# Selected Perspectives on Platforms in E-Commerce Ecosystems

Tobias Wulfert

# Selected Perspectives on Platforms in E-Commerce Ecosystems

## Recommendations for the Design and Management of Boundary Resources and Guidance on the Orchestration of Ecosystem Participants

Springer Vieweg

Tobias Wulfert
Universität Duisburg-Essen
Bottrop, Germany

Selected Perspectives on Platforms in E-Commerce Ecosystems: Recommendations for the Design and Management of Boundary Resources and Guidance on the Orchestration of Ecosystem Participants
Dissertation zur Erlangung des Doktorgrades „Dr. rer. pol.“ der Fakultät für Informatik der Universität Duisburg-Essen vorgelegt von Tobias Wulfert aus Gladbeck, Deutschland.
Betreuer: Prof. Dr. Reinhard Schütte
Lehrstuhl für Wirtschaftsinformatik und integrierte Informationssysteme
Essen, den 01.09.2023
Gutachterinnen und Gutachter: Prof. Dr. Reinhard Schütte, Prof. Dr. Frederik Ahlemann
Tag der mündlichen Prüfung: 01.02.2024

ISBN 978-3-658-45197-4 ISBN 978-3-658-45198-1 (eBook)
https://doi.org/10.1007/978-3-658-45198-1



This Springer Vieweg imprint is published by the registered company Springer Fachmedien Wiesbaden GmbH, part of Springer Nature.
The registered company address is: Abraham-Lincoln-Str. 46, 65189 Wiesbaden, Germany

*Dedicated with great respect and deep affection to my family, my friends, and those loved ones who are no longer with us.*

# Foreword

Global e-commerce sales have reached enormous proportions and are projected to continue to grow. 25 years ago, I published a book on the Internet and the retail industry, and I have been following the development of retail companies and pure-play e-commerce providers ever since. It is still fascinating to see how the development has accelerated—again due to the pandemic. Technological developments at the customer interface are not even that severe. However, the habitualization of buying behavior will contribute to the fact that the long-standing experience of users and customers in this sales channel will lead to further increases in sales in the future and that the expectations of the customers will in turn shape their expectations of the performance of the providers.

A major change in the last 25 years has been the emergence and effectiveness of platforms in e-commerce. Thanks to the impact of network effects and the enormous scalability of the information technology base, they have become perhaps the lasting achievement of the digital economy. They have established themselves as transaction or innovation platforms and are increasingly dominating the competition, e.g. Amazon, eBay, Zalando. From a retail perspective, this could also be formulated as follows: The new department store is the e-commerce platform. The successful companies in e-commerce have harnessed instead of this idea of the department store, which is attractive from the customer's point of view, via digital platforms and, thus, realized economies of scale that have reached unprecedented economic dimensions compared to a traditional retail reality. The underlying transaction platforms enable and represent enormous sales through large ecosystems of participating partners, resulting in the sales are not generated by one company. Value creation is not only achieved with a vertical value chain, but the interaction between the participants involved, such as logistics service providers, payment providers and many suppliers, generates additional value for customers and, thus, for the value creation network.

Tobias Wulfert has dedicated his work to a subject area that is both relevant and intellectually interesting. He has published an unusual number of papers at conferences and in academic journals for a cumulative dissertation. These contributions can be divided into four coherent blocks.

The first block contains contributions that describe and structure the phenomenon. The reader is given a good insight into e-commerce platforms, their categorization in a frame of reference and the possibility of a taxonomy of transaction platforms in e-commerce. An interesting and readable introduction for those who are eager to understand the problem of platforms in e-commerce.

The second part of the book deals with the architecture of e-commerce platforms, in particular the roles of the retailer and the platform. This chapter of the book takes an architectural perspective and will be particularly useful for those who are also confronted with design problems in operational reality.

The third block deals with boundary resources in e-commerce ecosystems. The ecosystems of transaction and innovation platforms in e-commerce require a large number of boundary resources. Tobias Wulfert presents the current state of research, develops design principles for the standardization of e-commerce ecosystems taking boundary objects into account and develops a model for the management of boundary resources.

The fourth block examines platforms as orchestrators of e-commerce ecosystems. It focuses on how the owner of the central platform of an e-commerce ecosystem can increase the value of the ecosystem by properly orchestrating the participants involved in commercial transactions and services. This is an exciting and economically important question that offers avenues for further research.

Tobias Wulfert's thesis is an outstanding piece of work in the field of information systems. It deals with a topic whose importance cannot be overestimated. At the same time, the topic has not been sufficiently analyzed theoretically. While there is some basic work on the economic effects of platforms as multi-sided markets, especially in economics, there is a lack of direct justification for the connection between technical properties or those of ecosystems and economic effects. The conceptual considerations on architectural categorization are therefore helpful and can be used for further research. The considerations are based on an extremely broad and deep literature foundation, which makes the work a sustainable work for platform research in the context of e-commerce ecosystems. The limitation of the work to e-commerce ecosystems is a scientific strength, as it allows for empirical verification in the sense of real-world research.

Because of the fundamental importance of the work to the information systems community and the many interesting ideas it contains, I hope that the work will be widely recognized and well received.

Essen, Germany

Prof. Dr. Reinhard Schütte
Professor at the University
of Duisburg-Essen

# Acknowledgements

The realm of platforms and their surrounding ecosystems has dominated both economic discourse and academic inquiry in recent times. Conversations concerning devices like smartphones, gaming consoles, and automobiles have evolved beyond mere hardware specifications to encompass broader ecosystem considerations, such as participant networks, app availability, and long-term viability. Within this landscape, both startups and established players now vie not only for optimal business models but also for the cultivation of robust platform-centered ecosystems.

Over the past few years, my research has delved deeply into the dynamics of platforms and business ecosystems in the context of e-commerce. Collaborating with my esteemed co-authors, we explored these subjects extensively, grounding our work in theory while engaging with industry practitioners. Beyond the intellectual journey that has manifested in this cumulative dissertation, my tenure as a research associate at the University of Duisburg-Essen allowed me to interact with a diverse array of students, guiding them through project theses, seminar papers, bachelor's theses, and master's theses.

This dissertation marks the culmination of a journey that would not have been possible without the support of numerous colleagues, friends, and my family. I extend my gratitude to my colleagues at the Chair for Integrated Information Systems, whose willingness to engage in substantive discussions and navigate the rigors of academia alongside me has been invaluable. Special recognition is owed to my longtime office mate, Dr. Sarah Seufert, for her unwavering support and empathetic ear.

I am grateful to all my co-authors, whose expertise and insights enriched the research underpinning the research articles comprising this cumulative dissertation. In particular, I would like to thank Dr. Gero Strobel and Robert Woroch for their enduring encouragement and contributions to many of the research endeavors.

A heartfelt acknowledgment goes to my supervisor, Prof. Dr. Reinhard Schütte, whose steadfast support, invaluable guidance, and constructive feedback steered me through this dissertation journey. His expertise and encouragement played a pivotal role in shaping the trajectory of my research, from its inception during a collaborative preparation for a speech on platforms in general to its final culmination in this book. I also extend my sincere appreciation to Prof. Dr. Frederik Ahlemann for graciously agreeing to serve as the second supervisor of my dissertation.

Lastly, I express my deepest gratitude to my family and all my friends for their unwavering support throughout this research journey. Their boundless love, encouragement, and understanding sustained me through the challenges of this endeavor. Thanks to my parents for their endless support and the enablement of this entire dissertation journey. I extend my gratitude to the members of the Pegelbrüder team for their support, which provided me with the opportunity to downshift and recover during this challenging period. I offer a heartfelt appreciation to my girlfriend Laura for her invaluable assistance, unwavering encouragement, and moral support throughout this demanding phase.

Grafenwald Tobias Wulfert
June, 2024

# Abstract

Digital platforms hold significant economic value, enriching various industry sectors and facilitating multi-sided markets, exemplified by industry giants like Apple, Amazon, and Alibaba. While they are pivotal in driving value creation and capture, it is not solely their technological prowess that underpins their economic significance. Rather, their effectiveness lies in their integration within business ecosystems. However, the terminology surrounding platforms remains ambiguous, lacking formalization, and sometimes struggles to delineate real phenomena from conceptual frameworks.

In this study, we examine digital platforms in the context of e-commerce through five lenses: enterprise architecture, boundary resources, application lifecycle, dynamic capabilities, and value creation, across ten research papers presented in this cumulative dissertation. Our analysis focuses on different e-commerce platform types, developing an architectural framework and a taxonomy for classification. We also delineate e-commerce ecosystem architectures, emphasizing the role of platforms as central components and the involvement of peripheral ecosystem participants. Additionally, we explore boundary resources as crucial interfaces between platforms and ecosystem participants. We further suggest a reference procedure model for the management of these boundary resources. Furthermore, we delve into the dynamic capabilities necessary for effective ecosystem orchestration, identifying configurations of dynamic capabilities that can confer sustainable competitive advantages. Our investigation extends to payment ecosystems, where we propose a value creation map to elucidate their role in the broader platform landscape.

In conclusion, we position platforms in e-commerce as enablers of the multi-lateral value creation of multiple participant types for the final customer, ensure an efficient use of resources through proper ecosystem orchestration, and provide the necessary boundary resources for digital exchanges.

# Contents

# Abbreviations

| | |
|---|---|
| ALM | Application Lifecycle Management |
| API | Application Programming Interface |
| ARIS | Architektur integrierter Informationssysteme |
| ARTS | Association for Retail Technology Standards |
| B2B | Business-to-Business |
| B2C | Business-to-Customer |
| B2G | Business-to-Government |
| BPO | Business Process Owner |
| C2B | Customer-to-Business |
| C2C | Customer-to-Customer |
| C2G | Customer-to-Government |
| csQCA | crisp-set Qualitative Comparative Analysis |
| CSV | Comma-Separated Values |
| e-commerce | Electronic Commerce |
| EC-SCP | E-Commerce Service Composition Platform |
| EU | European Union |
| fintech | Financial Technology |
| G2B | Government-to-Business |
| G2C | Government-to-Customer |
| G2G | Government-to-Government |
| GEMS | Global Electronic Market Stands |
| GMV | Gross Merchandising Volume |
| IBM | International Business Machines Corporation |
| IDE | Integrated Development Environment |

| | |
|---|---|
| IoT | Internet of Things |
| IPD | Internal Platform Developer |
| IS | Information Systems |
| ISO | International Organization for Standardization |
| IT | Information Technology |
| ITIL | Information Technology Infrastructure Library |
| MEMO | Multi-Perspective Enterprise Modelling |
| NGECP | Next-Generation E-Commerce Platform |
| OECD | Organization for Economic Co-operation and Development |
| P2P | Peer-to-Peer |
| PFM | Personal Finance Management |
| PLM | Product Lifecycle Management |
| QCA | Qualitative Comparative Analysis |
| SDK | Software Development Kit |
| TAM | Technology Acceptance Model |
| TOGAF | The Open Group Architecture Framework |
| UI | User Interface |
| UID | User Interface Designer |
| UML | Unified Modeling Language |
| US | United States |
| UTAUT | Unified Theory of Acceptance and Use of Technology |
| XML | Extensible Markup Language |

# List of Figures

# List of Tables

# Introduction

# 1

This introductory chapter contains the motivation to investigate platforms in e-commerce ecosystems as objects under consideration. We emphasize the economic significance of platforms in e-commerce with their surrounding ecosystems and clarify challenges in forming e-commerce ecosystems. Moreover, we refer to the body of literature on platforms in e-commerce to carve out a research gap in the context of ecosystem orchestration and related boundary resources. Against the identified research gaps, we formulate our problem statement and introduce our research questions. We also sketch the structure of the research papers integrated within this cumulative dissertation and the remainder of this thesis.

## 1.1 Motivation

In this section, we elaborate on the relevance of platforms and business ecosystems in the context of e-commerce and refer to limitations in current research with regard to focal platforms in e-commerce ecosystems.

### 1.1.1 The Relevance of Platforms in E-Commerce

The development and global dissemination of new technologies offer a wide range of economic opportunities (e.g., Arora, Longo, et al., 2022; Brackmann et al., 2023; Grewal, Roggeveen, et al., 2017; Muschkiet et al., 2023; Schütte,


T. Wulfert, *Selected Perspectives on Platforms in E-Commerce Ecosystems*, https://doi.org/10.1007/978-3-658-45198-1_1

Ahlemann, et al., 2022; Weber and Schütte, 2019b;).[1] The prevalence of the Internet, for instance, enabled new forms of commerce in the digital environment (Choi et al., 1997; Nath et al., 1998; Zwass, 1996). The global Electronic Commerce (e-commerce) revenue is expected to reach $ 7.4 trillion in 2025 (EMarketer, 2022),[2] covering 23.6% of overall retail sales (Lebow, 2021).[3] The share of e-commerce transactions in total retail sales is likely to increase in the upcoming years. This is due to the amplification of electronic transactions during the COVID-19 pandemic and increased customer convenience (Bhatti et al., 2020; Dietz et al., 2020; Eggert, 2020; Stüber et al., 2019). Since e-commerce allows electronically supported retail transactions without spatial and temporal restrictions (Nath et al., 1998; Zwass, 1996), e-commerce is likely to be conducted globally "without borders" (Stallkamp and Schotter, 2021, p. 58). Increasing globalization and tremendous digitalization efforts at the societal, company, and individual levels have resulted in complex systems of independent but intertwined actors (Adner, 2017).

The digital transformation[4] of the retail sector leads to a penetration of stationary retail with additional digital services (Muschkiet et al., 2023; Schütte, Muschkiet, et al., 2022; Wulfert, Betzing, et al., 2019) and an establishment of additional online sales channels resulting in multi-channel and omni-channel constellations (Saghiri et al., 2017; Verhoef, Kannan, et al., 2015). Besides the establishment of own online stores as additional online sales channels, retailers can participate in an existing transaction platform[5] and establish transaction platforms themselves (e.g., Kaufland, Otto, Zalando) (Schütte and Wulfert, 2022a, p. 213). Joining an established transaction platform allows "to target customers at their preferred sites" (Eden et al., 2021, p. 6) increasing the number of potential customers. The development of own transaction platforms allows for a diversification of the assortment offered and enables to

[1] These opportunities include business model innovations as well as benefits for existing economic principles.

[2] The revenues in this cumulative dissertation are indicated in US$. If another national dollar is applied, it is mentioned accordingly.

[3] Overall retail sales involves online and offline retail transactions in business-to-consumer mode. E-commerce revenue includes products or services ordered using the Internet without differentiation for the method of payment or fulfillment. E-commerce revenue excludes travel and event tickets; payments such as bill pay, taxes or money transfers; food services and drinking place sales; gambling; and other vice goods sales.

[4] Vial (2019, p. 121) defined digital transformation as "a process that aims to improve an entity by triggering a significant change to its properties through a combination of information, computing, communication, and connectivity technologies."

[5] Thirty-five percent of the top 500 retailers globally also offer (parts of) their assortment via established transaction platforms (e.g., Amazon Marketplace, eBay) (Eden et al., 2021, p. 8).

benefit and capture value from orchestrated transactions between third-party sellers and customers (Schütte and Wulfert, 2022a, p. 213).

**Observation 1:** E-commerce is an economically important sector, an important sales channel for retailers, and a convenient shopping environment for customers. Transaction platforms are a preferred option for online sales.[6]

In general, transaction platforms are of tremendous economic importance and have enriched many industry sectors that comprise some of the most valuable companies (e.g., Apple, Amazon, and Alibaba), orchestrating multi-sided markets (Gawer, 2021b, p. 7 ff.).[7] Platforms are important levers for increased value creation and value capture (Zhao, von Delft, et al., 2020, p. 1).[8] Miremadi et al. (2023, p. 1) even stated that the "structures of many industries have changed dramatically in recent years with the implementation of digital platforms."[9] Owing to their widespread dissemination, platforms have a huge impact on economics (Stallkamp and Schotter, 2021, p. 59). Abdelkafi et al. (2019, p. 553) described transaction platforms as "the most powerful business models in the digital economy."[10] Across industries, seven organizations conducting platform business models[11]—such as Apple, Microsoft,

[6] At this point, we have to restrict that the observations elaborated within this introduction originate from our interpretation of the developments described. Without our focus on the analysis of platforms in e-commerce ecosystems, one might come to different conclusions.

[7] Another fact that emphasizes the economic success of platforms is that around two-thirds of unicorns in the United States (US) operate platform business models (Cusumano, Gawer, and Yoffie, 2019, p. 3). A unicorn is defined as a startup that exceeds a valuation of $1 billion.

[8] Murakami Wood and Monahan (2019, p. 1) even claimed that "digital platforms fundamentally transform social practices and relations," with the threat of digitally mediated relations producing exploitable data. MurakamiWood and Monahan (2019, p. 1) described this phenomenon as "platform surveillance."

[9] The number of users of selected transaction platforms across industries, such as Facebook, Whatsapp, and YouTube, pars or even exceeds the number of inhabitants of the most populated countries in the world (Manyika et al., 2016, p. 6).

[10] Stark and Pais (2020, p. 48) even formulated that "much of our daily life is conducted on or with platforms. It is on platforms that we can access information, watch movies, listen to music, read books, shop for all kinds of products, archive documents, and find partners. We go to platforms to get a ride, book travel, support a cause, order food, or finance a project. Do you want to know what the people you care about are doing and thinking? Are you looking for someone for housekeeping, home repairs, medical advice, legal or accounting services, or lectures on any kind of topic? Or perhaps someone to organize your closets or coach you in organizing your life? There are platforms for all of these and more." We will focus on e-commerce as the domain of analysis within this cumulative dissertation.

[11] The business model concept is widely adopted in the digital economy, combining technological innovation with the realization of economic value on the one hand, and harmonizing

Alphabet, Amazon, and Meta (Companiesmarketcap.com, 2023)—are among the ten most valuable companies compared by market capitalization (Constantinides et al., 2018, p. 381).[12] By 2025, more than 30 % of the global economic revenue will be conducted via platforms (Atluri et al., 2017, p. 6).[13]

The success of transaction platforms across industries is rooted in the economic effects of two-sided markets (e.g., matching and subsidizing selected participants) (Merton, 2021; Parker, Van Alstyne, and Choundary, 2016; Wulfert, Seufert, et al., 2021). When it comes to network effects, the platform's value for affiliated participants increases when additional participants joining.[14] These network effects lead to winner-takes-all tendencies, allowing only a reduced amount of platforms to dominate a certain industry (Hagiu, 2007; Karhu and Ritala, 2021; Ruutu et al., 2017), and amplify the evolution of "gigantic" platform owners in terms of revenue and affiliated participants (Parker, Van Alstyne, and Choundary, 2016; Spaeth and Niederhöfer, 2022). Platforms are the "origin" (Miremadi et al., 2023, p. 1) and form the "core" (Staykova and Damsgaard, 2015, p. 10) of platform-centered ecosystems (Bresnahan and Greenstein, 1999).[15]

The most successful organizations in e-commerce operate transaction platform business models exploiting the economic effects of two-sided markets (e.g., matching and subsidizing selected participants) (Merton, 2021; Parker, Van Alstyne, and

business strategy with business processes on the other hand (Al-Debei and Avison, 2010, p. 365 ff.). Although the business model concept considers the collaboration of multiple actors in performing transactions, its focus lies on the value proposition of single organizations (Adner, 2017; Al-Debei and Avison, 2010; Zott et al., 2011).

[12] Some of the aforementioned organizations also employ linear business models besides their platform strategies.

[13] However, we would like to stress that successful platforms are mostly established in the US and China (Hermes, Pfab, et al., 2020, p. 9), while platforms are only considered a future opportunity in the European Union (EU) (Riemensperger and Falk, 2020, p. 62).

[14] If the additional participants are of the same type (e.g., additional customers for other demand-side participants), the demand-side participants record direct or same-side network effects. If the additional participants are of another type (e.g., additional third-party seller for demand-side participants), the demand-side participants record indirect or cross-side network effects (Armstrong, 2006; Briscoe, Odlyzko, et al., 2006; Dou and Wu, 2021; Hagiu and Rothman, 2016; Wulfert, Seufert, et al., 2021). We introduce network effects in more detail in Sect. 3.2.

[15] We also describe the central platform of a platform-centered ecosystem as "pivotal" and "focal" in the context of e-commerce ecosystems as platform owners govern and influence other ecosystem participants (e.g., Wulfert, Woroch, Strobel, Schoormann, et al., 2023; Wulfert, Woroch, Strobel, Seufert, et al., 2022). We apply the terms central, pivotal, and focal synonymously throughout this thesis.

Choundary, 2016).[16] Evans and Schmalensee (2016, p. 3) described platforms as "powerful engines of commerce" in the context of e-commerce. More than 60% of the world's global e-commerce revenue is generated via transaction platforms (e.g., Amazon, Alibaba, eBay, Mercado Libre)[17] conducting business internationally (Eden et al., 2021, p. 27).[18] In the case of Amazon, the number of third-party sellers' transactions accounts for more than 60 percent of the overall marketplace revenue (Amazon, 2019, 2021, 2023).[19] In recent years, a lot of transaction platforms have emerged in e-commerce (Prothmann, 2017). [20] Transaction platforms observed an overall increase in the number of affiliated third-party sellers (Eden et al., 2021, p. 6) and benefit from this development by offering a variety of (paid) services[21] to them, such as marketing services, analytic services, and fulfillment services (Amazon, 2023). Moreover, transaction platforms offer local businesses and individual shops the possibility to offer their products online benefiting from the user base of an existing platform (Berendes et al., 2020).[22] Transaction platforms further increase customer convenience by providing access to the assortment of multiple sellers through a unified user interface (Teller and Elms, 2010; Wulfert and Schütte, 2022). Eden et al. (2021, p. 6) reported that transaction platforms are the target for casual browsing, are used for produced searches, and are the preferred option for repeated purchases.

[16] Companies owning transaction platforms in the retail sector have up to four times higher valuations than traditional retailers and online resellers (Libert et al., 2014).

[17] The three transaction platforms with the highest third-party Gross Merchandising Volume (GMV), namely Taobao, Tmall, and Amazon, account for 53% of the total GMV of the top 100 transaction platforms (Eden et al., 2021, p. 31). It needs to be noted that—both—Taobao and Tmall belong to the Alibaba Group(Ackroyd, 2023).

[18] This share is independently supported by DigitalCommerce360 (2023) reporting a share of 53 % accounted to transaction platforms in e-commerce. DigitalCommerce360 (2023) added JD and Shopee in the top five transaction platforms in e-commerce measured by third-party GMV.

[19] Network effects lead to winner-takes-all-tendencies in e-commerce (Hagiu, 2007; Karhu and Ritala, 2021; Ruutu et al., 2017). Hence, platforms need to position themselves in market niches (Cennamo, 2021).

[20] Prothmann (2017) identified more than 900 across the world in his report. Eden et al. (2021) and DigitalCommerce360 (2023) reported more transaction platforms that have been established by retailers that opened their online stores for third-party sellers.

[21] We follow a service-dominant logic and define a service as "the application of competences (knowledge and skills) by one entity for the benefit of another" (Vargo and Lusch, 2004, 2006; Vargo, Maglio, et al., 2008).

[22] Already in 2016, a report published by the McKinsey Global Institute summarized that transaction platforms, such as Alibaba, eBay, and Etsy, empowered small- and medium-sized enterprises to reach a global market (Manyika et al., 2016, p. 44 f.).

**Observation 2**: Transaction platforms are involved in a large proportion of e-commerce transactions. Successful transaction platforms have grown large ecosystems with affiliated participants.

Owing to the transformative and disruptive impact of digital platforms and surrounding digital business ecosystems on realizing the value proposition, value is no longer created in isolation but co-created by multiple ecosystem participants (de Reuver et al., 2017; Iansiti and Levien, 2004b). Digital business ecosystems are formed by independent but intertwined and (digitally) connected participants multilaterally co-creating a value for other ecosystem participants (Adner, 2017; Hein, Weking, et al., 2019; Tsujimoto et al., 2018). Besides other ecosystem benefits, companies can leverage value among participants and profit from an increased number of participants and resulting network externalities propelling their revenues (Gao, Zhang, et al., 2013; Hiller et al., 2022; Shapiro and Varian, 1998).[23] Cennamo (2021, p. 279 ff.) stated that the level of competition has lifted from the company level to the ecosystem level with focal platforms and their surrounding ecosystem participants.[24] Across industries, the ecosystem of affiliated participants has up to three times higher revenue compared with the revenue of the focal platform (Delteil et al., 2020, p. 6).[25] In e-commerce, transaction platforms form the center of e-commerce ecosystems, which are surrounded by networks of independent participants (Wulfert, Woroch, Strobel, Seufert, et al., 2022).[26] The owner of the focal platform in e-commerce ecosystems orchestrates the ecosystem-wide value co-creation[27] across organizational boundaries by setting and enforcing governance

[23] Although an increasing number of ecosystem participants amplifies intra-ecosystem competition and diminishes exclusive offers, the overall ecosystem value for each participant on the same and different connected markets (inter-ecosystem) is on the increase (Gawer, 2021b; Wulfert, Woroch, Strobel, Seufert, et al., 2022).

[24] Early examples for ecosystem-level competitions were identified by Moore (1996) and Iansiti and Levien (2004a). While Moore (1996) described the competition between Microsoft and Netscape for the future of the Internet, Iansiti and Levien (2004a) reported on the competition between Microsoft and Intel on the control of computing ecosystems (Corallo, 2007, p. 19).

[25] The high relevance of business ecosystems in practice is underpinned by the term ecosystem having increasingly appeared in industrial and corporate news over the past years (Kapoor, 2018, p. 1 f.).

[26] E-commerce ecosystems are manifestations of digital business ecosystems in the context of e-commerce with the purpose of providing (final) customers with products and services (Adner, 2017; Wulfert, Woroch, Strobel, Seufert, et al., 2022). Grönroos (2012, p. 1520) described the participant for which the value is generated as the "beneficiary."

[27] Making use of research on the service-dominant logic (e.g., Vargo and Lusch, 2004, 2008; Vargo, Maglio, et al., 2008), the (final) customer is (more or less) involved in the process of

rules (Hein, Schreieck, Riasanow, et al., 2020; Santos and Eisenhardt, 2005). Organizing the value co-creation in ecosystems results in increased outputs compared to disintermediated scenarios (Autio, 2022; Hiller et al., 2022).

Applying Adner's (2017) ontology describing ecosystems, e-commerce ecosystems are identified as a special subtype of digital business ecosystems. E-commerce ecosystems enable the exchange of products and services in order to achieve the (final) customer's objective (Wulfert, Woroch, Strobel, Seufert, et al., 2022). These ecosystems evolve around a focal transaction platform (Wulfert and Karger, 2022b; Wulfert, Woroch, Strobel, Seufert, et al., 2022).[28] The conduct of e-commerce transactions is orchestrated by the owner of the focal transaction platform (Koch, 2010; Täuscher and Laudien, 2018; Wulfert, Woroch, and Strobel, 2024) and supported by a variety of actors, such as content and software providers (Böttcher, Rickling, et al., 2021; Wulfert, Woroch, Strobel, Schoormann, et al., 2023). Gao, Han, et al. (2019, p. 143) identified four characteristics of e-commerce ecosystems, namely "high system renewal rate, the leading position of core enterprises, the fuzziness of system boundary and high environmental threat."[29]

**Observation 3:** Transaction platforms in e-commerce form e-commerce ecosystems surrounded by multiple independent participants, such as third-party sellers, logistics service providers, and payment providers. These participants co-create value which is the exchange of products and services for the (final) customers.

The emergence of platforms amplified the joint value co-creation by multiple ecosystem participants (de Reuver et al., 2017; Iansiti and Levien, 2004a).[30] The owner of the focal platform's active orchestration of the ecosystem participants' actions can increase the multilateral value co-creation compared with that of disintermediated scenarios (Autio, 2022; Hiller et al., 2022). Proper ecosystem orchestration

value creation. However, we take the concept more broadly referring to multiple ecosystem participants' collaboration for a joint value creation (Adner, 2006, 2017; Wulfert, Woroch, Strobel, Seufert, et al., 2022). The customer does not necessarily have to be involved in creating the value (Grönroos, 2012, p. 1521) but experiences the "value in use" (Lim, Kim, et al., 2018, p. 126). For a more detailed elaboration on value co-creation in e-commerce ecosystems, please refer to Sect. 3.4.

[28] The owner of the focal platform can behave competitively to third-party sellers providing its own assortment to customers (Wulfert and Schütte, 2021, 2022).

[29] For a more detailed elaboration on the specific characteristics of e-commerce ecosystems, please refer to Sect. 3.4.2.

[30] This fact also resulted in a shift of competition from single companies to competition among business ecosystems with their affiliated participants (e.g., Amit and Zott, 2021; Cennamo, 2021; Rietveld and Schilling, 2021; Zhao, von Delft et al., 2020).

can therefore be the foundation of sustainable competitive advantage (Burkhalter, 2020; Rietveld and Schilling, 2021; Sirmon et al., 2011). Consequently, we need to investigate the focal platform owner's ecosystem strategy[31] where partners' business models are also considered (Adner, 2017) and coordinated in a network (Arreola González, et al., 2019; Bharadwaj et al., 2013).[32] Successful e-commerce ecosystems involve multiple companies co-creating a value proposition based on multilateral relationships. Therefore, the consideration, orchestration, and alignment of different business models are essential aspects of an ecosystem strategy (Adner, 2017; Vargo and Lusch, 2016). Focal platform owners in e-commerce ecosystems (e.g., Amazon Marketplace, Walmart Marketplace, and Alibaba.com) actively orchestrate[33] ecosystem participants by executing governance rules (Hein, Schreieck, Riasanow, et al., 2020). They match supply- and demand-side participants to enable retail transactions involving additional supporting actors (e.g., content providers) (Böttcher, Rickling, et al., 2021; Wulfert, Seufert, et al., 2021).

> **Observation 4:** The owner of a focal platform in e-commerce ecosystems can increase the ecosystem's overall value creation by properly orchestrating the different participants involved in retail transactions and involved services (e.g., third-party seller, customer, service provider).

With increasing revenues in the e-commerce sector and customers becoming increasingly likely to shop in virtual environments (EMarketer, 2022), former brick-and-mortar retailers initiate additional online sales channels and new startups begin to sell online (Rudolph et al., 2015; Wulfert, Betzing, et al., 2019). The creation of online sales channels requires the implementation of Information Systems (IS) supporting online sales (Schütte and Wulfert, 2022a). In e-commerce, these IS are often

---

[31] An "ecosystem strategy is defined by the way in which a focal firm approaches the alignment of partners and secures its role in a competitive ecosystem" (Adner, 2017, p. 47).

[32] While the focal platform owner is described by Iansiti and Levien (2004a, p. 3) as a "keystone" and by Jacobides et al. (2018, p. 2257) as a "hub," Moore (1996, p. 4) conceptualized the focal platform owner in an ecosystem as an "ecosystem leader."

[33] The orchestration of ecosystem participants resembles enablement and "facilitation of economic exchanges between two or more [independent but related] user groups" (Zhao, von Delft, et al., 2020, p. 1). We understand orchestration in e-commerce ecosystems "as an undertaking by a focal [platform owner] that intends to integrate and leverage resources and capabilities of others within a business ecosystem" (Mann et al., 2022, p. 3). Mann et al. (2022, p. 3) used the more neutral term "focal actor" who is responsible for orchestration in business ecosystems. However, within this dissertation, we refer to e-commerce ecosystems surrounding a focal platform (Wulfert, Woroch, Strobel, Seufert, et al., 2022, p. 1816 f.). Hence, the focal actor in e-commerce ecosystems is/are the platform owner/s.

provided as innovation platforms.[34] Since the implementation of new IS is not a core competence of most retailers (Wulfert, Strobel, et al., 2024, interview #2, interview #6), they rely on external software providers for providing their products online. Hence, the total market for e-commerce applications is expected to grow from $ 6.1 billion in 2020 to $ 7.4 trillion in 2025 with an annual average growth rate of 3.9 % (AppsRunTheWorld, 2021). Shopify, Adobe Magento Commerce, Salesforce, and SAP Commerce Cloud are among the leading e-commerce application providers (AppsRunTheWorld, 2021). Datanyze (2021) identified more than 600 different innovation platforms (e.g., WooCommerce) enabling e-commerce businesses. As a technological artifact, innovation platforms can be the foundation upon which (third-party) developers can create additional modules (e.g., localizations, themes) based on the digital platform services provided (Baldwin and Woodard, 2009; Tiwana, Konsynski, and Bush, 2010). Most of these innovation platforms offer boundary resources for third-party developers extending the feature set of these platforms with dedicated complements (Baldwin and Woodard, 2009; Jansen, 2013; Staykova and Damsgaard, 2015).

In this cumulative dissertation, we describe innovation platforms[35] as the operating systems for e-shops, digital marketplaces, and other business models in e-commerce, that provide the application services necessary for executing the business processes and environments for developing external modules (Aulkemeier, Paramartha, et al., 2016; Baldwin and Woodard, 2009; Hanseth and Lyytinen, 2010; Tiwana, Konsynski, and Bush, 2010).[36] Blaschke, Haki, et al. (2019, p. 4) described innovation platforms as the "technical core." In e-commerce, innovation platforms need to cope with a variety of specific peculiarities (e.g., Rudolph et al., 2015; Schütte and Wulfert, 2022a; Teller and Elms, 2010; Turban, Whiteside, et al., 2017; Wulfert and Schütte, 2022). Innovation platforms need to deal with a diverse set of ecosystem participants (Böttcher, Rickling, et al., 2021; Eisenmann, Parker, et al., 2009; Wulfert, Woroch, Strobel, Schoormann, et al., 2023). They need to support the retailer's business model, such as allowing the affiliation of third-party sellers for transaction platforms and routing instructions to external service providers (Kawa

[34] In general, innovation platforms are an important source of innovation. The nine largest US-based innovation platforms measured by annual revenue were awarded more than 11,000 patents in 2014 (Evans and Gawer, 2016).

[35] We refer to innovation platforms as an "extensible codebase of a software-based system that provides core functionality shared by the modules that interoperate with it and the interfaces through which they operate" (Tiwana, Konsynski, and Bush, 2010, p. 676).

[36] These external modules can extend the functionality of the innovation platform, add integrations to other platforms, and include new user interface themes (Aulkemeier, Paramartha, et al., 2016; Eaton et al., 2015; Wulfert and Karger, 2022b).

and Walesiak, 2019; Täuscher and Laudien, 2018). In most cases, third-party developers provide additional platform features and shop themes (Wulfert and Karger, 2022b, p. 4). The third-party sellers' different product and service assortments need to be adequately represented in the digital environment (Teller and Elms, 2010, p. 28). Moreover, the innovation platform needs to deal with high fluctuations in the number of transactions (e.g., holiday season) (e.g., Chiu et al., 2021; Goswami et al., 2015). To support new shopping trends (e.g., virtual shopping) and cope with customers' convenience (e.g., same-day delivery), innovation platforms in e-commerce are developed highly flexible (Aulkemeier, Iacob, et al., 2019; Aulkemeier, Paramartha, et al., 2016; Wulfert and Karger, 2022b).

**Observation 5**: Innovation platforms provide the application services necessary for conducting transactions in e-commerce (e.g., via transaction platforms). Innovation platforms in e-commerce attract third-party developers providing extensions (e.g., shop themes, feature add-ins).

In electronic environments, such as e-commerce, a set of boundary resources (e.g., Application Programming Interfaces (APIs), Software Development Kits (SDKs), documentation, support) is required for collaboration and value creation.[37] These boundary resources translate diverging viewpoints among actors and interface between different information (Ghazawneh and Henfridsson, 2013; Star, 2010; Wulfert, Woroch, Strobel, Seufert, et al., 2022). Boudreau (2010, p. 1853 ff.) as well as Yoo, Henfridsson, et al. (2010, p. 728 ff.) described boundary resources as strategically relevant. In a recent study, Evans and Basole (2016) showed that boundary resources are especially relevant for retailers in e-commerce, and that boundary resources are also increasingly implemented by brick-and-mortar incumbents. Boundary resources enable a "tight integration" (Wang, Zheng, et al., 2008, p. 565) of affiliated participants' IS in e-commerce ecosystems. Collaboration and value co-creation in e-commerce ecosystems require not only application boundary resources for linking application systems, but also development boundary resources extending the feature set of the focal platform and accompanying social boundary resources (Dal Bianco et al., 2014, p. 15 f.).

While application boundary resources are especially relevant for linking third-party sellers' and service providers' application systems with the focal platform in e-commerce ecosystems (Wulfert, Strobel, et al., 2024, p. 24), the provision of sophisticated development boundary resources is necessary for the creation of

[37] In general, the deployment of boundary resources has become a strategic option across industries (Zapadka et al., 2022, p. 2). These resources allow external actors to engage in joint value creation.

platform extensions (Dal Bianco et al., 2014; Jansen, 2013; Wulfert, 2023).[38] Since development boundary resources are a prerequisite for developing platform extensions, they are also required to attract third-party developers to an innovation platform. Making use of these boundary resources, third-party developers can increase the initial feature set of the innovation platform (Dal Bianco et al., 2014; Eaton et al., 2015; Jansen, 2013). The provision of sophisticated boundary resources and the attraction of third-party developers are likely to increase the ecosystem's overall generativity (Blaschke and Brosius, 2018; Fürstenau, Anisimova, et al., 2019; Fürstenau, Baiyere, et al., 2023; Miremadi et al., 2023; Wulfert, 2023).

> **Observation 6:** Innovation and transaction platforms in e-commerce need to provide a set of boundary resources to enable the co-creation of value among different participants (e.g., third-party sellers, service providers, third-party developers).

### 1.1.2 Current Research on Platforms in E-Commerce

Platforms' high relevance in practice goes along with an increasing research interest in platforms and surrounding ecosystems (e.g., Asadullah et al., 2018a; Liu, Li, et al., 2021; Schreieck et al., 2016; Wulfert and Karger, 2022a). The number of scientific publications that focus on platforms and surrounding business ecosystems has been on the increase since the last decade (Asadullah et al., 2018a; Baumann and Leerhoff, 2022; Schreieck et al., 2016; Tsujimoto et al., 2018). Liu, Li, et al. (2021, p. 4) conducted a bibliometric analysis of research on platforms and identified an increase in publications between 1996 and 2019. Rietveld and Schilling (2021, p. 1532) provided a systematic literature review across disciplines for competition among transaction platforms. The authors revealed a significant increase in the number of publications between 2015 and 2019 (Rietveld and Schilling, 2021, p. 1531 f.).[39] Schreieck et al. (2016, p. 5) conducted a focused literature review in IS research and also revealed an increase in the number of studies on the "platform topic," with the majority of research endeavors using qualitative research approaches.[40]

---

[38] We describe social boundary resources as supplementary boundary resources supporting technical boundary resources (i.e., application boundary resources and development boundary resources) (Wulfert, 2023). For instance, social boundary resources provide the documentation for using APIs.

[39] In a similar vein, Schüßler et al. (2021, p. 1218) also reported an exponential increase of publications with the term platform economy since 2015.

[40] It should be noted here that the number of studies on platforms in IS research (Schreieck et al., 2016, p. 5) is significantly smaller compared to more general investigations (Liu, Li, et al., 2021; Rietveld and Schilling, 2021; Wulfert and Karger, 2022a).

More particularly, there is a growing interest in research on platform boundary resources, which is reflected by recent publications (e.g., Eaton et al., 2015; Engert et al., 2022; Ghazawneh and Henfridsson, 2015; Petrik, Model, et al., 2021). These publications addressed boundary resources in the industrial Internet of Things (IoT) (Petrik, 2022), in industrial ecosystems led by consortia of multiple organizations (Mohagheghzadeh and Lindman, 2022), and in content management systems for commerce (Engert et al., 2022).

Following an increase in research with a focus on platforms, investigations on business ecosystems and their digital counterparts (i.e., digital business ecosystems) are of increasing interest in research (e.g., Adner, 2017; Baumann, 2022a; Jacobides et al., 2018; Tsujimoto et al., 2018). Since Moore's (1993; 1996 introduction of the ecosystem concept to the economics literature, Tsujimoto et al. (2018, p. 50) revealed an increase in publications on the concept of ecosystems since 2008. The rising research interest in ecosystems is also supported by Bogers et al. (2019, p. 3) who determined an increase in scientific publications on the Social Science Citation Index, mentioning the term ecosystem as a topic between 2007 and 2018. Baumann and Leerhoff (2022, p. 16) also found a steadily increasing research interest in digital business ecosystems since 2015.[41]

Following the general increase in the scientific interest in platforms and business ecosystems, we address a number of aspects scholars raised in recent scientific publications with our research on platforms in the context of e-commerce (e.g., de Reuver et al., 2017; McIntyre and Srinivasan, 2017; Wulfert and Karger, 2022a). De Reuver et al. (2017, p. 129 f.) called for research on domain-specific manifestations of platforms and their economic implications. Wulfert and Karger (2022a, p. 194 ff.) recommended future research to investigate configurations of successful platforms to inform research and practice. Moreover, the authors uncovered different actors' affiliation with the platform and their impact on value co-creation as a potential research gap (Wulfert and Karger, 2022a, p. 195). McIntyre and Srinivasan (2017, p. 152 ff.) suggested research on the design of platforms and complementors' involvement in the platform's value creation. In a similar vein, Yoo, Henfridsson et al. (2010, p. 731) called for research on platform owners' requirements for providing successful platforms.

We further address calls for research on the interaction between platform owners and ecosystem participants in general and third-party developers in particular (e.g.,

[41] Baumann and Leerhoff (2022, p. 17 f.) identified four disciplines dealing with digital business ecosystems, namely business administration and management, business informatics and computer science, industrial engineering, and economics. By far, the most contributions (60) between 2000 and 2020 have been attributed to the business administration and management discipline.

Huber, Kude, et al., 2017; Hurni, Huber, Dibbern, and Krancher, 2021; Karhu, Gustafsson, et al., 2018). More specifically, recent research called for extensive research on the design and control of boundary resources in the platform context (e.g., de Reuver et al., 2017; Hein, Schreieck, Riasanow, et al., 2020; Yoo, Henfridsson, et al., 2010). Although "the technological architecture, software, as well as platform governance aspects and rules are crucial [for the design of platforms, TW], literature on generally valid design principles and criteria is scarce" (Abdelkafi et al., 2019, p. 555). Schüler and Petrik (2021, p. 25) called for "specific design guidelines" for the design and implementation of boundary resources in platform-centered ecosystems. Yoo, Henfridsson, et al. (2010, p. 733) asked for future research on "the strategic roles of technical boundary resources such as APIs and SDKs and social boundary resources such as incentives, intellectual property rights, and control within a layered modular architecture." De Reuver et al. (2017, p. 132 f.) suggested the boundary resource concept as a possible unit of analysis "[...] for the theoretical treatment of the fundamental characteristics of digital platforms and innovation arrangements."[42] Applying the boundary resource concept might result in new theoretical categories (Tilson, Lyytinen, et al., 2010, p. 11). Hein, Schreieck, Riasanow, et al. (2020, p. 93 f.) called for research endeavors on the balancing of platform control and generativity and the role of boundary resource standardization.[43]

Jacobides et al. (2018, p. 2268), Thomas et al. (2014, p. 22), and Yoo, Henfridsson, et al. (2010, p. 730 ff.) called for an in-depth analysis of value creation in the digital business ecosystem and an investigation of the specific roles of different ecosystem participants. Rietveld and Schilling (2021, p. 1544) called for research on ecosystem orchestration and management in dynamic environments such as e-commerce. Thus far, research has proposed a variety of ignition strategies that can support platform owners in establishing an ecosystem by reaching a critical mass of participants to benefit from network effects (e.g., Evans and Schmalensee, 2016; Ojanperä and Vuori, 2021; Parker and Van Alstyne, 2016; Parker, Van Alstyne, and

[42] Digital innovations can be defined "as the creation of (and consequent change in) market offerings, business processes, or models that result from the use of digital technology" (Nambisan et al., 2017, p. 224).

[43] Hein, Schreieck, Riasanow, et al. (2020, p. 93 f.) stated the following with regard to understanding the trade-off between boundary resource standardization and flexibility in using platform capabilities: "By understanding this trade-off, we can elaborate what measures platform owners can take to control the generativity of a digital platform ecosystem by either reducing entrance barriers to a variety of complementors through standardization or by granting flexibility to come up with innovative complements" (Hein, Schreieck, Riasanow, et al. 2020, p. 93 f.).

Choundary, 2016; Reillier and Reillier, 2017; Schirrmacher et al., 2017; Wanner et al., 2019; Wulfert and Dennhardt, 2023).[44] Nevertheless, research on how to further develop transaction platforms in e-commerce after reaching critical mass and after passing the ignition stage in lifecycle models (Muzellec et al., 2015; Teece, 2017) remains sparse (e.g., Amit and Zott, 2021; Cennamo, 2021; Rietveld and Schilling, 2021; Zhao and Luo, 2022). In line with Priem et al.'s (2013, p. 478) suggestion for extending strategy research beyond the boundaries of single companies and supply chains, and for also investigating ecosystem-level strategies, an investigation of focal platform owners' "alignment of partners" (Adner, 2012, p. 4) and capabilities required for ecosystem orchestration seem valid. Following recent scientific publications (e.g., Amit and Zott, 2021; Koch, 2010; Mann et al., 2022; Rietveld and Schilling, 2021), we argue that proper orchestration is a critical factor for the success of a business ecosystem and, thus, a potential source of competitive advantage in competition among ecosystems.

Analogous to a general increase in the research interest in platforms and surrounding business ecosystems, we identified an increase in the relevance of platforms in research on e-commerce since 2015. The number of publications focusing on platforms in e-commerce increased from 213 in 2015 to 1,226 in 2021 (Wulfert and Karger, 2022a, p. 189). In a recent study on the impact of digital transformation in retail, Reinartz et al. (2019, p. 355 ff.) emphasized the intermediary role of transaction platforms as central touchpoints with a uniform user interface for customers (Kawa and Walesiak, 2019; Teller and Elms, 2010). The scientific community is merely focused on platforms with a business-to-consumer focus (Van der Aalst et al., 2019; Wulfert, Seufert, et al., 2021). E-commerce research also focuses on the platform business models[45] (i.e., transaction platforms). However, Zhao, von Delft, et al. (2020, p. 2 f.) argued that the phenomenon of transaction platforms—orchestrating independent participants in a network—"remains undertheorized." Although innovation platforms as technological "core" provide the necessary application services for e-commerce transactions and simultaneously form the center of innovation ecosystems, they are rarely considered as objects of reflection in e-commerce research (Aulkemeier, Paramartha, et al., 2016; Baldwin and

[44] These ignition strategies can either attract the different markets sides sequentially or simultaneously (Schirrmacher et al., 2017). However, the aforementioned publications did not evaluate the applicability of these strategies to e-commerce ecosystems (Wulfert and Dennhardt, 2023, p. 131).

[45] Platform business models are also referenced as digital marketplaces (Kawa and Walesiak, 2019; Wulfert, Seufert, et al., 2021) or transaction platforms (Evans and Gawer, 2016; Gawer, 2021a) in recent literature. We prefer the term transaction platform in this thesis, but also use the term platform business model when appropriate.

Woodard, 2009; Blaschke, Haki, et al., 2019). While research on innovation platforms is already sparse in e-commerce, research on boundary resources providing necessary development features for third-party developers and connectivity for other ecosystem participants is even rarer.

## 1.2 Problem Statement and Research Questions

We have observed a high relevance of platforms in the domain of e-commerce, both in practice and in research. Although successful transaction platforms (e.g., Amazon Marketplace, Walmart Marketplace, AliExpress) and innovation platforms (e.g., WooCommerce, Shopify) have been established, research on platforms in e-commerce, in general, and on the success factors of these platforms has remained elusive since the first publications on e-commerce (e.g., Nath et al., 1998; Zwass, 1996). This observation is supported by recent calls for domain-specific research on platforms (e.g., de Reuver et al., 2017; McIntyre and Srinivasan, 2017; Wulfert and Karger, 2022a).[46] Since a number of perspectives exist for describing platforms (e.g., Gawer, 2014; Gawer, 2021b; Sun, Gregor, et al., 2015; Thomas et al., 2014), an ambiguity prevailed in the e-commerce literature, with the ambiguity pertaining to the terminology and concepts applied—often interchangeably—for platforms.[47] As potential success factors for platforms in e-commerce, the symbiotic relationship between the governance and the architecture of platforms (e.g., Evans and Schmalensee, 2016; Tiwana, 2014) remains under-theorized in e-commerce. Although the importance of digital interfaces for collaboration and the conduct of transactions prevails in e-commerce, domain-specific analyses remain scarce (e.g., Engert et al., 2022). Similarly, it is undisputed that multiple actors jointly create

[46] More generally, More generally, Penrose (1989, p. 11) suggested that the value of "universal comprehensive generalization purporting to explain all aspects of economic reality in one grand model" is limited.

[47] Platforms are termed very ambiguously in existing research and proper denominations are also often lacking (Thomas et al., 2014, p. 203). While innovation platforms are often described as technological infrastructure, digital platforms, and software engine (e.g., Aulkemeier, Paramartha, et al., 2016; Cichocki, 2018; Engert et al., 2022; Ghazawneh and Henfridsson, 2015; Halckenhaeusser et al., 2022; Wulf and Blohm, 2020), transaction platforms are also termed multi-sided market, electronic marketplace, digital marketplace, and electronic environment (e.g., Aulkemeier, Schramm, et al., 2016; Hagiu and Wright, 2015a; Hagiu and-Wright, 2015b; Kawa and Walesiak, 2019; Täuscher and Laudien, 2018). We apply the terms innovation platform and transaction platform throughout this cumulative dissertation. A platform incorporating characteristics of both types is described as a hybrid platform, which is in line with existing publications (e.g., Evans and Gawer, 2016; Gawer, 2021b).

value in e-commerce ecosystems (Adner, 2017; Aulkemeier, Iacob, et al., 2019; Gui et al., 2018; Yi and Ming, 2011) and that the value creation is more efficient if it is properly orchestrated by a focal participant (Autio, 2022; Dhanasai and Parkhe, 2006; Mann et al., 2022). However, means for ensuring a proper ecosystem orchestration have not yet been identified in the context of e-commerce. Adner (2017, p. 43) identified activities, actors, positions, and links as elements for an ecosystem's value proposition to materialize. Hence, the prerequisites for and amplifiers of multilateral value creation in e-commerce ecosystems need to be further investigated. We summarize the aforementioned elaboration in the research problem of this doctoral thesis in Fig. 1.1

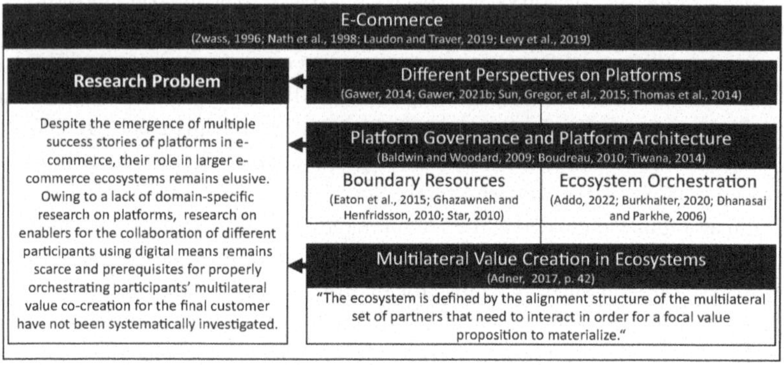


**Fig. 1.1** Perspectives on the Research Problem

This thesis addresses the overarching research problem with a set of accumulated research endeavors on platforms in the context of e-commerce. We seek to describe platforms and digital business ecosystems (Benedict, 2018; Jacobides et al., 2018) in the context of e-commerce from an affiliation perspective and a structural perspective (Adner, 2017). This conceptualization comprises the different types of participants involved in e-commerce ecosystems and the participants' contributions to the overall multilateral value creation for the (final) customer (Adner, 2006; Doherty and Ellis-Chadwick, 2006; Wulfert, Woroch, Strobel, Seufert, et al., 2022). Throughout this cumulative dissertation, we adopt the perspective of the owner of the focal platform and investigate its role in orchestrating and shaping e-commerce ecosystems. In order to conduct the research for this dissertation, the aforementioned research problem is translated into the following research question:

> **Research Question:** How do owners of focal platforms in e-commerce ecosystems shape ecosystem structure and participants?

To operationalize this research question, it is hierarchically subdivided into three investigative questions (Cooper and Schindler, 2014; Recker, 2013). These investigative questions focus on particular aspects of platforms and their owners' role in e-commerce ecosystems. First, we draw on Observation 1, Observation 2, and Observation 5 (i.e., the economic success of innovation platforms and transaction platforms in e-commerce), the ambiguity in the terminology on platforms in e-commerce (e.g., Aulkemeier, Paramartha, et al., 2016; Aulkemeier, Schramm, et al., 2016; Engert et al., 2022; Hagiu and Wright, 2015a; Hagiu and Wright, 2015b; Kawa and Walesiak, 2019; Täuscher and Laudien, 2018), and the numerous calls for domain-specific research on platforms (de Reuver et al., 2017; McIntyre and Srinivasan, 2017; Wulfert and Karger, 2022a). In general, platform business models as multi-sided markets that benefit from network effects have been introduced (Armstrong, 2006; Rochet and Tirole, 2003; West, 2003). These transaction platforms could be seen as additional intermediaries in e-commerce ecosystems (Hagiu and Wright, 2015b). Moreover, innovation platforms could form the technological "core" (Blaschke, Haki, et al., 2019, p. 4) of e-commerce ecosystems, providing the necessary application services for conducting transactions in e-commerce. From an IS perspective, it is also worthwhile investigating which application services are necessary for an innovation platform in the context of e-commerce (Aulkemeier, Paramartha, et al., 2016; Baldwin and Woodard, 2009). To provide clarity on the concept of platforms in e-commerce in general and e-commerce ecosystems in particular, and also to reduce ambiguity in the terminology, we pose the following investigative question:

> **Investigative Question 1:** How can platforms be conceptualized in e-commerce?

Second, we consider Observation 3 (i.e., the relevance of e-commerce ecosystems) and Observation 6 (i.e., the necessity of boundary resources in e-commerce ecosystems) as well as calls for design guidelines for boundary resources (de Reuver et al., 2017; Hein, Schreieck, Riasanow, et al. 2020; Schüler and Petrik, 2021; Yoo, Henfridsson, et al., 2010). Transactions in e-commerce ecosystems demand boundary resources as demarcation points between the focal platform and different types of ecosystem participants and their IS. In this regard, boundary resources are the backbone of successful e-commerce ecosystems and need to be designed adequately (Engert et al., 2022; Star, 2010; Wulfert, Woroch, Strobel, Seufert, et al., 2022). Furthermore, we acknowledge the role of third-party developers in increasing the degree of generativity in e-commerce ecosystems. This generativity is required to provide the necessary application services for a variety of different business models, companies, and articles extending the service portfolio of the innovation platform by third-party extensions. These extensions can involve, among others, shop themes, platform integrations, and service integrations (Aulkemeier, Paramartha, et al., 2016; Jansen and van Capelleveen, 2013; Wulfert, Woroch, Strobel, Seufert, et al., 2022). The design of the boundary resources aims at attracting additional ecosystem participants in general and third-party developers in particular (Dal Bianco et al., 2014; Ghazawneh and Henfridsson, 2013). While the former is likely to increase network effects and the value of the ecosystem for participants, the latter seeks to increase the degree of generativity of the overall ecosystem. Since boundary resources have strategic relevance for the owners of focal platforms in e-commerce ecosystems (Boudreau, 2010; Yoo, Henfridsson, et al., 2010), they need to be managed carefully beyond the design phase. To prescribe the design and management of boundary resources in e-commerce ecosystems, we pose the following investigative question:

**Investigative Question 2:** How should the boundary resources of a focal platform in e-commerce ecosystems be designed and managed to attract and maintain ecosystem participants?

Third, we enclose Observation 4 (i.e., ecosystem orchestration propelling ecosystem value) and research on factors amplifying the value creation in business ecosystems. Ecosystem orchestration by a focal actor was described as a potential amplifier for more efficient resource usage in business ecosystems creating more value for the ecosystem participants (Autio, 2022; Dhanasai and Parkhe, 2006; Mann et al., 2022). Proper ecosystem orchestration can therefore be a source of advantage in the inter-ecosystem competition (Amit and Zott, 2021; Parker, Van Alstyne, and Choundary, 2016; Rietveld and Schilling, 2021). However, antecedents for a proper orchestration of e-commerce ecosystems, such as environmental conditions and platform owner's capabilities, have not been identified thus far. To further investigate the prerequisites for a proper ecosystem orchestration in e-commerce, we pose the following investigative question:

**Investigative Question 3:** How can platform owners in e-commerce ecosystems successfully orchestrate participants' multilateral value creation?

## 1.3 Integration of Research Papers

To address these investigative questions and answer the overall research question, this cumulative dissertation includes a total of ten research papers arranged in chapters four to seven. The ten academic publications have been or are currently subject to double-blind peer-reviewing processes.[48] The research papers have been published or are under consideration for publication in leading IS journals, such as Electronic Markets and Electronic Commerce Research, and results have been presented recently at renowned IS conferences, such as the International Conference on Information Systems, Pacific Asia Conference on Information Systems, and the Hawaii International Conference on System Sciences. In Table 1.1, we provide the bibliographic information on the publications included in this cumulative dissertation referenced by their IDs in consecutive order. We also indicate whether the publication is already published (P) or under review for publication (U) and the ranking of the (potential) publication outlet. For the identification of the publications' rankings, we either refer to the VHB-JOURQUAL3 (VHB, 2022) or the CORE-2018 (Core, 2023).[49]

**Table 1.1** List of Publications

| ID | Bibliographic Information | Status | Ranking |
|---|---|---|---|
| P1 | T. Wulfert and E. Karger (2022a). "A Bibliometric Analysis of Platform Research in E-Commerce: Past, Present, and Future Research Agenda." In: *Corporate Ownership&Control* 20.1, pp. 185–200 | P | VHB: C |
| P2 | T. Wulfert and E. Karger (2022b). "Shaping Digital Platforms in E-Commerce: Developing an Architecture Framework." In: *Proceedings of the 26th Pacific Asia Conference on Information Systems (PACIS2022).* Virtual Conference, pp. 1–17 | P | VHB: C |

(Continued)

[48] Eight of the research papers have either been published in renowned IS journals or presented at leading IS conferences, while two of the research papers are currently under review for publication.

[49] The *SN Computer Science* (P4) has been established in 2020 and is not yet included in any of the mentioned rankings for scientific publications. Nevertheless, an intermediary version of this paper has been presented at the International Conference on Enterprise Information Systems ranked C in the CORE-2018 ranking (Wulfert and Schütte, 2021). It was invited with a fast-track to the *SN Computer Science* journal following its initial presentation.

**Table 1.1** (Continued)

| ID | Bibliographic Information | Status | Ranking |
|---|---|---|---|
| P3 | T. Wulfert, S. Seufert, et al. (2021). "Developing Multi-Sided Markets in Dynamic Electronic Commerce Ecosystems—Towards a Taxonomy of Digital Marketplaces." In: *Proceedings of the 54th Hawaii International Conference on System Sciences (HICSS 2021)*. Maui, Hawaii, pp. 6133–6142 | P | VHB: C |
| P4 | T. Wulfert and R. Schütte (2022). "Retailer's Dual Role in Digital Marketplaces—Reference Architecture for Retail Information Systems". In: *SN Computer Science 3.206*, pp. 1–17. | P | n.a. |
| P5 | T. Wulfert and J. Busch (2022). "Reference Architectures Facilitating a Retailer's Dual Role on DigitalMarketplaces: A Literature Review." In: *Proceedings of the 24th International Conference on Enterprise Information Systems (ICEIS2022) Volume 2*. Online Event, pp. 494–505 | P | Core: C |
| P6 | T. Wulfert, R.Woroch, G. Strobel, S. Seufert, et al. (2022). "Developing Design Principles to Standardize E-Commerce Ecosystems: A Systematic Literature Review and Multi-Case Study of Boundary Resources." In: *Electronic Markets* 32, pp. 1813–1842 | P | VHB: B |
| P7 | T. Wulfert, G. Strobel, et al. (2024). "Let's Join Forces: Boundary Resource as Enablers of Value Cocreation in E-Commerce Ecosystems." In: *Electronic Commerce Research* Forthcoming | U | VHB: C |
| P8 | T. Wulfert (2023). "Boundary ResourceManagement in Innovation Ecosystems: The Case of E-Commerce." In: *Electronic Markets* 33.27, pp. 1–27 | P | VHB: B |
| P9 | T. Wulfert, R.Woroch, and G. Strobel (2024). "How Giants Rule the World: A Qualitative Comparative Analysis of Dynamic Capabilities for E-Commerce Ecosystem Orchestration." In: *Journal of Strategic Information Systems* Forthcoming | U | VHB: A |
| P10 | R.Woroch et al. (2022). "Four Shades of Customer: How Value Flows in Fintech Ecosystems." In: *Proceedings of the 43rd International Conference on Information Systems (ICIS2022)*. Copenhagen, Denmark, pp. 1–17 | P | VHB: A |

Status: published (P), under review (U); Ranking refers to the outlet ranking VHB-JOURQUAL3 (VHB, 2022) and CORE-2018 (Core, 2023) as indicated.

The published publications in this cumulative dissertation have not been altered regarding content, but have been adapted to the layout of this thesis. If modifications are made, they involve corrections of minor spelling mistakes, and the standardization of labels, tables, footnotes, and abbreviations.[50] We also annotate linguistic differences across the research papers published with footnotes to increase the overall consistency of this cumulative dissertation.[51]

We integrate the ten research endeavors in an overarching design science research process following Peffers et al. (2007, p. 54).[52] The research papers are positioned within the concept of e-commerce ecosystems centered around a focal platform that orchestrates the ecosystems' participants (Wulfert, Woroch, Strobel, Seufert, et al., 2022; Yi and Ming, 2011).[53] We investigate platforms in e-commerce from the enterprise architecture (Ahlemann, 2012; Angelov, Grefen, et al., 2012; Cloutier et al., 2010), the boundary resource (Dal Bianco et al., 2014; Eaton et al., 2015; Ghazawneh and Henfridsson, 2013), the application lifecycle (AXELOS, 2012; Kääriäinen, 2011; Stark, 2011), dynamic capabilities (Eisenhardt and Martin, 2000; Teece et al., 1997), and the value creation perspectives (Adner, 2017; Vargo and Lusch, 2014; Vargo, Maglio, et al., 2008). In the following, we will describe the grouping of the research papers, the relations among the four groups, and briefly introduce the single research papers as well as their positioning within the overarching design science research process depicted in Fig. 1.2 (Peffers et al., 2007, p. 54).

The first group consists of three research papers and is concerned with platforms in e-commerce (Chap. 4). This chapter forms the outer environment of our research artifacts focusing on more detailed aspects of platforms (Simon, 1996). These platforms form the center of e-commerce ecosystems. In the first research paper, we introduced different platform types in e-commerce (i.e., innovation platforms, transaction platforms, and hybrid platforms) and developed a future research agenda for platforms in e-commerce (Evans and Gawer, 2016; Gawer, 2021b; Wulfert and Karger, 2022a). In general, platforms form the center of an e-commerce ecosystem

[50] We would like to stress that the size and positions of figures and tables may differ slightly from their original publications because of these layout alterations.

[51] These footnotes did not appear in the original publications.

[52] Peffers et al.'s (2007) design science research process consists of six phases, namely problem identification and motivation, the definition of the objectives of the solution, design and implementation, demonstration, evaluation, and communication. The process allows for iterations across selected phases. We will provide additional details on the overarching design science research process and the meta-theoretical assumptions involved in Sect. 2.2.

[53] For a more detailed introduction of e-commerce ecosystems, we refer to Sect. 3.4 (Wulfert, Woroch, Strobel, Schoormann, et al., 2023; Wulfert, Woroch, Strobel, Seufert, et al., 2022).

and orchestrate the other ecosystem participants (Wulfert, Woroch, Strobel, Seufert, et al., 2022). As Staykova and Damsgaard (2015, p. 4) described the platform as the "core" of an ecosystem, the platform owner has the power to control and govern the overall ecosystem (Tiwana, 2014). The bibliometric analysis of 7,463 research papers in the first research paper (P1) revealed three periods of platform research in e-commerce, four thematic clusters of platform research in e-commerce, and numerous potential avenues for future research involving all types of platforms (Wulfert and Karger, 2022a). The thematic clusters and related avenues for future research motivated the overall design science research process and posed the first objectives for the artifacts designed in the following research papers. In the second research paper (P2), we developed a platform architecture framework for e-commerce as an industry-specific conceptualization of platforms and their surrounding ecosystems (Wulfert and Karger, 2022b). Taking an enterprise architecture perspective on platforms in e-commerce (Angelov, Grefen, et al., 2012; Winter, 2009; Wulfert and Karger, 2022b), we answered calls for domain-specific conceptualizations of platforms (de Reuver et al., 2017). In the third research paper (P3), we developed a taxonomy for describing transaction platforms (Wulfert, Seufert, et al., 2021).[54] The established differentiation informs the architecture analysis, the boundary resource analysis, and the value creation investigation. Besides the description objective (Gregor, 2006), the taxonomy can also be applied by practitioners to evolve their transaction platform in dynamic e-commerce environments (Wulfert, Seufert, et al., 2021). We demonstrated the applicability of the taxonomy using 44 transaction platforms in e-commerce.

We clarify the architecture of these platforms that partition the e-commerce ecosystem "into stable core components and variable periphery" (Baldwin and Woodard, 2009, p. 24) in the second group of research papers forming the fifth chapter. The architecture is considered an important building block of a platform in literature (Baldwin and Woodard, 2009; Blaschke, Haki, et al., 2019; Cloutier et al., 2010; Tiwana, 2014). Since several reference architectures[55] for e-commerce exist

[54] In the published research paper, the term transaction platform is not used. Instead, we described transaction platforms as digital marketplaces for reasons of consistency with the existing literature. As already introduced, we use these terms synonymously in this thesis because of their common objective and shared theoretical foundation (i.e., two-sided market theory) (Armstrong, 2006; Rochet and Tirole, 2003; Rochet and Tirole, 2006; West, 2003).

[55] Reference architectures can be described as templates for the design of company-specific architectures (Bass et al., 2003; Schütte and Wulfert, 2022b). Taylor et al. (2010, p. 58) defined a reference architecture as a "set of principal design decisions that are simultaneously applicable to multiple related systems, typically within an application domain, with explicitly defined points of variation."

in the literature, we derived specific requirements for platforms in e-commerce based on a retailer's dual role[56] on transaction platforms. The dual role results from the simultaneous role of a retailer as platform owner and reseller on a particular transaction platform (Wulfert and Schütte, 2021, 2022). In the fourth research paper (P4), we derived a total of nine architectural requirements (i.e., retailer's dual role, additional types of participants, affiliation with the platform, matching as core value proposition, diversity of services, innovation platform services, smart service provision, aggregated assortment, sophisticated boundary resources) for a retailer's dual role on transaction platforms. We also developed four architectural patterns (i.e., aggregated assortment, boundary resource, innovation platform, matching) as parts of a potential reference architecture addressing these requirements (Wulfert and Schütte, 2022).[57] We developed the patterns based on a design science research approach emphasizing problem identification, objectives, and solution design (Peffers et al., 2007, p. 54). Applying the derived architectural requirements, the developed architectural patterns, and additional reference architecture characteristics (Angelov, Grefen, et al., 2012) as analysis criteria in the fifth research paper (P5), we identified 13 reference architectures for retail and e-commerce in a systematic literature analysis (Wulfert and Busch, 2022). However, none of the 13 identified reference architectures fully supports the architectural requirements resulting from a retailer's dual role in e-commerce ecosystems.

Having introduced platforms in e-commerce as the center of surrounding ecosystems, we performed a drill-down to platform boundary resources as boundary objects between the focal platform and additional e-commerce ecosystem participants in Chapter six (Ghazawneh and Henfridsson, 2013; Santos and Eisenhardt, 2005). Boundary resources are an integral part of the platform's architecture for the attraction of and communication with participants in e-commerce ecosystems (Dal Bianco et al., 2014; Ghazawneh, 2012; Tiwana, 2014). Boundary resources can be characterized as social boundary resources, platform boundary resources, and development boundary resources (Dal Bianco et al., 2014). Following Baldwin and Woodard (2009, p. 41), boundary resources are "the only components that remain truly stable over long periods of time." This chapter involves three research papers in a

[56] Wulfert and Schütte (2022) argue that the dual role can also become a triple role if the platform owner also develops and provides its own application services and opens these services to other ecosystem participants (Jansen and van Capelleveen, 2013).

[57] Taylor et al. (2010, p. 73) defined an architectural pattern as "a named collection of architectural design decisions that are applicable to a recurring design problem, parameterized to account for different software development contexts in which that problem appears." We apply the ArchiMate notation as a well-established standard for the description of enterprise architectures within this thesis (Hansen and Neumann, 2007; Open Group, 2019).

multi-paper approach (Vom Brocke, Winter, et al., 2020). The sixth research paper (P6) aimed at developing design principles for boundary resources in e-commerce ecosystems (Chandra Kruse, Seidel, and Purao, 2016; Wulfert, Woroch, Strobel, Seufert, et al., 2022). We derived six meta-requirements for the design of a platform's boundary resources in e-commerce and developed 19 design principles for the development of concrete design instances of boundary resources in e-commerce ecosystems (Wulfert, Woroch, Strobel, Seufert, et al., 2022). We updated these design principles using the results of an interview study increasing the design principles' projectability and fitness in the seventh research paper (P7) (Iivari, Rotvit Perlt Hansen, et al., 2021; Vom Brocke, Winter, et al., 2020; Wulfert, Strobel, et al., 2024). We conducted a series of semi-structured interviews with twelve domain experts employed at innovation platforms and transaction platforms on the reusability of the design principles (Iivari, Rotvit Perlt Hansen, et al., 2021). We summarized and updated the design principles for boundary resources to seven types and assigned them to roles responsible for the implementation and management (Wulfert, 2023). We applied an application lifecycle management perspective in the eighth paper (P8) to develop a boundary resource management lifecycle for innovation platforms (Wulfert, 2023). The reference procedure model for the management of boundary resources consists of four layers (i.e., boundary resource governance, continuous communication, technical boundary resources, and supplementary boundary resources). The lifecycle for managing technical boundary resources has a total of five subsequent lifecycle stages (i.e., plan and design, development and test, release, maintenance and optimization, and deprecation). We also mapped major supplementary boundary resources to each lifecycle stage.

As platforms in e-commerce are likely to be the "core" (Staykova and Damsgaard, 2015, p. 4) of a surrounding ecosystem and the platform owner is described as the "ecosystem leader" (Moore, 1996, p. 126) and the "matchmaker" (Evans and Schmalensee, 2016, p. 2), we analyzed these e-commerce ecosystems in more detail from the dynamic capabilities (Eisenhardt and Martin, 2000; Teece et al., 1997) and the value creation (Vargo and Lusch, 2004; Vargo, Maglio, et al., 2008) perspectives in Chapter seven. As a platform's architecture restricts and enables the ecosystem's governance (Tiwana, 2014, p. 45), we further investigated platform governance in terms of capabilities required for an increased value creation (Burkhalter, 2020; Dhanasai and Parkhe, 2006; Koch, 2010). Starting from a structural perspective on e-commerce ecosystems involving participants, value creation activities, and relations among participants (Adner, 2017), we prescribed successful e-commerce ecosystems and investigated their capabilities for ecosystem orchestration in the ninth research paper (P9) (Wulfert, Woroch, and Strobel, 2024). We identified twelve major participants in the e-commerce ecosystem by conducting a systematic literature review (Wulfert, Woroch, and Strobel, 2024, p. 11). We extended this affiliation perspective to an ecosystem-as-structure perspective focusing on the platform owner's activities for orchestrating the value creation within e-commerce ecosystems (Adner, 2017), derived five dynamic capabilities for ecosystem orchestration involving 12 second-order themes and 44 first-order constructs (Gioia et al., 2013), and proposed 12 capability configurations for a successful ecosystem orchestration. Besides the holistic analysis of the e-commerce ecosystem, we further analyzed Financial Technology (fintech) ecosystems in the tenth research paper (P10), responsible for the payment processing in e-commerce (Woroch et al., 2022). We considered the payment ecosystem as a sub-ecosystem of e-commerce ecosystems. We proposed a reference model of fintech ecosystems using the e$^3$-value modeling notation and further investigated four roles customers can take on in fintech ecosystems. We depict the ten research papers within this cumulative dissertation along Peffers et al.'s (2007, p. 54) design science research process in Fig. 1.2.

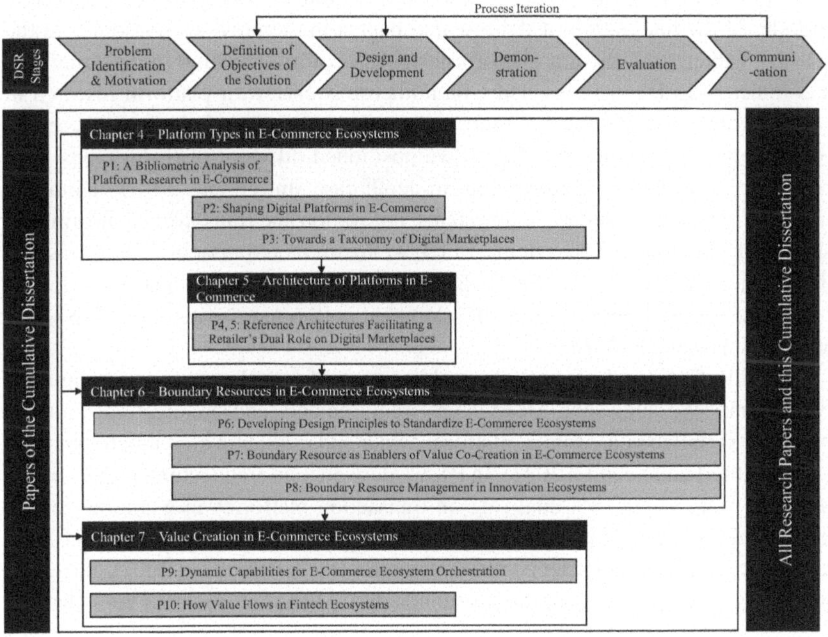


**Fig. 1.2** Integration of Research Papers (according to Peffers et al., 2007)

## 1.4 Structure of the Dissertation

Subsequent to this introductory chapter, the remainder of this dissertation is structured in the following way: a section on the scientific approach; a section on the related literature, unfolding the conceptual foundation of this cumulative dissertation; the main body containing the single research papers of this cumulative dissertation; a discussion section; and a brief summary of the main contributions of this thesis. In the second chapter, we introduce the theoretical stance taken and elaborate on the scientific approaches used for conducting the research endeavors. In the third chapter that follows this introductory chapter, we unfold e-commerce, platforms, business ecosystems, and boundary resources as key concepts of this cumulative dissertation.

In total, this cumulative dissertation consists of ten research papers. The research papers are organized in four chapters: platform types in e-commerce, the architecture of platforms in e-commerce, boundary resources in e-commerce ecosystems, and orchestration of and value creation in e-commerce ecosystems. We already introduced the integration of the research papers in the previous section.[58] In the fourth Chapter, we focus on platforms as objects of investigation in the context of e-commerce. In P1, we provided a bibliometric overview of platform research in e-commerce deriving topical clusters and developing an agenda for future research on platforms in e-commerce. In P2, we positioned different platform types in e-commerce ecosystems. Based on an in-depth case study (Yin, 2018) of innovation platforms in e-commerce ecosystems, innovation and transaction platforms are shaped in e-commerce ecosystems. In P3, we focused on transaction platforms and derived a taxonomy for describing different types of transaction platforms. In this paper, the approach for taxonomy development in IS research was applied (Nickerson, Varshney, et al., 2013).

The fifth Chapter deals with the architecture of focal platforms in e-commerce ecosystems. We suggested architectural requirements and architectural patterns based on a retailer's dual role as platform owner and reseller in e-commerce ecosystems (Wulfert and Schütte, 2022). In P4, we derived nine architectural requirements resulting from a retailer's dual role on transaction platforms in e-commerce and developed four design patterns for innovation platforms in e-commerce ecosystems, applying a design science research approach (Peffers et al., 2007). Based on these requirements and patterns, we conducted a literature review (Vom Brocke,

[58] For reasons of consistency, we will describe the remainder of this cumulative dissertation completely. However, we only refer briefly to the single research papers.

Simons, et al., 2009) in P5 to investigate the fulfillment of these requirements in existing reference architecture for e-commerce.

The sixth Chapter is concerned with boundary resources in e-commerce ecosystems. This chapter includes three research papers in a multi-paper design science research approach (Vom Brocke, Winter, et al., 2020) developing design principles for boundary resources of a focal platform in e-commerce ecosystems. A platform owner can implement these boundary resources to attract third-party developers, with the objective of increasing the overall generativity of the ecosystem and network effects for participants (e.g., Eaton et al., 2015; Ghazawneh and Henfridsson, 2010, 2015). While we derive design principles for boundary resources based on a dual approach involving a literature review (Vom Brocke, Simons, et al., 2009) and a multi-case study (Yin, 2018) in P6, we evaluated these design principles based on a semi-structured interview study (Misoch, 2015) and the framework for light reusability evaluation of design principles (Iivari, Rotvit Perlt Hansen, et al., 2021) in P7. Moreover, we developed a lifecycle for the management of boundary resources in P8. In this research paper, we followed the design science research approach by Vaishnavi, Kuechler, et al. (2019) informed by a literature review (Vom Brocke, Simons, et al., 2015), multi-case analysis(Yin, 2018), and an additional interview study (Misoch, 2015).

The seventh Chapter focuses on the orchestration of the diverse set of participants in e-commerce ecosystems for improved value creation (Koch, 2010; Mann et al., 2022). In this chapter, we explicitly take on the ecosystem-as-structure perspective emphasizing the multilateral value co-creation in e-commerce ecosystems (Adner, 2017) In P9, we identified major participants of an e-commerce ecosystem from the literature (Vom Brocke, Simons, et al., 2009) and derived dynamic capabilities for the owner of the focal platform necessary for an efficient ecosystem orchestration based on Qualitative Comparative Analysis (QCA) (Park, Fiss, et al., 2020). In P10, we dealt with value activities and value flows (i.e., structural perspective) in fintech ecosystems as a specific sub-ecosystem of e-commerce ecosystems responsible for payment processing. Value activities and value flows were deduced from a multi-case study (Yin, 2018) that formed the foundation of the value analysis (Peppard and Rylander, 2006).

In the discussion section, the research papers are summarized and their main contributions are explicated. This section closes with a depiction of the limitations involved in this thesis as an outlook on possible avenues for future research. Finally, the results of this thesis are briefly summarized. The overall structure of this dissertation is depicted in Fig. 1.3.

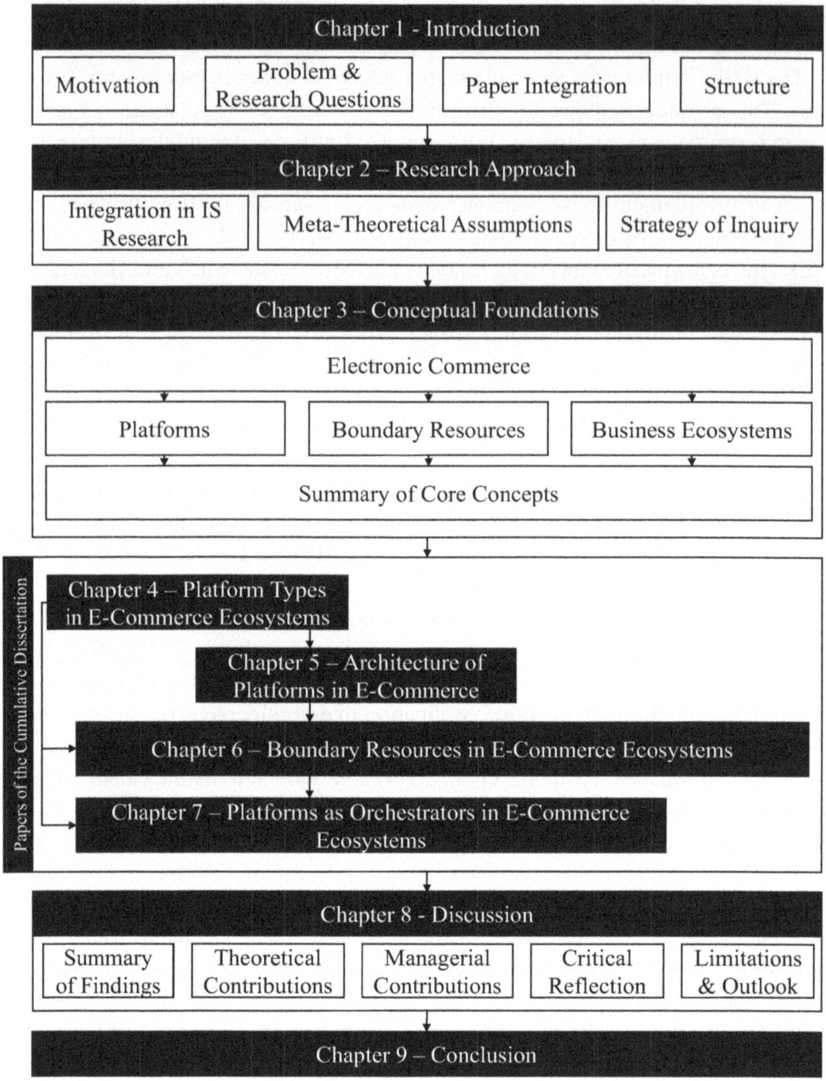


**Fig. 1.3** Overview of the Dissertation Structure

# Scientific Approach 2

In this section, we position our research endeavor within the IS discipline.[1] We also introduce our ontological assumptions and epistemological stance as the foundation of our research. Moreover, we present the methods of inquiry[2] applied in this cumulative dissertation with emphasis on the data collection and data analysis methods applied.

## 2.1 Integration in Information Systems Research

The research objectives of this cumulative dissertation match the object of inquiry and research agenda of design-oriented information systems research (Österle et al., 2011; Schütte, Ahlemann, et al., 2022). Our objective for prescribing mechanisms

[1] As this cumulative dissertation is written in a German context, we locate our research endeavor more in the tradition of the German Wirtschaftsinformatik following a design-oriented research approach developing (software) artifacts and deriving prescriptive knowledge "about technological innovations that are or can be useful for individuals, organizations, or society" (Drechsler and Hevner, 2018, p. 82). We intentionally discriminate our research from the positivist, behavioral research in the Anglo-American context (Benbasat and Zmud, 1999; Drechsler and Hevner, 2018; Gregor, 2006). We will not further differentiate between the German Wirtschaftsinformatik and the Anglo-American Information Systems (IS) discipline (e.g., Stahl, 2009).

[2] Iivari, Hirschheim, et al. (1998, p. 174) defined the inquiry as "the procedures or means by which [...] knowledge" is obtained.

**Supplementary Information** The online version contains supplementary material available at https://doi.org/10.1007/978-3-658-45198-1_2.



T. Wulfert, *Selected Perspectives on Platforms in E-Commerce Ecosystems*,
https://doi.org/10.1007/978-3-658-45198-1_2

for collaboration in e-commerce ecosystems and the orchestration of these ecosystems follows the tradition of design science research (Gregor, Müller, et al., 2013; Hevner, March, et al., 2004; Peffers et al., 2007).[3] The core subject matter in information systems research is information systems[4] as socio-technical systems (Hesse, Barkow, et al., 1994; Hevner, March, et al., 2004; Orlikowski and Baroudi, 1991; Schütte, 1998).[5] Österle et al. (2011, p. 8) defined the object of inquiry in information systems research involving "IS both for organizations and individuals in society and economy."[6] The authors further denoted the socio-technical nature of IS that consists "of three object types, namely people (i.e. human task bearers), information and communications technology (i.e. technical task bearers), and organizational concepts (i.e. functions, structures, processes), and the interrelationships between them" (Österle et al., 2011, p. 8).[7] Information systems research advocates "to address behavior and technology in unison" (Niehaves and Becker, 2006, p. 3580). The information systems discipline can be described as an application-oriented discipline whose results should be transferred into practice in a beneficial way (Frank, 2007, p. 155).[8] Kanellis and Papadopoulos (2009, p. 10 f.) also highlighted the applied nature of socio-technical systems in information systems research. Therefore, information systems research refers to the design of social science endeavors that are of high "relevancy to industry and commerce" (Kanellis and Papadopoulos, 2009, p. 10 f.).

Österle et al. (2011, p. 8) also acknowledged that information systems knowledge is not only comprised in scientific publications but also held to a greater extent

---

[3] Design research is relevant to information systems due to its focus on (IT) artifacts (Orlikowski and Iacono, 2001, p. 121). It addresses the socio-technical core of information systems research "by making systems and methods the unit of analysis and by evaluating research outcomes in an organizational context" (Arnott, 2006, p. 57).

[4] For an elaborated discussion on information systems definitions from various perspectives, we refer to Alter (2008, 449 f.).

[5] Walsham (2006, p. 91) described the pivotal vision of IS research even broader as the contribution to "a better world" by investigating socio-technical systems.

[6] Recently, Schütte, Ahlemann, et al. (2022, p. 530) extended information systems research's object of inquiry including economic efficiency when investigating socio-technical systems.

[7] This description of the research object in information systems research is the translated version of the original quote postulated in the *Memorandum zur gestaltungsorientierte Wirtschaftsinformatik* (Österle et al., 2010, p. 3).

[8] As such, information systems are considered as means for reaching organizational goals and need to be managed accordingly (Schütte, Seufert, et al., 2022). The origins of information systems research are rooted in computer science and economics (Becker, Holten, et al., 2003, p. 3) and, as a sub-discipline, information systems research also functions thematically as an interface between these two disciplines (Winter, 2009, p. 223).

by organizations in the form of "the experiences and knowledge accumulated in business concerning IS, software products, organizational concepts, methods, and tools." In general, information systems research follows description, explanation, and design objectives (Mertens et al., 2017; Schütte, Ahlemann, et al., 2022). In this cumulative dissertation, we describe innovation platforms and transaction platforms in the context of e-commerce. The investigated innovation platforms and transaction platforms in e-commerce resemble information systems (Aulkemeier, Paramartha, et al., 2016; Blaschke, Haki, et al., 2019; Tilson, Sorensen, et al., 2012; Wulfert and Karger, 2022b).[9] These platforms consist of a number of loosely-coupled modules accessible by ecosystem participants using boundary resources (Baldwin and Woodard, 2009; Tiwana, Konsynski, and Bush, 2010; Yoo, Henfridsson, et al., 2010). The surrounding e-commerce ecosystems exhibit socio-technical characteristics with the focal platform and the platform owner as well as a number of surrounding individuals and organizations (Gui et al., 2018; Wulfert, Woroch, Strobel, Schoormann, et al., 2023; Wulfert, Woroch, Strobel, Seufert, et al., 2022; Yi and Ming, 2011).

For the partitioning of e-commerce ecosystems and the differentiation of platform types, we took on an enterprise architecture perspective. We also developed architectural patterns for information systems supporting transaction platforms in e-commerce. The description, modeling, and analysis of enterprise architectures is considered a core topic in information systems research (Winter, 2009, p. 226).[10] Moreover, we provide a set of design principles for boundary resources that enable the collaboration and value creation of ecosystem participants with the focal platform. These design principles prescribe the design and implementation of a platform's boundary resources in e-commerce and are augmented by a reference procedure model that guides the management of these boundary resources. We also prescribe capabilities required from a focal platform owner for a successful orchestration of e-commerce ecosystems. Furthermore, we shed light on the value creation in the e-commerce ecosystem in general and payment sub-ecosystems in particular.

Our research results are compliant with the research objectives of design-oriented information systems research that are described as "instructions for action (i.e., normative, practically applicable means-ends conclusions) that allow the design and operation of IS and innovative concepts within IS (instances)" (Österle et al.,

[9] Several authors also described innovation platforms and transaction platforms as socio-technical systems (e.g., Hermes and Schreieck, 2021; Kapoor, et al., 2021; Wulfert and Karger, 2022b).

[10] Winter (2009, p. 226) especially focused on information systems architectures and respective sub architectures.

2011, p. 8).[11] The architectural patterns, the design principles, the capability sets, and the value maps resemble normative design guidelines for the establishment and development of successful platforms in e-commerce.[12]

## 2.2 Meta-Theoretical Assumptions

### 2.2.1 Research Paradigms in Information Systems Research

As the information systems discipline is concerned with economic and technical systems (i.e., socio-technical) as man-made entities, it can be described as a social science (Schütte, 1998, p. 214). In social sciences, such as information systems research, researchers need to explicate their underlying philosophical paradigm[13] to make the results of their research projects inter-subjectively comprehensible (Guba and Lincoln, 1982; Heron and Reason, 1997).[14] Because of the multidisciplinary and international context as well as a developed methodological pluralism, "the extensive publication of epistemological assumptions is [...] almost mandatory" (Becker and Niehaves, 2007, p. 197) in IS research.[15]

These ontological and epistemological assumptions are grounded in the philosophy of science that informs the nature of the object and phenomena under investigation and the methods of inquiry.[16] As the research endeavors and related findings

[11] We focus on the development and provisionof instructions for action guiding the design, implementation and operation of (future) Information Technology (IT) artifacts. In general, design-oriented information systems research enables the improvement of existing information systems and the development of future IT artifacts as well as the derivation of adjacent (design) knowledge.

[12] Feenberg (1991, p. 173) noted that design per se entails "normative implications and is not simply a matter of efficiency."

[13] A research paradigm resembles "the most fundamental set of assumptions adopted by a professional community which allow them to share similar perceptions and engage in common practices" (Hirschheim and Klein, 1989, p. 1201).

[14] If the philosophical assumptions of a researcher are not explicitly disclosed, they must be brought to light through careful interpretations (Albert, 1978, p. 3 f.).

[15] In general, Guba and Lincoln (1994, p. 116) emphasized that "paradigm issues are crucial; no inquirer, we maintain, ought to go about the business of inquiry without being clear about just what paradigm informs and guides his or her approach."

[16] Although an explicit articulation of the philosophical stance taken is often missing in research publications, any mindful and responsible scientific investigation must comprehend the implications of its choice of philosophical grounding (Bechara and Van de Ven, 2007; Guba and Lincoln, 1982).

underlie the researcher's inquiry[17] paradigm,[18] we elaborate on the choice of the philosophy underlying the investigations within this cumulative dissertation (Guba and Lincoln, 1982, p. 363). A paradigm refers to the fundamental worldview that directs the inquirer as well as the ontological, epistemological, and methodological grounds of study (Greene, 2007; Guba and Lincoln, 1994). Guba and Lincoln (1994, p. 107) summarized the research paradigms on three research questions that reflect the levels of scientific philosophy. These questions are the ontological question ("What is the form and nature of reality?"), the epistemological question ("What is the nature of the relationship between the knower or would-be knower and what can be known?"), and methodological question ("How can the inquirer [...] go about finding out whatever he or she believes can be known?") (Guba and Lincoln, 1994, p. 107). The assumptions formulated towards these three questions form the philosophical paradigm underlying scientific research.

Historically, IS research was dominated by positivism and interpretivism as research paradigms (Alharahsheh and Pius, 2020; Chen and Hirschheim, 2004). While positivism considers an objective reality beyond the human mind, interpretivism implies an intersubjectively perceived reality based on meanings and an understanding of social and experiential levels resulting in (potentially) multiple realities (Alharahsheh and Pius, 2020; Chen and Hirschheim, 2004).[19] In general, four major research paradigms have been introduced to IS research (e.g., Becker and Niehaves, 2007; Guba and Lincoln, 1994; Niehaves, 2007a; Thapa and Haj-Bolouri, 2023; Walsham, 1995). These paradigms are positivism, interpretivism, critical realism, and pragmatism.[20] Morgan (2007, p. 52) justified the emergence of the pragmatic paradigm with his criticism of the practicability and usefulness of a detailed elaboration on philosophical assumptions. The author criticized that

[17] We use the terms research and inquiry synonymously in this section.

[18] The inquiry paradigm can be understood as a researcher's grasp of the world (Ogilvy, 1996, p. 31). Creswell and Plano Clark (2018, p. 86) described paradigms as "worldviews." Kuhn (1962, p. 23) defined a paradigm in scientific research as an "accepted model or pattern" informing positions and approaches in a discipline. For Morgan (2007, p. 66), it is more important that scholars "share a consensus about which questions are most important to study and which methods are most appropriate for conducting those studies."

[19] Mertens (1998, p. 67) detailed the interpretive paradigm suggesting that "knowledge is socially constructed by people active in the research process, and that research is a product of the values of the researchers conducting it."

[20] For an overview of the characteristics of these four paradigms along the dimensions of ontology, epistemology, methodology, axiology, methods, and vocabulary, we refer to Thapa and Haj-Bolouri (2023, p. 6). We also need to note here that these paradigms are not undisputed in the context of IS research (e.g., Burrel and Morgan, 2019; Goles and Hirschheim, 2000; Hirschheim and Klein, 1989; Hirschheim, Klein, and Lyytinen, 1995).

these assumptions may detail the understanding of the inquiry made, but these abstract elaborations "tell us little about the more substantive decisions such as what to study and how to do so" (Morgan, 2007, p. 52).[21] We summarize these four research paradigms in Table 2.1. Several scholars have argued for design science[22] as a potential third research paradigm in information systems research (e.g., Goldkuhl, 2012a; Hevner, March, et al., 2004; Vaishnavi, Kuechler, et al., 2019).[23]

**Table 2.1** Summary of the Paradigms in Information System Research (Creswell and Creswell, 2022; Creswell and Plano Clark, 2018; Thapa and Haj-Bolouri, 2023)

| | Positivism | Interpretivism | Critical Realism | Pragmatism |
|---|---|---|---|---|
| Ontology | Realist | Non-realist | Realist | Pluralist view |
| Epistemology | Positivist | Interpretivist | Relativist | Relativist/ Interpretivist |
| Methodology | Quantitative or Qualitative | Qualitative or Quantitative | Qualitative or Quantitative | Qualitative or Quantitative |
| Approaches | Hypothesis testing, experiments | Qualitative case studies, ethnography | case studies | action-oriented approaches |
| Vocabulary | Explanation, prediction, causality, generalization | Understand, thick-description, context | explore, explain, context, mechanism, retrodiction | Practical solution, action, context |

For this cumulative dissertation, we follow the pragmatic paradigm (Creswell and Creswell, 2022; Creswell and Plano Clark, 2018; Goldkuhl, 2012a,b; Thapa and Haj-Bolouri, 2023). Pragmatism as a philosophical paradigm emphasizes "what

[21] Morgan (2007, p. 52) further elaborated that "this combination of strong demands for self-conscious allegiance to one paradigm but less advice about how that should play out in the practices of "workaday" researchers created ongoing difficulties for the metaphysical paradigm."

[22] Historically, design science has been described as "sciences of the artificial" (Simon, 1996, p. 1), "systemeering" (Iivari, 1983), "a constructive approach" (Iivari, 2007, p. 40), and "systems development" (Nunamaker et al., 1990, p. 90). As no recognized definition of design science research was established up to now, we refer to Deng and Ji (2018, p. 5) for an overview of possible definitions.

[23] Vaishnavi, Vaishnavi, et al. (2015, p. 30 ff.) described design science as a philosophical paradigm besides positivism and interpretivism. We will subsume this "paradigm" under the umbrella of pragmatism.

works while abstaining from the use of metaphysical concepts such as truth and reality" (Goles and Hirschheim, 2000, p. 260).[24] In general, pragmatism has "an orientation towards a prospective, not yet realized world" (Goldkuhl, 2012b, p. 140). Philosophers have identified several kinds of pragmatism[25] (Goldkuhl, 2012a; Lovejoy, 1908; Rescher, 2000). Within the pragmatic paradigm in IS research, Goldkuhl (2008) initially identified three sub-types and later Goldkuhl (2012a) described four sub-types, namely local functional pragmatism, general functional pragmatism, referential pragmatism, and methodological pragmatism.[26] We take a general functional pragmatism stance aiming at the development of abstract design knowledge for a class of systems (Goldkuhl, 2012a,b; Gregor, Müller, et al., 2013). We summarize the pragmatic approach in Table 2.2.

**Table 2.2** Characteristics of the Pragmatic Approach

| | Pragmatic approach |
|---|---|
| Ontology | Multiple realities exist. Current truth, meaning, and knowledge are changing. |
| Epistemology | Knowledge is both constructed and based on the reality of the world we experience and live in. Practicality in "what works" for addressing research questions. Justification of knowledge comes from warranted assertions. |
| Methodology | Quantitative, qualitative, and mixed-methods approaches can form the basis of inquiry. |
| Knowledge accumulation | The researcher constantly tries to improve upon past understandings in a way that fits and works in the world in which he or she operates. |
| Communication and shared meaning | A degree of mutual understanding should be achieved between research colleagues and participants. |
| Transferability | The research findings of one study should be explored to understand whether they are useful in other circumstances. |

[24] Nevertheless, we will unfold our ontological, epistemological, and methodological assumptions in the following section.

[25] Lovejoy (1908) described 13 different types of pragmatism.

[26] Goldkuhl (2012a, p. 91) further differentiated local functional pragmatism and general functional pragmatism. While the former focuses on the development of artifacts for problem-solving and the derivation of knowledge from these artifacts, the latter is concerned with the creation of "abstract knowledge that has value beyond local practice[s]" (Goldkuhl, 2012a, p. 91).

Following this approach, we acknowledge that the research objectives of this cumulative dissertation are addressed pragmatically taking the types of research questions and related scientific approaches into account. In the selection of research approaches, we followed existing research in the domain of e-commerce and research endeavors on platforms to the greatest possible extent. We draw on an increase in research on platforms in general (e.g., Asadullah et al., 2018a; Cole et al., 2005; Liu, Li, et al., 2021; Schreieck et al., 2016; Wulfert and Karger, 2022a) and calls for domain-specific research on platforms (e.g., de Reuver et al., 2017; McIntyre and Srinivasan, 2017; Wulfert and Karger, 2022a) suggesting that ecosystem orchestration and boundary resources as enablers of collaboration in e-commerce ecosystems are worth investigating and of interest to researchers and practitioners alike. Furthermore, our bibliometric analysis of platform research in e-commerce indicates a plethora of different scientific approaches for their analysis (Wulfert and Karger, 2022a). This suggestes that a pragmatic approach to platform research in e-commerce would be an appropriate strategy with agreed-upon philosophical assumptions for the researchers within this domain of interest. The pragmatic assumptions also support the design science research approaches applied in several of our research papers (Cole et al., 2005; Deng and Ji, 2018; Goldkuhl, 2012a,b; Hovorka, 2009) and the design-oriented implications of the remaining research papers (e.g., Wulfert, Seufert, et al., 2021; Wulfert, Woroch, and Strobel, 2024).[27]

In general, design science research aims at problem-solving, providing solutions that transform an as-is situation into a desired to-be state (March and Storey, 2008, p. 726).[28] We focus on design science research following a pragmatic approach focusing on the development of new artifacts or the augmentation of existing ones "to [improve] social and individual factors" (Gregg et al., 2001, p. 172 f.).[29] Design science research can be described as a process of "knowing through making" (Carlsson,

[27] Deng and Ji (2018, p. 7) argued that researchers usually apply pragmatic assumptions in design science research. However, Niehaves (2007b) tried to position design science research within an interpretivist paradigm, and Carlsson (2010) positioned design science research within critical realism. In an alternative attempt not further detailed in this thesis, Iivari (2007) introduced Popper's (1978) three worlds model as a potential ontology for design science research.

[28] Simon (1996, p. 130) stated that "everyone designs who devises courses of action aimed at changing existing situations into preferred ones."

[29] Following Gregg et al.'s (2001, p. 172) introduction, a "[s]ocio-technologist/[d]evelopmentalist [...] addresses the valuable contribution of software systems and associated processes to scientific knowledge building".

2010, p. 14).[30] We will detail our ontological, epistemological, and methodological assumptions following the overarching pragmatic paradigm in the following section.

### 2.2.2 Ontological, Epistemological, and Methodological Assumptions

Creswell and Plano Clark (2018, p. 87) referred to "the nature of reality" with ontological assumptions that both enable and limit research endeavors. Following Iivari, Hirschheim, et al. (1998, p. 172), ontology "is concerned with the structure and properties of 'what is assumed to exist', i.e., the basic building blocks that make up the phenomena or objects to be investigated."[31] Niehaves and Becker (2006, p. 3581) described that ontological assumptions are concerned "with the question of a "'real world' existence." The focus of the philosophy of pragmatism is on pragmatic approaches for research on societal problems that guide action highlighting "what works" (Creswell and Plano Clark, 2018, p. 86).[32] Following the pragmatic paradigm, multiple alternative realities and world-states are accepted that are "technologically created [...] upon a single stable underlying physical reality that constrains the multiple world-states" (Deng and Ji, 2018, p. 7). The reality, knowledge, and truth can be changed through the creation and implementation of "novel artifacts" (Deng and Ji, 2018, p. 7).[33]

Following the ontological assumptions about the objects and reality under investigation, "epistemology is concerned with the question of how humans can achieve 'true knowledge'" (Niehaves and Becker, 2006, p. 3581). This involves the "nature of knowledge" (Iivari, Hirschheim, et al., 1998, p. 174) and considerations on validity, truth, and scope (Lee and Nickerson, 2010; Levy and Hirschheim, 2012). Moreover,

[30] However, the "scientific foundations underlying this critical area of the IS field—design [science] research—have remained largely undeveloped. [...] Over the years, in spite of important writing about design [science] research [...] philosophical underpinnings of this form of research have been largely unexplored" (Purao, 2002, p. 1). As this observation has changed over time, we followed the suggestion by numerous researchers and took on the pragmatic paradigm.

[31] Schütte (1998, p. 14) summarized ontology in the question "[w]hat is knowledge?"

[32] In the pragmatic paradigm, a "top-down privileging of ontological assumptions in the metaphysical paradigm [is denied] as simply too narrow an approach to issues in the philosophy of knowledge" (Morgan, 2007, p. 68).

[33] Hirschheim, Klein, and Lyytinen (1995) exemplified that socially constructed elements, such as system requirements, can be adapted by changing perceptions and assumptions.

epistemology describes the constituents of knowledge, possibilities for the acquisition of knowledge, and the transferability of knowledge (Creswell and Creswell, 2022; Creswell and Plano Clark, 2018; Guba and Lincoln, 1994).[34] Blaikie (1993, p. 6) explicates that epistemology refers to "the claims or assumptions made about the ways in which it is possible to gain knowledge about social reality."[35] It is important to note that the epistemological assumptions are already restricted by the ontological stance taken (Guba and Lincoln, 1994, p. 108).

As we follow pragmatism as a research paradigm in this cumulative dissertation, knowledge is considered to be constructed and based on the reality experienced (Morgan, 2007; Teddie and Tashakkori, 2009; Thapa and Haj-Bolouri, 2023; Vaishnavi, Vaishnavi, et al., 2015). Knowledge is obtained "through making" (Deng and Ji, 2018, p. 7).[36] Pragmatism entails "an instrumental view on knowledge" (Goldkuhl, 2012b, p. 140) applicable in practice for changing a current state.[37] Hence, knowledge generation in pragmatism is future-oriented. Goldkuhl (2012b, p. 140) claimed that the character of knowledge in pragmatism goes beyond explaining and understanding.[38] The author subsumed "prescriptive [knowledge] (giving guidelines), normative [knowledge] (exhibiting values), and prospective [knowledge] (suggesting possibilities)" under what he called "constructive knowledge" (Goldkuhl, 2012b, p. 140).[39]

For pragmatists, truth is defined by "what works at the time" (Creswell and Creswell, 2022, p. 294).[40] Truth is achieved in terms of its practical relevance

34 Iivari, Hirschheim, et al. (1998, p. 174) also included "methods of inquiry" within epistemological assumptions, which we discuss distinctly in the remainder of this section.

35 Kanellis and Papadopoulos (2009, p. 8) claimed for the information systems discipline, that "[p]roven avenues for the acquisition of scientific knowledge exist and can be borrowed from a number of disciplines."

36 With reference to potential artifacts created and implement through the making, pragmatic research can be characterized as instrumental because of the predetermined functions of the artifacts (Vaishnavi, Vaishnavi, et al., 2015, p. 32).

37 Dewey (1931) described the instrumental view on knowledge as follows: "If we form general ideas and if we put them into action, consequences are produced which could not have been produced otherwise. Under these conditions, the world will be different from what it would have been if thought had not intervened. This consideration confirms the human and moral importance of thought and of its reflective operation in experience."

38 Goldkuhl (2012a, p. 88) emphasized that pragmatists not only propose ideas for changing reality but also try to implement them "through action."

39 Following Goldkuhl (2012b, p. 140), constructive knowledge can also include descriptive and explanatory knowledge.

40 Lee and Nickerson (2010, p. 4) described truthfulness as "whether the predictions or other observational consequences of a scientific theory are upheld by actual observations."

and consequences (Lee and Nickerson, 2010, p. 3). Besides the meaning of truth, "[p]ragmatism elevates [...] usefulness and moral rightness to the same level of importance as truthfulness" (Lee and Nickerson, 2010, p. 4). Rescher (2000, p. 175 f.) reported on appropriateness as an important nominator of truth "which is a matter of combining a whole range of evaluative factors not efficiency and effectiveness alone but also their broader normative nature" (Rescher, 2000, p. 175 f.). Following the pragmatic paradigm, research strives for the best possible understanding of a research problem for the given time of the inquiry (Creswell and Creswell, 2022; Creswell and Plano Clark, 2018; Goldkuhl, 2012b).[41] Pragmatism is a "school of thought that considers practical consequences or real effects to be vital components of both meaning and truth" (Hevner, 2007, p. 93)

Besides the pragmatic knowledge conceptualization of "knowledge through making" and the truthfulness of "what works," several scholars reported on changing philosophical assumptions in iterative research endeavors (e.g., Deng and Ji, 2018; Gregg et al., 2001; Purao, 2002; Vaishnavi, Vaishnavi, et al., 2015).[42] Hence, functional pragmatism can be informed by the interpretation of the reality surrounding the research. Borrowing the knowledge assumptions from interpretivism, meaning and truth are regarded as embodied in human and social interactions that need to be investigated, learned, and interpreted (Chen and Hirschheim, 2004; Creswell and Plano Clark, 2018; Orlikowski and Baroudi, 1991). Understanding human and social interactions, a subjective meaning of reality can be constructed (Chen and Hirschheim, 2004; Walsham, 1995). Therefore, we describe an alternating mode and integration of a pragmatic and interpretive epistemology.[43]

Following the epistemological assumptions about knowledge acquisition and the truth, methodological assumptions denote how a researcher can find out what is believed to be known (Guba and Lincoln, 1994, p. 108).[44] Crotty (1998, p. 3) described the methodological assumptions as the "strategy, plan of action, process or design" shaping the selection and application of research approaches. Method-

[41] In this regard, we will not further unfold the discussion whether pragmatic inquiries lead to "good" results that can be valued as "true" (Goldkuhl, 2008; Niehaves, 2007a; Niehaves and Becker, 2006).

[42] Goldkuhl (2012b, p. 143 f.) showcased the combination of pragmatism and interpretivism reporting on an e-government development project incorporating action and design science research approaches.

[43] Gregg et al. (2001, p. 181) claimed that the ontological assumptions of positivism, interpretivism, and design science are even interdependent.

[44] The research methodology describes "processes, methods, and tools" (Nunamaker et al., 1990, p. 91) and their underlying general assumptions usable to conduct research in a field in a general and abstract manner.

ological assumptions also denote potential sources of knowledge (e.g., dialectic method, experience, a combination of both). The methodology is informed by the ontological and epistemological assumptions made (Crotty, 1998, p. 4).

Pragmatic research "emphasize[s] the importance of research problems" (Creswell and Creswell, 2022, p. 294) and focuses "on the consequences of research, on the primary importance of the question asked" (Creswell and Plano Clark, 2018, p. 86). The pragmatic paradigm explicitly regards knowledge hold by practitioners (e.g., managers, executives) as relevant and seeks for approaches to benefit from this knowledge (Lee and Nickerson, 2010, p. 3). The adoption of a pragmatic paradigm in research "opens the door for multiple methods [...] as well as different forms of data collection and analysis" (Creswell and Creswell, 2022, p. 12). In pragmatic research endeavors, usually, multiple methods of data collection are applied to inform the problem under investigation (Creswell and Creswell, 2022, p. 294). Hence, it is pluralistic regarding research methods applied "and oriented toward 'what works' and real-world practice" (Creswell and Plano Clark, 2018, p. 86). Burrel and Morgan (2019, p. 6 ff.) made a distinction between nomothetic and idiographic research methods.[45] Iivari, Hirschheim, et al. (1998, p. 175) added constructive methods concerned with the design and implementation of artifacts (Simon, 1996). In a pragmatic stance, the methodologies applied are often "constructive, starting from problem identification to the artifact description or formal implementation" (Deng and Ji, 2018, p. 7) as potential research outcomes. Iivari, Hirschheim, et al. (1998, p. 175) described purely conceptual artifacts and technical artifacts (e.g., software) as potential research outcomes.[46]

## 2.3 Strategy of Inquiry

The strategy of inquiry summarizes the types of quantitative and qualitative methods or a combination of them applied in a particular research endeavor (Creswell and Plano Clark, 2018, p. 84).[47] The research design resembles the individual orchestra-

[45] For an unfolding of these types of research methods, we refer to Burrel and Morgan (2019, p. 6 ff.). Iivari, Hirschheim, et al. (1998, p. 175) introduced "constructive methods to information systems research because of its nature as an "applied discipline."

[46] Constructive research methods are emphasized by Brooks (1996); Hevner, March, et al. (2004); March and Smith (1995).

[47] Creswell and Plano Clark (2018, p. 5 f.) highlighted that the research paradigm presupposed, the strategy of inquiry chosen, and the research methods applied are highly intertwined and need to match each other's assumptions and peculiarities.

tion of methods[48] and the specific research process of a research endeavor (Nunamaker et al., 1990; Recker, 2013).[49]

### 2.3.1 Design Science Research Process and Integrated Methods

In line with the meta-theoretical assumptions introduced,[50] we followed a design science research approach within this cumulative dissertation, aiming at describing and differentiating different platform types, prescribing the platforms' roles in e-commerce ecosystems, and proposing design principles for boundary resources involved in the orchestration of ecosystem participants. Our overarching design science research endeavor involved several design iterations addressing more nuanced research problems as we progressed with our research on platforms in e-commerce (Peffers et al., 2007). For the illustration of our overarching design science research approach, we applied the design science research process proposed by Peffers et al. (2007, p. 54). In Fig. 2.1, we depict the methods applied, the sources used for data collection and data analysis, and stress the artifacts developed within each research paper. We also indicate the perspectives taken for the analysis of platforms in e-commerce.

We informed our research on platforms in e-commerce and our overarching design science research process with a bibliometric overview of the literature on platform research in e-commerce involving 7,463 research papers in total (P1) (Wulfert and Karger, 2022a). The research papers were collected from Scopus and quantitatively analyzed involving performance and co-citation analyses (Forliano et al., 2021). We identified three phases of platform research in e-commerce and

48 Zelewski (2008, p. 31) defined a method as "an intersubjectively comprehensible and systematically described procedure that is used to solve problems or achieve goals." This quote was translated from German to increase the overall linguistic consistency.

49 We use the terms strategy of inquiry and research design interchangeably in this cumulative dissertation.

50 With reference to Lee and Nickerson (2010); Levy and Hirschheim (2012), Deng and Ji (2018, p. 7) argued that the pragmatic paradigm is more suitable for design science research in information systems because of three major reasons. First, the truth and validity of research results is not only dependent on truthfulness but also on the practical consequences of the artifacts developed (Lee and Nickerson, 2010, p. 3). Second, pragmatism conceptualizes the researchers as an active and integrative part of the research process (Deng and Ji, 2018; Sein et al., 2011). Third, pragmatism allows the researchers to alter their philosophical assumptions during the progress of the research process (Deng and Ji, 2018; Gregg et al., 2001; Purao, 2002; Vaishnavi, Vaishnavi, et al., 2015).

four topical clusters on platform research in e-commerce. We also investigated the most cited articles qualitatively to derive a future research agenda that also motivated our overall research (Aria and Cuccurullo, 2017; Moral-Muñoz et al., 2020). We used this knowledge base with an additional multi-case analysis (Yin, 2018) of three innovation platforms to provide clarity on the role of platforms in e-commerce, emphasize the intertwined nature of innovation platforms and transaction platforms in e-commerce, and develop a platform architecture framework for e-commerce (P2) (Wulfert and Karger, 2022b). The framework was developed using a design science research approach (Hevner, March, et al., 2004) informed by an analysis of three innovation platforms (Yin, 2018) and a review of the literature on platform extensions (Jansen, 2013). We took on an enterprise architecture perspective for the development of the platform architecture framework in P2 (Ahlemann, 2012; Schekkerman, 2004; Winter and Fischer, 2006). The platform architecture framework positions innovation platforms and transaction platforms in e-commerce ecosystems. Drawing on the differentiation of platforms in e-commerce, we developed a taxonomy for describing and developing transaction platforms in e-commerce based on an analysis of existing taxonomies from the knowledge base and an analysis of 44 transaction platforms in e-commerce (P3) (Nickerson, Varshney, et al., 2013; Wulfert, Seufert, et al., 2021).[51] The taxonomy classifies transaction platforms in e-commerce involving 21 dimensions and 99 characteristics in total integrating the concepts of multi-sided markets and network effects with platform governance dimensions in the context of e-commerce (Armstrong, 2006; Hein, Schreieck, Wiesche, et al., 2016; Rochet and Tirole, 2003; Shapiro and Varian, 1998; Tiwana, 2014). Furthermore, we applied our taxonomy to 44 transaction platforms and derived possible avenues for the evolution of selected platforms (Wulfert, Seufert, et al., 2021).

Based upon the differentiation of innovation platforms and transaction platforms in e-commerce, we developed four architectural patterns on a conceptual basis to inform the development of innovation platforms that support a retailer's dual role on transaction platforms (Wulfert and Schütte, 2022). We developed the architectural patterns against nine architectural requirements for transaction platforms in e-commerce derived from the literature in a design science research approach (P4) (Peffers et al., 2007). We further applied the architectural requirements identified and architectural patterns developed in a systematic literature review on reference architectures in e-commerce (P5) (Vom Brocke, Simons, et al., 2009). We presented

[51] In line with our integration of the taxonomy development in an overarching design science research process, Kundisch, et al. (2022) presented an approach to integrating the taxonomy development in design science research.

our literature analysis in a concept matrix (Webster and Watson, 2002). The literature review revealed reference architectures potentially suitable for the implementation of platforms in e-commerce (Wulfert and Busch, 2022).

For the subsequent development of design principles for boundary resources on innovation platforms and transaction platforms in e-commerce, we followed a multi-paper design science research approach (Vom Brocke, Winter, et al., 2020). We took on the perspective of boundary resources as boundary objects between the focal platform of an ecosystem and the further ecosystem's participants (Eaton et al., 2015; Ghazawneh and Henfridsson, 2015; Star and Griesemer, 1989). We developed design principles for a platform's boundary resources in e-commerce based upon a systematic literature review on design requirements for boundary resources and a multi-case analysis on five innovation platforms and five transaction platforms to further extract design requirements (Wulfert, Woroch, Strobel, Seufert, et al., 2022). From the design requirements, we synthesized a total of six meta-requirements and 19 design principles for the design and implementation for a platform's boundary resources in e-commerce (P6) (Chandra et al., 2015; Walls et al., 1992; Wulfert, Woroch, Strobel, Seufert, et al., 2022). The design principles developed resemble prescriptive design knowledge for platform development (Chandra et al., 2015; Gregory and Muntermann, 2020). To improve the fitness and transferability of the design principles, we evaluated them in 12 interviews with domain experts in e-commerce (Wulfert, Strobel, et al., 2024). The semi-structured interviews followed the structure of the minimum reusability framework (Iivari, Rotvit Perlt Hansen, et al., 2021) and were analyzed using a qualitative content analysis (Mayring, 2014) of the transcribed and anonymized interviews in MAXQDA.[52] The experts' feedback allowed us to assign the design principles to the platform types introduced and three internal roles involved in the design, implementation, and management of boundary resources (P7). As boundary resources need to be managed carefully (Boudreau, 2010; Yoo, Henfridsson, et al., 2010), we also proposed a boundary resource management lifecycle within the multi-paper design science research approach (Wulfert, 2023). The development of this reference procedure model followed a design science research approach (Vaishnavi, Kuechler, et al., 2019) informed by a systematic literature review on application lifecycle management (Vom Brocke, Simons, et al., 2009) and the multi-case analysis of eight innovation platforms in e-commerce (P8) (Yin, 2018). We evaluated the boundary resource management lifecycle in nine expert interviews and adapted the lifecycle following the experts' suggestions (Wulfert, 2023).

[52] We refer to the company's website for an overview of MAXQDA's features: *https://www.maxqda.com/de/*.

Having established the technological foundation for collaboration in e-commerce ecosystems (i.e., boundary resources), we further analyzed a platform owner's capabilities required for a successful orchestration of ecosystem participants (P9) (Wulfert, Woroch, and Strobel, 2024). For the capability analysis, we drew on the concept of dynamic capabilities (Eisenhardt and Martin, 2000; Teece et al., 1997). The development of capability configurations for a successful orchestration of e-commerce ecosystems is informed by a systematic literature review (Vom Brocke, Simons, et al., 2015; Webster and Watson, 2002) on potential ecosystem participants, multi-case analysis (Yin, 2018) of five large and successful transaction platforms, and a case survey (Larsson, 1993) of 100 transaction platforms in e-commerce (P9). Moreover, we further analyzed the input data from the case survey in a crisp-set Qualitative Comparative Analysis (csQCA) (Park, Fiss, et al., 2020) to derive 12 capability configurations for ecosystem orchestration. Further investigating these capability configurations in an ecology of configurations (Park, Fiss, et al., 2020), we proposed six orchestration types for successful e-commerce ecosystems depending on the ecosystem's structure. We derived five dynamic capabilities for ecosystem orchestration involving 12 second-order themes and 44 first-order constructs (Gioia et al., 2013). Following the orchestrating role of platforms in e-commerce ecosystems, we analyzed payment sub-ecosystems and developed a value map depicting value activities and value flows in Financial Technology (fintech) ecosystems in $e^3$-value modeling notation (P10) (Woroch et al., 2022). Moreover, we identified four different roles that a customer can take on in fintech ecosystems. We conducted a network value analysis, as suggested by Peppard and Rylander (2006), drawing on a systematic literature review (Vom Brocke, Simons, et al., 2015) and a multi-case analysis of 171 fintech platforms (Yin, 2018). The value analysis is driven by research on value creation (Gleiss et al., 2021; Peppard and Rylander, 2006; Pousttchi, 2008).

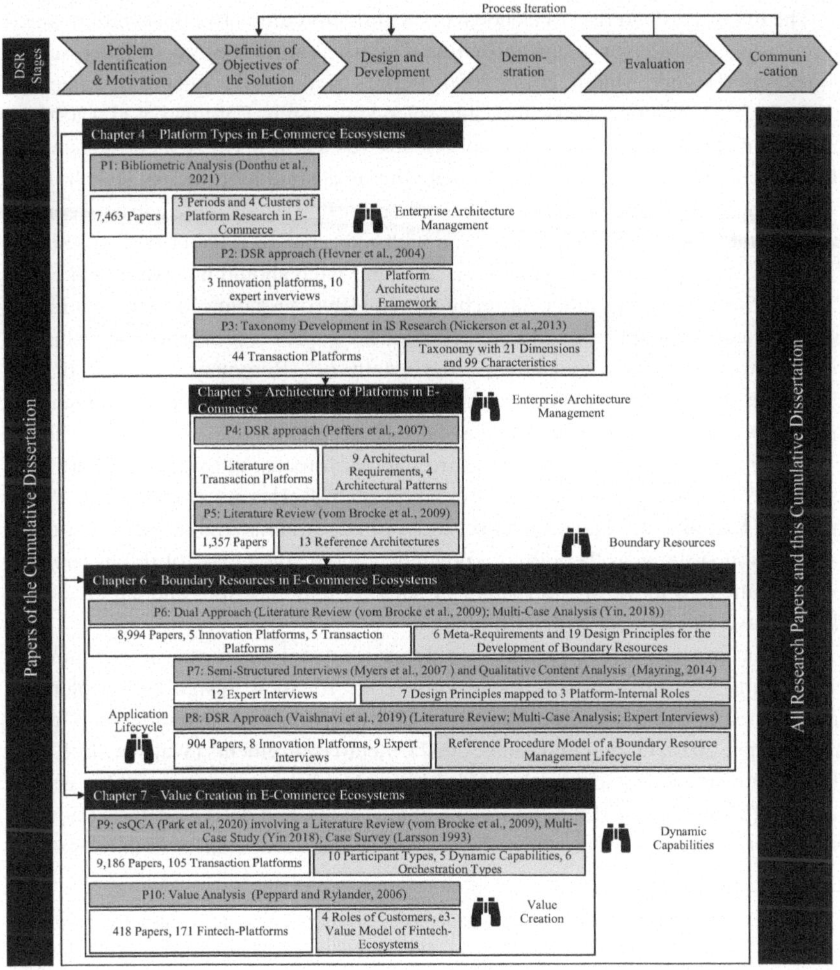


**Fig. 2.1** Overarching Design Science Research Approach

The majority of our research endeavors within this cumulative dissertation aimed at providing theories for design and action (Gregor, 2006) in design science research endeavors (Hevner, March, et al., 2004; Peffers et al., 2007; Vaishnavi, Kuechler, et al., 2019).[53] In the context of design science research, Bergman et al. (2007, p. 548) defined a design problem metaphorically "as a gulf between a current state of the chosen target system [...] and the state that is desired by principal stakeholders within one or more social worlds." The authors further detailed the gap between the current state and the desired future state with reference to Hage (1980) and Simon (1996). Design gaps exist between "what is and what should be" (Bergman et al., 2007, p. 566). The conceptualization by Simon (1996) pronounced "a more directed and concerted thought behind the selection of the gaps, i.e. problems, the resources required to address them, and the expected benefits of" (Bergman et al., 2007, p. 566) closing the gap. Design science research can be described as "a stepwise refinement process where the designer seeks the solution that satisfies the constraints" (Takeda et al., 1990, p. 39). In design science, the researcher "creates and evaluates IT artifacts intended to solve identified organizational problems" (Hevner, March, et al., 2004, p. 77).[54] Kanellis and Papadopoulos (2009, p. 22) summarized design research in information systems as "a body of knowledge about some class of things (objects or phenomenon) in the world (nature or society) that describes and explains how they behave and interact with each other."

For the provision of boundary resources for focal platforms in e-commerce ecosystems, we provide prescriptive knowledge for their design and implementation. Prescriptive design knowledge can take the form of design theories (Gregor and Hevner, 2013; Gregor and Jones, 2007), meta-requirements (Walls et al., 1992), design principles (Chandra et al., 2015; Möller, Guggenberger, et al., 2020), design features (Meth et al., 2015; Schoormann et al., 2023), and design artifacts (Hevner, March, et al., 2004).[55] The types of prescriptive knowledge differ in the degree of

[53] We especially refer to the research papers developing design principles for boundary resources in e-commerce ecosystems and the boundary resource management lifecycle (Wulfert, 2023; Wulfert, Strobel, et al., 2024; Wulfert, Woroch, Strobel, Seufert, et al., 2022). Theories for design and action investigate "underlying principles, methods and frameworks that indicate how to transform the current reality into an improved reality by producing new artifacts (Gregor and Jones, 2007; Simon, 1996; Walls et al., 1992)" (Bergman et al., 2007, p. 547).

[54] Bergman et al. (2007, p. 547) further specified that design is "the practice of inventing, creating, and implementing (technical) artifacts that depends upon, integrates, and transforms heterogeneous and uncertain domains of knowledge."

[55] Meta-requirements are abstract requirements that must be met by the design principles and resulting artifacts that will result as instances of the design (Meth et al., 2015; Schoormann et al., 2023; Walls et al., 1992). Meth et al. (2015, p. 807) described design features as specific

abstraction from the actual design instances implemented (Fig. 2.2). We focus our design knowledge derivation on abstract design principles with meta-requirements[56] as a prerequisite and make reference to exemplary design instances when applicable. The concept of design principles has entered design science research within IS research progressively to "supplement constructs, methods, models, and instantiations" (Iivari, 2020, p. 287).[57] Design principles "define the structure, organization, and functioning of the design product or design method" (Gregor and Jones, 2007, p. 325). Chandra et al. (2015, p. 4044) highlighted that design principles include "prescriptive knowledge for a class of systems."[58] Möller, Guggenberger, et al. (2020, p. 3) defined design principles as "fundamental propositions that aid designers in achieving a successful transfer of requirements to design".

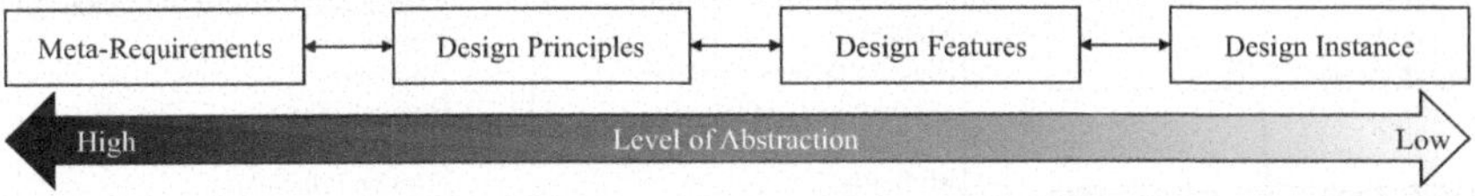


**Fig. 2.2** Level of Abstraction for Prescriptive Design Knowledge

As the objective of design science research is the design of (novel) artifacts to improve the current reality and derive knowledge from the design process (Goldkuhl, 2012a; Hevner, March, et al., 2004; Vaishnavi, Vaishnavi, et al., 2015), we will briefly introduce different artifact types that can result from design science research endeavors (e.g., Hevner, March, et al., 2004; March and Smith, 1995; Vaishnavi, Vaishnavi, et al., 2015). In general terms, Simon (1996, p. 4 f.) described an artifact as an artificially created object (usually) under human involvement and contrasted it

procedures for implementing design principles in resulting design instances that resemble the actual artifact.

[56] Meta-requirements are also described as "design requirements" in the literature (e.g., Dellermann et al., 2016; Möller, Schoormann, et al., 2021; Siering et al., 2021; Walls et al., 1992).

[57] Multiple design principles can be accumulated to form a broader design theory. Walls et al. (1992, p. 36-37) described a design theory as "a prescriptive theory based on theoretical underpinnings which says how a design process can be carried out in a way which is both effective and feasible."

[58] For the actual formulation of our design principles, we followed the design principle formulation framework by Chandra et al. (2015, p. 4044). The authors proposed the following scheme for formulating design principles: "Provide the system with [material property-in terms of form and function] in order for users to [activity of user/group of users-in terms of action], given that [boundary conditions -user group's characteristics or implementation settings]."

with objects provided by nature.[59] In design science research, Hevner, March, et al. (2004, p. 82 ff.) described constructs, models, methods, and instantiations as possible artifact types with reference to March and Smith (1995). Prat et al. (2015, p. 234) introduced and further specified four artifact types in design science research, drawing on March and Smith (1995) and Offermann et al. (2010). Vaishnavi, Kuechler, et al. (2019, p. 16) also included frameworks, architectures, design principles, and design theories as additional output types of design science research. We summarize the sets of potential artifacts developed in design science research endeavors in Fig. 2.3.

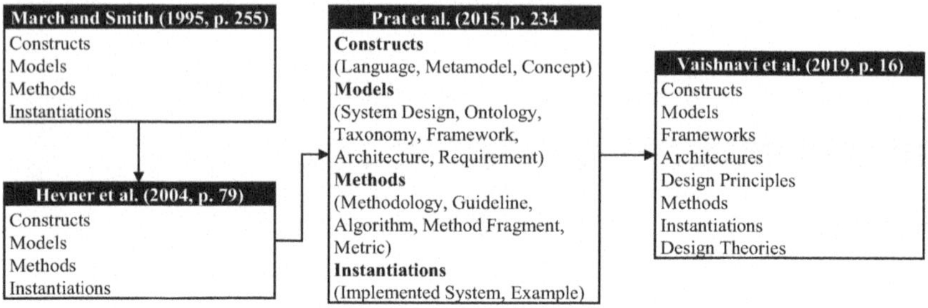


**Fig. 2.3** Output Artifacts in Design Science Research

The single design science research endeavors involved multiple research methods for data collection, data analysis, and artifact evaluation indicated in Fig. 2.1. Combining multiple forms of inquiry allowed us to derive knowledge that is not accessible using a single approach (Shannon-Baker, 2015, p. 36). We mostly inform our design endeavors with qualitative measures extracting knowledge from the existing literature or abstracting knowledge from organizations and practitioners.[60] Qualitative methods aim at describing and understanding the situation of a phenomenon under investigation (Chen and Hirschheim, 2004, p. 205).

To inform our research endeavors with information from the literature, we applied quantitative and qualitative approaches to analyze the existing body of literature in

---

[59] Following the Oxford English Dictionary (2023), the term artifact stems from the Latin terms "factum" (something made) and "arti" (by or using art). Hence, an artifact is not created by nature but created artificially involving human action.

[60] We make use of the changing conceptualization of knowledge and truth as well as the methodological pluralism described in the meta-theoretical assumption to combine these research approaches.

specific areas. For a quantitative overview on existing research on platforms in e-commerce and the derivation of a research agenda (Wulfert and Karger, 2022a), we conducted a bibliometric analysis of the literature (Forliano et al., 2021; Moral-Muñoz et al., 2020). Moreover, we qualitatively analyzed prior literature relevant to our research endeavors depending on the research question posed (Benbasat and Weber, 1996; Vom Brocke, Simons, et al., 2015).[61] The identified literature formed the informed foundation for our research projects (Vom Brocke, Simons, et al., 2009; Vom Brocke, Simons, et al., 2015), helped to focus our investigations as a theoretical lens (Niederman and March, 2019), and uncovered potential blind spots for our research objectives (Webster and Watson, 2002, p. xiii).[62] A systematic literature review process ensures reliability and reproducibility of the research endeavor (Vom Brocke, Simons, et al., 2015; Webster and Watson, 2002). We usually apply the approach by Webster and Watson (2002, p. xvi ff.) to present and synthesize the identified literature in a structured manner. To improve the relevance and appropriateness of identified research papers with the research question posed, we applied additional formal and content-related quality criteria to our literature reviews (Bandara, et al., 2015; Cram et al., 2020).

To incorporate knowledge from organizations operating innovation platforms and transaction platforms, we applied several multi-case analyses (Benbasat, Goldstein, et al., 1987; Yin, 2018).[63] The multi-cases analyses either involve innovation platforms (Wulfert, 2023; Wulfert and Karger, 2022b) and transaction platforms (Wulfert, Woroch, and Strobel, 2024) exclusively or both platform types simultaneously (Wulfert, Woroch, Strobel, Seufert, et al., 2022). Case study research refers to an "empirical inquiry that investigates a contemporary phenomenon within its real-life context, especially when the boundaries between phenomenon and context are not clearly evident" (Yin, 2018, p. 15). For the selection of case companies, we followed a purposive sampling approach (e.g., Wulfert, 2023; Wulfert and Karger, 2022b; Wulfert, Woroch, Strobel, Seufert, et al., 2022), which allows researchers to identify cases with rich information related to the research topic in question (Eisenhardt, 1989b; Patton, 1990). We also considered rankings on website visits, number of third-party sellers, revenue, and Gross Merchandising Volume (GMV) to select

[61] Webster and Watson (2002, p. xiii) emphasized that a review of the literature "is an essential feature of any academic project."

[62] Another reason for conducting structured literature reviews "is to describe available knowledge for professional practice" (Okoli, 2015, p. 82).

[63] Case studies usually involve a single or only a few objects of analysis (e.g., organizations, projects) over a predefined period of time (Chen and Hirschheim, 2004, p. 206). We usually consulted multiple platforms to abstract from the peculiarities of single cases and inform our research questions with a rich set of information.

the most successful platforms (Wulfert and Karger, 2022b; Wulfert, Woroch, and Strobel, 2024). For the analysis of the case information, we extracted and triangulated knowledge from "multiple sources of evidence" (Yin, 2018, p. 13), such as the company website, press releases, and developer information and aggregated them in rich single case reports. We compared these single case reports and extracted information confirmed by a majority of the cases (Eisenhardt, 1989b; Van Aken, 2004; Yin, 2018).

Besides the case analyses, we conducted a number of interviews with domain experts. Walsham (1995, p. 78) suggested that interviews are a primary data source in information systems research because they provide the best interpretations that participants have regarding the actions and events that have taken and are taking place. These semi-structured expert interviews were an additional source of information that aimed to grasp information on an individual level (Misoch, 2015, p. 13). Schultze and Avital (2011, p. 2) described an interview as an "exchange of views between two people [...] talking about a topic of common interest." In information systems research, usually, one person is the researcher and the other is the interviewee providing his or her personal experiences on the topic of interest (Polkinghorne, 2005, p. 138). Interviewees' experiences are beneficial to inform research endeavors and evaluate design science research artifacts (Hevner, March, et al., 2004; Vaishnavi, Vaishnavi, et al., 2015; Verschuren and Hartog, 2005). The interviews were transcribed (Rädiker and Kuckartz, 2019) and formally analyzed using tool-supported qualitative content analyses (Krippendorff, 2013; Mayring, 2014).[64]

### 2.3.2 Tactics for Increasing Reliability and Validity

As mentioned before, the research papers forming the body of this cumulative dissertation involve a number of qualitative research methods applied from an interpretivist's perspective (e.g., systematic literature review, multi-case study, semi-structured interviews). Even the design science research approaches applied involve qualitative research methods for data collection and data analysis. As the results of qualitative research are often criticized as subjective and the interpretations may be biased by the investigator (e.g., Beuving and De Vries, 2015; Guba and Lincoln, 1982; Miles et al., 2014; Straub and Gefen, 2004), research rigor with regards to reliability and validity needs to be ensured for qualitative research endeavors (Golaf-

[64] In particular, we used the expert interviews to partition e-commerce ecosystems, evaluate the set of design principles, and discuss the reference procedure model (Wulfert, 2023; Wulfert and Karger, 2022b; Wulfert, Woroch, and Strobel, 2024).

shani, 2003; Miles et al., 2014; Recker, 2013).[65] Moreover, reliability and validity indicators and measures from quantitative research "do not apply to non-numerical data" (Recker, 2013, p. 94) interpreted in qualitative research (Merrick, 1999). To account for these problems, the existing research proposes a number of guidelines to address the subjective and interpretive nature of qualitative research and increase research rigor (e.g., Beuving and De Vries, 2015; Lincoln and Guba, 1985; Merrick, 1999; Myers and Newman, 2007; Walsham, 1995; Walsham, 2006; Wulfert, 2023).[66] We refer to Guba's (1981) and Miles et al.'s (2014) criteria for addressing subjectivity and investigator bias and, thus, improving reliability and validity.

Internal validity (i.e., credibility) of findings "concerns whether the researcher has been able to provide sufficient substantiated evidence for the interpretations offered in qualitative data analysis" (Recker, 2013, p. 94). We ensured internal validity by adapting well-established research methods in information systems research. We provided transparency on the actual process of research as far as possible. When possible, we also applied data triangulation from different sources (Shenton, 2004, p. 64). We used a variety of secondary data sources for our multi-case analyses (Hänninen et al., 2018, p. 156 ff.). Moreover, we tried to match patterns derived from existing literature with our case observations (Yin, 2018, p. 91). In interview studies, we applied a random sampling approach for the selection of interviewees. In line with suggestions from the literature, we informed the interviewees a priori to the interview to allow for preparation on the interviewee's side. We were also always transparent on the purpose of the interviews and the use of the transcripts for reasons of an improved analysis (Guba, 1981). As suggested by Shenton (2004, p. 65), each interviewee had the chance to refuse participation in the interview studies. With regards to the number of interviewees, we applied sufficiency and saturation as indicators for information sufficiency (Guest et al., 2006; Seidman, 2019). Sufficiency was achieved by contacting a variety of different interviewees and applying random sampling. Data saturation was measured by the degree of new information added and changes to the coding scheme caused by the consecutive interviews (Guest et al., 2006, p. 65). Guest et al. (2006, p. 65) conceptualized information sat-

---

[65] Although the following measures do not seem to matter following a pragmatic paradigm, we applied these measures as additional quality criteria for paper reviews in peer-review processes.

[66] To distinguish reliability and validity concerns in qualitative research from those in quantitative research, these researchers refer to confirmability, dependability, credibility, and transferability as quality criteria instead of reliability and validity (e.g., Miles et al., 2014; Shenton, 2004). As the terms reliability and validity are more often used in information systems research, we use those well-established terms. We also refer to the terms used in qualitative research in brackets.

uration "as the point in data collection and analysis when new information produces little or no change to the codebook."

External validity (i.e., transferability) is concerned with the extent to which the results of a particular research endeavor can be applied in other settings, domains, or cases (Recker, 2013, p. 94). To increase the transferability of the research results, we provided information on the boundaries of our research endeavors but left the test for transferability to other domains of application for future research. We always provided information on the methods applied for data collection and analysis in general and provided information on the number, length, and time period of the expert interviews. We provided demographic information on our interviewees if applicable. In the context of platforms in e-commerce ecosystems, we described the interviewees' roles within their organization according to Zutshi and Grilo (2019, p. 551) to enable the researcher to better assess the interviewees' statements. Conducting our interview studies and multi-case analyses, we abstracted from the peculiarities of single organizations to enable a wider application of our research results. Although we conducted our multi-case analyses using a replication logic (Yin, 2018, p. 91), the external validity of our studies could be increased by analyzing platforms in domains other than e-commerce. This would also increase the projectability of our research results (Vom Brocke, Winter, et al., 2020). To ensure construct validity for the multi-cases analysis, we established a chain of evidence triangulating multiple data sources (Yin, 2018, p. 179). In the same vein, we asked our interviewees to verify the interview transcripts to avoid misunderstandings and misinterpretations (Shenton, 2004, p. 66 f.).

Reliability (i.e., dependability) describes whether the process of the study is consistent and reasonably stable over time and across researchers. Similar conclusions should be drawn by other researchers. We provided rigorous information on our research design and also provided our data sources if possible (e.g., concept matrix). Among the teams of authors of the single research papers, we periodically reflected on the research design, execution, and results (Guba, 1981, p. 81). Reliability was achieved by keeping track of intermediary results and preparing research diaries over the single research projects (Yin, 2018, p. 180).

# Conceptual Foundation

# 3

Since an unambiguous use of concepts and terms is one of the core requirements of scientific research (Töpfer, 2009), we unfold the conceptual framework and core terms of this thesis in the following section. We introduce focal platforms as nexus of dynamic e-commerce ecosystems (Wulfert, Seufert, et al., 2021), and boundary object theory as a theoretical lens (Niederman and March, 2019). We refer to related literature introducing the core concepts. The order of the introduced concepts already hints at the thesis's research papers' structure and integration.

## 3.1 Electronic Commerce

Within this section, we introduce e-commerce as the environment of our research endeavor. We describe e-commerce as a subset of electronic business, present a broad and narrow perspective on e-commerce, and elaborate on selected e-commerce definitions. Moreover, we elicit digital products, contact principles, and transaction phases in the context of e-commerce.

**Supplementary Information** The online version contains supplementary material available at https://doi.org/10.1007/978-3-658-45198-1_3.



T. Wulfert, *Selected Perspectives on Platforms in E-Commerce Ecosystems*,
https://doi.org/10.1007/978-3-658-45198-1_3

### 3.1.1 Electronic Business Environment

Following the emergence and increasing dissemination of the Internet (Berners-Lee et al., 1994), transactions on societal, organizational, and private levels are increasingly supported electronically. Electronic business[1] (e-business)[2] includes "all electronic transactions[3] within an organi[z]ation" (Chaffey, 2015, p. xv). Laudon and Traver (2020, p. 45) described electronic business as "the digital enabling of transactions and processes within a firm, involving information systems under the control of the firm."[4] In a similar vein, Combe (2006, p. xv) explained that electronic business includes "all electronic transactions within an organization." Wirtz (2020, p. 23) took the term electronic business more broadly, also considering inter-company relationships referring "to the initiation, partial or complete support, processing and maintenance of [transactions] between economic partners by means of information technology (electronic networks)." In a nutshell, electronic business involves internal and external as well as cross-company and inter-company business processes using modern communication technologies, such as the Internet. E-commerce can be distinguished from this rather broad conceptualization of electronic business. Figure 3.1 illustrates the hierarchical structure of the concepts surrounding electronic business and presents them in an overall context as dimen-

[1] Although some scholars claim that digital business is synonymous with electronic business (e.g., Chaffey, 2015; Kollmann, 2019, 2022), we will continue to use electronic business for reasons of consistency, because electronic business is the more disseminated term in literature (Wirtz, 2020). We will not put forth the discussion of the terms electronic business and digital business in this thesis. Following Chaffey (2015, p. 14) "[d]igital business is just a new label—there is no distinction between the role of digital business and traditional information systems management." Bican and Brem (2020, p. 2f.) discussed different terms under the digital business umbrella.

[2] The acronym e-business was coined in a commercial campaign by International Business Machines Corporation (IBM). IBM preferred the notation "eBusiness" (Barton, 2014, p. 3). However, in the literature, the notation e-business is more common. IBM defined e-business as "a secure, flexible and integrated approach to delivering differentiated business value by combining the systems and processes that run core business operations with the simplicity and reach made possible by internet technology" (IBM, 2003).

[3] A transaction can be defined as "the exchange of values between two parties" (Kotler, 1972, p. 48). Williamson (1985, p. 1) denoted that a "transaction occurs when a good or service is transferred across a technologically separable interface." With reference to commerce, Picot (1982, p. 269) described a transaction as the "process of clarifying and agreeing on an exchange of products and services."

[4] For Laudon and Traver (2020, p. 45), electronic transactions involving products, services, and an "exchange of value across organizational boundaries" are not part of electronic business.

sions of the information society as described by Wirtz (2020, p. 15 ff.). In line with Chaffey's (2015) explanations, e-commerce[5] is depicted as a subset of electronic business focusing on an electronically supported transaction involving products and services. Transactions in e-commerce involve the overall purchasing process from need-recognition to after-sales services (Kotler and Keller, 2016, p. 195). Wirtz (2020, p. 15) described electronic collaboration, electronic communication, electronic education, and electronic entertainment as additional subsets of electronic business.[6]

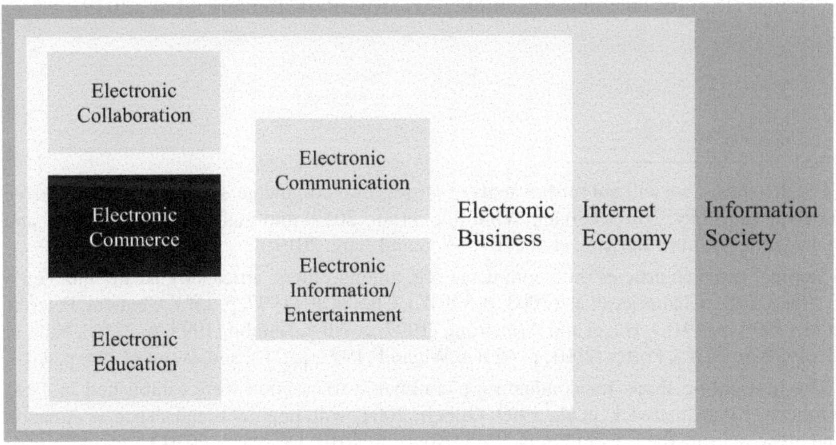


**Fig. 3.1** Dimensions of the Information Society (Wirtz, 2020, p. 15)

### 3.1.2 Toward a Conceptualization of Electronic Commerce

We will further explain the concept of e-commerce that forms the environment of our investigation on platforms, boundary resources, and digital business ecosystems.

[5] E-business is the super-ordinate concept, as it includes processes that go beyond the purchasing and selling of products (Wirtz, 2020, p. 32).

[6] However, it seems that a clear delimitation of the mentioned terms is not possible due to their overlapping and intertwined nature. Several authors use the terms electronic business and electronic commerce as synonyms (Wirtz, 2020, p. 20).

E-commerce[7] is a topic of interest in research[8] and practice[9] since the mid-1990s[10] (Dan, 2014, p. 135 f.). The concept of e-commerce is composed of the two components electronic and commerce.[11] The former component emphasizes the use of electronic means for conducting commerce.[12] Important to note here is the use of electronic networks, such as the Internet, for communication among parties involved in the transaction. A prerequisite for exchanging information about the commerce transaction via electronic networks is the availability of that information in a digital representation. This is usually achieved by using an electronic shop system for sellers (Treutner and Ostermann, 2011; Wirtz, 2020) and corresponding Information Systemss (ISs) for customers (Schütte, 2017; Wulfert, Betzing, et al., 2019).

---

[7] For this thesis, we will not further analyze subsets of e-commerce, such as social commerce (Grewal and Levy, 2007; Turban, Whiteside, et al., 2017) and mobile commerce (Wu and Hisa, 2008; Wulfert, Betzing, et al., 2019; Yoo and Jang, 2019).

[8] Seminal research articles on e-commerce are, among others, articles by Adam and Yesha (1996, p. 5 ff.), Chatterjee et al. (2002, p. 520 ff.), Choi et al. (1997, p. 1 ff.), Clement, Peteres, et al. (1999, p. 49 ff.), Hagel and Armstrong (1997, p. 3 ff.), Kambil (1995, p. 27 ff.), Nath et al. (1998, p. 92 ff.), Porter (2001, p. 62 ff.), Wigand (1997, p. 2 f.), and Zwass (1996, p. 4 ff.).

[9] The first online shops for conducting e-commerce transaction were established in 1994 (Doherty, Ellis-Chadwick, et al., 1999; Gilbert, 2004) with popular brands, such as Amazon and eBay, starting their operation in 1995 (Amazon, 2019; Donszem, 2020.)

[10] The first approaches for electronically-mediated transaction in commerce date back to the 1970s with the emergence of electronic data interchanges for purchase orders and invoice processing for business-to-business transactions (Dan, 2014, p. 136). Illik (1998, p. 13) even described the first Business-to-Customer (B2C) e-commerce applications drawing on Viewdata, Teletel, and Minitel as electronic networks. We date the beginnings of e-commerce to the 1990s when the Internet was commercialized and its number of private customers was on the increase (Von Reden, 2015, p. 166 f.).

[11] The acronym e-commerce implies that commercial action activities are carried out via electronic media, leading to the digitalization of traditional trade and may involve monetary or intangible added value (Meier and Stormer, 2008; Zheng et al., 2009).

[12] Cater-Steel and Grist (2008, p. 2096) introduced the electronic component of e-commerce as another dimension for the broad and narrow perspective of e-commerce. The authors described the broad electronic perspective as composed of all kinds of electronic media, such as electronic data interchange and telecommunication systems. The narrow perspective is restricted to the Internet as electronic media. The OECD also reflected different types of electronic networks in their broad and narrow definition of e-commerce (OECD, 2002, p. 89). Since we focus on the Internet as an electronic network for e-commerce transactions, we did not further consider this additional dimension further.

The latter component refers to the activities for the exchange of goods and services among market participants. In general, a functional definition and a institutional definition for commerce[13] have been established in research (e.g., Becker and Schütte, 2004; Müller-Hagedorn et al., 2012). The functional perspective describes the major activities attributed to trading companies.[14] In the literature, these activities concerning the distribution in an economy are summarized in a number of catalogs (e.g., Buddeberg, 1959; Seÿffert, 1972). Buddeberg (1959, p. 23 ff.) introduced a total of eight trading functions from which the trading company needs to select a bundle of activities according to its strategy (Müller-Hagedorn et al., 2012, p. 99). The functions are the contact function, information function, consulting function, goods regrouping function, quantity regrouping function, spatial balancing function, temporal balancing function, and refinement function. Schütte (2017, p. 14) referred to activities bridging the discrepancies in space, time, volume, and quality between manufacturers and customers as major functions of trading companies. In the context of e-commerce, (parts of) the retail and wholesale functions (e.g., bridging, information, storing) (Laudon and Traver, 2019; Schütte, 2017) are conducted via electronic means (Zwass, 1996, p. 5 f.). E-commerce is likely to mitigate spatial and temporal constraints of traditional brick-and-mortar commerce, allowing for conducting global transactions at any time (Gunasekaran et al., 2002; Lefebvre and Lefebvre, 2002; OECD, 1999).

[13] Historically, trade, commerce, and market exchange were considered significant as long ago as the eighth century BC (Von Reden, 2015, p. 161). Zentes (2006, p. 5) even sees commerce as one of the early central points in the structuring of human life.

[14] According to Müller-Hagedorn et al. (2012, p. 39 f.) "trading in a narrow functional sense occurs when market participants procure goods from other market participants, do not process these goods, and sell them to third parties or participate in the procurement and sales. In practice, the term generally refers to the exchange of tangible goods, and more frequently to the exchange of movable tangible goods."

The institutional perspective refers to trading companies[15] including retailers[16] and wholesalers[17] in an institutional sense. Trading companies have been established because they involve relatively lower transaction costs than those for direct trade scenarios (Coase, 1937; Picot, 1986; Williamson, 1985).[18] Traditionally, these economic effects led to distribution systems often consisting of multiple stages (Specht, 1998, p. 16). The application of electronic media in e-commerce leads to disintermediation that diminishes existing trade structures,[19] enables direct contact between manufacturers as well as customers, and favors the integration of additional distribution supports (Schütte and Vetter, 2016, p. 81 ff.).[20] Transaction platforms agglomerating the assortment of a diverse set of sellers and providing it to (end) customers via a uniform user experience (Teller and Elms, 2010) can be, among others, such additional distribution supports in existing and new distribution systems.

Despite intense research on the topic of e-commerce, no agreed-upon definition for e-commerce has been established (e.g., Cater-Steel and Grist, 2008; Deutsch, 1998; Hermanns and Sauter, 1999; Schmeken, 2007; Schubert, 1999; Wamser,

---

[15] Müller-Hagedorn et al. (2012, p. 40 f.) defined trading companies as follows: "Trading companies are companies which mainly purchase and sell movable tangible goods without processing them in a technical sense (apart from customary manipulations) and often combine this with the offer of services. The goods to be sold can be procured by purchase (own-account trading), but they can also be sold in their own name and on behalf of third parties (commission) and for the account of third parties (commission trade) or in the name of third parties and for the account of and for the account of a third party (agency trade)." The authors further denote that, in more narrow definitions, transactions need to be made on the trading company's own account, with economic independence, and with autonomy in procurement (Müller-Hagedorn et al., 2012, p. 41 ff.).

[16] Retailers usually operate in B2C mode, with smaller packet sizes and value-added tax indicated on shelf labels.

[17] Wholesalers usually operate in business-to-business mode selling articles to other companies. Wholesalers and retailers are present in trading companies that comprise multiple levels of trade (Specht, 1998, p. 16).

[18] For further information on transaction costs, we refer to Williamson (1985). The relatively low transaction costs were illustrated by Picot (1986, p. 3 f.) and Wigand (1997, p. 8).

[19] Disintermediation tendencies and the increase in direct distribution from manufacturer to customers are also observed by Kollmann (2019, p. 115).

[20] Literature on marketing and supply chain management often refers to the direct interaction between manufacturer and customers as "direct-to-customer" (e.g., De Bont, 2018; Priyanka, 2016). However, these transaction relations can be characterized as business-to-consumer involving manufacturers and customers.

2001).[21] We therefore present a broad and a narrow perspective[22] on e-commerce and discuss some of the definitions of e-commerce put forward in the literature. In a broad perspective on e-commerce, several authors treat the concept as synonymous with electronic business (e.g., Chaffey, 2015; Gunasekaran et al., 2002; Kalakota and Whinston, 1996; Wirtz, 2020; Zwass, 1996). It is considered as the electronicization and electronic support of all business processes within and across companies (Zheng et al., 2009, p. 8). From the broad perspective, it is important to note that the electronic support of business processes, in general, is emphasized. In a narrow perspective on e-commerce, e-commerce includes "all activities associated with a commercial transaction" (Cater-Steel and Grist, 2008, p. 2096) and can be seen as synonymous with electronic shopping, online shopping, and digital shopping (Hermanns and Sauter, 1999, p. 14).[23]

Based on an in-depth screening of the existing literature synthesizing 28 prior definitions of e-commerce, Holsapple and Singh (2000, p. 152 ff.) identified five perspectives on e-commerce. The authors clustered existing definitions along the perspectives activity view, effects view, information exchange view, trading view, and value chain view. Based on their analysis of these perspectives, Holsapple and Singh (2000, p. 161) defined e-commerce as "an approach to achieving business goals in which technology for information exchange enables or facilitates execution of activities in and across value chains as well as supporting decision making that underlies those activities." In their synthesized definition, Holsapple and Singh (2000, p. 161)

[21] We also refer to the literature provided by these authors. Wigand (1997, p. 5) noted that e-commerce "is poorly understood and frequently used to denote different meanings, very often depending on the individual's job function, professional orientation and background, focal product or service, and type of information technology deployed" and involves many different technologies. Cater-Steel and Grist (2008, p. 2093) even described attempts for defining e-commerce as "dilemma." Owing to the technologies that hae been developed and that have emerged since 1997, Wigand (1997, p. 5) is still correct in abstracting from the concrete technology and instead focusing on the business goal achieved in e-commerce.

[22] The understanding of the e-commerce concept also seems to differ across languages. While the *Wörterbuch der deutschen Sprache* following a narrow understanding of e-commerce, describes e-commerce as the "[d]istribution of goods or services via the Internet" (DWDS, 2023), the *Oxford English Dictionary* defines e-commerce more broadly as "business that is transacted by transferring data electronically, especially over the Internet" (Lexico.com, 2023).

[23] As described later, e-commerce is concerned with the overall purchasing process including pre-sales and after-sales (Kotler and Keller, 2016). We indicate the broad and narrow perspective on e-commerce in Table A.1 in Appendix A.1 in the Electronic Supplementary Material.

documented a broad perspective on e-commerce. Although the authors identified the trading view as a building block of e-commerce (Holsapple and Singh, 2000, p. 152), they did not consider it further in their definition. Schmeken (2007, p. 12) identified four core topics of e-commerce, namely digital transaction (Choi et al., 1997, p. 12), coverage of the entire (Adam and Yesha, 1996, p. 5), purchasing process (Kotler and Bliemel, 2001, p. 1220 ff.), transaction partners as economic entities (Clement, Peteres, et al., 1999, p. 49), and focus on electronic media such as the Internet (Picot et al., 2001, p. 23). Kalakota and Whinston (1996, p. 1) also described e-commerce from a broad perspective as "a modern business methodology." In this regard, they introduced their generic framework for e-commerce, consisting of the two pillars public policy and technical standards, as well as the five layers e-commerce applications, business services, messaging services, multimedia content, and network infrastructure (Kalakota and Whinston, 1996, p. 3 ff.).[24] Gunasekaran et al. (2002, p. 186) also defined e-commerce from a broad perspective as "conducting business electronically." In their definition of e-commerce, Gunasekaran et al. (2002, p. 186) emphasized that e-commerce involves different types of entities[25] and that it serves an "organizational or individual objective."[26] Wigand (1997, p. 2) also applied a broad perspective on e-commerce, including "any form of economic activity conducted via electronic connections" and emphasized the "accomplishment of a business goal" (Wigand, 1997, p. 5) as the objective of e-commerce besides the application of a particular technology. Following Wigand (1997, p. 5), e-commerce "denotes the seamless application of information and communication technology from its point of origin to its endpoint along the entire value chain of business processes conducted electronically and designed to enable the accomplishment of a business goal." The Organization for Economic Co-operation and Development (OECD) (2002, p. 89) described a more narrow definition of e-commerce and denoted it as "the sale or purchase of goods or services, whether between businesses, households, individuals, governments, and other

---

[24] While Kalakota and Whinston (1996, p. 4) denoted the "telecom, cable TV, wireless" connection as network infrastructure, we will follow a narrow approach and focus on the Internet as the major infrastructure of contemporary e-commerce.

[25] An overview of potential entities involved in e-commerce transactions is presented by Böttcher, Rickling, et al. (2021, p. 5). In the remainder of this thesis, we also introduce ten participant types in e-commerce ecosystems (Wulfert, Woroch, and Strobel, 2024; Wulfert, Woroch, Strobel, Schoormann, et al., 2023).

[26] The presence of an objective in e-commerce is in line with the ecosystem-as-a-structure perspective introduced by Adner (2017). This perspective is applied in the further course of this thesis for the introduction of the e-commerce ecosystem as an instantiation of digital business ecosystems in the context of e-commerce.

public or private organi[z]ations, conducted over computer-mediated networks."[27] With reference to the OECD (2002), Numberger and Rennhak (2005, p. 237) proposed their narrow definition of e-commerce, emphasizing the "spatial separation of transaction and delivery" (Numberger and Rennhak, 2005, p. 273) in automated transactions.[28] Combe (2006, p. 1) narrowed e-commerce down to "the buying and selling, marketing, and servicing of products and services via computer networks." Deges (2020, p. 2) also applied a narrow perspective on e-commerce, focusing on the "electronically supported commerce of goods and services." Following the narrow perspective on e-commerce, Laudon and Traver (2020, p. 45) defined e-commerce as "digitally enabled commercial transactions between and among organizations and individuals." The authors hinted at the different parties involved in e-commerce transactions.[29] We present exemplary definitions of the term e-commerce in Table A.1 in Appendix A.1 in the Electronic Supplementary Material.

Following the discussion on the broad (i.e., electronic business) and the narrow perspective on e-commerce (Zheng et al., 2009, p. 8), Combe (2006, p. 2) presented a comparison of the two constructs (Fig. 3.2). The author subsumed electronic shopping activities, such as buying, selling, and procurement, under the term e-commerce and only identified an overlap in "the buying and selling of products and services" (Combe, 2006, p. 1). With regards to the transaction process in e-commerce, we argue that the electronic distribution of digital articles and online customer services (Butler and Peppard, 1998; Luxem and Ehlers, 2000) are also relevant in e-commerce, as they are subsequent activities in the transaction process (Kotler, 1972). We therefore updated Fig. 3.2 to include the distribution of digital products and electronic pre-sales and after-sales service[30] on the e-commerce side. Moreover, we depicted the broad and the narrow perspective on e-commerce within

---

[27] It is important to note, that the OECD (2002, p. 89) amends that "the payment and the ultimate delivery of the good or service may be conducted on- or off-line." The OECD (2002, p. 89) further denoted a more narrow definition of the component electronic, narrowing the "computer-mediated networks" down to the "internet." We also apply this focus throughout this thesis, using the Internet as an underlying network for transactions in e-commerce.

[28] We would like to point out that the addition of the second characteristic "automated transaction" (Numberger and Rennhak, 2005, p. 273) is especially relevant in this context. Otherwise, orders in showrooms with home delivery would be considered as e-commerce although no network mediation seems to be involved (Choi et al., 1997; Zwass, 1996).

[29] In the further course of this section, we present different types of participants involved in e-commerce transactions and describe possible combinations of actors on the demand and supply side (Aichele and Schönberger, 2016; Lefebvre and Lefebvre, 2002).

[30] With regard to the pre-sales and after-sales services, we did not further differentiate whether the services are provided by humans or whether they are automated by machines (Ferstl and Sinz, 2013, p. 4 f.).

the illustration and implied further activities subsumed under the broad perspective in line with the broad definitions of "doing business electronically" (Choi et al., 1997, p. xxiv).

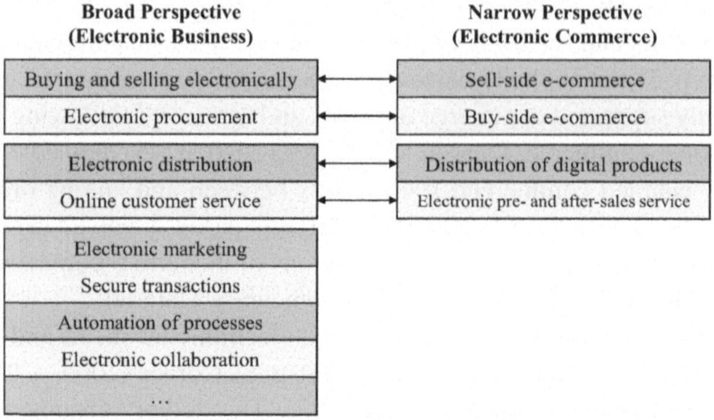


**Fig. 3.2** The Relationship Between E-Business and E-Commerce (according to Combe, 2006, p. 2)

For the progress of this thesis, we apply a narrow perspective on e-commerce. Therefore, we focus on electronically supported commerce transactions entailing physical and digital products and services via the Internet, involving customers and businesses as major transaction partners. In this context, e-commerce means not only the exchange of goods and services, but also all activities and technologies that support this goal (Riggins and Rhee, 1998, p. 90).

### 3.1.3 Features of Electronic Commerce

We apply the aforementioned perspective for the following elaboration on the phases of e-commerce, the e-commerce cube, transaction phases, and transaction partners. Heinemann (2018, p. 41 f.) divided the development of e-commerce into three phases.[31] In the first phase ranging from 1993 to 1999, simple e-commerce concepts

[31] In the course of this thesis, we will introduce three phases of research on platforms in e-commerce (Wulfert and Karger, 2022a, p. 192 f.). The first phase (1993–2004) is concerned with various e-commerce business models. The second phase (2005–2014) introduces transaction platforms, discusses advanced architectures, and describes social commerce situations. The third phase (2015–2021) consists of an in-depth analysis of social commerce

were implemented and investments were made in data traffic. During this period, well-known companies, such as Amazon and eBay, were founded. The early forms of e-commerce were extremely complex in the use of Information Technology (IT) and were reserved only for large companies at high prices. This changed over time such that small entrepreneurs also had the chance to start their business in e-commerce at a cost of only a few thousand euros (OECD, 1999, p. 10). Heinemann (2018, p. 41 f.) described the second phase, lasting from 1999 to 2005, as the comparison portal phase. In this phase, comparison websites were implemented, search engines were founded, and customer ratings were written online for the first time. The third phase started in 2005, and consists of optimization and scaling. Online stores are continuously optimized and improved to further meet customer requirements, including premium membership services, responsive implementations for mobile commerce, and omni-channel efforts (Heinemann, 2018, pp. 33–35).

In e-commerce, transactions involving information, products, and services are executed using information and communication technology (Kalakota and Whinston, 1996; Nath et al., 1998). Choi (1997, p. 13)) defined e-commerce as the "use of electronic means and technologies to conduct commerce."[32] The authors described the "core" (Choi et al., 1997, p. 18) of e-commerce along the three dimensions agent, process, and product.[33] While non-e-commerce[34] is about physical agents, processes, and products, pure e-commerce deals with digital agents, processes, and products.[35] In Fig. 3.3, partial e-commerce is depicted with unshaded cubes, traditional brick-and-mortar commerce is indicated with light grey shading, and pure e-commerce is introduced as the "core" is indicated with dark grey shading. Choi et al. (1997, p. 20) extended the cube with axes along the three dimensions agent,

---

and the integration of evolving technologies, such as blockchain and artificial intelligence, in e-commerce.

[32] The authors applied a narrow perspective on e-commerce. In the preface of their book entitled *The Economics of Electronic Commerce*, Choi et al. (1997, p. xxiv) denoted that e-commerce "go[oes] far beyond simply 'doing business electronically'." The authors aimed at innovative and disruptive business models and business processes in the context of e-commerce in a narrow sense (Choi et al., 1997, p. 15 ff.).

[33] In light of technological developments in extended reality applications, a number of scholars introduced the degree of virtuality as another dimension describing products in e-commerce (e.g., Reisberger et al., 2008). Schultz (2004, p. 7) argued that the choice of the transaction medium (e.g., internet, television system, mobile system) can also be an additional dimension.

[34] Choi et al. (1997, p. 18) described non-e-commerce as traditional commerce.

[35] Illlik (1998, p. 15 f.) further separated the product dimension into physical products, semi-physical products, semi-digital products, and digital goods. The author further denoted the former three product types as non-digital as they still contain physical components (Illik, 1998, p. 16).

process, and product to indicate that the digital manifestation is an intermediary step toward virtual agents, virtual processes, and virtual products.

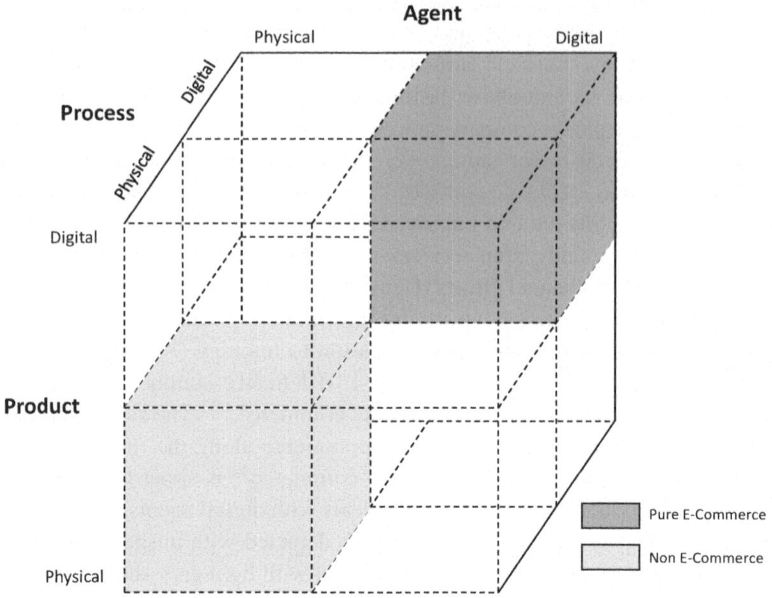


**Fig. 3.3** The Core of E-Commerce (Choi et al., 1997, p. 18)

The degree to which the transaction in e-commerce must be digital varies in the literature and can be illustrated using different purchasing phases. The purchasing phases are organized in an overall purchasing process that is understood as comprising a series of criteria that must be fulfilled from the customer's point of view to make a purchase decision. The purchasing process can be subdivided into several phases that take place before, during, and after the actual purchase of an article (Khalifa and Shen, 2005; Kotler and Keller, 2016; Schmid, 1999). Kotler and Keller (2016, p. 195) proposed the following five steps: need recognition, information search, evaluation of alternatives, purchase phase, and post-purchase phase. With reference to Schmid (1999), Risch (2008, p. 36 f.) introduced a stimulation phase, information phase, agreement phase, settlement phase, and loyalty phase as transaction phases in e-commerce.[36] The former and the latter phases were added

[36] We translated the German phase denominators for readability reasons.

to Schmid's (1999) core phases to emphasize the benefits of customer relationship management systems in e-commerce.[37] Turban, Whiteside, et al. (2017, p. 6) used the purchasing phases of ordering and payment, order fulfillment, and delivery as phases for their classifications of e-commerce. The degree of e-commerce can be described on a continuum between a completely digital transaction (Laudon and Traver, 2019) and a transaction where only a small part[38] of the procurement process is digital (Timmers, 2000). In this thesis, we associate retail transactions with e-commerce if they are purely and partially e-commerce in Turban et al.'s (2017, p. 6) classification. Turban et al.'s (2017, p. 6) classification, which involves eight combinations (i.e., 1–8) of their three commerce phases, is illustrated in Table 3.1.[39]

**Table 3.1** Classifications of E-Commerce (Turban, Whiteside, et al., 2017, p. 6)

| Activity | 1 | 2 | 3 | 4 | 5 | 6 | 7 | 8 |
|---|---|---|---|---|---|---|---|---|
| Ordering and payment | p | p | p | p | p | d | d | d |
| Order fulfillment | p | p | d | p | d | d | p | d |
| Delivery | p | d | d | d | d | p | p | d |
| Types of e-commerce | **non** | **partial** | **partial** | **partial** | **partial** | **partial** | **partial** | **pure** |
| physical (p), digital (d) | | | | | | | | |

[37] The transaction phases introduced have a generic character and are not only relevant in e-commerce but can also be applied to brick-and-mortar scenarios. Nevertheless, they are applied as an overview for the transaction phases supported by transaction platforms (Brenner and Breuer, 2001, p. 142 f.). Transaction platforms can support and take on selected transaction phases and even the entire transaction (Schmid, 1993, p. 468).

[38] According to Turban, Whiteside, et al. (2017, p. 6), a completely digital transaction is "pure" e-commerce, while a traditional retail transaction is non-e-commerce. The continuum in between is called "partial e-commerce."

[39] Following the differentiation of transaction phases (Khalid, et al., 2014; Kotler and Keller, 2016; Turban, Whiteside, et al., 2017), their increasing penetration with IT (Arora Longo, et al., 2022; Grewal and Levy, 2007; Grewal, Roggeveen, et al., 2017), and the discussion on digital business (e.g., Bican and Brem, 2020; Chaffey, 2015; Kollmann, 2022), it is highly debatable whether the differentiation between electronic commerce—including its various sub-forms—and traditional or stationary commerce is still useful. Even stationary commerce is increasingly penetrated with IT (Arora Longo, et al., 2022; Brackmann et al., 2023; Kollmann, 2022; Muschkiet et al., 2023; Wulfert, Betzing, et al., 2019). Because of the blurring boundaries between the forms of commerce, McKinsey introduced the acronym NeXT commerce (Arora Longo, et al., 2022).

As indicated in the e-commerce cube (Choi et al., 1997, p. 18), the digitalization of commerce enables the creation and trade of entirely digital products and services, such as e-books, software, and banking applications, and are non-physical objects (Lefebvre and Lefebvre, 2002, p. 313). As stated above, pure e-commerce involves digital products (Choi et al., 1997; Luxem, 2000; Turban, Whiteside, et al., 2017).[40] Loebbecke (1999, p. 1) defined digital products[41] as "data, information, and knowledge traded on the Internet [...] that consist only of content and whose total value, therefore, can [be], and usually is, produced, traded, and delivered online." Choi et al. (1997, p. 69–73) identified indestructibility, transmutability, and reproducibility as basic characteristics of digital products. Indestructibility indicates that the usage of the product does not diminish its quality (Choi et al., 1997). While indestructibility means that there is no wear and tear for digital products, transmutability refers to the digital product's ability to be perceptible to change. Compared to traditional products, it is relatively easy to customize digital products to a customer's preferences (Luxem, 2000, p. 2). Finally, reproducibility refers to a digital product's ability to be duplicated with less effort and less cost compared to traditional products. For instance, storage costs are related to the costs of physical hard drives and cloud storage required for the initial copy of the digital product from which all copies can be reproduced subsequently (Loebbecke, 1999).[42] To produce an additional copy of a digital product, the marginal costs are almost zero (Luxem, 2000, p. 2). Referring to Choi et al. (1997), Loebbecke (1999, p. 4 f.) distinguished digital products along the four dimensions of customization, timeliness, intensity in use, and network effects.[43] For a high degree of customization, Loebbecke (1999) suggested pull-based delivery modes. Pull-based delivered products can be customized to the preferences of the specific customer pulling the product. Timeliness refers to the time dependence of a digital product (time dependence vs. time independence). Time-independent products are likely to be used more frequently. Intensity in use describes how often a digital product can be or is likely to be used

---

[40] Since the further analysis within this thesis takes a narrow perspective on e-commerce involving commercial transactions, transaction platforms trading physical and digital products are considered. Innovation platforms need to provide the necessary application services for storing, selling, and delivering digital products (Aulkemeier, Schramm, et al., 2016; Zwass, 1996).

[41] Originally, Loebbecke (1999, p. 1) termed digital products as "online delivered content."

[42] Compared with physical products, the reproducibility characteristic significantly reduces the need for storage capacity in terms of warehouses.

[43] Originally, Loebbecke (1999) used the term "externality" to describe network effects. However, we prefer to term these network effects for increased consistency throughout this thesis (Shapiro and Varian, 1998). Network effects are further unfolded in the following sections.

(single-use vs. multiple-use). Network effects positively or negatively influence the digital product's value[44] for customers, depending on the number of other customers.[45]

Since e-commerce does not only mean "'doing [existing] business electronically'" (Choi et al., 1997, p. xxiv), e-commerce also bears the potential for altered and entirely new business models. New business models benefit from the potential leveraged by new technologies, and because of that, a number of different business models have emerged in e-commerce (Alt, 2020; Deutsch, 1998; Kollmann, 2022; Timmers, 1998). In the editorial to the journal entitled *Electronic Markets*, writing on business model development, Alt (2020, p. 407) provided an overview of different electronic business types with reference to other researchers.[46] Timmers (1998, p. 5 f.) mentioned ten different business models in the context of e-commerce.[47] Tapscott et al. (2000, p. 28 ff.) introduced a typology consisting of five types of business models in e-business. These business models are agora, aggregation, value chain, alliance, and distributive networks. While Täuscher and Laudien (2018, p. 324 ff.) introduced six business model types,[48] Cennamo (2021, p. 269 f.) described three platform business models[49] in e-commerce. E-commerce requires dedicated infrastructure and services (Aulkemeier, Paramartha, et al., 2016; Zwass, 1996). In this research, we will especially focus on platform business models and innovation platforms that also enable other business models with adequate application services (Aulkemeier, Schramm, et al., 2016).

---

[44] The value here involves "economic consequences that are not fully accounted for by the price or market system" (Loebbecke, 1999, p. 4).

[45] A further differentiation in direct and indirect network effects will be undertaken in the following section. We will also dive deeper into the impacts of and prerequisites for network effects (Armstrong, 2006; Briscoe, Odlyzko, et al., 2006; Rochet and Tirole, 2003; Shapiro and Varian, 1998).

[46] Following Stähler (2002, p. 60), the different business model typologies exist due to an evolution of the electronic business and e-commerce and possible uncertainties regarding their definition.

[47] Timmers (1998, p. 5 f.) introduced e-shop, e-procurement, e-auction, e-mall, third-party marketplace, virtual communities, value chain service provider, value chain integrator, collaboration platforms, and information brokers as business models in e-commerce.

[48] Täuscher and Laudien (2018, p. 324 ff.) mentioned efficient product transactions, digital product community, product aficionados, on-demand offline services, online services, and peer-to-peer offline services as business model types in e-commerce.

[49] Cennamo (2021, p. 269 f.) referred to multi-sided transaction markets (e.g., Amazon Marketplace, eBay), complementary innovation markets (e.g., SAP NetWeaver, Nike+, iPhone), and information markets (e.g., GoogleSearch, Facebook, TripAdvisor).

These e-commerce business models can involve three groups of participants (i.e., customers, businesses, governmental organizations[50]) on the demand side and supply side, which are frequently mentioned in the literature (e.g., Lefebvre and Lefebvre, 2002, p. 313). The first form of e-commerce considered here is Business-to-Business (B2B).[51] B2B includes financial transactions such as insurance and loans, but the focus of B2B e-commerce is more on wholesale trade between companies in the services, resources, and technologies sectors (Dan, 2014; Große Holtforth, 2018). Aichele and Schönberger (2016, p. 5 ff.) described this form of e-commerce as an electronic distribution between companies, where a company can take the position of the supplier or the buyer. The scope ranges from digital marketplaces to the customer's and supplier's complete integration into the company's own value creation processes (Aichele and Schönberger, 2016, p. 7). Another form of e-commerce is B2C.[52] B2C e-commerce refers to the traditional sales of products or services to end customers. The focus of B2C transactions is on the exchange of digital and physical goods and services from companies to customers (Aichele and Schönberger, 2016; Turban, Whiteside, et al., 2017). B2C e-commerce "applies to any company or organization that sells products or services to customers over the Internet for their own benefit" (Nemat, 2011, p. 101).[53] Customer-to-Customer (C2C) e-commerce involves transactions among customers usually facilitated by a third party (e.g., eBay, eBay Classifieds). To make this possible, companies take an intermediary position and usually charge a fee for this help in matching supply and demand (Nemat, 2011, p. 102).[54] Customer-to-Business (C2B) e-commerce is mainly characterized by the customers transferring information to

---

[50] Governmental organizations as participants in e-commerce transactions are not considered in detail in our investigation on platforms in e-commerce. We argue that they are included if they act as companies in our analysis.

[51] According to Nemat (2011), B2B originally referred to electronic communication between companies in order to differentiate it from the company's communication with the customer. Over time, the term evolved into the previously mentioned understanding, namely, B2B e-commerce. The volume of B2B transactions is much higher compared to other manifestations of e-commerce, as many components and resources are purchased from other companies within the value chain via electronic means (Combe, 2006).

[52] Nemat (2011, p. 101) defined B2C as "activities with which the company provides the customer with products or services."

[53] Furthermore, B2C can also refer to other offerings, such as online banking, travel services, online auctions, health information, or the real estate market (Nemat, 2011, p. 101). For the further analysis of platforms in e-commerce ecosystems, we focus on the exchange of tradable products and services in a narrow sense.

[54] A purchasing process involving a transaction does not necessarily have to exist in this form. It can also be an exchange of goods (Aichele and Schönberger, 2016, p. 68–69). For

companies.[55] Other forms of e-commerce involve governmental organizations as actors.[56] Business-to-Government (B2G) e-commerce describes the execution of administrative processes between businesses and government units. Customer-to-Government (C2G) e-commerce consists mainly of the use of digital technologies to transfer information from customers to governmental units (Wirtz, 2020, p. 63–64). Interactions with governmental organizations are also possible the other way around through the manifestations of Government-to-Business (G2B), Government-to-Government (G2G), and Government-to-Customer (G2C) (Nemat, 2011, p. 103). In this thesis, we mainly focus on businesses and customers as major transaction partners in e-commerce. Therefore, we investigate platforms supporting B2C, B2B, C2C, and C2B transactions. In Fig. 3.4, we summarize the transaction alternatives with regard to entity types involved, with the transaction partner combinations this thesis focuses on shaded in grey.

| To: Consumer of service | From: Supplier of product/service | | |
|---|---|---|---|
| | Customer (C) | Business (B) | Government (G) |
| C | Customer-to-Consumer (C2C) | Business-to-Customer (B2C) | Government-to-Customer (G2C) |
| B | Customer-to-Business (C2B) | Business-to-Business (B2B) | Government-to-Business (G2B) |
| G | Customer-to-Government (C2G) | Business-to-Government (B2G) | Government-to-Government (G2G) |

**Fig. 3.4** Summary of Transaction Alternatives Between Customer, Business, and Government (Chaffey, 2015, p. 26)

According to Heinemann et al. (2019, p. 4), commerce can be classified into several types, each of which can be differentiated from the other, based on the amount of contact the transaction partners have during the business transaction, and in what direction the contact is made. In total, five transaction principles according to which

this reason, C2C e-commerce usually involves auctions or second-hand products that are traded among customers (Dan, 2014; Lefebvre and Lefebvre, 2002).

[55] C2B does not represent a main focus of this dissertation. We consider information exchanged for e-commerce transactions as mandatory in electronic environments.

[56] E-commerce transactions involving governmental organizations are not the focus of this thesis. However, we would like to note that these organizations can provide transaction platforms and can also participate in e-commerce ecosystems.

the trade can be characterized, based on the contact direction in trade are identified (Schröder, 2005, p. 9 ff.). These principles are the residence principle, the domicile principle, the meeting point principle, the distance principle, and hybrid forms. The residence principle (1) resembles brick-and-mortar retailing. The customers go to the seller and conduct the transaction at the retail store. In the domicile principle (2), the transaction takes place at the customers' locations, which means that the seller goes to the customer. The meeting point principle (3) entails the seller and customer going to a "neutral" location owned by a third party (e.g., a weekend market). The distance principle (4) is applied in e-commerce since the transaction partners do not meet physically and the goods are transported to the customers' domiciles making use of logistic service providers. These transactions are often conducted in a virtual environment supported by IT. Hybrid principles (5) involve an arbitrary combination of the aforementioned principles (Heinemann et al., 2019, p. 4). The different contact principles are illustrated in Fig. 3.5 (Heinemann et al., 2019; Schröder, 2005).

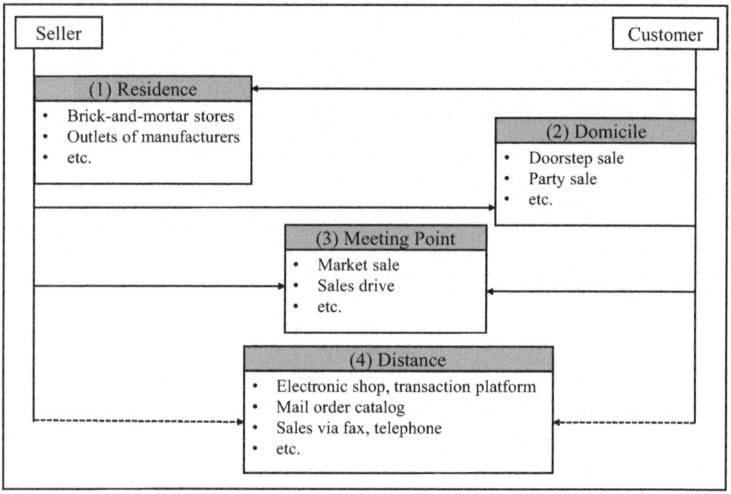


**Fig. 3.5** Contact Principles in Commerce (Schröder, 2005, p. 10)

In the following section, we unfold different notions for understanding platforms and major platform types in the context of e-commerce. These platforms are likely to form platform-centered ecosystems (e.g., Adner, 2017; Benedict, 2018; Wulfert, Woroch, Strobel, Seufert, et al., 2022).

## 3.2 Platforms in E-Commerce

In this section, we clarify different facets of the platform concept (Sun, Gregor, et al., 2015; Thomas et al., 2014). Particularly, we focus on innovation platforms and transaction platforms including a combination of the two concepts called hybrid platforms. First, we provide a brief overview of the etymology of the term platform and introduce different perspectives on the investigation of platforms. Then, we describe innovation platforms in e-commerce, introduce transaction platforms in e-commerce, and integrate the two platforms types in a hybrid platform (Gawer, 2021b; Wulfert and Karger, 2022a). We also elaborate on platform governance as an overarching activity across platform types (Boudreau, 2010; Tiwana, 2014).

### 3.2.1 Platform Etymology and Research Perspectives

Depending on the application context, the term platform has various meanings. These meanings can be caused by the term's etymology. First coined in German-language usage in the mid-16th century, the origins of the term platform stem from the French term *plateforme*.[57] The French term initially described a flat plain or surface used in fortification. From the mid-19th century on, the term platform also established itself as a synonym for a political standpoint. Owing to English-language influences, the current spelling eventually became established (Pfeifer, 1993) and, since these influences stemming from the English language, the term platform also acquired additional meanings. The *Oxford English Dictionary* thereby introduced the term platform as a synonym for a platform in train stations, a standard for the hardware of computers that can run certain software, and platform shoes (Lexico.com, 2019). The term platform "is deployed as a metaphor for infrastructural operations in multiple areas" (Murakami Wood and Monahan, 2019, p. 2).[58] According to Gillespie (2010, p. 349 f.), the four current associations of the term platform are computational, architectural, figurative, and political. The computational association involves hardware and software environments as infrastructure for the "design and use of particular applications" (Gillespie, 2010, p. 349).[59] These environments

[57] We consider the etymology of the term platform from a German language perspective.

[58] Murakami Wood and Monahan (2019, p. 3) even argued that a number of established platforms, such as Google, Meta, Microsoft, and Amazon, have become commonly used "infrastructures."

[59] The computational association described by Gillespie (2010) refers to the concept of innovation platforms that will be introduced in the remainder of this section.

formulate de facto standards for the engagement of third-party developers [60] who provide the complements. In the computational association, a platform is seen as a "framework" (Murakami Wood and Monahan, 2019, p. 2) or "foundation upon which other computing processes are built or the environment within which such processes run" (Murakami Wood and Monahan, 2019, p. 2). The architectural association describes "human-built or naturally formed physical structures, whether generic or dedicated to a specific use" (Gillespie, 2010, p. 349).[61] The figurative association abstracts from the physical structure, preserving its metaphorical meaning. It describes a "ground, foundation, or basis of an action calculation, condition, etc." (Gillespie, 2010, p. 350). The political association refers to the political basis of a politician, program, or party in an abstract meaning.[62]

The multiple meanings of the term platform already give rise to different perspectives in research on platforms, both in the scientific literature and in entrepreneurial practice. Different notions for understanding platforms (e.g., organizational capabilities, technology set, two-sided market, extendable system) and various research areas (e.g., construction or management literature) investigating platforms have been established (e.g., Gawer, 2014; McIntyre and Srinivasan, 2017; Reillier and Reillier, 2017; Sun, Gregor, et al., 2015; Thomas et al., 2014).[63]

Platforms are mostly analyzed in four research streams (Gawer, 2014; McIntyre and Srinivasan, 2017; Sun, Gregor, et al., 2015; Thomas et al., 2014). Thomas et al. (2014, p. 29) identified organizational capabilities, product family, market intermediary, and technology system as research streams on platforms.[64] The organizational capability stream deals with research on dynamic capabilities (e.g., Eisenhardt and Martin, 2000; Teece et al., 1997; Teece, 2007) and describes platforms as "a

---

[60] Third-party developers are also named "complementors" developing and providing complements to a platform (e.g., Benlian et al., 2015; Boudreau and Jeppesen, 2015; Engert et al., 2023; Hurni, Huber, Dibbern, and Krancher, 2021; Petrik and Herzwurm, 2020; Rickmann et al., 2014). For reasons of consistency, we will use the term third-party developer throughout this thesis.

[61] In this sense, it also includes the above-mentioned platforms in train stations.

[62] Although a platform is typically conceived as "flat, featureless and open" (Gillespie, 2010, p. 350), it is coined by a specific position in the political association.

[63] Besides the different perspectives for analyzing platforms, Kenney and Zysman (2016, p. 62) described a platform's overall objective as "mobilizing human beings to contribute." Kenney and Zysman (2016, p. 62) referred to Google, Facebook, LinkedIn, and Uber as examples to support their proposition.

[64] Thomas et al. (2014, p. 4) emphasized that "[e]ach of the four streams is characterized by a distinctive, although usually implied, theoretical logic" (Thomas et al., 2014, p. 4) and that the streams consider the platform as a lever for "organizational advantage and value creation" (Thomas et al., 2014, p. 4).

collection or specific architecture of resources and capabilities that have been realized and deployed by dynamic capabilities" (Thomas et al., 2014, p. 6). In this research stream, the platform concept is used to describe a structure for organizational capabilities and resources (Kogut and Kulatilaka, 1994; Rai et al., 2006). The product family research stream describes a product platform as the foundation for the derivation of several products (Thomas et al., 2014, p. 7).[65] Following Meyer and Lehnerd (1997, p. 7), a product platform enables the efficient creation of derivative products. The product platform concept was first applied in production companies, more specifically in the automotive industry (Wheelwright and Clark, 1992). In this context, a "platform strategy" (Gebhart et al., 2016, p. 128) in production describes a "long-term, systematic bundling of product-inherent elements into platforms and the management of these platforms within the framework of long-term product planning" (Stauss, 2006, p. 322). When using the platform strategy in production, the same components and modules are used for different models of a company or even across companies (Wallentowitz et al., 2009, p. 144). The main benefit of this strategy lies in the exploration and exploitation of economies of scale and scope (Wheelwright and Clark, 1992). In production, the platform strategy represents the intermediate step from the common parts strategy[66] to a module strategy.[67] The market intermediary stream focuses on platforms as orchestrators of two-sided or multi-sided markets (Armstrong, 2006; Rochet and Tirole, 2006).[68] Thomas et al. (2014, p. 7 f.) identified network effects, value capture, pricing, and matching as major topics of concern. The technology system stream investigates innovation

[65] A product platform can be defined as a "collection of assets that are shared by a set of products where assets may include components, processes, knowledge, and people" (Robertson and Ulrich, 1998, p. 5).

[66] When a common parts strategy is used in production, a small number of standardized parts, in relation to the total number of components used, are used for several variants and product families (Gebhart et al., 2016, p. 131).

[67] In analogy to the common parts strategy, a module strategy involves the use of autonomous, functional units, also referred to as modules, in multiple product variants (Gebhart et al., 2016, p. 133). The product platform "is a set of common components, modules, or parts from which a stream of derivative products can be efficiently created and launched" (Meyer and Lehnerd, 1997, p. 7).

[68] At this point, it should be noted that the use of the term market appears superficially misleading. The market represents a price-based form of coordination of supply and demand (Reiss and Günther, 2010). However, multi-sided markets are not primarily about price-based coordination, but about the orchestration of different groups of participants. Some authors therefore use the more neutral term network (e.g., Sun and Tse, 2009). Owing to the more established terminology, this thesis refers to two-sided and multi-sided markets (e.g., Armstrong, 2006; Rochet and Tirole, 2003; Rochet and Tirole, 2006).

platforms as "the hub of a technology system" (Thomas et al., 2014, p. 8) incorporating the "logic of modularity, standards, and product differentiation to a product or service system broader than an internal or supply-chain level product family" (Thomas et al., 2014, p. 8). In the operation management literature, platforms form a system or architecture for extension development and execution (Cusumano and Gawer, 2002; Taudes et al., 2000; Tiwana, Konsynski, and Bush, 2010; West, 2003).[69] In the technology system, the innovation platform's owner intentionally relinquishes the "control of the overall finished product" (Thomas et al., 2014, p. 8) to independent platform participants developing and providing complementary products and services. While we adopt the market intermediary stream as a transaction platform, we consider the technology system stream as an innovation platform within this cumulative dissertation

Reillier and Reillier (2017, p. 21 f.) indicated three scientific areas that deal with platforms. These areas are product development, software development, and economic transactions. The product development area deals with the idea of an already introduced product platform and focuses on the development of products that is supported by a platform a number of products share (e.g., vehicles and aircraft) (Reillier and Reillier, 2017, p. 21). The product platform approach is particularly researched in the area of technology and software in which technologies and software form the foundation for the development of new products and services complementing these platforms (Reillier and Reillier, 2017, p. 21). The economic transactions perspective focuses on multi-sided markets and the transactions enabled between the different market sides (Reillier and Reillier, 2017, p. 22).

McIntyre and Srinivasan (2017, p. 157) introduced technology management, industrial organization economics, and strategic management as major research perspectives on platforms. The technology management perspective highlights the innovative potential of platforms in engineering and design (McIntyre and Srinivasan, 2017, p. 150 ff.). Innovation platforms are described "as modular systems [...that break] up a complex system[70] into discrete components that interact through standardized interfaces" (McIntyre and Srinivasan, 2017, p. 150 ff.). From an industrial organization economics perspective, the authors describe platforms as loci "that

[69] Baldwin and Woodard (2009, p. 7) described architecture as an abstract description of the entities of a system and the relationships between them. The IS architecture includes a "[d]escription of the logical and physical arrangement of the building blocks of a complex system, as well as the relationships between these building blocks" (Hansen and Neumann, 2007, p. 24) as well as the process sequences within an organizations (Ahlemann, 2012; Schekkerman, 2004).

[70] Simon (1962, p. 468) described a system as complex if it is "made up of a large number of parts that interact in a non-simple way."

facilitate exchange between two or more categories of users" (McIntyre and Srinivasan, 2017, p. 144). This perspective considers platforms as two-sided and multi-sided markets including network effects and winner-takes-all tendencies.[71] As a third research perspective on platforms, McIntyre and Srinivasan (2017, p. 148 ff.) introduced strategic management. This perspective is concerned with the time of entry in an already established platform and with the timing for the establishment of an own platform. The strategic management perspective also includes discussions on appropriate ignition strategies and platform competition (Amit and Zott, 2021; Rietveld and Schilling, 2021).

In the context of e-commerce, the economic perspective and the technological perspective (Bughin and van Zeebroeck, 2017; Gawer, 2014; Reillier and Reillier, 2017; Thomas et al., 2014) are especially relevant (Hagberg et al., 2016; Hagiu and Wright, 2015a; Hänninen et al., 2018; Wulfert and Karger, 2022a).[72] In a special issue focusing on multi-sided platforms, Abdelkafi et al. (2019) emphasized in their editorial introduction to the peer-reviewed journal *Electronic Markets* that transaction platforms (economic perspective), innovation platforms (technological perspective), and hybrid platforms (combination of both perspectives) are especially relevant to research in e-commerce. The aforementioned differentiation is supported by recent research in platforms in general and in e-commerce in particular (Gawer, 2021b; Schütte and Wulfert, 2022a). Gawer (2014, p. 1240) also focused on the economics perspective and the engineering design perspective for her integrative framework for platforms.[73] In a similar vein, Schreieck et al. (2016, p. 5 ff.) focused on a technology-oriented perspective and a market-oriented perspective.[74] Schütte and Wulfert (2022a, p. 213) also highlighted that platforms are an important concept in e-commerce on both infrastructure and business model levels. Constantinides et al. (2018, p. 381) defined platforms in an abstract manner as "a set of digital resources—including services and content—that enable value-creating interactions

[71] The industrial organization economics perspective aligns with the concept of transaction platforms explained in the remainder of this section.

[72] We provide an overview of platform definitions in Appendix A.2 in the Electronic Supplementary Material. We also assigned the perspective taken by the authors of each definition.

[73] This integrative framework for platforms resembles a classification scheme along the dimensions level of analysis, constitutive agents, technological architecture, interfaces, innovative capabilities, and coordination mechanisms (Gawer, 2014, p. 1244).

[74] These perspectives resemble the innovation platform and transaction platform concepts unfolded in the remainder of this section.

between external producers and consumers."[75] This abstract definition includes the economic and technological perspectives.

Besides the economic and technological perspectives on platforms and the resulting differentiation between innovation platforms and transaction platforms in e-commerce, Gawer and Cusumano (2014, p. 418 ff.) differentiated between internal and external platforms. Internal platforms are defined "as a set of assets organized in a common structure from which a company can efficiently develop and produce a stream of derivative products" (Gawer and Cusumano, 2014, p. 419). In contrast, external platforms allow others outside the focal organization's boundaries to develop complementary products and services (Gawer, 2014; Porch et al., 2015). In a related study, Gawer and Cusumano (2014, p. 143 ff.) introduced supply-chain platforms as an intermediary type of internal and external platform. Affiliated organizations from the supply chain can use supply-chain platforms as a foundation for product and service development (Gawer and Cusumano, 2014, p. 144).[76] For our investigation on platforms in the context of e-commerce in this cumulative dissertation, we focus on external platforms creating a surrounding business ecosystem in e-commerce (Schreieck et al., 2016, p. 5).[77]

Focusing on external platforms, we present selected platform concepts in line with the economic and technological perspectives on platforms introduced in Table A.2 in the Electronic Supplementary Material. In the following sections, we introduce innovation platforms and transaction platforms in further detail, adopting selected definitions and disentangling their characteristics while highlighting their intertwined nature resulting in hybrid platforms.

### 3.2.2 Innovation Platforms

Innovation platforms are defined as "a set of stable components that supports variety and evolvability in a system by constraining the linkages among the other components" (Baldwin and Woodard, 2009, p. 19).[78] Tiwana, Konsynski, and Bush (2010,

[75] In line with this definition and aside from the understanding of research perspectives on platforms, Baldwin and Woodard (2009, p. 25) questioned whether the "various platform concepts are synonymous or simply evocative uses of the same word to mean different things".

[76] In supply-chain platforms, the focal company usually controls the development of products and services.

[77] We consider the external features as requirements for the attraction and matching of a number of participants (Ghazawneh and Henfridsson, 2011).

[78] Ciborra already noted the potential of innovation platforms in (1996, p. 104). He characterized innovation platforms as an "unrecognized source of productivity in the high-tech

p. 675) emphasized that third-party developers can use these stable components as an "extensible codebase [...] that provides core functionality shared by the modules that interoperate with it and the interfaces through which they interoperate." While the stable components form the core of the innovation platform, the third-party modules[79] can be described as the periphery (Staykova and Damsgaard, 2015, p. 4 f.). In this cumulative dissertation, we refer to the periphery as software modules extending the innovation platform.[80] The platform core and periphery are evolvable, with only the interface specifications that should remain stable over a longer period of time to avoid necessary module adjustments for the developers (Baldwin and Woodard, 2009; Staykova and Damsgaard, 2015; Tiwana, Konsynski, and Bush, 2010) (Figs. 3.6 and 3.7).

Whereas the platform owner develops and operates the platform core,[81] third-party developers provide platform extensions (i.e., the periphery) (Jansen and Cusumano, 2013; Tiwana, Konsynski, and Bush, 2010; Wulfert and Karger, 2022b). Third-party developers are "actors who contribute to an open platform by programming code for applications" (Von Krogh et al., 2003, p. 1219).[82] To provide platform users with a wide variety of applications, the platform owner sets incentives to support external innovation. The relationship between the platform owner and third-party developers is therefore not in "command and order" form (Tiwana, 2014). Instead, the platform owners should motivate and support the third-party developers (Huber, Kude, et al., 2017; Mohagheghzadeh and Lindman, 2022).[83]

---

industries." Innovation platforms have the "intrinsic potential to efficiently generate new combinations of resources, routines, and structures which are able to match the present, turbulent circumstances" (Ciborra, 1996, p. 104). Innovation platforms allow for "an exciting mixture of ready-made arrangements and interpretations, and of half-realized, not-yet-made solutions and visions" (Ciborra, 1996, p. 104).

[79] Modules are defined as "components of a system whose elements are powerfully connected among themselves and relatively weakly connected to elements in other components" (Baldwin and Clark, 2000, p. 63).

[80] Innovation platforms build upon the concept of complementary assets (Teece, 1986). In this case, the complementary assets are represented by (software) extensions provided by third-party developers (Dal Bianco et al., 2014; Eaton et al., 2015; Ghazawneh and Henfridsson, 2013; Saloner, 1990).

[81] The "platform owner" (Baldwin and Woodard, 2009, p. 38) role adopts the roles of the platform provider and platform sponsor (Eisenmann, Parker, et al., 2009, p. 134 f.). Throughout this cumulative dissertation, we use the term platform owner and do not differentiate between platform provider and platform sponsor roles.

[82] In contrast to internal developers, the platform owner does not employ external developers and does not have any direct command over them.

[83] Engert et al. (2023, p. 5) defined third-party developers' engagement in an innovation platform as "state-based partly volitional resource contribution in its interactions, activities,

The main idea of innovation platforms is the reuse of core (software) modules and their recombination for the creation of new products and services "to achieve economies of scale while reducing the cost of creating a wide variety complementary components" (Baldwin and Woodard, 2009, p. 19). Boudreau (2010, p. 1851) added that an innovation platform "may include physical components, tools and rules to facilitate development, a collection of technical standards to support interoperability or any combination of these things." Kenney and Zysman (2016, p. 64) defined innovation platforms as "complicated mixtures of software, hardware, operations, and networks. The key aspect is that they provide a set of shared techniques, technologies, and interfaces to a broad set of users who can build what they want on a stable substrate".[84] Innovation platforms "consist of technological building blocks that are used as a foundation on top of which a large number of innovators can develop complementary services or products" (Evans and Gawer, 2016, p. 6).[85] Gawer and Cusumano (2002, p. 2) hinted at the platform's internal architecture, when stating that it is "made [up] of interdependent pieces that can each be innovated upon."[86] Perks et al. (2017, p. 107) described innovation platforms "as dynamic configurations of (tangible and intangible) resources that act as a foundation upon which network members co-create value through a set of specific practices." Kapoor et al. (2021, p. 95 ff.) highlighted the socio-technical nature of innovation platforms not only describing the composition of modules, but also the involvement of platform owners and third-party developers. Sangiovanni-Vincentelli and Martin (2001, p. 26) emphasized that an innovation platform "is an abstraction that covers several

---

and relationships." Third-party developers' contribution is subject to continuous change that requires careful platform governance (Engert et al., 2023, p. 13 ff.). Platform governance is further emphasized in the remainder of this section.

[84] In the literature, innovation platforms are also described as software platforms (e.g., Dal Bianco et al., 2014; Evans, Hagiu, et al., 2008; Jansen and Cusumano, 2013; Messerschmitt and Szyperski, 2003), while Evans and Schmalensee (Evans and Schmalensee (2016), p. 209 f.) introduced the innovation platform as "invisible engine," Moazed and Johnson (2016, p. 41 ff.) as "maker platforms," and Evans and Schmalensee (2007, p. 7 ff.) as "cost minimizer." However, we prefer to use the term innovation platform, as Evans and Gawer (2016, p. 5) noted that "most of today's platforms are digital" and thus entail software per definition, regardless of whether or not they focus on extension development.

[85] It is important to emphasize here that Evans and Gawer (2016, p. 4) focused their survey not on platforms, but on entire organizations.

[86] We describe the "interdependent pieces" (Gawer and Cusumano, 2002, p. 2) and "technological building blocks" (Evans and Gawer, 2016, p. 6) of an innovation platforms as "modules" (Fig. 3.8). While modules developed by the platform owner are described as "internal modules," modules forming platform extensions developed by third-party developers are named "external modules."

possible lower-level refinements" and, thus, indicates the encapsulation of platform features through boundary resources.

In the literature on innovation platforms, several dimensions have been introduced for describing them (e.g., Blaschke, Haki, et al., 2019; Boudreau, 2010; Eisenmann, 2008). Eisenmann (2008, p. 33 ff.) described innovation platforms along the dimensions of platform provider and platform sponsor and differentiated, for each case, if the provider and sponsor are a single organization or a conglomerate of companies. Boudreau (2010, p. 1856) augmented the owner-centered description by Eisenmann (2008) with an investigation of the opening of the focal platform's hardware for third-party developers. Economides and Katsamakas (2006, p. 1060 ff.) took this dimension and differentiated between proprietary and open-source innovation platforms.[87] Knoll and RinderleMa (2015, p. 324 ff.) differentiated innovation platforms into hardware platforms, operating system platforms, and software platforms.

The architecture of innovation platforms is often described as layered and modular (e.g., Baldwin and Clark, 2000; Baldwin and Woodard, 2009; Blaschke, Haki, et al., 2019; Kazan, Tan, et al., 2018). Owing to an innovation platform's layered architecture (Yoo, Henfridsson, et al., 2010, p. 726 f.), new extensions can be added to the "platform to extend its functionality" (Mohagheghzadeh and Svahn, 2016, p. 2).[88] In the literature about platforms, a layered approach has been established for describing a platform's architecture (e.g., Choudary, 2015; Yoo, Henfridsson, et al., 2010; Zutshi and Grilo, 2019). [89]

Choudary (2015, p. 60 ff.) analyzed platforms, using so-called "platform stacks" consisting of the layers of the network-marketplace community, infrastructure, and

[87] For the development of their model, the authors focused on operating systems as platforms. While the authors chose Microsoft Windows as an example of a proprietary platform, Economides and Katsamakas (2006, p. 1060 ff.) analyzed Linux as an example of an open-source platform. At this juncture, it is necessary to emphasize that the authors only presented two extreme cases and did not provide any further differentiating characteristics of proprietary and open-source platforms.

[88] Kapoor (2018, p. 6) defined the architecture of an ecosystem as "[t]echnological interactions between offers and input-output flow interactions between actors." Thus, the ecosystem architecture prescribes the relations among ecosystem participants and with the focal platform owner. Moreover, it highlights the importance of complementors in platform-centered ecosystems (Huber, Kude, et al., 2017) and the relationship between the platform owner and complementors as a source of extended supply (Hurni, Huber, and Dibbern, 2022; Hurni, Huber, Dibbern, and Krancher, 2021).

[89] In general, Yoo, Henfridsson, et al. (2010, p. 727) introduced a layered architecture for digital technology, consisting of a contents layer, a service layer, a network layer, and a device layer.

data.[90] The network-marketplace community layer depicts the platform participants and their relations with the platform and among each other. The infrastructure layer involves "tools, services, and rules [...third-party] producers build on" (Choudary, 2015, p. 61). Choudary (2015, p. 62) described the infrastructure as a layer upon which supply-side participants can create value for customers. The data layer stores and analyzes the data generated through interactions with the platform. The main objective of data analysis is to improve the matching between the platform's supply side and demand side (Choudary, 2015, p. 62).[91]

Sun, Gregor, et al. (2015, p. 8) described a governance layer, add-ons, standards, and a technological base in their platform conceptual model. The governance layer entails rules and guidelines for extension development. For Sun, Gregor, et al. (2015, p. 9), the add-on resembles platform extensions "following defined standards and using stable interfaces" of the focal platform. Standards mainly refer to technical boundary resources and supplementary rules for the access of core platform services. They enable interoperability with the platform core and transactionality among platform participants.[92] The technological base refers to the platform core providing application services for extension development and operation as well as boundary resources for accessing these services (Sun, Gregor, et al., 2015, p. 16).

Blaschke, Haki, et al. (2019, p. 8) included four dimensions in their taxonomy for describing platforms. The service dimension includes a variety of platform-powered services provided by one participant for the benefit of another. While the ecosystem dimension includes the various platform participants (i.e., supply-side participants, platform owner, demand-side participants), the core dimension includes "a set of stable technical core [modules] of software and hardware" (Blaschke, Haki, et al., 2019, p. 8). The authors described this core dimension as the technical foundation upon which third parties can provide their value-added services. The authors included computing and network resources in the infrastructure dimension.[93]

---

[90] Choudary (2015, p. 61) introduced his platform stack as a "framework to explain all platforms." The layers can be adapted in size for specific platform types.

[91] Making use of the platform stack, Choudary (2015, p. 62 ff.) described marketplaces, infrastructure platforms, and data platforms as basic configurations of platforms. While the infrastructure and data platforms resemble our understanding of innovation platforms, the marketplace inhibits transaction characteristics unfolded in the remainder of this section.

[92] Thompson (1967, p. 51 ff.) already differentiated an organization into its technical core and boundary-spanning buffers that protect an organization against influences from outside. Scott (2014, p. 186) also emphasized the concept of "buffering units" protecting an organization against influences from the environment.

[93] Based on these four layers, Blaschke, Haki, et al. (2019, p. 11 ff.) introduced three platform types, namely orchestration platform, amalgamation platform, and innovation platform.

Zutshi and Grilo (2019, p. 551) described six layers of digital platforms. The business model is defined in the business layer including its major value proposition. In the case of a platform business model with multiple market sides, Zutshi and Grilo (2019, p. 551) identified the discovery of collaborators, price discovery, and the elimination of other gatekeepers as major activities for the formulation of a business model. In the user interface layer, the participants' interactions with the platform are defined. While the development layer provides Software Development Kits (SDKs) for third-party developers' extensions, the integration layer allows the connection with external service providers (e.g., data analytics) through predefined Application Programming Interfaces (APIs) (Zutshi and Grilo, 2019, p. 551 f.). The data layer involves data discovery and data storage systems as well as features for data analysis. The IT layer resembles the platform core defining the architecture for the data layer and for designing data analysis processes.

Wulfert and Karger (2022b, p. 8) applied an enterprise architecture perspective for the derivation of their platform architecture framework and identified the business actor layer, business layer, application layer, and infrastructure layer in e-commerce ecosystems. The business actor layer includes participant types in e-commerce ecosystems. While the business layer described the possible business models in e-commerce (Timmers, 1998), the application layer involved innovation platforms providing the necessary application services for the e-commerce business models. The infrastructure layer provides the necessary computing power for the innovation platform. We provide an overview of possible platform layers in Fig. 3.6.

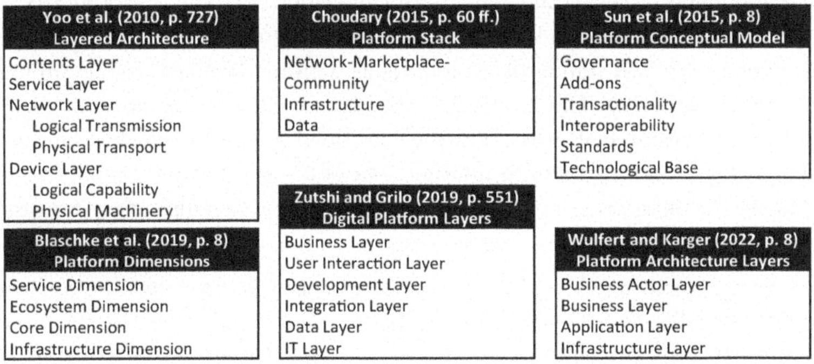


**Fig. 3.6** Selected Platform Layers

Summarizing the considerations of an innovation platform's layered architecture, several authors differentiated the platform core from peripheral third-party extensions linked by boundary resources (e.g., Baldwin and Clark, 2000; Baldwin and Woodard, 2009; Gawer, 2009a, 2014; Tiwana, 2014). With reference to Baldwin (2008); Gawer (2014); Tiwana, Konsynski, and Bush (2010), McIntyre and Srinivasan (2017, Appendix A) defined the architecture of innovation platforms as "[t]he conceptual specification of interfaces that allows an ecosystem to be partitioned into the relatively stable platform and a complementary set of modules and governs the interactions among these different parts." The architecture of an innovation platform is recognized as a primary driver of the nature and distribution of third-party developers' complements. Cusumano and Gawer (2002, p. 55) revealed that the platform core's architecture is evolvable and is evolved by successful platform owners.[94] Baldwin and Woodard (2009, p. 24) further emphasized the platform core's stable nature and the peripheral components' evolvability and variability. The authors highlighted the role of boundary resources[95] that link the platform core with the periphery. Since the boundary resources enable access to the platform core, "they may be the only components that remain truly stable over long periods of time" (Baldwin and Woodard, 2009, p. 41). Baldwin and Woodard (2009, p. 41) added that the platform core's relative stability and the periphery's variety are "accomplished via stable, yet versatile interfaces which govern the interactions of components." These interfaces also delimit the single platform-internal modules (Baldwin and Woodard, 2009, p. 23).

The power of innovation platforms not only comes from their layered design but rests on their architectural modularity (Baldwin and Clark, 2000; Tiwana, Konsynski, and Bush, 2010), catalyzing the reconfigurability of technical and organizational components to accelerate generativity and value creation.[96] Modularity helps manage complexity "by breaking up a complex system into discrete components, which interact through standardized interfaces within a standardized architecture" (Langlois, 2002, p. 19). The modularity enables the (re)configuration of platform components for use in platform extensions provided by third-party developers. The modularization involves a strong internal cohesion[97] of the single modules, weak coupling across module boundaries, and interface abstraction through encapsulation

[94] Baldwin and Woodard (2009, p. 35) also pointed to this fact: "even the core components can evolve—only the interfaces need to be stable."

[95] Baldwin and Woodard (2009, p. 24) termed the boundary resources as interfaces. However, for reasons of consistency, we use the term boundary resource.

[96] The concept of modularity is rooted in early works on complex systems. For instance, Simon (1962, p. 473 f.) described modular systems as "hierarchical and decomposable systems that consist of interacting and inter-dependable subsystems."

[97] If the cohesion within a module is not high, it could be further decoupled (Messerschmitt and Szyperski, 2003, p. 88).

of the technological details (Messerschmitt and Szyperski, 2003, p. 88 f.). In modular environments, the single modules can be exchanged and operated under different circumstances (Baldwin and Clark, 2000; Schilling, 2000; Ulrich, 1995). The interfaces only contain information that is necessary for third-party developers to make their platform extensions compatible with the platform boundary resources (Gawer, 2014, p. 1243). Hence, modularity and the possibility for (re)configuration can lead to significant cost savings if a high degree of reuse can be achieved (Buxmann, Diefenbach, et al., 2011). Tiwana and Konsynski (2010, p. 294) introduced architecture modularity in the context of innovation platforms as the "degree of decomposition of an organization's IT portfolio into loosely coupled subsystems that communicate through standardized interfaces." Jacobides et al. (2018, p. 2260) described modularity as "separability along a production (or production and consumption) chain." Dai (2023, p. 3 f.) introduced independent and loosely coupled modules that are structured hierarchically and that are connected via boundary resources as major building blocks of an innovation platform's modular architecture (Baldwin and Clark, 2000; Simon, 1962; Ulrich, 1995).[98] Modularity can lead to combinatorial innovation (de Reuver et al., 2017; Tiwana, Konsynski, and Bush, 2010). Baldwin and Clark (2000, p. 13) described this innovation as a platform's reconfigurability. The components of single modules are strongly interconnected, but weakly connected to the platform through technical boundary resources (Baldwin and Clark, 2000; Tiwana, 2015a). The clear separability of single modules allows the platform owner to restrict access to some modules while allowing third-party developers to use others and recombine them for their own complements.[99] This modularization results in a system "with intentionally minimized interdependencies between the single subsystems" (Sanchez and Mahoney, 1996, p. 65). This enables high extensibility and evolvability of the platform and its periphery (Dai, 2023, p. 4). Baldwin and Clark (2000, p. 11) emphasized that the platform extensions in the periphery need to be loosely integrated among themselves and the platform core. The linkages between the platform and the third-party extension should be minimized to the predefined technical boundary resources (Dellermann et al., 2016, p. 4).

The concept of modularity results in a high degree of generativity for innovation platforms (Yoo, Boland, et al., 2012), allowing technical modules and business models that the platform owner had not initially intended. This is caused by the

[98] Dai (2023, p. 12 ff.) identified positive effects of an innovation platform's modular architecture on network effects for participants.

[99] In this regard, Jacobides et al. (2018, p. 2260) emphasized that although modularity is a prerequisite for platform openness, it does not necessarily result in platform openness.

shared innovation and contributions of third-party developers whose actions are only partially controllable (Witte and Zarnekow, 2018). Innovation platforms generate generativity from the interaction between the platform owner and third-party developers (Hein, Schreieck, Riasanow, et al., 2020). Generativity is the characteristic of a platform to let third-party developers implement modules with functionalities that platform owners had not even intended (Henfridsson, Nandhakumar, et al., 2018) Innovation platforms are generative systems that allow for unbounded and unforeseeable growth (Constantinides et al., 2018; Foerderer, Kude, Mithas, et al., 2018; Hein, Schreieck, Riasanow, et al., 2020; Zittrain, 2009).[100] Witte and Zarnekow (2018, p. 734) defined generativity in the context of innovation platforms as "the platform's ability to generate change that was not anticipated." While platform owners open their platform to third-party developers, these developers use their own knowledge and expertise to create a variety of unforeseeable platform extensions (Fürstenau, Baiyere, et al., 2023, p. 3).[101]

Fürstenau, Baiyere, et al. (2023, p. 3) introduced two perspectives on generativity in innovation platforms, namely the product view and the social interaction view. The product view of generativity is introduced as a platform's ability "to expand its product boundaries beyond the current categories" (Fürstenau, Baiyere, et al., 2023, p. 3). Based upon the services and components provided by the innovation platform, third-party developers can produce "unanticipated change" (Zittrain, 2008, p. 70) and develop "a seemingly infinite number of variations and specifications" (Yoo, 2013, p. 232). Fürstenau, Baiyere, et al. (2023, p. 4) summarize the unanticipated development of platform complements as the expansion of product boundaries.[102] From a social interaction view, generativity describes the expansion of ecosystem boundaries if third-party developers are involved in the development of platform extensions (Fürstenau, Baiyere, et al., 2023, p. 4). This view considers platforms as a "trading zone" (Fürstenau, Baiyere, et al., 2023, p. 4) in which a variety of actors

[100] In general, Zittrain (2006, p. 1980) introduced generativity as "a technology's overall capacity to produce unprompted change driven by large, varied and uncoordinated audiences."

[101] Although we described third-party developers' involvement in the development of the innovation platform's periphery as largely positive, it needs to be noted that complements can be harmful to the focal innovation platform, for instance, if the provided complements are of bad quality and negatively affect the users of the complements (Brandwijk and de Reuver, 2023, p. 2 f.). Moreover, third-party developers may interfere with the control of the focal innovation platform owner resulting in continuous tensions requiring a balance between generativity and platform control (Benlian et al., 2015; Blaschke and Brosius, 2018; Boudreau, 2010; O'Mahony and Karp, 2022; Schreieck et al., 2016).

[102] In section 3.3, we will further denote the boundary of organizations in general and platforms in particular. Langley, Lindberg, et al. (2019, p. 704) described the reconfiguration of platform components and extension with own developments as "configurational boundary work."

engage in collaborative work.[103] Formerly independent actors become ecosystem participants if they engage "in generative or constraining discussion" (Fürstenau, Baiyere, et al., 2023, p. 4) with the platform owner or other ecosystem participants expanding the ecosystem boundary. Wareham et al. (2014, p. 1203) describes this relationship as "self-selection." From their panel vector auto-regressive approach, Fürstenau, Baiyere, et al. (2023, p. 15 ff.) further derived that growth in the participant base can lead to an expansion of product boundaries (inverse generativity). Moreover, growth in the participant base can lead to stabilized ecosystem boundaries (bounded generativity) (Fürstenau, Baiyere, et al., 2023, p. 15 ff.). We summarize the two perspectives on generativity put forth by Fürstenau, Baiyere, et al. (2023, p. 3 f.) in Table 3.2.

**Table 3.2** Views on Generativity on Digital Platforms (Fürstenau, Baiyere, et al., 2023, p. 3)

| View | Definition | Reference |
|---|---|---|
| Product View | Generativity refers to the capacity of a platform to enable unbounded growth of new components that expand the product boundaries of a platform beyond its initial conception and further attract more users. | Zittrain (2008), Yoo, Henfridsson, et al. (2010), Yoo (2013) |
| Social Interaction View | Generativity refers to the capacity of a platform to produce unbounded growth of interactions between human and technology resources that expand the ecosystem boundaries and leads to user growth by resolving emerging conflicts. | Faraj et al. (2011), Shaikh and Vaast (2016), Jarvenpaa and Standaert (2018) |

Making use of the modular and layered architecture of innovation platforms, third-party developers provide extensions to the platform.[104] Tiwana, Konsynski, and Bush (2010, p. 675) characterized third-party extensions as an "add-on software subsystem." Ghazawneh and Henfridsson (2013, p. 175) described third-party extensions as "executable pieces of software that are offered as applications, services, or systems to end-users of the platform." A (software) extension is any kind of functional component added by third-party developers (Jansen, 2013, p. 188). An innovation platform can grow and generate more value for its participants by being

[103] Langley, Lindberg, et al. (2019) denoted the collaboration of different actors across organizational boundaries as collaborative boundary work.

[104] Engert et al. (2022, p. 4) described the engagement of third-party developers in innovation platforms as a "symbiotic relationship" beneficial for the generativity of the overall ecosystem.

expanded by third-party developers. Jansen (2013, p. 191) introduced three scenarios for how extensions can interact with each other in innovation platforms. These extension types are independent extensions, extensions that can communicate with each other, and interdependent extensions that build upon other extensions. Independent extensions (a) make use of technical boundary resources to connect to the focal innovation platform and benefit from its provided services (Jansen, 2013, p. 191). The independent extensions act independently from other ecosystem participants and other third-party extensions. Communicative extensions (b) provide their own technical boundary resources to allow additional extensions access to information. Interdependent extensions provide technical boundary resources to enable third-party developers to develop additional extensions based on the extensions already affiliated with the focal platform.[105] Kenney and Zysman (2016, p. 64) described extension type (c) (i.e., interdependent extensions) as "[p]latforms [that] can grow on platforms." The different types of platform extensions provided by third-party developers are depicted in Fig. 3.7.

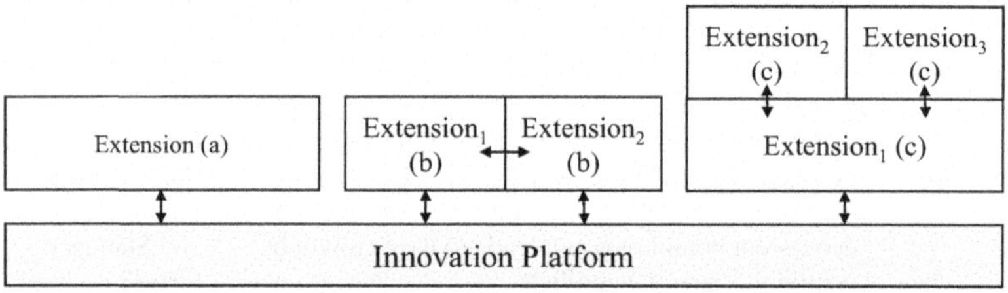


**Fig. 3.7** Extension Model for Platforms (Jansen, 2013, p. 192) *Note. Independent extension: (a); communicative extension: (b); interdependent extensions: (c)*

Miremadi et al. (2023, p. 8 ff.) observed a positive effect of platform openness[106] on a platform's generativity and platform adoption by third-party developers.[107] The platform owner can control the platform openness actively by opening (parts of) the platform, for instance, through the provision of boundary resources and through

---

[105] Wulfert and Karger (2022b) further elaborated on the extension types and their possible application in innovation platforms in the context of e-commerce.

[106] The openness of a platform refers to "the easing of restrictions on the use, development, and commercialization of a technology" (Boudreau, 2010, p. 1851).

[107] Miremadi et al. (2023) based their analysis on data from the video game industry involving six different innovation platforms. The authors also revealed that openness leads to higher generativity for complex platforms.

the provision of an extension marketplace for platform users (Wulfert and Karger, 2022b; Wulfert and Schütte, 2022). Homscheid et al. (2015, p. 193 ff.) highlighted that the openness of a platform can be dynamically adjusted to govern third-party developers on innovation platforms. To distribute and deploy third-party developers' extensions, either the innovation platform' owner or independent third parties can establish extension marketplaces (Jansen and Cusumano, 2013, p. 20 ff.). A digital extension marketplace can be defined as a "platform component that offers a venue for exchanging applications between developers and end-users belonging to single or multiple ecosystems" (Ghazawneh and Henfridsson, 2015, p. 200).[108] The extension marketplace can add value the platform's users, and the value can be captured by the owner of the innovation platform. Beverungen, Kundisch, et al. (2020, p. 508) widened the concept of an extension marketplace for "service providers and their customers to directly interact and co-create service, fueled by information."

Expected and unexpected platform extensions provided by third-party developers increase the platform core's value for customers (McAfee and Brynjolfsson, 2017, p. 160 ff.). Additional platform extensions deployed via extension marketplaces have their own demand curves as complementary products creating potential consumer surplus.[109] This consumer surplus for the extension influences the demand curve for the core platform positively and pushes the demand curve outwards (McAfee and Brynjolfsson, 2017, p. 160 ff.). McIntyre and Srinivasan (2017, Appendix A) emphasized this characteristic of platform extensions affecting the platform core and described a platform's complements broadly as "[g]oods and services built upon a platform which enhance the value of the core good to its network, such that the value of the core good to adopters is higher in tandem with the complement than without it."

In e-commerce, innovation platforms provide the technological infrastructure for the conduct of transactions in a multitude of e-commerce business models (Aulkemeier, Paramartha, et al., 2016; Aulkemeier, Schramm, et al., 2016; Wulfert and Karger, 2022b).[110] As described in general above, innovation platforms form the "technical core" (Blaschke, Haki, et al., 2019, p. 1) of e-commerce (Hanseth and Lyytinen, 2010). They provide the application services necessary for executing the business model-related business routines (Wulfert and Schütte, 2021). In this regard,

[108] The extension marketplace resembles a transaction platform for the extensions of an innovation platform. Transaction platforms are described in further detail in the remainder of this section.

[109] The effect can be further increased if platform extensions are provided free of charge (McAfee and Brynjolfsson, 2017, p. 161).

[110] For this cumulative dissertation, we focus on platform business models (i.e., transaction platforms) in particular

Aulkemeier, Paramartha, et al. (2016, p. 471 ff.) suggested a "lightweight platform architecture" that provides pivotal e-commerce application services that can be extended with external services and extensions. For the provision of external services and extensions, we proposed an extension marketplace (Wulfert and Karger, 2022b, p. 8 f.) in line with recent research on software platforms (e.g., Jansen, 2013; Jansen and van Capelleveen, 2013).

In the following, we summarize the description of innovation platforms, with a meta-model for innovation platforms in e-commerce.[111] The innovation platform consists of several internal modules with a high internal cohesiveness weakly coupled with other modules (Messerschmitt and Szyperski, 2003, p. 88). Hence, it is depicted as a structure of multiple internal modules developed by the innovation platform owner. The functionality of the internal modules is encapsulated by abstraction in the form of boundary resources. The innovation platforms have an extension marketplace attached, on which third-party developers' platform extensions are provided to users and deployed to their platform instances (Jansen and Cusumano, 2013; Wulfert, Woroch, Strobel, Seufert, et al., 2022). Sine multiple organizations can own and develop the innovation platform, we also depict a structure for the innovation platform owner. Besides the innovation platform as an extensible codebase upon which third-party developers design their extensions, the platform owner also governs the extension marketplace by enforcing governance rules.

We model the platform extensions developed by third-party developers in analogy to the focal innovation platform. A platform extension consists of a number of external modules that can make use of the functions provided by an innovation platform's internal modules via technical boundary resources (Eaton et al., 2015; Ghazawneh and Henfridsson, 2015). As technical boundary resources, APIs and SDKs are provided (Ghazawneh and Henfridsson, 2015; Wulfert, Woroch, Strobel, Seufert, et al., 2022). These technical boundary resources can also be used to access the core services provided by the innovation platform's internal modules. The technical boundary resources are accompanied and described by supplementary social boundary resources (Wulfert, 2023, p. 10 ff.). The third-party extensions can be multiple types (e.g., modules for enhanced analytics, shop themes, and merchandise management system integrations (Wulfert and Karger, 2022b, p. 9).

Independent third parties and retailers that want to establish a transaction platform instantiate the innovation platform and make use of the provided application

---

[111] It needs to be noted that the meta-model can be applied to innovation platforms in general. It is only customized for e-commerce with retailers as potential users instantiating the innovation platform for the support of their business models (Wulfert and Schütte, 2021, 2022). We also detailed enhanced analytics, shop themes, and merchandise management system integrations as possible platform extensions in the e-commerce context (Wulfert and Karger, 2022b, p. 9).

services to support their business model. These platform instances can be further customized with third-party developers' extensions provided via an extension marketplace. Following the elaboration on the layered and modular architecture of innovation platforms, we provide a simplified meta-model of an innovation platform in e-commerce in Fig. 3.8.[112]

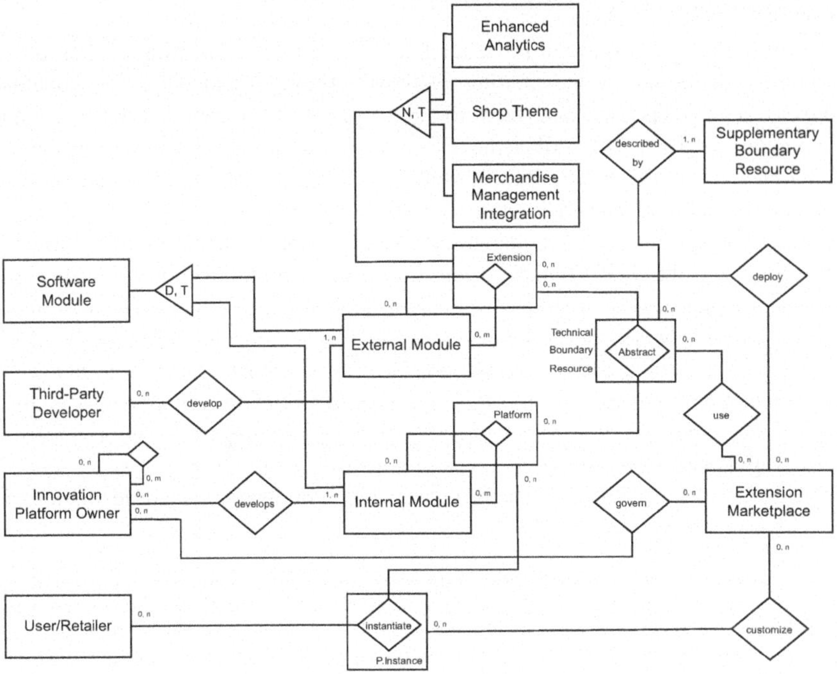


**Fig. 3.8** Simplified Meta-Model for Innovation Platforms in E-Commerce

[112] We apply the entity-relationship notation according to Chen (1976) for modeling the meta-models. Deviating from Chen's (1976) original notation, we used reinterpreted relationship types as an extension of the notation. The reinterpreted relationship types are relationships reinterpreted to an entity type and are represented with a rectangle around the diamond of the relationship type. The reinterpretation is necessary because a direct connection of two relationship types is not possible following Chen's (1976) entity-relationship notation, but it is necessary for the modeling of some circumstances (Becker and Schütte, 2004, p. 89).

### 3.2.3 Transaction Platform

In contrast to innovation platforms that provide the technical infrastructure for e-commerce transactions (Aulkemeier, Paramartha, et al., 2016; Wulfert and Karger, 2022b) and extensible codebase for the development of e-commerce applications (Baldwin and Woodard, 2009; Tiwana, Konsynski, and Bush, 2010), transaction platforms[113] enable and orchestrate transactions between previously independent actors (Evans, 2003; Hagiu and Wright, 2015a; Rochet and Tirole, 2003).

Transaction platforms can be understood as a business model that is based on enabling efficient and value-creating relations between providers of complements and customers (Haselhoff and Harwardt, 2022, p. 138).[114] Parker, Van Alstyne, and Choundary (2016, p. 5) described the objective of a transaction platform as the facilitation of goods, services, and social currency exchanges "enabling value creation for all participants" and the value capture from the orchestration of these exchanges.[115] Transaction platforms "attract, match and connect people to enable them to transact" (Reillier and Reillier, 2017, p. 6). Eisenmann, Parker, et al. (2011, p. 1270) defined transaction platforms as "markets, where [participants'] interactions with each other are subject to network effects and are facilitated by a common platform provided by one or more intermediaries." Transaction platforms "facilitate transactions between different types of individuals and organizations that would otherwise have difficulty to find each other" (Evans and Gawer, 2016, p. 5). Facilitating exchanges by "otherwise fragmented groups of consumers and/or firms" (Teece, 2016, p. 19), Abdelkafi et al. (2019, p. 553) specified that transaction platforms entail "an intermediation or

[113] In the literature, transaction platforms are also described as "economic catalyst" (Evans and Schmalensee, 2007, p. 3), "multi-sided platform" (Boudreau and Hagiu, 2009; Evans and Schmalensee, 2016; Hagiu, 2007), and "matchmaker" (Evans and Schmalensee, 2016, p. 210). Using the term matchmaker, Evans and Schmalensee (2016, p. 210) stressed the platform's main purpose of matching participants from different market sides to enable transactions between them.

[114] Zhao, von Delft, et al. (2020, p. 4 f.) differentiated transaction platforms matching multiple market sides from product platforms in which an original equipment manufacturer (e.g., Mercedes, Airbus) receives parts and components from a network of suppliers.

[115] Platform owners may take a share of ecosystem participants' revenues, for instance, through service fees (Wulfert, Seufert, et al., 2021, p. 8). The transaction platform owner's major source of revenue are transaction fees charged for matching participants and enabling transactions (Wulfert, Seufert, et al., 2021, p. 6136). The transaction fee depends on the price of the product or service exchanged. Armstrong (2006, p. 668 f.) identified the volume of network effects, payment types, and multi-homing possibilities as determinants of the price structure in two-sided markets. Rochet and Tirole (2006, p. 646 f.) also referred to fixed fees and usage fees as their price structure.

matchmaking role." Gawer (2014, p. 1241) argued that "[p]latforms fundamentally create value by acting as conduits between two (or more) categories of consumers who would not have been able to connect or transact without the platform." Hagiu (2007, p. 115) and Hagiu andWright (2015b, p. 163) emphasized that the groups of participants need to be affiliated with the transaction platform for transactions among them. In summary, transaction platforms "adopt platform business models that use information and communication technologies to facilitate interactions (including commercial transactions) between users, collection and use of data about these interactions, and [are subject to] network effects which make the use of the platforms with most users most valuable to other users" (KOM, 2018).[116]

Besides the facilitation of single transactions, Rach and Tschöpel (2011, p. 3) noted that transaction platforms can also take on the role of interorganizational ISs establishing long-term relationships between supply-side participants and demand-side participants. Constantinides et al. (2018, p. 381) described that transaction platforms consist of "a set of digital resources." These resources are further differentiated into the technological infrastructure and rules[117] for collaboration amplifying the collaboration among different participant types (Van Alstyne, Parker, and Paul Choudary, 2016, p. 4). Parker, Van Alstyne, and Choundary (2016, p. 5)s described this infrastructure as "open [and] participative" to propel the interactions among participants.[118] Indeed, innovation platforms such as those described above provide the infrastructure, with software modules specifically tailored for the amplification of exchanges among formerly independent actors. Hence, we focus the description of transaction platforms on business models reflecting multi-sided markets and their orchestration for increased value creation for the platform participants and amplified value capture for the transaction platform's owner.

[116] The European Commission proposed this definition for online platforms (KOM, 2018). This definition is in line with our understanding of transaction platforms in the e-commerce context.

[117] Parker, Van Alstyne, and Choundary (2016, p. 5) described the set of accompanying rules as "governance conditions."

[118] The openness of the infrastructure is achieved by providing boundary resources that allow access to the transaction platform's core (Eaton et al., 2015; Engert et al., 2022; Ghazawneh and Henfridsson, 2013). In section 3.3.3, we provide further details on the role of boundary resources in the context of platforms.

Transaction platforms incorporate characteristics of two-sided markets[119] as they orchestrate at least two different market sides.[120] A two-sided market "refers to a market with two distinct sides that benefit from network effects by interacting on a common platform" (Rochet and Tirole, 2003, p. 990).[121] The two distinct sides are addressed with distinct products, whereas the sales of one product affect the demand for the other product (i.e., network effect). In their definition, Rochet and Tirole (2006, p. 647) also described the price structure of a two-sided market. They defined a two-sided market as "one in which the volume of transactions between end-users depends on the structure and not only on the overall level of the fees charged by the platform" (Rochet and Tirole, 2006, p. 647). The authors pointed out that in a two-sided market, the volume of transactions depends not only on the level of fees but also on their structure.[122] Rochet and Tirole (2006, p. 646 f.) mentioned usage fees and fixed fees for the execution of transactions in two-sided markets. Eisenmann (2008, p. 41) noted that the more price-sensitive market side should be subsidized and the growth-valuing side should be charged. Multi-sided markets describe platforms with more than two independent types of actors (Boudreau and Hagiu, 2009).[123] In multi-sided markets, a single actor can also adopt multiple roles in subsequent transactions (Adner, 2017; Wulfert, Seufert, et al., 2021).

[119] A two-sided market is the "original name" (Evans and Schmalensee, 2016, p. 212) introduced by Rochet and Tirole (2003) for industries with platforms that match two or multiple market sides. We incorporate multi-sided market characteristics in transaction platforms and will use this term in the remainder of this cumulative dissertation.

[120] A market side resembles a group of participants with the same characteristics and objectives (Evans and Schmalensee, 2016, p. 210). Armstrong (2006, p. 66) stressed that the interaction of the two market sides takes place on a (transaction) platform. In e-commerce, these essential market sides are third-party sellers and customers.

[121] Adner (2017, p. 50) emphasized that a two-sided market entails two major presuppositions. First, there must be a clear separation and demarcation of the market sides. Second, the market participants must be clearly assignable to one (or more) of the market sides. Hagiu and Wright (2015b, p. 163) emphasized the presence of each participant's affiliation with the focal platform as a third presupposition. Evans (2003, p. 192) added the more efficient coordination of transactions by an intermediary compared with bi-lateral scenarios as a fourth presupposition.

[122] Following Rochet and Tirole (2006, p. 650), the fee structure defines the allocation and apportionment of the fees on the market sides. The authors referred to the sum of fees as the "price level" (Rochet and Tirole, 2006, p. 650).

[123] For instance, Reiss and Günther (2010, p. 167 ff.) discussed focal platforms that have three distinct market sides.

Network effects[124] for participants (e.g., lower transaction costs for search and initiation) and economies of scale for market owners (e.g., marginal costs for adding another supplier and article are almost zero) are strong drivers for the success of transaction platforms (Briscoe, Odlyzko, et al., 2006; Wulfert and Schütte, 2021). Transaction platforms benefit from the concept of two-sided markets, matching two or more previously independent market sides and generating value from transactions between the participants (Armstrong, 2006; Rochet and Tirole, 2003). Direct and indirect network effects are exploited to further propel the transaction platform (Rochet and Tirole, 2003; Shapiro and Varian, 1998). Network effects describe the benefits for platform participants, caused by the number of participants on the same (i.e., direct) or other (i.e., indirect) market side (Rochet and Tirole, 2003; Shapiro and Varian, 1998).[125] Shapiro and Varian (1998, p. 183) preferred the term network externality to highlight that single participants affect the value of others "without […] compensation being paid" (Shapiro and Varian, 1998, p. 183). Network effects arise between two or more sides of a market orchestrated by a transaction platform (Evans, 2003; Rochet and Tirole, 2003; Rochet and Tirole, Rochet and Tirole, 2006).[126] These effects are seen as exogenous and fixed and are fundamental drivers of platform competition. Network effects can trigger a self-reinforcing positive feedback loop that magnifies incumbents' early advantages (Eisenmann, Parker, et al., 2006).

[124] Farrell and Saloner (1985) and Shapiro and Varian (1998) were the first to describe network effects. In scientific research, network effects are also called network externalities or demand-side economies of scale (e.g., Armstrong, 2006; Briscoe, Odlyzko, et al., 2006; Farrell and Saloner, 1985; Shapiro and Varian, 1998).

[125] Network effects rely on the phenomena described in Metcalfe's law (Metcalfe, 1995). In a nutshell, Metcalfe's law described the increase of the network's value as "the square of the number of users joining it" (Shapiro and Varian, 1998, p. 184). The additional network benefits from the addition of more users for existing users (Gilder, 1993). Even if there is disagreement in the scientific community about the actual degree of benefits, the basic tendency to increase benefits is undisputed. For further information on this discussion, the interested reader is referred to Briscoe, Odlyzko, et al. (2006).

[126] Network effects require a minimum threshold of participants on all sides of the platform to materialize (Evans and Schmalensee, 2010; Wulfert and Dennhardt, 2023). In the literature, this minimum threshold is also described as a critical mass (e.g., Alt and Zimmermann, 2019; Evans and Schmalensee, 2010; Oren and Smith, 1981; Ruutu et al., 2017; Wanner et al., 2019). Evans and Schmalensee (2016, p. 207) defined critical mass as "[t]he minimum sets of participants for the platform sides that are necessary for the platform to ignite—that is, have self-sustaining growth."

For direct network effects,[127] the network participants' benefit "depends on the number of other network users with whom they can interact" (McIntyre and Srinivasan, 2017, Appendix A). The value of a platform with direct network effects increases for participants when there is a significant number of participants on the same side of the platform "because of the higher potential of exchange (e.g., of words, of files, of knowledge)" (Eisenmann, Parker, et al., 2006, p. 93).[128] Direct network effects can result in a "positive feedback loop" (Eisenmann, Parker, et al., 2006, p. 93), as the number of potentially new participants depends on the number of participants already in the network.

For indirect network effects,[129] the ecosystem's different market sides "mutually benefit from the size and characteristics of the other side" (McIntyre and Srinivasan, 2017, Appendix A) (e.g., sellers and customers). Indirect network effects are especially relevant in markets with complementary products (e.g., smartphones and apps). In this case, the other market side is responsible for providing additional complements to the focal product or service and, therefore, additional complementors may provide more complements, which increase the value for the other market side (Hagiu and Wright, 2015a, p. 163). Hence, "[i]ndirect network effects [...] reflect a pre-existing underlying inter-dependency (and complementarity) between the demands from two or more types of" ecosystem participants (Gawer, 2014, p. 1241). As with direct network effects, indirect network effects can also result in a positive feedback loop (Eisenmann, Parker, et al., 2006, p. 94). This feedback loop can be accelerated with incentives to join the platform for a particular market side and with subsidies for a particular market side in scenarios with indirect network effects (Evans and Gawer, 2016, p. 6). We depict direct and indirect network effects in Fig. 3.9

Besides the differentiation between direct network effects and indirect network effects, Tiwana (2014, p. 36) emphasized that network effects are not always beneficial for the network participants, but can also be negative in nature (Boudreau and Hagiu, 2009).[130] In this regard, an additional network participant can decrease the

[127] Direct network effects are also described as within-group and same-side network effects to emphasize the focus on one market side (Constantinides et al., 2018; Gawer, 2014; Sun and Tse, 2009).

[128] Gawer (2014, p. 1241) described the same phenomenon for technology-based ecosystems.

[129] Indirect network effects are also described as cross-group and cross-side network effects to highlight the involvement of different market sides (e.g., Constantinides et al., 2018; Eisenmann, Parker, et al., 2006; Gawer, 2014; Sun and Tse, 2009).

[130] Network effects are only present for platforms with an ecosystem that is open for a larger community (Gawer and Cusumano, 2014, p. 423). Indirect network effects also require at least two clearly distinct market sides.

network's appeal for participants on the same side and for those on the other market side. Transaction platforms that act globally with geographically distinct markets can benefit from local network effects (McAfee and Brynjolfsson, 2017, p. 219). The offering of local products and services can therefore generate geographically restricted network effects independent of the larger global network. With strong network effects in the platform competition, "[t]here's a tendency toward winner-takes-all markets" (McAfee and Brynjolfsson, 2017, p. 217). These winner-takes-all tendencies result in markets with one or only a limited number of transaction platforms competing for participants (Eisenmann, Parker, et al., 2006, p. 7 ff.). Following these winner-takes-all tendencies, a leading platform can become an industry standard for a particular technology, among competing alternative standards (Gawer and Cusumano, 2014, p. 425).

In analogy to the architectural considerations of innovation platforms, Staykova and Damsgaard (2015, p. 5 ff.) applied the duopoly of core and periphery to transaction platforms. West and O'Mahony (2008, p. 147) described this analogy as "participation architecture" for ecosystems surrounding a transaction platform. [131] Staykova and Damsgaard (2015, p. 6 ff.) used the metaphor of core and periphery to depict network effects and describe the different market sides orchestrated by a transaction platform. While the initial conceptualization (Baldwin and Clark, 2000; Baldwin and Woodard, 2009; Blaschke, Haki, et al., 2019) of the core involves "the main features and functionalities offered by the platform owner" (Staykova and Damsgaard, 2015, p. 4), the authors also include the initial setup of market sides in the platform core. Staykova and Damsgaard (2015) used the platform's periphery to illustrate the evolution of the platform, with the periphery resembling market sides added with alterations of the business model over time. In e-commerce, transaction platforms are likely to be established with an existing customer base from a reseller business (e.g., Amazon, Zalando) (e.g., Hagiu and Wright, 2015b). If the retailer opens its reseller business model to third-party sellers as the periphery, a transaction platform with customers as the core market side is established (Fig. 3.9).

[131] O'Reilly (2005, p. 474 ff.) introduced the concept of an "architecture of participation" as a heuristic for encouraging participation in innovation.

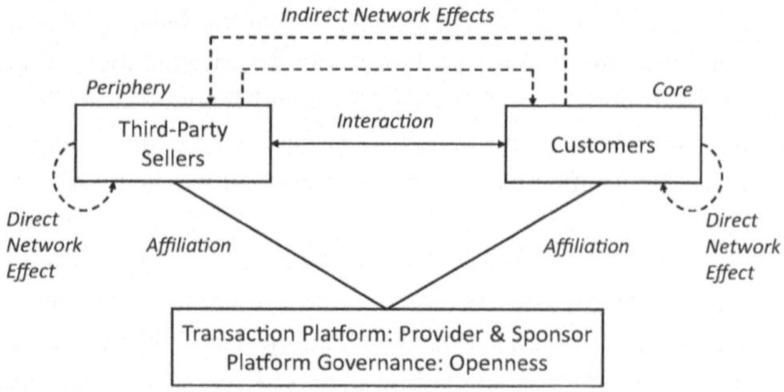


**Fig. 3.9** Network Effects on Transaction Platforms in E-Commerce

Wulfert and Schütte (2022, p. 206) explicated and depicted network effects by applying the concept of triadic closures from network theory (Song, Tang, et al., 2019, p. 914).[132] As already described, for direct network effects, the value of a transaction platform and its surrounding digital business ecosystem for participants increases with the size of the network on the same side. The increase of the value is caused by the higher potential of exchanges (e.g., of product review, of knowledge) between these actors (Eisenmann, Parker, et al., 2006; McIntyre and Srinivasan, 2017). Complementarily, an indirect network effect arises if the benefit to participants depends on the number of participants on another market side (e.g., the supply side of the digital business ecosystem subsidizing the demand side) (Armstrong and Wright, 2007; Rochet and Tirole, 2003). Wulfert and Schütte (2022, p. 206) applied graph theory to depict network effects as triadic closures[133] in social affiliation networks on one (direct: black circles) or multiple sides (indirect: dashed circles) on a transaction platform (Fig. 3.10). Enabling triadic closures, the focal transaction platform is likely to establish a larger ecosystem of independent participants surrounding the platform (Easley and Kleinberg, 2010).[134]

---

[132] In this context, a network can be described as a "[s]ystem of entities or nodes which are interconnected; such nodes can be either individuals or 'collective' participants, such as organizations" (McIntyre and Srinivasan, 2017, Appendix A).

[133] Following Easley and Kleinberg (2010, p. 97 ff.), we differentiated focal and membership closures. While a focal closure is orchestrated by the focal platform in the ecosystem, a membership closure is propelled by another member on the same side of the platform.

[134] The ecosystem evolves dynamically as actors join or depart while creating new or interrupting prior stable triadic closures (Schütte and Wulfert, 2022a; Wulfert, Seufert, et al., 2021).

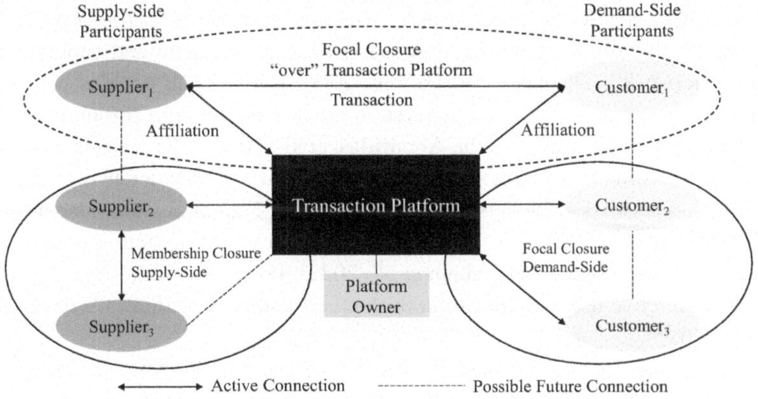


**Fig. 3.10** Triadic Closures via Transaction Platforms (Wulfert and Schütte, 2022, p. 206)

The concept of orchestration is inherently linked with the concept of platforms in ISs research (Kazan, Tan, et al., 2018).[135] It involves matching different participant types with diverging desires in two-sided market constellations (Nambisan et al., 2017; Rochet and Tirole, 2003). Orchestration is conceptualized as an organization's "leverage over dispersed resources and capabilities of others" (Mann et al., 2022, p. 3) and involves new "centers of gravity" (Ivarsson and Svahn, 2020a, p. 1) if organizations establish platform-centered ecosystems. Mann et al. (2022, p. 3) defined orchestration in the context of transaction platforms "as an undertaking by a focal actor that intends to integrate and leverage resources and capabilities of others within a business ecosystem." The authors referred to Breidbach et al. (2016), Dhanasai and Parkhe (2006), and Ivarsson and Svahn (2020a) for their definition of orchestration.[136] Orchestrating value exchanges, the platform owner "coordinate[s] the demand of distinct groups of customers" (Evans, 2003, p. 191).

With reference to Linde et al. (2021), Addo (2022, p. 362) described ecosystem orchestration as "a dynamic process that requires a range of capabilities from an

[135] The concept of orchestration has also been established in strategy (e.g., Hinterhuber, 2002), management (e.g., Dhanasai and Parkhe, 2006), supply chain management (e.g., Wind et al., 2009), entrepreneurship (e.g., Boudreau, 2017), service science (e.g., Breidbach et al., 2016), and marketing (e.g., Perks et al., 2017) research. Orchestration is understood as "actions of dedicated human actors facilitating value-creating interactions for and between others outside of a traditional customer-firm dyad" (Mann et al., 2022, p. 3) in these disciplines.

[136] Mann et al. (2022, p. 3) purposefully delimited orchestration from organizational planning, coordination, and open innovation.

orchestrator to enable the creation and extraction of value from the network."[137] Orchestration involves "centralized decision making to create and capture value within a network" (Dhanasai and Parkhe, 2006, p. 661) and enables ecosystem participants to contribute to the ecosystem's value co-creation through enabling (Guggenberger et al., 2020; Ritala, Agouridas, et al., 2013). The concept of orchestration stems from research on resource orchestration and knowledge orchestration in innovation networks (Dhanasai and Parkhe, 2006; Foss et al., 2023; Isckia et al., 2020; Mann et al., 2022). It is also used in the context of transaction platforms in IS research (Burkhalter, 2020; Mann et al., 2022). Proper orchestration guarantees "the most effective use and allocation of the [ecosystem's] resources" (Burkhalter, 2020, p. 10).[138]

The orchestration can be described as a process of attracting, matching, and connecting participants to enable transactions between them and subsequent optimization of the orchestration process (Reillier and Reillier, 2017, p. 46 ff.). Holzmann et al. (2014, p. 603) conceptualized matching as a "process of complex resource-allocation" among ecosystem participants in multi-sided market environments.[139] Since platform owners lack explicit market control mechanism, they "need to explore the extrinsic and intrinsic motivations of actors to incentivize participation in economic activity" (Ivarsson and Svahn, 2020b, p. 5933).[140] Ivarsson and Svahn (2020b, p. 5933) described the practice of orchestration as an undertaking by the focal platform owner to encourage interactions between participants from previously independent market sides.

---

[137] In the literature, orchestration is not only described in the context of transaction platforms, but is also investigated for innovation platforms. In this context, orchestration "is the processes through which platform providers ensure a constant development and commercialization of innovations" (Isckia et al., 2020, p. 198).

[138] Since Addo (2022, p. 2022) highlighted that successful ecosystem orchestration "requires a range of capabilities from an orchestrator," we analyzed dynamic capabilities required for the orchestration of the e-commerce ecosystem in the ninth research paper (Wulfert, Woroch, and Strobel, 2024).

[139] In contrast to Holzmann et al.'s Holzmann et al. (2014, p. 603 f.) description of matchmaking as a "complex resource allocation," matchmaking in the context of e-commerce is focused on the active management of single transactions. The transaction platform matching participants from two separate market sides is likely to keep these participants separated, requiring further intermediation.

[140] Ivarsson and Svahn (2020b, p. 5938) provided a detailed overview of the differences between conventional matchmaking and digital matchmaking via transaction platforms. Kapoor (2018, p. 6) further described such economic activities as "[t]asks that underlie the different offers that contribute to" the (final) customers value proposition. The ecosystem participants jointly create value for the customers. The joint value co-creation is also reflected in Adner's (2017) ecosystem-as-a-structure perspective.

In traditional markets, customers collect information on their own until, after their own evaluation, they have a sufficient amount of information to make a purchase decision (Strader and Shaw, 1997, p. 189). These processes are changed by the introduction of transaction platforms in e-commerce.[141] This change mainly involves the transformation of the search process and the digitization of products and services. First, the reduction of search costs leads to cheaper and more uniform prices in the market because price comparisons are easier. Second, products and their distribution become digitized to share, reducing both the cost and time of transaction processing (Strader and Shaw, 1997, p. 189 ff.). The aspect of lower search costs is considered across the literature as one of the key developments in the context of electronic markets (Bakos, 1998; Strader and Shaw, 1997; Turban, Whiteside, et al., 2017). This development refers to the more favorable purchase of product and price information for the customer and easier communication of this information for the seller (Bakos, 1998, p. 39).

Transaction platforms resemble virtual loci on which participants can conduct retail transactions (Turban, Whiteside, et al., 2017, p. 39 ff.). The transaction platform owner orchestrates individual participants from various markets and social groups to form a dynamic ecosystem (Corallo, 2007; Gawer, 2021b). The objective of a transaction platform in e-commerce is the electronic facilitation of transactions between groups of participants Derave (2019, p. 291). With regard to the main trading functions in e-commerce (Becker and Schütte, 2004; Levy, Weitz, et al., 2019; Müller-Hagedorn et al., 2012), some or even huge amounts of them may be performed by participants other than the platform owner.[142] Kenney and Zysman (2016, p. 65) described transaction platforms in e-commerce as "a set of digital frameworks for [...] marketplace interactions." In e-commerce, transaction platforms "facilitate the interaction and the seamless exchange of products [and services] between consumers and independent suppliers" (Hänninen et al., 2018, p. 156). Schreieck et al. (2016, p. 6) emphasized that the purpose of transaction platforms is not only to facilitate the interaction but also "to match supply and demand." Conducting

[141] In the context of e-commerce, transaction platforms are also called "electronic marketplaces" (e.g., Bakos, 1991; Schenkel et al., 2013; Soh and Markus, 2002; Timmers, 1998; Wang, Zheng, et al., 2008; Wang and Archer, 2007) and "digital marketplaces" (e.g., Derave, 2019; Derave et al., 2020; Kollmann, 2022; Linnhoff-Popien et al., 2018; Wulfert and Busch, 2022; Wulfert and Schütte, 2022). We apply the term transaction platform throughout this cumulative dissertation.

[142] Transaction platforms also provide an environment for customer-created content (e.g. customer feedback) that is integrated into several transaction phases influencing customers' decisions (Riggins, 1999; Rudolph et al., 2015).

transactions on transaction platforms also involves "the exchange of information about goods and prices" (Rach and Tschöpel, 2011, p. 3).[143].

A transaction platform supports the identification of potential trading partners and the execution of retail transactions (Love et al., 2010, p. 1). Christiaanse et al. (2004, p. 152) described that transaction platforms in e-commerce not only involve third-party sellers and potential buyers in a single transaction but also support "other activities, such as collaborative planning, logistics, transportation arrangements, and fulfillment." Transaction platforms offer participants a variety of retail-related services, for example, fulfillment, marketing, and payment (Wulfert, Seufert, et al., 2021). These trading partners involve "suppliers, producers, intermediaries, customers, and producers of complementary products and services" (Tan, Lu, et al., 2015, p. 249) and are arranged in a network structure.[144] Tan, Lu, et al. (2015, p. 249) described these actors as "complementors" in analogy to our elaboration on innovation platforms. The participants of a transaction platform in e-commerce "are held together through formal contracting and/or mutual dependency" (Tan, Lu, et al., 2015, p. 249). From a customer point of view, transaction platforms in e-commerce "resemble retail agglomerations" (Hänninen et al., 2018, p. 155), integrating the range of articles of participating suppliers, retailers, and wholesalers through a single digital channel (Teller and Elms, 2010; Wulfert and Schütte, 2021). Intermediating independent participants from separate market sides (e.g., advertising partner, developer, or seller) that can be assigned with unambiguous roles (Böttcher, Rickling, et al., 2021, p. 8 f.), transaction platforms in e-commerce can be characterized as multi-sided (Hagiu and Wright, 2015a; Wulfert, Seufert, et al., 2021). As such, they form the center of larger e-commerce ecosystems (Wulfert and Karger, 2022b; Wulfert, Woroch, Strobel, Seufert, et al., 2022).[145]

Transaction platforms in e-commerce can inhibit several different characteristics.[146] Hänninen et al. (2018, p. 157 f.) differentiated between pure transaction platforms and hybrid business models in e-commerce. While pure transaction

---

[143] This quote was translated from German.

[144] Bakos (1991, p. 296) characterized transaction platforms as inter-organizational ISs that allow "the participating buyers and sellers to exchange information about prices and product offerings." Transaction platforms in e-commerce are described "as a bundle of standard components around which buyers and sellers coordinate efforts" (Bresnahan and Greenstein, 1999, p. 4).

[145] We elaborate in further detail on e-commerce ecosystems in section 3.4.

[146] Transaction platforms in e-commerce can be described with taxonomies specifically developed for the context of e-commerce (e.g., Abdollahi and Leimstoll, 2011; Clement et al., 2019; Derave, 2019; Kawa and Walesiak, 2019; Lavassani, et al., 2011; Petersen et al., 2007; Soh and Markus, 2002; Täuscher and Laudien, 2018; Wang and Archer, 2007).

platforms focused on the facilitation of transactions between third-party sellers and customers, hybrid business models are vertically integrated, complement third-party sellers' assortments with their own inventory, and often offer supply chain services. Pure transaction platforms can be considered as "open, neutral, egalitarian and progressive support" (Gillespie, 2010, p. 352) for e-commerce transactions. Another differentiating dimension can be the ownership dimension. Transaction platforms can be owned by a single organization (e.g., retailer), independent third parties, or a consortium of multiple organizations (Bakos, 1991, p. 296). For transaction platforms owned by a single organization, Kollmann (2022, p. 512 ff.) differentiated demand-side and supply-side dominated transaction platforms.

Following the explanations of the Bundeskartellamt (2015, p. 11) a decisive demarcation between retail and a two-sided market is the control of important variables of the transactions. While this control, such as pricing sovereignty, remains with the retailer for retail transactions, huge parts of the transaction control are beyond the influence of the transaction platform owner in two-sided market constellations.[147] Hence, traditional retail and wholesale are not considered two-sided markets (Rysman, 2009).[148] Rochet and Tirole (2003, p. 994 f.) added that, from the customer's point of view, each market side is only knowledgeable about one price and cannot consider the price demanded from the other market side.[149]

---

[147] "For example, consider a farmer who sells a product to a grocery store once and does not otherwise interact with the grocer. The grocer then picks a retail price based on inventory and demand. In this one-sided market, the farmer collects the wholesale price and is then indifferent to the success of the grocer in actually selling the good. The distinguishing feature in this case is whether the seller is paid "based on the success of the platform with the buying side" (Rysman, 2009, p. 126).

[148] In a two-sided market model of traditional retail and wholesale (Armstrong, 2006, p. 677 ff.), the retailer sets the price per customer, and the wholesaler is paid based on how many customers show up. This conceptualization is contrary to traditional retail, where a wholesaler sells a product for a known sum, the retailer takes possession of the product and sells the product as it wishes, and the wholesaler has no concern about how many units the retailer is able to sell.

[149] Rochet and Tirole (2003, p. 994 f.) denoted this fact as a key differentiation from the purchase of complementary products bought by the same customer.

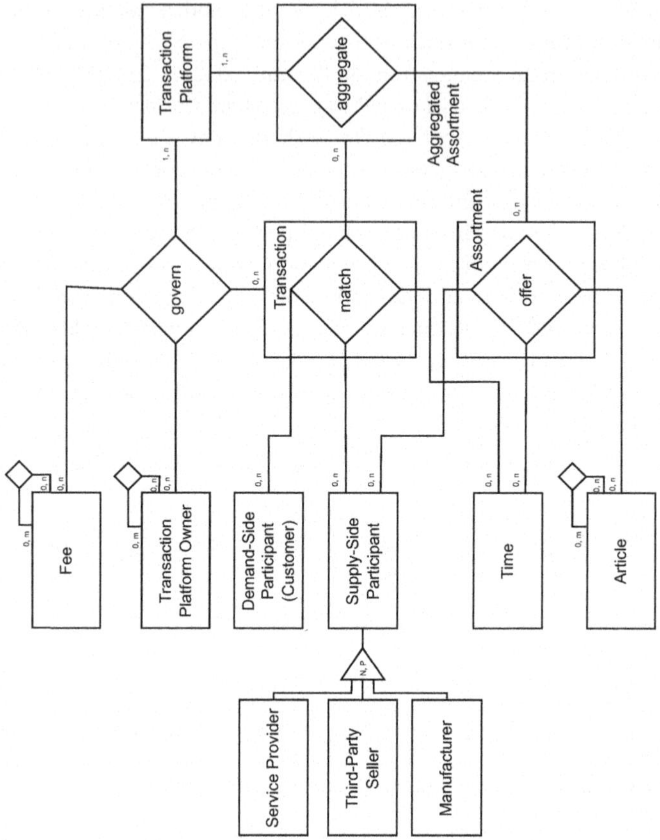


**Fig. 3.11** Simplified Meta-Model for Transaction Platforms in E-Commerce

In the following, we summarize the description of transaction platforms in a meta-model for transaction platforms in e-commerce.[150] The transaction platform is governed by a platform owner that can either by a single organization or a consortium of organizations (Hein, Schreieck, Riasanow, et al., 2020; Helfat and Raubitschek, 2018). The governance involves, among others, a fee structure that defines the transaction-dependent fees. Transaction platforms match supply-side participants, such as manufacturers, third-party sellers, and additional service providers with demand-side participants (i.e., customers). These participants can be augmented by additional ecosystem actors, such as opinion researchers, advertising partners, developers, and infrastructure providers (Wulfert and Schütte, 2021). The matching also involves articles offered by supply-side participants.[151] The assortments of single demand-side participants are aggregated by the transaction platform accessible to customers.[152] The matching also involves an article via the aggregated assortment of the transaction platform. If the customer, the supply-side participants, and the article are matched successfully at a particular point in time,[153] the transaction can be executed. In Fig. 3.11, we provide a simplified meta-model of a transaction platform in e-commerce.

### 3.2.4 Hybrid Platforms

Although we introduced innovation platforms and transaction platforms separately in this cumulative dissertation, Cusumano, Gawer, and Yoffie (2019, p. 23) revealed a general tendency for mature transaction platforms to develop innovation functions. Hybrid platforms incorporate features of the aforementioned platform types and offer both transaction and innovation services (Gawer, 2021b). Gawer (2021b, p. 10) argued that the "main reason transaction platforms add an innovation side is that more apps or features will make the transaction platform a more compelling

[150] A transaction platform can be modeled as the location on which the matching and (parts of) the transaction is executed (Turban, Whiteside, et al., 2017). Unlike internal and supply-chain platforms, transaction platforms are open to and accessible for interested participants (Gawer and Cusumano, 2014).

[151] The article offered can be a product or a service.

[152] The assortments can be further specified as full assortments and sub-assortments (Wulfert, Woroch, and Strobel, 2024).

[153] The entity type time allows the reference of a specific point in time to a relation. In a database, this time is usually part of the primary key.

experience for users." Hence, the innovation services should further drive the fees generated through the transactions. The addition of another market side enables additional opportunities for value capture (Gawer, 2021b, p. 10). In e-commerce, there are several examples (e.g., Amazon Marketplace, Alibaba) for transaction platforms that provided additional innovation services attracting third-party developers in more mature phases of their lifecycle (Alibaba Group, 2022; Wittig et al., 2016).[154] In general, the formation of hybrid platforms increases the reach of the platform owner and the value of the ecosystem for incumbents and new participants (Schütte and Wulfert, 2022a).

Hybrid platforms are innovation platforms and transaction platforms simultaneously.[155] Cusumano, Gawer, and Yoffie (2019, p. 13) defined platforms broadly with a focus on joint innovation or interaction without a further focus on either innovation or transaction. In a similar vein, McIntyre and Srinivasan (2017, Appendix A) defined platforms as "[i]nterfaces that serve to mediate transactions among firms and/or individuals that may not be able to transact otherwise, often two or more distinct 'sides'; building blocks that serve as the foundation upon which other firms can build related products or services."[156] The two parts of their definition describe transaction platforms and innovation platforms. With reference to Hagiu (2007), Boudreau and Hagiu (2009, p. 188) described hybrid platforms as innovation platforms with transaction platform characteristics. These platforms provide "foundations upon which other parties can build complementary products, services, or technologies" (Boudreau and Hagiu, 2009, p. 188) through which "distinct groups of consumer and 'complementors'" (Boudreau and Hagiu, 2009, p. 188) can interact.

---

[154] Muzellec et al. (2015, p. 142 ff.) introduced four maturity stages for transaction platforms, namely the embryonic stage, the emergent stage, the growing stage, and the maturity stage. Moore (1993, p. 77) reported birth, expansion, leadership, and self-renewal as possible lifecycle stages for business ecosystems that can also include platforms. Teece (2017, p. 19 f.) adopted Moore's (1993) lifecycle stages for the context of transaction platforms. Parker, Van Alstyne, and Choundary (2016, p. 187) proposed a lifecycle for platforms that covered a startup phase, a growth phase, and a maturity phase.

[155] Evans and Gawer (2016, p. 6) and Schütte and Wulfert (2022a, p. 215) described hybrid platforms as "integrated platforms." Evans and Gawer (2016, p. 6) further introduced investment platforms that bundle a number of investing companies in a holding and active investment approach (Evans and Gawer, 2016, p. 9). However, this type is not of special relevance in the context of e-commerce.

[156] For their definition of transaction platforms, McIntyre and Srinivasan (2017, Appendix A) referred to Evans (2003); Gawer (2014); Hagiu and Wright (2015a); Rochet and Tirole (2003); Rysman (2009).

Besides the simultaneous provision of transaction-related and innovation-related services, other scholars reported on additional services combined by hybrid platforms in e-commerce in addition to the orchestration of transactions. Qin et al. (2020, p. 4 ff.) reported on the provision of logistics services by the platform and developed a mathematical model for sharing logistics services among affiliated participants. Schütte and Wulfert (2022a, p. 215) reported on the integration of reseller services in a hybrid platform combining innovation and transaction services. In this approach, the hybrid platform has its own assortment that is sold to customers and, thus, behaves competitively to other third-party sellers.

In the following, we summarize the description of hybrid platforms in a meta-model combining innovation platforms and transaction platforms in e-commerce.[157] The hybrid platform owner develops the internal modules that are the foundation for the innovation platform providing the necessary application services for the transaction platform. In Fig. 3.8, we further depicted the technological foundation the innovation platform provides for the development of third-party developers' platform extensions (Baldwin and Woodard, 2009; Tiwana, Konsynski, and Bush, 2010). The hybrid platform owner instantiates and customizes the provided application services to form its transaction platform as a virtual environment for e-commerce transactions involving a number of ecosystem participants (Aulkemeier, Paramartha, et al., 2016; Wulfert and Karger, 2022b; Wulfert, Woroch, Strobel, Seufert, et al., 2022). Besides the instantiating of the transaction platform for the fulfillment of the business requirements, the hybrid platform owner also governs the e-commerce transactions between the transaction partners and orchestrates them enabling an efficient transaction execution. In Fig. 3.12, we depict the simultaneous provision of innovation and transaction services in a hybrid platform in e-commerce.[158]

[157] For the illustration of hybrid platforms, we make use of the platform architecture framework by Wulfert and Karger (2022b, p. 8). In section 3.4.2, we further detail the integration of innovation platforms and transaction platforms in the unfolding of e-commerce ecosystems.

[158] The other elements of Fig. 3.8 and Fig. 3.11 could be added to Fig. 3.12. However, we decided to focus on the integration between innovation platforms and transaction platforms as building blocks for hybrid platforms. Therefore, we kept the two platforms separate for reasons of integrability between Figs. 3.8, 3.11, and 3.12.

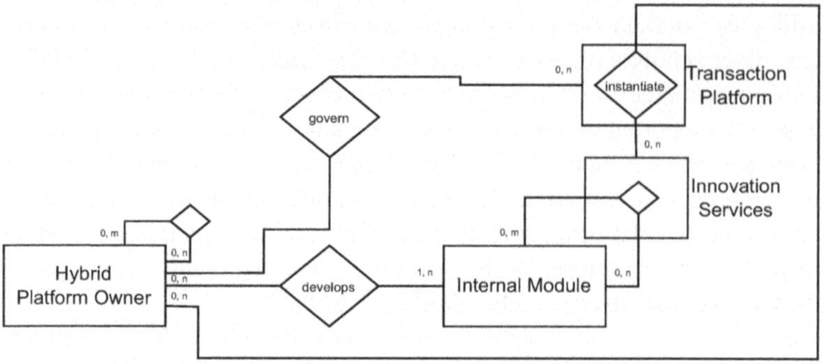


**Fig. 3.12** Simplified Model for Hybrid Platforms in E-Commerce (Wulfert and Karger, 2022b, p. 8)

Since the governance of platforms is relevant for innovation platforms and transaction platforms alike, we describe platform governance and major governance dimensions in the following.[159] The platform owner "sets, and often enforces the governance rules, determines timing" (Adner, 2017, p. 48), and captures a large amount of the ecosystem's revenues. Boudreau and Hagiu (2009, p. 183 f.) described the platform owner as a "private regulator" that invokes "an unusually rich set of strategic instruments to influence the behavior of complementors and users [...], while benefiting from privileged information and a privileged position within the ecosystem." Governance has been identified as an important activity for a platform owner, and a determinant for a platform's success (Alves et al., 2017; Tiwana, 2014). As Tiwana (2014, p. 39) described it, "[g]overnance broadly refers to who decides what in a platform's ecosystem". Boudreau and Hagiu (2009, p. 185) also remained vague and described platform governance as "a series of sometimes rather nuanced controls and inducements." Governance and control mechanisms can be formal and informal (Tiwana, 2014, p. 39).[160] According to Schreieck et al. (2016, p. 9), the above-mentioned mechanisms refer "to how the platform owner governs the processes within his platform ecosystem."

[159] Platform governance can be seen as an overarching activity not only for defining the modularity and generativity of an innovation platform , but also for providing orchestration guidelines.

[160] More general, Wiesche et al. (2011, p. 135) described that control is applied to "direct attention, motivate, and encourage organizational members to act according to organizational goals and objectives".

Platform governance enables platform owners to "control interactions between multiple stakeholders without jeopardizing their incentives for value creation" (Constantinides et al., 2018, p. 383). Platform governance "is the set of rules, norms, and policies that the platform adhere to in order to build its ecosystem" (Reillier and Reillier, 2017, p. 165). Baldwin and Woodard (2009, p. 23) described platform governance as a "set of stable constraints, or design rules, that govern the relationships among components." Governance rules are set and enforced by the owner of the focal platform in a business ecosystem (Adner, 2017, p. 48).[161] Evans and Schmalensee (2016, p. 209) summarized the set of governance rules as a "governance system." The governance system includes "[r]rules that prohibit bad behavior [...] by platform participants and mechanisms for enforcing those rules, including methods of detection and punishment" (Evans and Schmalensee, 2016, p. 209). An adequate level of platform governance ensures "appropriately bound participant behavior without excessively constraining the desired level of generativity" (Wareham et al., 2014, p. 1195 f.). Boudreau and Hagiu (2009, p. 185) argued that platform governance serves to minimize negative network effects.

Extant literature has structured and aggregated governance rules and decisions along several governance dimensions (e.g., Hein, Schreieck, Riasanow, et al., 2020; Hein, Schreieck, Wiesche, et al., 2016; Reillier and Reillier, 2017; Tiwana, 2014). Economides and Katsamakas (2006, p. 1058) described the degree of collaboration and competition with providers of complementary goods and services as a "core strategic question for a platform" owner. Evans and Gawer (2016, p. 7) included decisions on platform access, division of value capture, and conflict resolution in platform governance. Hein, Schreieck, Wiesche, et al. (2016, p. 3 f.) described six dimensions of platform governance for transaction platforms. The authors introduced governance structure, resources and documentation, accessibility and control, trust and perceived risk, pricing, and external relationships as part of a platform's governance. Boudreau and Hagiu (2009, p. 187) stressed that price setting as a governance mechanism is augmented by "nuanced combinations of legal, technological, informational and other instruments." The authors further differentiated between regulation of access and interactions among participants that are already on the multi-sided platform (Boudreau and Hagiu, 2009, p. 187).[162] Staykova and Damsgaard (2015, p. 3 f.) emphasized the determination of subsidy and revenue side as important governance decisions by the platform owner. Gawer (2021b,

[161] In contrast, Adner (2017, p. 48) described an ecosystem follower as "a firm that agrees to these terms, and cedes the leadership role."

[162] Access regulations are applied to have "the 'right'" (Boudreau and Hagiu, 2009, p. 184) participants on all sides of the platform.

p. 10) described the definition of the platform's boundary and boundary resources as important instruments in platform governance. Besides these rather general dimensions, Schüler and Petrik (2021, p. 15) argued that it is crucial to consider "the specific context of each market [...] for understanding [platform] governance mechanisms and their implementation."[163]

A platform owner can "define openness (for access and resources) and facilitate complementors' work" (Karhu, Gustafsson, et al., 2018, p.) with the deployment of boundary resources. Obtaining control over the boundary resources of an open platform is a key element of a platform owner's ecosystem governance (Baldwin and Woodard, 2009; Boudreau, 2010; Pon et al., 2014; Schilling, 2000). The platform owner influences the openness of the digital platform and the data exchange between the platform and ecosystem participants through the selection of boundary resources (Gawer, 2021b). Thus, the platform owner decides via the boundary resources how open the platform is and how the data exchange takes place. Ghazawneh and Henfridsson (2010, p. 4) considered the control of the platform provider as another impact of boundary resources. Boundary resources are an important building block of platform governance (Foerderer, Kude, Schuetz, et al., 2019). By implementing a set of boundary resources as part of its governance function (Boudreau, 2010), the platform owner defines the openness of and the access to the innovation platform (Boudreau, 2010; Karhu, Gustafsson, et al., 2018; Tiwana, 2014).[164] Mohagheghzadeh and Lindman (2022, p. 9 ff.) argued that joint development of boundary resources by the platform owner and third-party developers can also be a favorable option for developing the focal platform. The authors identified a decrease in process uncertainty, an increase in developer engagement, and an increase in boundary resource usage when designing boundary resources jointly (Mohagheghzadeh and Lindman, 2022, p. 6 ff.). [165] Increased coordination within and across organizations can be achieved by standardization of boundary resources (Thompson, 1967, p. 56).[166] The opening of (parts of) the platform is also described

[163] Summarizing a platform's instantiation of the governance dimensions, metaphors, such as open and closed, are often used (e.g., Boudreau and Hagiu, 2009; Hagiu, 2007; Parker and Van Alstyne, 2018).

[164] A certain degree of platform openness invokes positive network effects and increases generativity of the ecosystem (Blaschke and Brosius, 2018; Karhu, Gustafsson, et al., 2018).

[165] Mohagheghzadeh and Lindman (2022) investigated the joint design and development of boundary resources by platforms owned by consortia of multiple organizations and third-party developers in the maritime industry.

[166] Thompson (1967, p. 56) introduced standardization as a mode of intra-organizational coordination involving agreed-upon "routines or rules which constrain the action of each unit or position into paths consistent with those taken by others in interdependent relationships."

as "selective revealing" in the literature by West (2003, p. 20). Selective revealing can be described as "a combination of free revealing and various means of protecting one's code" (Henkel, 2006, p. 972).[167] Following Henkel's (2006, p. 972 f.) argumentation, boundary resources can be a means of selectively providing insights to ecosystem participants to the platform's core while consciously protecting other parts of the platform. Using boundary resources, a platform owner can provide third-party developers with selected design capabilities (Ghazawneh and Henfridsson, 2013,p. 174).

## 3.3 Boundary Resources Connecting Platforms and Participants

In this section, we introduce boundary resources as boundary-spanning organizational artifacts in digital business ecosystems in general and e-commerce ecosystems in particular (Ghazawneh and Henfridsson, 2013; Tsujimoto et al., 2018). We describe a variety of perspectives on the organizational boundary (e.g., Santos and Eisenhardt, 2005; Scott, 2014) and introduce boundary objects as potential measures in boundary-spanning endeavors (Fennell and Alexander, 1987; Star and Griesemer, 1989). This section closes with an overview of different types of boundary resources (e.g., Dal Bianco et al., 2014; Engert et al., 2022; Foerderer, Kude, Schuetz, et al., 2019) and concrete manifestations of boundary resources (e.g., Karhu, Gustafsson, et al., 2018; Petrik, 2022; Wulfert, Woroch, Strobel, Seufert, et al., 2022) in the context of innovation platforms and transaction platforms.

### 3.3.1 Organizational Boundaries

The organizational boundary[168] is a multifaceted concept widely studied in organizational theory (Santos and Eisenhardt, 2005; Scott, 2014), management research (Carlile, 2002, 2004), innovation management (Tushman, 1977), and social sciences (Bhatt et al., 2022).[169] Pelizza (2021, p. 96) described organizational, cultural, and

[167] For a more focused elaboration on the process of selective revealing, please refer to Henkel et al. (2014) and O'Mahony and Karp (2022).

[168] For this thesis, and in line with pivotal research contributions, we use the term organizational boundary (e.g., Star, 2010; Star and Griesemer, 1989).

[169] Boundaries are often discussed in ecology for delimiting ecological units. For the delimitation of boundary resources, Jax (2006) described topographical and functional boundaries. While the former is introduced as boundaries based on "discontinuities in space" (Jax, 2006,

functional boundaries in the context of organizations.[170] The organizational boundary can also be described by applying social identity theory.[171] The boundary of an organization is referred to as "the demarcation between the organization and its environment" (Santos and Eisenhardt, 2005, p. 491). It encompasses several dimensions that reflect the nature and extent of an organization's interactions with its external environment, such as its social structure and the organization's influence on resources and partners (e.g., Aldrich, 1979; Berne, 1966; Helfat, 1997; Santos and Eisenhardt, 2005; Scott, 2014). The organizational boundary is a critical determinant of organizational identity, not only defining the extent of an organization's activities and resources, but also its relationships with its stakeholders (Santos and Eisenhardt, 2005). Zelewski (2008, p. 62 ff.) made use of systems theory to depict an organization's boundaries when positioning an organization in its peripheral system (Schütte, 1998; Von Bertalanffy, 1950). In this case, the boundary is described as an area that is situated between an organization and its environment (Carlile, 2002, p. 443). In the following, we clarify several perspectives on the organizational boundary and different organizational boundary types (e.g., Berne, 1966; Santos and Eisenhardt, 2005; Scott, 2014).

The concept of organizational boundaries can be dated back to Weber's (1947, p. 139 ff.) definition of a formal organization.[172] Weber (1947) described a "corporate group" that clearly delimits "outsiders." Therefore, "[f]ormal organizations admit some persons while excluding others, thus allowing an observer to draw a boundary around the organization" (Aldrich, 1979, p. 221).[173] Aldrich (1971, p. 283) added

p. 241), the latter are set "on the basis of the extension of functional relationships between the elements of the unit" (Jax, 2006, p. 241).

[170] IS research began investigating organizational boundaries and the altering of boundaries in the context of IT outsourcing endeavors (Loh and Venkatraman, 1992, p. 8).

[171] According to social identity theory, a boundary between a group and its environment exists assuming in-group homogeneity while being aware of an out-group (Ashforth and Mael, 1989, p. 25).

[172] Weber (1947, p. 139 ff.) described closed social relationships as follows: "[…] a [social] relationship will […] be called 'closed' against outsiders so far as, according to its subjective meaning and the binding rules of its order, participation of certain persons is excluded, limited or subjected to conditions […]. A social relationship which is either closed or limits the outsiders by rules will be called a 'corporate group' so far as its order is enforced by the action of specific individuals […]."

[173] Excluding the perception of organizational boundaries follows Laumann's (1983, p. 21) realist approach. Laumann, et al. (1983, p. 21) identified two approaches for the investigation of organizational boundaries. The realist approach considered the "vantage point of the actors themselves in defining the boundaries" (Laumann, et al., 1983, p. 21). The organization "is treated as a social fact only in that it is consciously experienced as such by the actors com-

that the entry to and exit from organizations are only possible to a limited degree for individuals.

With reference to an innovation process, Tushman (1977, p. 590 ff.) identified three types of organizational boundaries. Extra-organization boundaries denoted the boundary between the organization and its external environment including adjacent organizations (Tushman, 1977, p. 590). ). Delimiting a laboratory from a larger organization requires an internal boundary within the organization, i.e., an internal boundary between research departments and other departments (Tushman, 1977, p. 590 f.). For innovation processes, Tushman (1977, p. 591) identified intra-laboratory boundaries between independent research departments. In the context of knowledge management, Carlile (2004, p. 559 ff.) developed an integrative framework of boundaries for knowledge transfer among organizations. The author differentiated pragmatic boundaries, semantic boundaries, and syntactic boundaries along the dimension increasing novelty for involved actors. Syntactic boundaries arise in situations "in which actors at the boundary lack a shared terminology (e.g., a communication device) for integrating knowledge" (Foerderer, Kude, Schuetz, et al., 2019, p. 129). While actors do not share a common meaning for knowledge, leading to semantic boundaries, actors do not even have any "shared interests" (Foerderer, Kude, Schuetz, et al., 2019, p. 129) in a situation of pragmatic boundaries. Baldwin (2008, p. 156) described the boundaries in production processes as "crossing points" between organizations. She differentiated between thin crossing points and thick crossing points. While thin crossing points are located at the boundary of a platform's complements and modules, thick crossing points can be found within these modules.[174]

Schwalbe et al. (2000, p. 430) introduced symbolic and spatial boundaries in the social context for individuals "to differentiate themselves from others" (Bhatt et al., 2022, p. 1686). By maintaining these boundaries, "dominant groups protect the material and cultural capital they have acquired and upon which they rely to preserve their dominance" (Schwalbe et al., 2000, p. 430). The aforementioned boundaries are often institutionally maintained by, for instance, governments and organizations (Schwalbe et al., 2000, p. 430). Symbolic boundaries are further specified by Bhatt et al. (2022, p. 1686) as "conceptual distinctions, interpretive strategies, and cultural traditions."

posing it" (Laumann, et al., 1983, p. 20 f.). In contrast, an investigator "imposes a conceptual framework constructed to serve his own analytic purposes" (Laumann, et al., 1983, p. 21) in the nominalist approach. In this approach, the boundaries are "analytically relative to the purpose of the investigator" (Laumann, et al., 1983, p. 21).

[174] Baldwin (2008) investigated that thick boundary resources involve higher transaction costs compared with thin crossing points.

Scott (2014, p. 232) described spatial boundaries and temporal boundaries of an organization. While spatial boundaries denote a geographical dimension,[175] temporal boundaries[176] refer to an organization's degree and speed of change affecting the organizational boundary (Scott, 2014, p. 232 ff.). Temporal boundaries reflect the temporal aspects of an organization's relationships with its stakeholders, such as the duration of relationships and the timing of interactions.

Boundaries of organizations are also investigated in psychiatry as the transition zone between a group and its environment (e.g., Berne, 1966).[177] Berne (1966, p. 56) described that the "external boundary, or boundary zone, separates the external environment from the group space."[178] Following Berne's (1966, p. 19) hypothesis on the scalability of his observations, a boundary zone also exists for organizations in e-commerce.[179] Since larger organizations in e-commerce can be considered as complex groups (Berne, 1966, p. 93), an internal boundary and several secondary boundaries within subgroups may exist. While the internal boundary separates an organization's leadership group from a membership group, secondary boundaries subdivide the aforementioned groups into subgroups. In Fig. 3.13, we summarize the organizational structure put forth by Berne (1966, p. 56 ff.).

---

[175] Scott (2014, p. 233) pointed out, that geographical proximity is not always the best way of conducting business. Especially with regard to virtual collaboration, the geographical proximity does not seem to be substantial for organizational success. Spatial boundaries can therefore be bridged by digital technologies. In e-commerce, spatial constraints are annulled using IT.

[176] Temporal boundaries can also be grasped as a characteristic of organizational boundaries. Organizational boundaries are not static in nature but change dynamically over time due to intra-organizational impacts and environmental influences.

[177] Although Berne (1966, p. 19) mostly analyzed smaller group constellations, he hypothesized that the "formulations are valid for all groups—regardless of their size and objectives." His observations are therefore also applicable to organizations in the context of this cumulative dissertation.

[178] Besides the external boundary that is relevant in the context of platforms and ecosystems in e-commerce, Berne (1966, p. 56) also introduced internal boundaries, dividing "the group space into a membership region and a leadership region."

[179] From an organizational perspective, Scott (2014, p. 186) argued for the establishment of "buffering units" protecting the organizations from external influences.

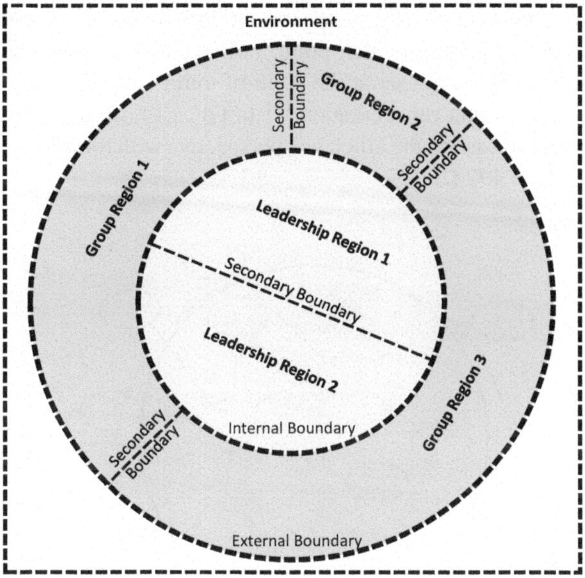


**Fig. 3.13** Boundary Types within an Organization (Berne, 1966, p. 93)

Santos and Eisenhardt (2005, p. 492) differentiated between horizontal and vertical boundaries.[180] While the former is "defined by the scope of product/markets addressed" (Santos and Eisenhardt, 2005, p. 492), the latter involves "the scope of activities undertaken in the industry value chain" (Santos and Eisenhardt, 2005, p. 492). Moreover, the authors identified four perspectives on organizational boundaries, namely boundaries of efficiency, boundaries of power, boundaries of competency, and boundaries of identity (Santos and Eisenhardt, 2005). The different perspectives on the organizational boundary deal with dedicated organizational challenges.

Boundaries of efficiency are determined by the cost of value creation, and the main decision in this instance involves internal (hierarchical) and external (market) value creation (Santos and Eisenhardt, 2005, p. 492 f.). For the efficiency perspective, Santos and Eisenhardt (2005, p. 492) argued for setting the demarcation of an organization "at the point that minimizes the cost of governing activities,"

---

[180] Santos and Eisenhardt (2005, p. 506) took an organizational-level perspective when analyzing boundaries. They did not consider the implications of boundaries for industries, groups, and individuals.

and referred to "transaction costs, measurement difficulties, and knowledge differences" for the determination of that point. Following the efficiency perspective, organizational boundaries are set at the point of minimum costs, comparing costs of hierarchical and market value creation.[181] In Fig. 3.14, we depict organizational boundary decisions based on the efficiency perspective with reference to Santos and Eisenhardt (2005, p. 492 f.)

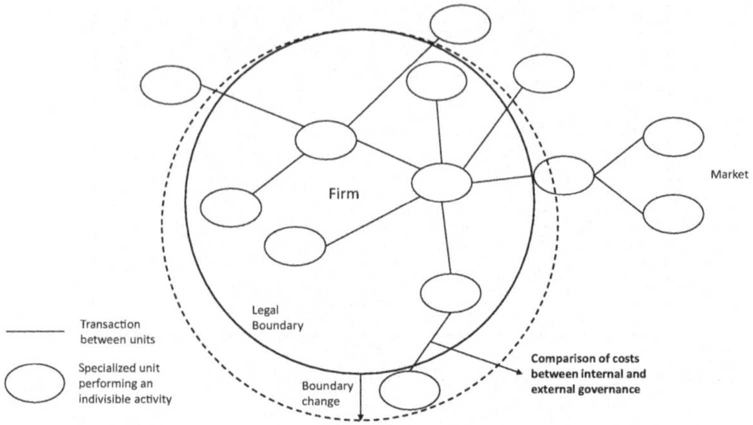


**Fig. 3.14** Boundaries of Efficiency (Santos and Eisenhardt, 2005, p. 492)

Boundaries of power are determined by the degree of control an organization can exercise over exchange relations if the organization is directly or indirectly involved (Santos and Eisenhardt, 2005, p. 495 f.). For this perspective, Santos and Eisenhardt (2005, p. 495) argued for setting the demarcation of an organization "at the point that maximizes strategic control over crucial external forces" and referred to the conception of "resource dependence (Pfeffer and Salancik, 1978; Thompson, 1967) and industrial organization (Porter, 1980)" for the determination of that point. Following the power perspective, organizational boundaries are set at the point of maximum control over critical resources and partners. For the determination of their boundary from a power perspective, organizations need to evaluate strategic

[181] While hierarchical value creation refers to production and value creation within the boundaries and governance of an organization, market value creation resembles external production (Coase, 1937). In section 3.4, we introduced multilateral value creation in ecosystems as "meta-organization" (Kretschmer, et al., 2022, p. 407) as an alternative between organization-internal and market value creation.

dependencies and select influence mechanisms for their external environment (Santos and Eisenhardt, 2005, p. 495). The maintenance of "a distinction between members and non-members involves establishing an authority, either exercised collectively or delegated to a specific role, empowered to admit some and exclude others" (Aldrich, 1979, p. 5). Authorities in organizations entail the power of "applying organizational rules in making decisions about entry and expulsion of members" (Aldrich, 1971, p. 283). In Fig. 3.15, we depict organizational boundary decisions based on the power perspective with reference to Santos and Eisenhardt (2005, p. 495 f.).

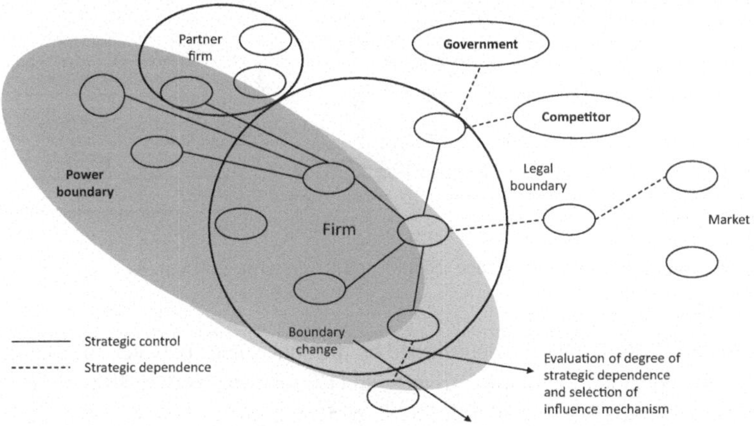


**Fig. 3.15** Boundaries of Power (Santos and Eisenhardt, 2005, p. 495)

Boundaries of competence are determined by decisions about which resources an organization needs to possess to gain competitive advantage matching internal resources with market opportunities (Santos and Eisenhardt, 2005, p. 497 f.).[182] For this perspective, Santos and Eisenhardt (2005, p. 497) argued for setting the demarcation of an organization "at the point that maximizes the value of the firm's resource portfolio" and referred to the two conceptions contingency theory (Chandler, 1962) and the resource-based view of the firm (Barney, 2015; Wernerfelt, 1984).[183] Following the competence perspective, organizational boundaries are set at the point of

[182] For the competence-oriented perspective on the organizational boundary, Santos and Eisenhardt (2005, p. 497) conceptualized organizations as "a unique bundle of resources."

[183] The dynamic determination of competencies not only involves physical resources and competencies, but also includes dynamic capabilities (Teece, 2007).

the resource portfolio' maximized opportunity value (Santos and Eisenhardt, 2005, p. 497). From the competence perspective, boundary management is a "process of aligning [internal] resource configurations with market opportunities" (Santos and Eisenhardt, 2005, p. 497). In Fig. 3.16, we depict organizational boundary decisions based on the competence perspective with reference to Santos and Eisenhardt (2005, p. 497 f.).

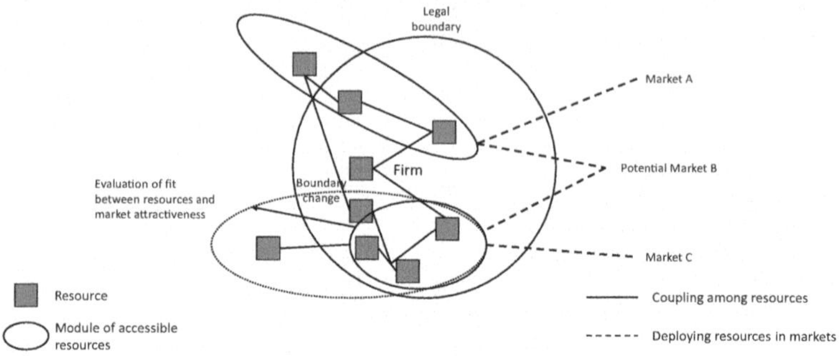


**Fig. 3.16** Boundaries of Competence (Santos and Eisenhardt, 2005, p. 492)

Boundaries of identity are determined by the coherence "between the identity of the organization and its activities" (Santos and Eisenhardt, 2005, p. 500). [184] For this perspective, Santos and Eisenhardt (2005, p. 500) argued for setting the demarcation of an organization at the point of maximum consistency between organizational identity and organizational activities, referring to research on managerial cognition and organizational identity (Albert and Whetten, 1995; Dutton and Dukerich, 1991; Walsh, 1995; Weick, 1995). Following the identity perspective, the organizational boundary is determined by a mapping of the organization's identity and its activities (Santos and Eisenhardt, 2005, p. 500). From the identity perspective, boundary management is "a process of resolving inconsistencies between identity and organizational activities and markets" (Santos and Eisenhardt, 2005, p. 500). In Fig. 3.17, we depict organizational boundary decisions based on the identity perspective with reference to Santos and Eisenhardt (2005, p. 500 f.).

[184] For the identity perspective on organizational boundaries, Santos and Eisenhardt (2005, p. 500) conceptualized an organization as "social contexts for sensemaking."

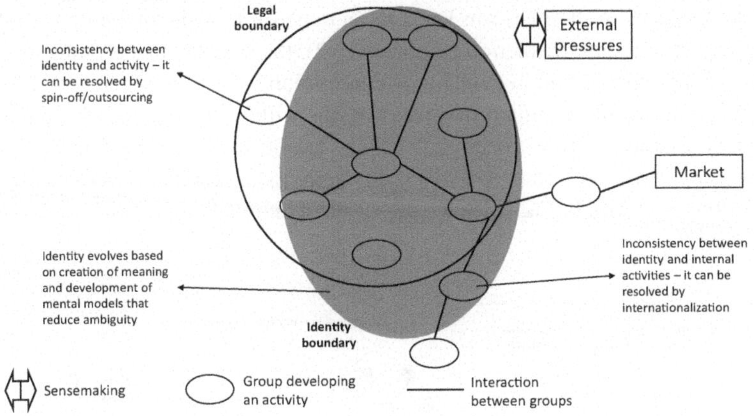


**Fig. 3.17** Boundaries of Identity (Santos and Eisenhardt, 2005, p. 500)

In their attempt to develop an extended generativity theory on transaction platforms, Fürstenau, Baiyere, et al. (2023, p. 5 f.) introduced the concept of a dynamic boundary[185] for transaction platforms integrating the product view and social interaction view on generativity.[186] Fürstenau, Baiyere, et al. (2023, p. 5) described the dynamic boundary as "the dynamic set of components that result from a continuous process in which providers and complementors negotiate conflicting expectations for functions on the service layer of a digital platform." The dynamic boundary concept includes the configurational boundary work (expansion of product boundaries) and the collaborative boundary work (expansion of the ecosystem boundary) (Fürstenau, Baiyere, et al., 2023; Langley, Lindberg, et al., 2019). From a product perspective on generativity (1), adding and opening new service categories and ser-

[185] Kellogg et al. (2006, p. 39) described the dynamic boundary as a "trading zone" in which internal developers negotiate with third-party developers on components design and openness.

[186] Although we previously described the concept of generativity for innovation platforms, Fürstenau, Baiyere, et al. (2023) claimed to investigate generativity in the context of transaction platforms in e-commerce. However, the authors did not investigate the relationship between the platform and third-party *sellers*, but the relationship between the platform and third-party *developers*. Fürstenau, Baiyere, et al. (2023, p. 2) investigated the opening of a transaction platform's service layer (Yoo, Boland, et al., 2012) to third-party developers and allowed them to provide complements to the platform. Hence, their investigation is more focused on the innovation perspective on digital platforms (e.g., Aulkemeier, Paramartha, et al., 2016; Tiwana, Konsynski, and Bush, 2010) than on the multi-sidedness and related economic concepts (e.g., Armstrong, 2006; Rochet and Tirole, 2003).

vice components to the platform leads to an expansion of the dynamic boundary. Indeed, from a social interaction perspective (2), these new service categories and service components result in conflicts in expectations of these services (2a, 2b) and joint negotiations (2). Conflicts and negotiations will further expand the platform's boundary. Fürstenau, Baiyere, et al. (2023, p. 13) further discovered that extended product boundaries and extended social interaction will result in an increasing user base of the platform. In Fig. 3.18, we depict the dynamic boundary of a transaction platform.[187]

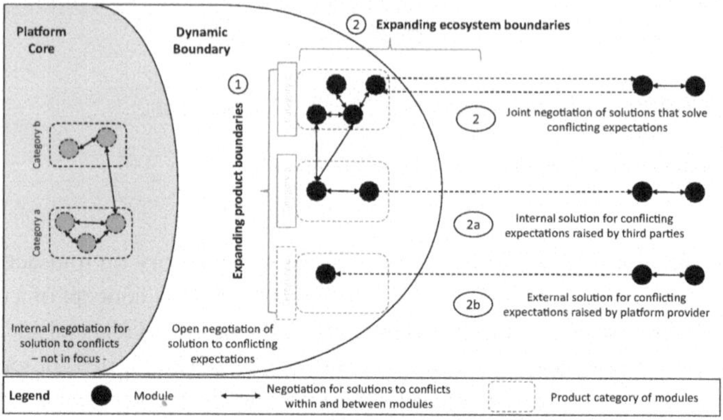


**Fig. 3.18** Dynamic Boundary and Generativity (Fürstenau, Baiyere, et al., 2023, p. 5)

### 3.3.2 Boundary Objects as a Means for Organizational Boundary-Spanning

In 1971, Aldrich (1971, p. 283) introduced organizations as "boundary-maintaining systems."[188] Following Aldrich's definition, an organization's "survival as an entity depends on its ability to control its boundaries" (Aldrich and Ruef, 2006, p. 5). In this regard, boundary resource theory denotes boundary management as "a central task of organizations" (Zhang, Viswanathan, et al., 2011, p. 319).[189] In sociology,

[187] In line with our conceptualization of platforms consisting of modules, we described a platform's "components" (Fürstenau, Baiyere, et al., 2023, p. 5) as "modules."

[188] Aldrich (1979, p. 4) defined organizations "as goal-directed, boundary-maintaining, and socially constructed systems" of human activity.

[189] Analyzing purchasing agents' role in trust development for buyer-supplier relationships, Zhang, Viswanathan, et al. (2011, p. 319) explained that "a central task of organizations is

boundary maintenance describes activities of group members that identify and "police the boundaries between themselves and purported outsiders" (Graham, 2019, p. 2034).[190] For the social sciences, Bhatt et al. (2022, p. 1687) further differentiated boundary-maintaining activities in boundary retention and boundary reinforcement. While boundary retention describes the continuation of existing boundaries across the members of a social group for a longer period of time (Bhatt et al., 2022; Durckheim et al., 2008), boundary reinforcement is described as the reinforcement of existing symbols and the innovation of new symbols (Bhatt et al., 2022, p. 1687) (Figs. 3.19 and 3.20).

In the course of increasing globalization and digitalization, organizations are "embedded in larger systems of action" (Thompson, 1967, p. 66). Since organizations are embedded in complex supply chain networks, dense competition, and loose relations (Adner, 2017; Cennamo, 2021; Corallo, 2007; Tiwana, 2015a), they need to manage their boundaries carefully (Zhang, Viswanathan, et al., 2011, p. 319). Langley, Lindberg, et al. (2019, p. 704) referred to the active and purposeful definition of organizational boundaries as "boundary work."[191] With reference to Lamont and Molnár (2002) and Phillips and Lawrence (2012), Langley, Lindberg, et al. (2019, p. 704) defined boundary work as the "purposeful individual and collective effort to influence the social, symbolic, material or temporal boundaries, demarcations, and distinctions affecting groups, occupations, and organizations." Langley, Lindberg, et al. (2019, p. 709 ff.) derived three major types of boundary work, namely competitive boundary work, collaborative boundary work, and configurational boundary work. Competitive boundary work aims at building boundaries around individuals and organizations for reasons of protection and market positioning in competition. In collaborative boundary work situations, (organizational) boundaries are realigned to enable inter-company collaboration. Configurational

---

to manage their boundaries with other organizations that supply critical resource inputs." In contrast, Laumann, et al. (1983, p. 20) noted that often "[...] the matter of boundary definition is one of no particular import" as boundaries "are presented so self-evident in a social situation studied to require no comment" (Laumann, et al., 1983, p. 20). We argue that, in the context of complex network structures in e-commerce ecosystems (Adner, 2017; Wulfert, Woroch, Strobel, Seufert, et al., 2022), the importance of organizational boundary management is even on the increase to retain and extend an ecosystem participant's position and to compete in the market (Wulfert, 2023).

[190] Fini et al. (2023) investigated boundary-maintaining tendencies in academia analyzing the example of an academic accreditation process.

[191] Langley, Lindberg, et al. (2019, p. 704) based their review of the literature on boundary work on the practice turn in organization and management theory. Gieryn (1983, p. 781 ff.) described the term boundary work in the context of science for scientists attempting to demarcate science from non-science.

boundary work can be described as a hybrid form of the former boundary work types. It allows for establishing boundaries simultaneously, enabling collaboration for selected areas and competition in others.[192]

Boundary management in digital business ecosystem contexts often requires extending organizational boundaries and cooperating with other ecosystem participants for value co-creation. In contrast to boundary-maintaining activities, this activity is introduced as boundary-spanning in the literature (e.g., Doolin and McLeod, 2012; Fennell and Alexander, 1987; Thompson, 1967). Fennell and Alexander (1987, p. 458 f.) identified two boundary-spanning activities, namely boundary bridging and boundary buffering.[193] In boundary bridging activities, an organization's boundaries are redefined through the connection with other organizations. In doing so, the boundaries of the organization tethering up are expanded with the other organization' boundary. With boundary buffering activities, an organization protects "against disturbing environmental influences" (Fennell and Alexander, 1987, p. 458). The authors described an increase in internal administrative structures and a strengthening of boundary management as potential strategies for boundary buffering (Fennell and Alexander, 1987, p. 459). Boundary-spanning units of an organization need to adjust "to constraints and contingencies not controlled by the organization" (Thompson, 1967, p. 66).[194] Galison (1997, 46) described "trading zones" for boundary-spanning activities in fast-paced and volatile environments.[195]

Doolin and McLeod (2012, p. 572) noted that "[b]oundary objects [can] perform a range of boundary-spanning functions." Boundary objects[196] are "plastic enough to adapt to local needs and the constraints of several parties employing them, yet

---

[192] The concept of configurational boundary work can be attributed to focal platforms in digital business ecosystems (i.e., e-commerce ecosystems). It can be characterized as carefully managing boundary resources as a platform governance mechanism (Hein, Schreieck, Wiesche, et al., 2016, p. 6). In this regard, the platform owner allows ecosystem participants to access selected parts of the core platform while restricting access to other areas (Staykova and Damsgaard, 2015, p. 7 f.).

[193] Although Fennell and Alexander (1987) analyzed boundary-spanning activities for the particular case of hospitals, we consider the identified activities as probably applicable to organizations in general and e-commerce ecosystem settings in particular.

[194] Economists also called these constraints and contingencies that are not controlled by the organization under consideration "exogenous variables" (Thompson, 1967, p. 66).

[195] The trading zone metaphor should emphasize exchanges and negotiation between members of different communities "in exquisite local detail, without global agreement" (Galison, 1997, p. 46). Trading zones can be situated at an organization's boundary, representing (virtual) environments for boundary-spanning activities (Kellogg et al., 2006, p. 39).

[196] Boundary objects are also described as "boundary-spanning components" (Thompson, 1967, p. 71).

robust enough to maintain a common identity across sites" (Star and Griesemer, 1989, p. 393). Star and Griesemer (1989) introduced repositories, ideal type, coincident boundaries, and standardized forms as boundary object types.[197] Bergman et al. (2007, p. 550) applied a broad definition of boundary objects referring to "any artifact that is shared between two or more actors at the border of two social worlds." Boundary objects connect individuals from different social worlds that have the ability to contain different meanings in each world (Briers and Wai, 2001, p. 264 f.). Since boundary objects are "objects that are shared and shareable across different problem-solving contexts" (Carlile, 2002, p. 451), they can help overcome organizational boundaries and promote the exchange of information between different demarcation points (Carlile, 2002, p. 443). Boundary objects intermediate between problem-solving contexts and different understandings, establishing a shared context that "sits in the middle" (Star and Griesemer, 1989, p. 47). Star (2010, p. 603) further highlighted that even if boundary objects are defined at a specific point in time, they are not static but evolvable over time to adhere to internal and external contingencies (Subrahmanian et al., 2003).[198] Star and Griesemer (1989) attributed five characteristics to boundary objects. Following these characteristics, boundary objects are weakly structured in common use; are strongly structured in individual site use; have different meanings in different social worlds, with a structure that is common enough to all; are simultaneously concrete and abstract, specific, and general; conventionalized and customized; and are often internally heterogeneous. The purpose of boundary objects is to standardize information between different ecosystem participants in order to interact with each other in achieving a common goal (Star and Griesemer, 1989, p. 408 ff.). In Table 3.3, we summarize the boundary-spanning functions of boundary objects identified by Doolin and McLeod (2012, p. 572)

[197] The boundary object types introduced by Star and Griesemer (1989) are depicted with exemplary manifestations in Fig. 3.21.

[198] In the context of platforms, Baldwin and Woodard (2009, p. 23 f.) mentioned that boundary resources are relatively stable over time compared with platform extensions and internal platform modules.

**Table 3.3** Boundary-Spanning Function of Boundary Objects (Doolin and McLeod, 2012, p. 572)

| Boundary-Spanning Function | Source |
|---|---|
| Boundary objects enable knowledge transfer between groups through a shared syntax. | Carlile (2002, 2004) |
| Boundary objects enable a translation between groups and their perspectives. | Star and Griesemer (1989) |
| Boundary objects enable a shared understanding in negotiations to mitigate concerns and interests. | Carlile (2002, 2004); Lee (2007) |
| Boundary objects enable the alignment of interest for legitimization and adherence to design knowledge. | Bergman et al. (2007) |
| Boundary resources are evolvable over time to adhere to internal and external contingencies. | Subrahmanian et al. (2003) |

In addition to boundary objects as software artifacts and artificial objects, Wei et al. (2020, p. 533) highlighted the role of "human boundary spanners" in inter-firm collaboration and coordination, such as salespeople and account managers.[199] Human boundary spanners can propel innovations within and across organizations (Tushman, 1977, p. 587).[200] Wei et al. (2020, p. 533) further denoted that human boundary spanners "and non-human boundary resources have co-existed and complemented each other for a long time."

### 3.3.3 Boundary Resources in Platform Contexts

Ghazawneh (2012) introduced the concept of boundary objects to research on innovation platforms and coined the term "boundary resource" (Ghazawneh, 2012, p. 24 ff.) in the context of software development.[201] Petrik (2022, p. 68) emphasized that boundary resources resemble boundary objects in platform-centered ecosystems. The concept of boundary resources is not limited to innovation platforms,

[199] Zhang, Viswanathan, et al. (2011, 320 ff.) denoted three major characteristics of human boundary spanners. Human boundary spanners handle strategic communication with ecosystem participants, entail knowledge about available boundary resources, and have the capability of effective mediation (Aldrich and Aldrich, 1977; Zhang, Viswanathan, et al., 2011).

[200] Tushman (1977, p. 590) illustrated organizational boundaries and boundary spanners in innovation processes with the example of multiple research and development departments within an organization and their relations to the external environment.

[201] Bergman et al. (2007) applied the concept of boundary resources to distributed system design involving participants from different social worlds.

but is also applied to transaction platforms (e.g., Fürstenau, Baiyere, et al., 2023; Gawer, 2021b; Wulfert, Woroch, Strobel, Seufert, et al., 2022). Brenner and Breuer (2001, p. 145) described boundary resources in the context of transaction platforms as "integration requirements" for affiliating with sellers and customers. Boundary resources enable the transfer of information between the platform owner and additional ecosystem participants providing platform extensions (Ghazawneh and Henfridsson, 2010, p. 4). Boundary resources can be applied to translate between all types of organizational boundaries (e.g., Engert et al., 2022; Ghazawneh, 2012; Star and Griesemer, 1989). We define boundary resources as "software tools and regulations that serve as the interface for the arm's-length relationship between the platform owner and the application developer" (Ghazawneh and Henfridsson, 2013, p. 175). Perks et al. (2017, p. 536) defined boundary resources[202] in the context of platforms broadly "as those digital artifacts that connect two or more organizations or components in a solution." Boundary resources are "the means the platform owner introduces to third-party developers so that they can contribute to activities carried out in the platform" (Mohagheghzadeh and Lindman, 2022, p. 1). Digital interfaces act "as a demarcation between agents and as a connector or conduit of data" (Gawer, 2021b, p. 3). In the remainder of this section, we present abstract models describing boundary resources in the context of innovation platforms, boundary resource types, and manifestations of boundary resources.

Ghazawneh and Henfridsson (2013, p. 176 ff.) developed a model to analyze the relationship between a platform owner and third-party developers. The relationship is facilitated by the provision of appropriate boundary resources through the platform owner. The boundary resource model by Ghazawneh and Henfridsson (2013, p. 176) is illustrated in Fig. 3.19.[203] The boundary resources are designed and provided by a focal platform (a) and used by the applications developed by third-party developers (b). Besides the design and use of boundary resources, Ghazawneh and Henfridsson (2013, p. 176) emphasized the role of boundary resources in developing the platform and securing the platform owner's control. Resourcing (c) denoted the expansion in scope and diversity of the platform. By expanding the platform through new boundary resources, new participants can be obtained. Hence, the provision of a new

[202] Perks et al. (2017, p. 536) originally denoted boundary resources as "boundary interfaces." Since the authors ascribed the same meaning to boundary interfaces as we denote for boundary resources, we call them boundary resources, which is in line with the discussions in the rest of this section. However, Perks et al.'s (2017, p. 536) focus on "digital artifacts" excluded some social boundary resources that involve human interactions, such as hackathons and support hot-lines (Dal Bianco et al., 2014).

[203] The relations between platform, third-party applications, and boundary resources are indicated using small letters in brackets in Fig. 3.19.

boundary resource, such as an API[204] with documentation, can attract additional third-party developers and increase the generativity of the platform (Ghazawneh and Henfridsson, 2013, p. 177). Securing (d) referred to ensuring control over the platform and its related services. In doing so, the platform owners strive to retain and increase their power over the ecosystem. For instance, existing developer agreements can be made more stringent and approval processes for new or updated applications can be introduced by the platform owner (Ghazawneh and Henfridsson, 2013, p. 177). Petrik (2022, p. 71) emphasized that the platform owner is likely to perform resourcing and securing activities in an alternating manner. This alternation should ensure an appropriate degree of platform openness while retaining proper control over the platform and the ecosystem (Blaschke and Brosius, 2018; Boudreau, 2010; Karhu, Gustafsson, et al., 2018).

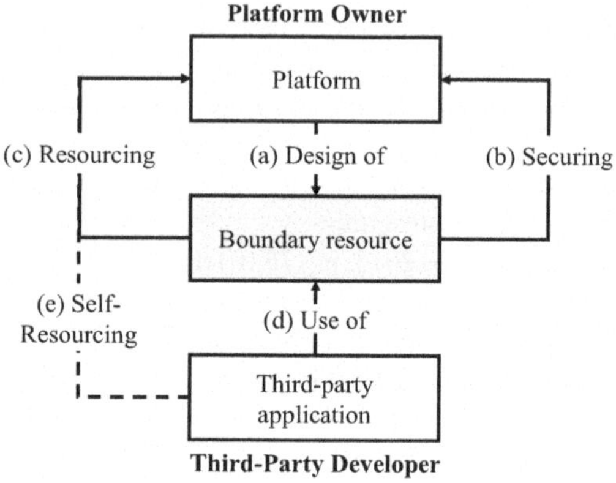


**Fig. 3.19** Platform Owner-Centric Boundary Resource Model (according to Ghazawneh and Henfridsson, 2013, p. 176)

Although Ghazawneh and Henfridsson's (2013, p. 176) model only depicted resourcing that is possible via the platform owner, the authors also described self-resourcing (e) situations (Ghazawneh and Henfridsson, 2013, p. 181). Referring to the first iPhone generation, Ghazawneh and Henfridsson (2013, p. 181) emphasized

[204] An API can be described as "a standardized machine-readable digital communication interface for a system which can be designed to have open or restricted access and be exposed both internally and externally of an organization's network" (Parker, Van Alstyne, and Choundary, 2016, p. 40).

that too restrictive resourcing and high securing by the platform owner may lead to situations in which third-party developers themselves introduce boundary resources for the platform.[205] Eaton et al. (2015, p. 235) called the creation and alteration of boundary resources by ecosystem participants "distributed tuning."[206]

Eaton et al. (2015, p. 235 f.) highlighted that boundary resources undergo a dynamic process of shaping and reshaping in the interaction between the platform owner and ecosystem participants.[207] Platform owners can speed up the tuning of boundary resources by increasing the contact with ecosystem participants and engaging them to participate in the tuning process (Mohagheghzadeh and Rudmark, 2017, p. 104 ff.). To highlight the pivotal role of boundary resources for innovation platforms, Eaton et al. (2015, p. 235 ff.) deviated from Ghazwaneh's and Henfridsson's platform owner-centric model and introduced a boundary resource-centric model for the distributed tuning of boundary resources. The conceptual model summarizing the distributed tuning of boundary resources is centered around the boundary resource as an artifact. Eaton et al. (2015, p. 237) introduced platform owners and third-party developers as major actors in the distributed tuning of boundary resources.[208] Reflecting the heterogeneity of third-party developers using innovation platforms, the authors depicted three groups of developers indicated by grey shading.[209] Besides the boundary resources already introduced in this thesis,

---

205 Therefore, we added a dashed line representing self-resourcing (e) relationships between the platform and extensions provided by third-party developers.

206 Pickering (1993, p. 567 ff.) introduced the concept of tuning as a generative process balancing resistance and accommodation. March and Simon (1958, p. 26) referred to the transmission of new information during the process of action as "coordination by feedback."

207 Eaton et al. (2015) analyzed the iPhone as an innovation platform with its surrounding ecosystem participants for the development of their conceptual model for the distributed tuning of boundary resources. The authors analyzed the iPhone platform from 2007 to 2011.

208 Although Eaton et al. (2015, p. 237) focused on platform owners and third-party developers as major actors in the distributed tuning of boundary resources, the authors added that the group of third-party developers is usually highly heterogeneous. Therefore, they extended the research by Pickering (1995) and Barrett et al. (2012). Pickering (1993) described the process of resource tuning as resistance and accommodation of single actors. Barrett et al. (2012, p. 1464) analyzed dyadic constellations of resource tuning in robotic innovations in the pharmaceutical industry. Eaton et al. (2015, p. 237) concluded that the process of boundary resource tuning "involves dialectical interrelating among distributed and heterogeneous actors and resources participating in" an ecosystem.

209 Eaton et al. (2015, p. 236) suggested a differentiation into large, medium, and small groups of third-party developers. Although the authors admitted that the "shading does not constitute a theoretical statement beyond the indication of the heterogeneity of actors" (Eaton et al., 2015, p. 236), the size of the third-party organization may reflect the degree of their influence on the tuning of boundary resources.

Eaton et al. (2015, p. 236) differentiated between technology resources and auxiliary boundary resources. While technology resources are "all sorts of hardware components, operating systems, middle-ware, software code, etc. that actors participating in a service system attempt to tune," auxiliary boundary resources are described as "peripheral to the focal platform." Technological boundary resources form the super-set of boundary resources and auxiliary boundary resources. Eaton et al. (2015, p. 236) observed that platform owners and third-party developers continuously change and adapt technology resources, auxiliary boundary resources, and boundary resources for their own purposes and benefits. In third-party developers' and the platform owner's attempts to tune boundary resources, "a boundary resource (BR1) can change over time (BR1') or new boundary resources can emerge (BR2)" (Eaton et al., 2015, p. 236). Third-party developers and platform owners are both influenced by other ecosystem participants such as partners and the government. In Fig. 3.20, we depict the conceptual model of the dynamic process of distributed tuning of boundary resources according to Eaton et al. (2015, p. 237).

Aside from these definitions for boundary resources identified in the literature, Gawer (2021b, p. 2) stated that there is no "holistic definition" for the boundaries of innovation platforms and transaction platforms. We will therefore present selected types of boundary resources in the context of innovation platforms and transaction platforms (e.g., Foerderer, Kude, Schuetz, et al., 2019; Gawer, 2021b; Ghazawneh and Henfridsson, 2013; Karhu, Gustafsson, et al., 2018). Moreover, we provide an overview of boundary resource instances (e.g., Karhu, Gustafsson, et al., 2018; Petrik, 2022; Wulfert, Woroch, Strobel, Seufert, et al., 2022).

Gawer (2021b, p. 2 ff.) identified three distinct streams of boundary decisions "to strategically demarcate [the platform's] resources and assets." These decisions are grounded in the literature on platform economics, platform strategy, and IS research. She defines digital platform boundaries as a combination of platform firm scope, platform sides' configuration and composition, and digital interfaces (Gawer, 2021b, p. 7). For the scope, companies "make decisions upon deciding what assets, activities, and resources they will own and what kind of labor they will employ" (Gawer, 2021b, p. 4). Scoping decisions become increasingly important in times of digitalization with altering "value creation, delivery, and capture processes" (Autio et al., 2018, p. 74). Decisions on the platform sides determine the number of participant types that can affiliate with a platform and specify which participants from the distinct types can access the platform.[210] Digital interfaces describe the

[210] While Gawer (2021b, p. 2) refers to the decision on affiliated participant types as a "configuration," she denoted the specific access rights for participants as a "configuration."

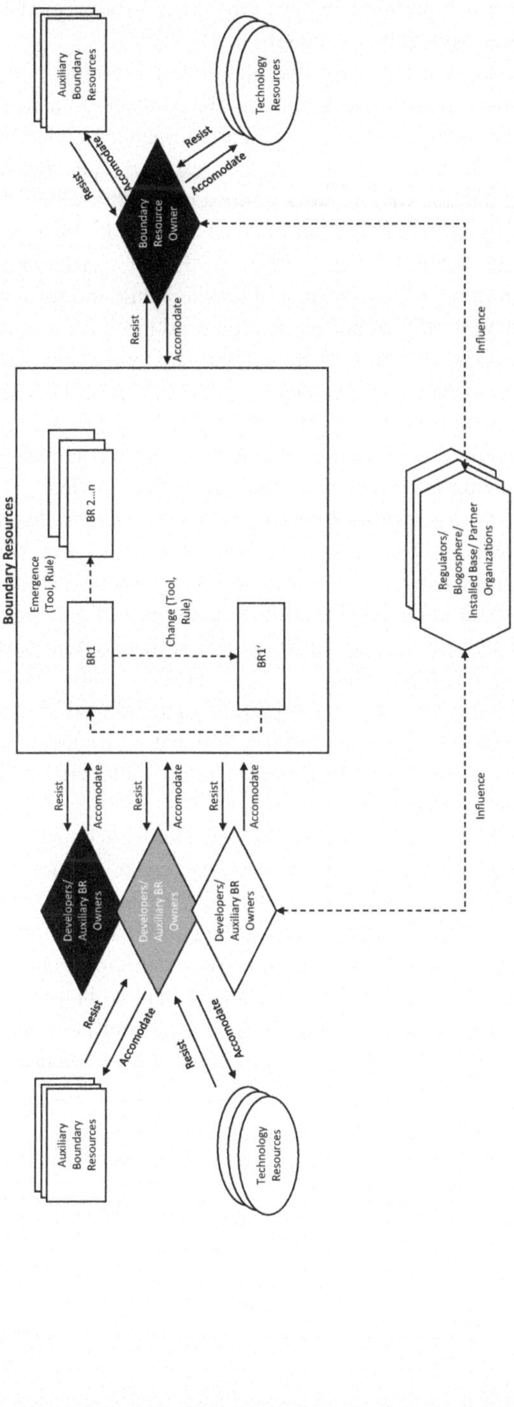


**Fig. 3.20** Distributed Tuning of Boundary Resources (Eaton et al., 2015, p. 237)

boundary resources used for the exchange of information between the focal platform and ecosystem participants (Gawer, 2021b, p. 2).

Karhu, Gustafsson, et al. (2018, p. 481) elaborated on boundary resources allowing participation on the platform and enabling the conduct of business. The authors referred to these boundary resources as "access openness" (Karhu, Gustafsson, et al., 2018, p. 481). As an additional boundary resource type, "resource openness" types provide access to internal resources and services "forfeiting the [intellectual property rights] of the resource" (Karhu, Gustafsson, et al., 2018, p. 481).[211]

Foerderer, Kude, Schuetz, et al. (2019, p. 130 ff.) differentiated boundary resources by different degrees of scale and scope. While the former describes the share of complementors the boundary resource addresses, the latter depicts the extent of information, skills, and organizational capability provided. Along these two dimensions, Foerderer, Kude, Schuetz, et al. (2019, p. 130 ff.) introduced three types of boundary resources addressing knowledge boundaries,[212] namely broadcasting boundary resources, brokering boundary resources, and bridging boundary resources. Broadcasting boundary resources "provide knowledge via transferable objects or through […] centralized databases" (Foerderer, Kude, Schuetz, et al., 2019, p. 130), are generally applicable, address all ecosystem participants, and do not require interaction with the platform owner for access. Brokering boundary resources provide meta-knowledge and address a subset of participants. They provide "meta-knowledge in terms of knowing where [ecosystem participants] may obtain help" (Foerderer, Kude, Schuetz, et al., 2019, p. 130). Bridging boundary resources address individual ecosystem participants and enable specific problem-solving. These boundary resources require "intensive, frequent, and sometimes informal exchanges between platform owners and developers" (Foerderer, Kude, Schuetz, et al., 2019, p. 130).

Engert et al. (2022, p. 7) differentiated between uniform boundary resources and individual boundary resources. Uniform boundary resources, such as general documentation, APIs, and Integrated Development Environments (IDEs), are standardized and provided to all or a group of ecosystem participants in the same specification. They can therefore be distributed to ecosystem participants in a "one-to-many" (Engert et al., 2022, p. 7) relation. In contrast, individual boundary resources are tailored to the specific requirements of single ecosystem participants. Individual boundary resources, such as individual promotions and personal contacts, are pro-

[211] While the former boundary resource type is more likely in transaction platforms, resource openness is a prerequisite for the development of extensions by third-party developers (Zapadka et al., 2022, p. 3).

[212] With reference to Jantsch (1980, p. 51), Carlile (2002, p. 443 ff.) introduced syntactic, semantic, and pragmatic boundaries in knowledge management.

vided on a "one-to-one" (Engert et al., 2022, p. 7) basis. Engert et al. (2022, p. 4) further differentiated between innovation-oriented boundary resources, governance-oriented boundary resources, and communication-oriented boundary resources. Innovation-oriented boundary resources provide access to a platform's services and enable the development of third-party extensions (Jansen and van Capelleveen, 2013; Tiwana, Konsynski, and Bush, 2010). While governance-oriented boundary resources aim at knowledge transfer for developers making available information on technical boundary resources, communication-oriented boundary resources facilitate the motivation and incentivize third-party developers (Engert et al., 2022, p. 4).

Dal Bianco et al. (2014, p. 15 f.) introduced three types of boundary resources, namely application boundary resources, development boundary resources, and social boundary resources.[213] Social boundary resources involve documentation, roadmaps, registration forms, forums, support, and hackathons. While application boundary resources include APIs and workflows, development boundary resources include SDKs, software libraries, software repositories, debugging tools, prototypes, and mockups. In Table 3.4, we summarize the boundary resource types derived by Dal Bianco et al. (2014, p. 15 f.).

**Table 3.4** Classification of Platform Boundary Resources Dal Bianco et al. (according to 2014, p. 15)

| Class | Definition |
|---|---|
| Application Boundary Resource | Program resources that enable the third-party application to interact with the platform. |
| Development Boundary Resource | Program resources or tools that enable the third-party developer to develop and evolve applications. |
| Social Boundary Resource | Resources that enable transferring knowledge about the third-party application development and the coordination thereof. |

[213] Dal Bianco et al. (2014, p. 16) accentuated that selected boundary resources, not being mutually exclusive, can be assigned to more than one group. The authors stressed the example of a method within an API that can simultaneously be an application boundary resource for accessing platform services and used by developers for their third-party extensions. Application boundary resources and development boundary resources can also be further summarized as technical boundary resources. In paper 8, we deal with this abstraction of boundary resources in technical boundary resources and supplementary social boundary resources (Wulfert, 2023).

As a further specialization of technical boundary resources, Wulf and Blohm (2020, p. 255 ff.) proposed three API archetypes, namely professional service APIs, mediation service APIs, and open asset APIs. Professional service APIs provide paid access to cloud-based IT services, such as the Amazon S3 API and the Google Maps API. Mediation service APIs provide access to the innovation platform's core services that third-party developers can use for extension development. Wulf and Blohm (2020, p. 256) mentioned the Facebook Graph API and the YouTube Live Streaming API as examples of this archetype. Open asset APIs, such as the New York Times API and the DB Open Data API, provide free access to organization-internal IT resources.

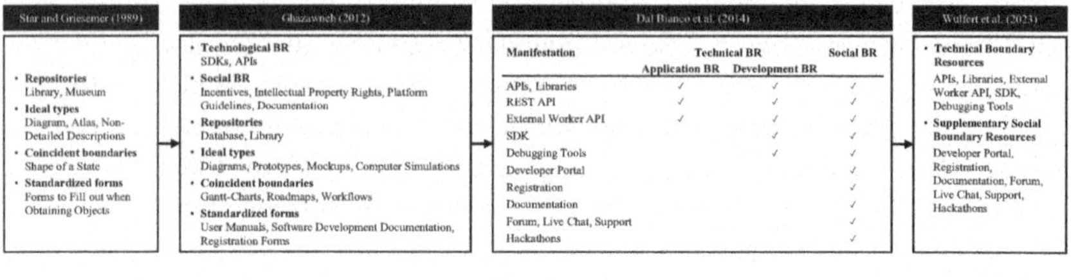


**Fig. 3.21** Overview of Major Boundary Resource Types (Wulfert, Woroch, Strobel, Seufert, et al., 2022, p. 1818)

Making use of these seminal works, a number of researchers developed sets of boundary resources for specific application scenarios. These sets include a variety of manifestations of the aforementioned boundary resource types (Eaton et al., 2015).[214] Dal Bianco et al. (2014, p. 15) presented a set of ten boundary resources for innovation platforms, mapped to their three boundary resource types. Mapped to his boundary resource types, Ghazawneh (2012, p. 20 ff.) introduced more than ten boundary resource manifestations.[215] Karhu, Gustafsson, et al. (2018, p. 495) developed a set of 16 boundary resources for open digital platforms. Foerderer, Kude, Schuetz, et al. (2019, p. 130 ff.) identified several empirical examples of boundary resources in a case study of two enterprise resource planning systems as innovation

[214] Please note that the overview of boundary resource manifestations is neither comprehensive nor complete. It should provide an overview of possible boundary resource implementations in different domains.

[215] Ghazwaneh's (2012) and Dal Bianco et al.'s (2014) boundary resources manifestations, mapped to the respective boundary resources types are shown in Fig. 3.22.

platform with surrounding ecosystems.[216] Petrik (2022, p. 109 ff.) derived a set of 15 boundary resources for the context of industrial Internet of Things (IoT) scenarios.[217] Wulfert, Woroch, Strobel, Seufert, et al. (2022) synthesized 14 distinct boundary resources for platforms in e-commerce ecosystems, making use of the publications by Ghazawneh (2012) and Dal Bianco et al. (2014). Schreieck et al. (2016, p. 12) investigated that data can adopt the role of a boundary resource. We summarize the aforementioned manifestations of boundary resources in chronological order in Fig. 3.22.[218]

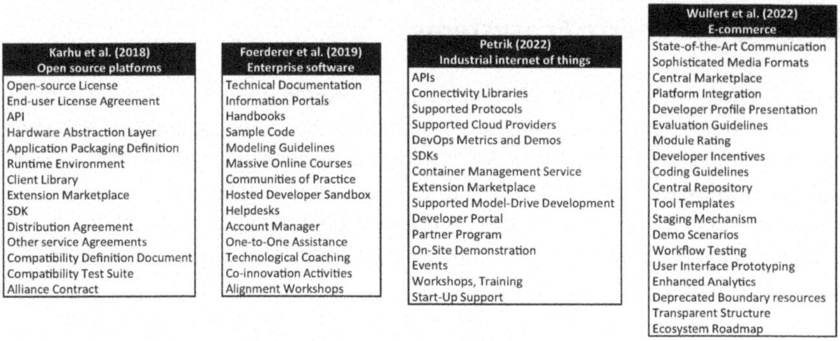


**Fig. 3.22** Overview of Boundary Resource Manifestations

In the remainder of this cumulative dissertation, we differentiate between technical boundary resources (i.e., application and development boundary resources) and supplementary social boundary resources. For this differentiation, we make use of the seminal works by Dal Bianco et al. (2014); Ghazawneh (2012); Star and

216 These are, among others, technical documentation, modeling guidelines, help desks, and technological coaching. Foerderer, Kude, Schuetz, et al. (2019, p. 130 ff.) assign these examples to the three boundary resource groups broadcasting, brokering, and bridging.

217 The 15 boundary resources developed by Petrik (2022, p. 109 ff.) were translated from German to English to achieve increased consistency within this thesis. Petrik, Model, et al. (2021) further analyzed third-party developers' satisfaction with technical boundary resources in IoT scenarios applying Kano's model of attractive quality (Kano et al., 1984).

218 We did not depict any relation between the boundary resource manifestations developed by Karhu, Gustafsson, et al. (2018), Foerderer, Kude, Schuetz, et al. (2019), Petrik (2022), and Wulfert, Woroch, Strobel, Seufert, et al. (2022), as they seem to be developed independently. They all address different areas of application.

Griesemer (1989) as depicted in Fig. 3.21.[219] While technical boundary resources are part of a platform's architecture (Baldwin and Woodard, 2009; Tiwana, 2014), the design of supplementary social boundary resources and the management of both types (Wulfert, 2023) are considered in a platform's governance. The boundary resources resemble the demarcation points between the focal platform and its surrounding ecosystem. We focus on platforms in e-commerce, where hybrid platforms form the center of broader e-commerce ecosystems and orchestrate a broad range of participants (Gawer, 2014; Wulfert, Woroch, Strobel, Seufert, et al., 2022).

## 3.4 E-Commerce Ecosystems

In this section, we develop an understanding of e-commerce ecosystems. We sketch the establishment of the ecosystem concept in the business context (i.e., business ecosystem) and emphasize the role of IS in digital business ecosystems. We conceptualize e-commerce ecosystems as manifestations of digital business ecosystems in the context of e-commerce (Wulfert, Woroch, Strobel, Seufert, et al., 2022, p. 1816 f.) and describe them from an ecosystem-as-affiliation and ecosystem-as-structure perspective (Adner, 2017).

### 3.4.1 Toward Digital Business Ecosystems

For the introduction of digital business ecosystems, we introduce the ecosystem metaphor to business research. We also elaborate on the characteristics of digital business ecosystems and introduce software ecosystems as a sub type of digital business ecosystems.

#### 3.4.1.1 Business Ecosystem

The ecosystem concept originated from biology and ecology (Basole and Russell, 2015, p. 3) and originally described biological systems[220] of interacting organisms

[219] We integrated both types of boundary resources in our boundary resource management lifecycle (Wulfert, 2023).

[220] Since this thesis focuses on e-commerce ecosystems as a sub type of (digital) business ecosystems, the biological concept will not be dealt with in further detail. Literature on biology provides a variety of research on ecosystems and a set of definitions. These definitions often include organisms of biotic (i.e., living organisms) and abiotic (i.e., physical environment) components (e.g., Chapin, 2011; Jax, 2006; Schaefer, 2002; Tansley, 1935). Initially, it was Tansley (1935) who proposed the ecosystem concept.

that populate a habitat.[221] Lindeman (1942, p. 400) further denoted that the ecosystem is a system composed of "physical-chemical-biological processes active within a space-time unit of any magnitude" and added that it is usually the "the biotic community plus its abiotic environment" Lindeman (1942, p. 400). Moore (1993, 1996) has introduced the ecosystem concept into the business literature in analogy to biology and ecology.[222] While single companies resemble the species of the business ecosystem (i.e., biotic community), the physical environment is mirrored by the legal environment in which these companies interact (i.e., abiotic environment) (Moore, 1996).[223] Hence, a business ecosystem can be described as a collection of independent (or even competing) participants[224] (natural or legal entities) who share a common interest, such as the success of the ecosystem (Corallo, 2007; McIntyre and Srinivasan, 2017; Wareham et al., 2014).[225] Ecosystem participants are autonomous actors.[226] This self-interest leads to the investment of ecosystem participants' resources, such as time and know-how, if the ecosystem participants can benefit from the value co-creation through value capture (Kude, et al., 2012; Rickmann et al., 2014).

---

[221] Tansley (1935, p. 299) defined an ecosystem as a "system (in the sense of physics), including not only the organism-complex, but also the whole complex of physical factors in the widest sense."

[222] It is important to note that several authors emphasized the necessity of providing meaning to the use of the ecosystem concept in non-ecologic contexts (e.g., Hyrynsalmi, Seppänen, et al., 2014; Ritala and Almpanopoulou, 2017; Suominen et al., 2019). Applying the ecosystem concept to a research area needs to provide additional insights and ensure consistent use of the concept (Hyrynsalmi and Hyrynsalmi, 2019, p. 2).

[223] Already before Moore's (1993; 1996) introduction of the ecosystem concept to the business literature, Hawley (1986) introduced the ecosystem concept in sociology. The author defined an ecosystem as an "arrangement of mutual dependencies in a population by which the whole operates as a unit and thereby maintains a viable environmental relationship" (Hawley, 1986, p. 26).

[224] Kapoor (2018, p. 6) described ecosystem participants as "[a]gents who undertake activities and produce the different offers" jointly, thereby forming the value proposition for the (final) customers.

[225] Ecosystem participants' shared objective is also reflected in Adner's (2017) ecosystem-as-a-structure perspective, which highlights the ecosystem participants' multilateral value-co creation.

[226] It is worthwhile mentioning that, compared with the focal platform owner, the ecosystem participants account for a majority of the value creation and innovations in an ecosystem (Delteil et al., 2020; Makinen and Dedehayir, 2012). However, the ecosystem participant' share of the value captured does not necessarily correspond to the same ecosystem participant's value-creating activities.

The biological and ecological metaphor ecosystem suggests extending the boundaries of a single company for strategic considerations (Adner and Kapoor, 2010; Dal Bianco et al., 2014), and emphasizes "the need for strategy to extend its consideration beyond rivals competing within industry boundaries" (Adner, 2017, p. 40–41).[227] Burkhalter et al. (2021, p. 264 ff.) described ecosystems as "allocentric business models" (Burkhalter et al., 2021, p. 264 ff.) that foster value co-creation beyond the organization's boundaries (Burkhalter et al., 2021, p. 264). Consequently, Kretschmer, et al. (2022, p. 407 ff.) described business ecosystems as "meta-organizations" positioning business ecosystem constellations between hierarchical, intra-firm structures and external markets. Moreover, the business ecosystem participants depend on each other for joint value creation (Blaschke and Brosius, 2018; Iansiti and Levien, 2004a).[228] Gossain and Kandiah (1998, p. 30) highlighted that the inclusion of supply-side participants (e.g., manufacturers, sellers) in business ecosystems "is the engine at the heart of the whole system." In a business ecosystem, companies form themselves into a competitive and cooperative collaboration, thus pursuing the goal of fulfilling customer needs as well as increasing innovation performance for new products (Corallo et al., 2007, p. 18). In this context, collaboration and competition form the basis for the success of a business ecosystem (Corallo et al., 2007, p. 19). Hence, the health of a business ecosystem is reflected by "a balance between co-operation and competition in a dynamic free market" (Stanley and Briscoe, 2010, p. 7). Other factors for the health or success of a business ecosystem are the ecosystem's robustness to changes in the environment, as well as the ability to handle the creation of new opportunities for new businesses (Korpela and Taipale, 2013, p. 3839).[229]

Aside from the biological and ecological metaphor, Iansiti and Levien (2004a, p. 50) emphasized that business ecosystems have fundamental differences compared to their biological counterparts. The authors elaborated that business ecosystems compete with others for external attention, such as customers' demands. Business ecosystems also need to compete for participants who can often choose between

[227] This is also supported by several scholars who emphasized a shift in competition from the company level to the ecosystem level (e.g., Amit and Zott, 2021; Jacobides et al., 2018; Rietveld and Schilling, 2021).

[228] In this regard, Moore et al. (2003) emphasized in analogy to biological ecosystems that "individual businesses cannot thrive alone—they must develop in clusters or economic ecosystems."

[229] Iansiti and Levien (2004a) described ecosystem health along the three parameters of productivity, robustness, and niche creation.

competing ecosystems or even join multiple ecosystems.[230] Moreover, Iansiti and Levien (2004a, p. 50) attribute intelligence in terms of the ability to anticipate future developments and planning to the participants of business ecosystems. Owing to their intertwined relations, the actions of a single ecosystem participant impact other participants and even the whole ecosystem (Adner and Kapoor, 2010). A business ecosystem therefore evolves dynamically and is subject to continuous change if, for instance, individual participants join and depart while creating new and interrupting existing relationships (Corallo et al., 2007; Wulfert and Schütte, 2022; Wulfert, Seufert, et al., 2021).[231] From this assumption it follows that the growth of an ecosystem and, thus, its value creation, depend to a large extent on its openness to new participants and, particularly, on the number of actors joining it (Adner, 2017).

Adner (2017) introduced two complementary perspectives for describing business ecosystems: the ecosystem-as-affiliation perspective and the ecosystem-as-structure perspective.[232] While the affiliation perspective is mainly concerned with the relationships between the individual ecosystem participants and their affiliation with a focal platform, the structural view of a business ecosystem focuses on the ecosystem's overarching purpose and an interdependent value proposition for a focal company's final customers. Making use of the concept of complementary assets introduced by Teece (1986), business ecosystems resemble value networks of independent participants multilaterally co-creating value that a single actor could otherwise not create. Following the ecosystem-as-structure perspective, Adner (2017, p. 42) stated that a business ecosystem is "defined by the alignment structure of the multilateral set of partners that need to interact in order for a focal value proposition to materialize."

Appropriate structures and the—not fully hierarchical—alignment of different participants (i.e., non-generic complementarities) are the basis of an ecosystem's value (Jacobides et al., 2018). While business ecosystems do not necessarily involve

[230] The phenomenon of being part of multiple business ecosystems simultaneously is described as "multihoming" in the literature (Idu et al., 2011; Koh and Fichman, 2014).

[231] Owing to increasingly interactive processes and global dependencies, even single actors have the power to influence the overall ecosystem (e.g., supply-chain shortages, shortage of specialists) (Schüßler et al., 2021).

[232] We will implicitly make use of the complementary nature of the ecosystem-as-affiliation perspective and the ecosystem-as-structure perspective focusing on affiliations with the focal platforms and network effects while also keeping in mind the ecosystem participants' value propositions (Adner, 2017).

a focal platform, ecosystems are likely to be formed around a focal actor.[233] In this regard, Iansiti and Levien (2004a) described the mutual dependencies of participants in business ecosystems and highlighted the focal participant's significance in business ecosystems.[234] The focal platform owner sets and enforces governance rules (e.g., access rights, fees, transaction structure) (Adner, 2017; Hein, Schreieck, Wiesche, et al., 2016). The configuration of the four characteristics of participants, activities, positions, and links characterize the necessary ecosystem structure for the realization of a value proposition (Adner, 2017, p. 43).[235] The participants comprise the ecosystem's population. The participants' activities jointly "materialize" (Adner, 2017, p. 40) the ecosystem's value for the (final) customer. The positions indicated the participants' location in creating the ecosystem value. The links describe transfers of tangible and non-tangible assets required for value co-creation. These transfers do not need to be done via the ecosystem's focal platform (Adner, 2017, p. 43).

In 1993, Moore (1993, p. 7) described that "[i]n a business ecosystem, companies co-evolve capabilities around a new innovation."[236] Following Moore's initial introduction to the business literature, research on ecosystems in the business literature has increased significantly, resulting in a number of IS research-related definitions (Hyrynsalmi and Hyrynsalmi, 2019, p. 4). We present pivotal business ecosystem definitions in the following.[237]

Jacobides et al. (2018, p. 2264) defined an ecosystem as "a set of actors with varying degrees of multilateral, non-generic complementarities that are not fully

---

[233] At this point, we would already like to highlight that we later on conceptualize e-commerce ecosystems that always surround a focal transaction platform (Wulfert and Karger, 2022b; Yi and Ming, 2011).

[234] The focal participant of an ecosystem is described as a "keystone firm" (Iansiti and Levien, 2004a). Other scholars denoted the focal participant as the "hub firm" (Nambisan, 2013), the "platform leader" (Gawer and Cusumano, 2008, 2014), the "ecosystem leader" (Moore, 1996), and the "platform owner" (Baldwin and Woodard, 2009).

[235] Adner (2017, p. 43) originally termed an ecosystem's participants as actors. For reasons of consistency, we will describe them as participants. The terms are used interchangeably throughout this thesis.

[236] Moore (1993, p. 7) further highlighted the relation of business ecosystems to innovations. He denoted that the participants of a business ecosystem "work cooperatively and competitively to support new products, satisfy customer needs, and eventually incorporate the new round of innovations."

[237] We provide an overview of ecosystem definitions in Appendix A.3 in the Electronic Supplementary Material. We also assigned the ecosystem types described by the authors' definitions. We differentiate between business ecosystems, platform-centered ecosystems, digital business ecosystems, and software ecosystems.

hierarchically controlled." Multilateral refers to the set of different participant types involved in an ecosystem and its value co-creation (Adner, 2017). Non-generic complementarities refer to platform extensions specifically provided by ecosystem participants for a common purpose (Jacobides et al., 2018, p. 2264).[238] The characteristic "not fully hierarchically controlled" (Jacobides et al., 2018, p. 2264) refers to the ecosystem participants' independence in providing the complementarities. Bogers et al. (2019, p. 4) defined an ecosystem as "an interdependent network of self-interested actors jointly creating value." The authors further emphasized that a defining characteristic of ecosystems is the independent actors' joint value co-creation that single actors would not be capable of creating (Adner, 2006). Business ecosystems can be described as cooperative forms of organization with multilateral relationships among the ecosystem participants (Petrik and Schüler, 2021, p. 97 ff.).[239] Following Iansiti and Levien (2004a, p. 35), features of a business ecosystem include fragmentation, interconnectedness, cooperation, and competition. Benedict (2018, p. 459 f.) detailed these features and described ecosystems along the following nine characteristics: asymmetric relationships, emergence and generativity, coevolution, cooperative behavior, competitive behavior, coopetition, openness, recombination, and self-organization.

While the aforementioned business ecosystem definitions and descriptions do not mention a dedicated focal ecosystem participant, other researchers already included a focal platform as a building block of business ecosystems (e.g., Ghazawneh and Henfridsson, 2015; Hein, Schreieck, Riasanow, et al., 2020; Iansiti and Levien, 2004a).[240] Kapoor (2018, p. 2) defined business ecosystems as "a set of actors that contribute to the focal offer's user value proposition." The authors emphasized the following: an explicit link between the supply side and the demand side, a joint value creation for the (final) customer, and different types of ecosystem participants involved in the value creation (Kapoor, 2018, p. 2). Moreover, Kapoor (2018, p. 3) also referred to "the simultaneous presence of complementarities and interdependencies between" ecosystem participants as the "main theoretical premise for ecosystem research" (Kapoor, 2018, p. 3) in the business context. While complementarities increase the value proposition for the (final) customer, interdependencies describe their value proposition's dependence on the focal platform and

[238] Initially, the concept of complements refers to innovation platforms and third-party developers who provide software-based platform extensions. In the case of e-commerce ecosystems, these complementarities can also be related to retail transactions, such as logistics services.

[239] Networks of organizations can also be characterized as cooperative forms of organizations. However, networks involve dyadic relationships among participating organizations (Petrik and Schüler, 2021, p. 97 ff.).

[240] We describe these business ecosystems as platform-centered ecosystems.

other ecosystem participant's complements (Kapoor, 2018, p. 3). Hein, Schreieck, Riasanow, et al. (2020, p. 90) defined platform-centered business ecosystems as consisting of "a platform owner that implements governance mechanisms to facilitate value-creating mechanisms on a digital platform between the platform owner and an ecosystem of autonomous complementors and consumers."

Following the seminal works by Iansiti and Levien (2004a); Iansiti and Levien 2004b, and Moore (1993, 1996), several authors have introduced additional and often overlapping ecosystem types (e.g., Benedict, 2018; Jacobides et al., 2018; Knodel and Manikas, 2015).[241] Knodel and Manikas (2015, p. 63 ff.) introduced four types of software ecosystems along the dimensions of leadership, structure, governance, and changeability. These ecosystem types are cornerstone ecosystems, standard-based ecosystems, protocol-based ecosystems, and infrastructure-based ecosystems. Seppänen et al. (2017, p. 5) identified 11 research communities as ecosystems. Benedict (2018, p. 455 ff.) differentiated seven ecosystem types along the dimensions of socio-technical systems and platform focus.[242] Jacobides et al. (2018, p. 2261) introduced three abstract types of value systems, with ecosystem-based value systems as the fundamental one. Faber et al. (2019, p. 4 ff.) identified 12 types of business ecosystems in a systematic mapping study. Hyrynsalmi and Hyrynsalmi (2019, p. 2 ff.) described 23 types of non-biological ecosystems. Among others, the authors included business ecosystems, innovation ecosystems, and platform-centered ecosystems. Based on a literature analysis, Guggenberger et al. (2020, p. 9 ff.) introduced five generic ecosystem types along the dimensions of population, ecosystem purpose, relationship structure, system configuration, and system dynamics. Guggenberger et al. (2020, p. 9) described sociocentric ecosystems, symbiotic collective ecosystems, centrally balanced ecosystems, orchestrating actor ecosystems, and structured resource-sharing ecosystems.[243]

---

[241] These ecosystem types go beyond our differentiation between business ecosystems, platform-centered ecosystems, digital business ecosystems, and software ecosystems.

[242] Benedict (2018, p. 455 ff.) described business ecosystems, service ecosystems, innovation ecosystems, digital business ecosystems, platform ecosystems, software ecosystems, and technology ecosystems. At this juncture, it must be noted, that the aforementioned ecosystem types do not involve a clear distinction with possible overlaps across the seven ecosystem types.

[243] While innovation platforms can form the center of the centrally balanced ecosystems, transaction platforms can be found in sociocentric ecosystems and orchestrating actor ecosystems.

Jacobides et al. (2018, p. 2260 ff.) differentiated business ecosystems from hierarchy-based (intra-company) value systems and market-based (inter-company) value systems. In business ecosystems, the final customer consumes the platform owner' product[244] and can individually augment the product with selected complements provided by other ecosystem participants. The ecosystem structure with individualized products based on end customers' free selection of complements requires "significant" (Jacobides et al., 2018, p. 2060) coordination. Jacobides et al. (2018, p. 2261) argued for different types of complementarities that amplify the need for coordination. The authors focus on two types of complementarities, namely unique complementarities and supermodular complementarities (Jacobides et al., 2018, p. 2261 f.). Jacobides et al. (2018, p. 2261 f.) described unique complementarities as dependency relationships between two items or activities. Unique complementarities can be described on a continuum between strict (item A requires item B) (Hart and Moore, 1990, p. 1135) and specific (item A needs a customized item B to be productive) (Jacobides et al., 2018, p. 2261 f.).[245] Supermodular complementarities are summarized as relationships maximizing the value between two items or activities (more of item A makes item B more valuable) (Milgrom and Roberts, 1990; Topkis, 1998). Supermodular complementarities can be found in production and consumption relationships (Jacobides et al., 2018, p. 2262).[246] In a consumption relationship, supermodular complementarities form the basis for direct and indirect network effects (Armstrong, 2006; Farrell and Saloner, 1985; Parker and Van Alstyne, 2005).[247] Building upon their conceptualization of complementarities, Jacobides et al. (2018, p. 2263) defined ecosystems as "groups of firms that must deal with either unique or supermodular complementarities that are non-generic, requiring the creation of a specific structure of relationships and alignment to create value." In Fig. 3.23, we present the ecosystem-based value system conceptualized by Jacobides et al. (2018, p. 2261).

[244] Jacobides et al. (2018, p. 2261) described the platform owner as the focal firm.

[245] This continuum of dependency relationships between two items or activities can even be expanded to generic (item A can use item B without any customization) (Helfat and Lieberman, 2002). Jacobides et al. (2018, p. 2262) provided the example of electricity having generic complementarities with activities such as water boiling or heating.

[246] Supermodular complementarities are also described as Edgeworth complementarities in the literature (Milgrom and Roberts, 1990; Weber, 2005).

[247] Direct and indirect network effects are discussed in further detail in section 3.2 in the context of transaction platforms.

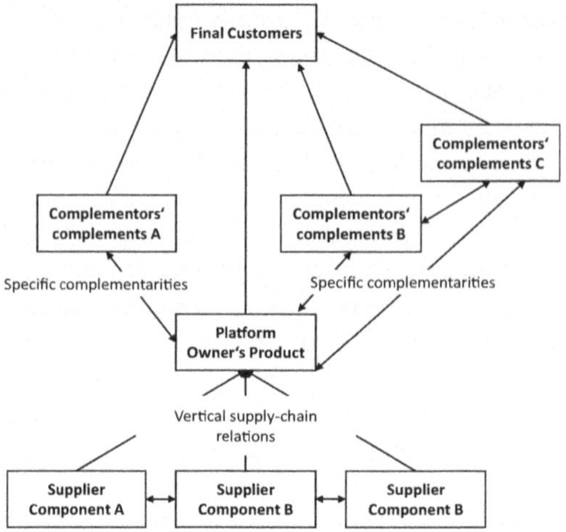


**Fig. 3.23** Ecosystem-Based Value System (Jacobides et al., 2018, p. 2261)

Following their definition of platform-centered digital business ecosystems, Hein, Schreieck, Riasanow, et al. (2020, p. 90 ff.) suggested three ecosystem building blocks, namely platform ownership, value-creating mechanisms, and complementor autonomy.[248] Hein, Schreieck, Riasanow, et al. (2020, p. 90 f.) summarized platform ownership as the entity owning the focal platform, with the platform owner wielding the distribution of power in the ecosystem. Hence, the authors identified single-ownership, consortium, and peer-to-peer ecosystems as potential platform owner archetypes.[249] Value-creating mechanisms make use of on the two major types of platforms in digital business ecosystems, i.e., innovation platforms and transaction platforms.[250] Following Hein, Schreieck, Riasanow, et al. (2020, p. 91

[248] Although Hein, Schreieck, Riasanow, et al. (2020, p. 91) argued that customers as demand-side participants in digital business ecosystems are not building blocks of the ecosystem, we indicated them as building blocks as ecosystems are likely to involve multiple different customer types who influence the entire ecosystem's configuration with their demand.

[249] The concentration of power among actors can be described as differentiating dimension for the platform ownership building block.

[250] Hein, Schreieck, Riasanow, et al. (2020) did not explicitly mention the possibility of providing both types of value-creating mechanisms simultaneously in hybrid platform constellations (e.g., Gawer, 2021b; Wulfert and Karger, 2022b). However, it is conceivable to have both types of value-creating mechanisms for a single company (e.g., Amazon, Alibaba). Please refer to section 3.2 for more information on the platform types.

f.), the value creation in digital business ecosystems relies either on the facilitation of transactions in multi-sided markets (Armstrong, 2006; Rochet and Tirole, 2003) or on the fostering of innovations made possible by a super-modular innovation platform (Tiwana, Konsynski, and Bush, 2010). In Fig. 3.24, we depict the adapted model originating from Hein, Schreieck, Riasanow, et al. (2020, p. 91).

Tsujimoto et al. (2018) introduced an ecosystem model integrating previous research streams identified in a literature review. The model integrates four perspectives on ecosystems, namely industrial ecology, business ecosystems, multi-actor networks, and platform management. The industrial ecology perspective is concerned with organizations and their supply chains (Tsujimoto et al., 2018, p. 51 f.). The focal organization with its supply chain of producers and customers is treated as a network and transformed "into a more integrated model: an industrial ecosystem" (Frosch and Gallopoulos, 1989, p. 144). The business ecosystem perspective and multi-actor network perspective describe types of ecosystem participants (Tsujimoto et al., 2018, p. 52 ff.). The business ecosystem perspective is grounded in organizational boundary theory (e.g., Santos and Eisenhardt, 2005) and involves ecosystem participants adjacent to the focal organization under consideration.[251] The multi-actor network perspective includes "the dynamic networks among actors with different attributes from private firms" (Tsujimoto et al., 2018, p. 54). Tsujimoto et al. (2018, p. 53 ff.) identified policy makers (e.g., Fabrizio and Hawn, 2013), consortia (e.g., Borgh et al., 2012; Leten, et al., 2013), complementors (e.g., Rohrbeck et al., 2009), entrepreneurs (e.g., Autio et al., 2018; Clarysse et al., 2014; Samila and Sorenson, 2010), and users (e.g., Hienerth et al., 2014; Khavul and Bruton, 2013; Ramachandran et al., 2012) as additional ecosystem participants from the multi-actor network perspective. The platform management perspective is concerned with managing the diverse ecosystem participants based on multi-sided market mechanisms (Armstrong, 2006; Gawer, 2014; Rochet and Tirole, 2003) and establishing potential leadership positions (Cusumano and Gawer, 2002; Gawer and Cusumano, 2008).[252] In Fig. 3.25, we depict the ecosystem constructs as introduced

[251] In section 3.3, we detail our elaboration on organizational boundaries in general and the seminal work by Santos and Eisenhardt (2005) in particular.

[252] Although Tsujimoto et al. (2018, p. 54) modeled the platform management only for private firms, such as suppliers, complementors, buyers, and competitors, adjacent to the focal organization, we would argue that platform management should consider the overall ecosystem with its participants also including these participants derived from the multi-actor network perspective. These peripheral ecosystem participants need to be governed in a holistic ecosystem approach as well (Gawer, 2021a; Staykova and Damsgaard, 2015; Wulfert and Karger, 2022a).

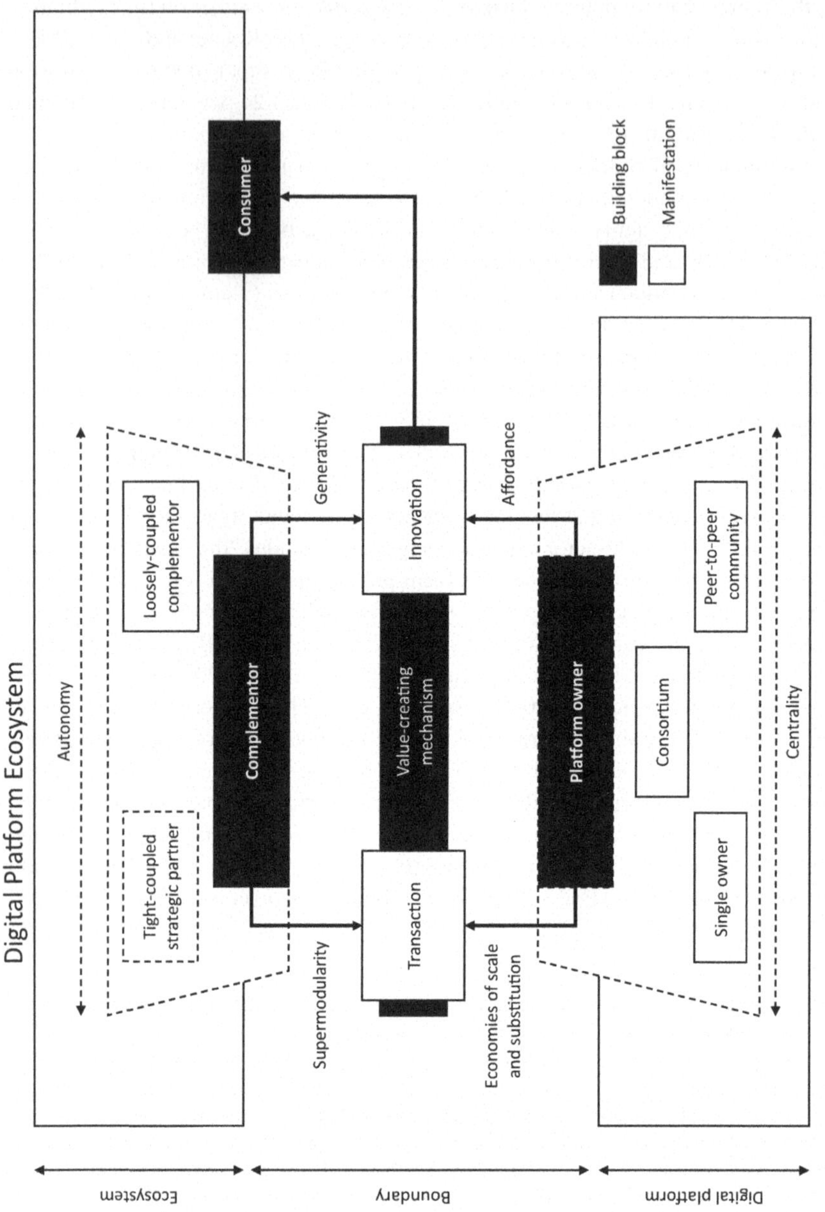


**Fig. 3.24** Building Blocks of Digital Business Ecosystems (Hein, Schreieck, Riasanow, et al., 2020, p. 91)

by Tsujimoto et al. (2018, p. 54), with the four above-mentioned perspectives on ecosystems printed in bold.

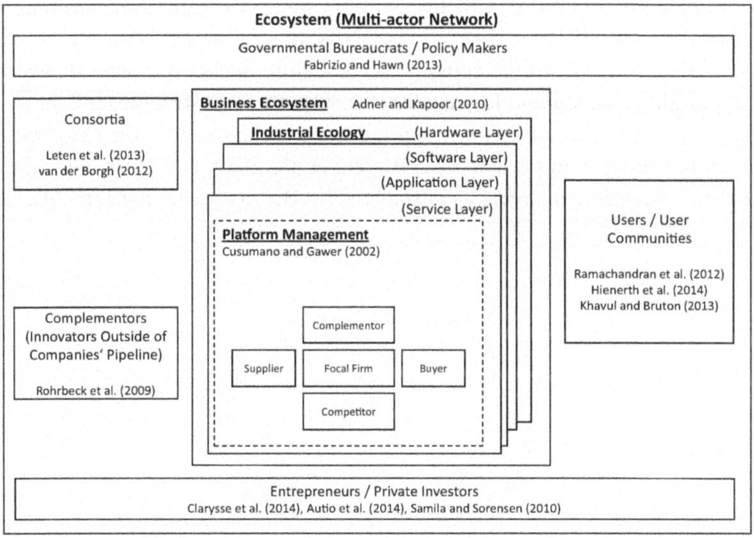


**Fig. 3.25** Integrated Model of Previous Ecosystem Research (Tsujimoto et al., 2018, p. 54)

From a literature analysis of research papers published in the field of economics, Bogers et al. (2019, p. 12 ff.) identified three constructs and the relations among them forming business ecosystems. The authors identified self-interested goals of participants, interdependence between participants, and the network of participant relationships as constructs forming business ecosystems (Bogers et al., 2019, p. 13 ff.).[253] Participants' self-interested goals emphasize that ecosystem participants have a common purpose and co-create value (Adner, 2017) while pursuing their own goals and implementing measures to capture their share of the value created (Chesbrough et al., 2018; Ritala, Agouridas, et al., 2013). Orchestrating the different goals of different ecosystem participant types (Eisenmann, Parker, et al., 2006) might be challenging for the owner of the focal platform, especially with two-sided market configurations in place (Bogers et al., 2019, p. 13). Ecosystem participants

[253] In their research paper, Bogers et al. (2019) described the ecosystem participants as (ecosystem) members. However, for reasons of consistency, we continue to use the term ecosystem participants throughout this cumulative dissertation.

depend upon each other for creating value for the (final) customer (Jacobides et al., 2018). Often, ecosystem participants only provide a defined proportion of the final product or service. The interdependence requires that single participants' interests are coordinated and controlled in line with the the ecosystem's overarching purpose (Bogers et al., 2019, p. 14).[254] The ecosystem participants are organized in a network that define both the ecosystem's structure and governance (Gawer and Henderson, 2007; Iansiti and Levien, 2004a). The network structure influences the complementary product and service configuration and prescribes the resource allocation among ecosystem participants (Bogers et al., 2019, p. 14). In Fig. 3.26, we depict the ecosystem constructs as introduced by Bogers et al. (2019, p. 24).

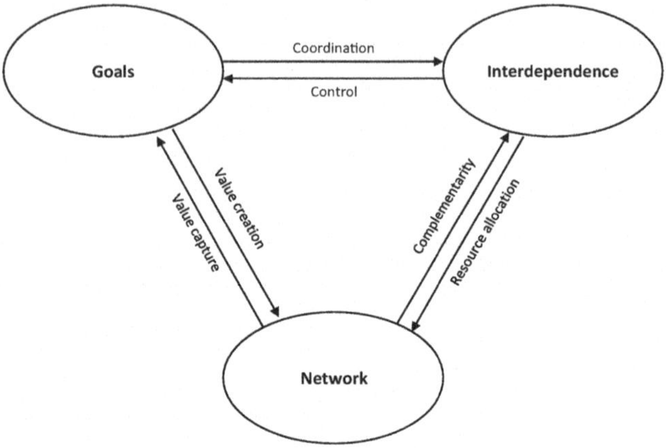


**Fig. 3.26** Constructs within the Ecosystem Concept (Bogers et al., 2019, p. 24)

### 3.4.1.2 Digital Business Ecosystem

Over time, the need for a new understanding of the business ecosystem has emerged and various research groups have begun to work on the topic of the digital business ecosystem (Razavi et al., 2010, p. 293). Based upon the aforementioned introduction of business ecosystems, we will focus on digital business ecosystems in which value

[254] Adner (2017, p. 40 ff.) described this overarching purpose in his ecosystem-as-a-structure perspective.

is created partially or fully via digital means (Baumann, 2022b; Nachira, 2002).[255] The digital business ecosystem concept was first described by the European Union (EU) in their Lisbon Agenda in 2000 (Khalil, et al., 2011; Nachira, 2002).[256] Nachira (2002, p. 5) synthesized the concept of business ecosystems as established by Moore (1993, 1996) with digital possibilities.

Digital business ecosystems form a digital environment that is populated by a diverse set of participants augmented with "digital species" (Nachira, 2002, p. 12).[257] Thus, it can be stated that digital business ecosystems are a combination of the business ecosystem and the support of technical infrastructure in a digital world (Nachira et al., 2007, p. 5).[258] Power and Jerjian (2001, p. 16) emphasized the integration of ecosystem participants and their real products and services in digital transaction environments in digital business ecosystems. Consequently, the digital business ecosystems' participants interact and co-create value, using IT (Tsujimoto et al., 2018). The participants in digital business ecosystems remain independent, but cooperate to achieve common objectives while competing for scarce resources (Corallo, 2007). Hence, the characteristic of digital business ecosystems is that they represent a loosely coupled and on-demand driven environment (Chang et al., 2006; Razavi et al., 2010). Thus, there is a free and open relationship between the participants (Chang et al., 2006, p. 34). A digital business ecosystem is a complex network of platform-mediated actor-to-actor interactions, becoming increasingly accessible to end users through third parties' platform derivatives (Corallo, 2007; Wareham et al., 2014).[259]

---

[255] Moore et al. (2003) even introduced the concept of the digital business ecosystem using IT in developing countries to amplify sustainable socio-economic development and poverty alleviation.

[256] The EU piloted the concept of digital business ecosystems in two framework programs to increase efficiency in the communication among small and medium companies (Korpela and Taipale, 2013, p. 2).

[257] The term digital species is broadly understood and includes a variety of things, such as "[…] software components, applications, services, knowledge, business models, training modules, contractual frameworks, and laws" (Nachira, 2002, p. 12).

[258] Nachira et al. (2007, p. 5) further explained that the technical infrastructure is also described in the literature as a digital ecosystem (Nachira et al., 2007, p. 5). However, from our point of view, this description seems misleading, as the technical infrastructure with its hardware components has a physical representation. They provide the foundation for digital environments, but are not fully digital in nature.

[259] Despite the presented characteristics of the introduced digital business ecosystems and the definitions summarized in Appendix A.3 in the Electronic Supplementary Material, we emphasize that business ecosystems already entail digital components in the course of increasing digitalization on the societal, organizational, and individual levels. Moreover, business

Digital business ecosystem "can be defined as a network of organizations such as manufacturers, suppliers, distributors, customers, competitors, and government agencies, that together create and deliver a specific product or service in a partly or fully digital environment" (Baumann, 2022b, p. 1).[260] Tsujimoto et al. (2018, p. 52) described digital business ecosystems as "IT-based complex relationships among firms in IT industries" (Tsujimoto et al., 2018, p. 52), analyzing them with a focus on the management of the relationships among organizations. Besides their additional interaction in a digital environment, digital business ecosystems inherit the basic characteristics of traditional business ecosystems. As such, participants of digital business ecosystems form a "non-homogeneous community of self-interested entities that [...] depend on each other for their survival" (Baumann, 2022b, p. 1). Information can be such a valuable and scarce resource in a digital business ecosystem in which ecosystem participants interact using IT (Wunck and Baumann, 2017, p. 2). The ecosystem is subject to constant evolution caused by participants joining and departing, creating new and releasing existing relations among participants (Briscoe and De Wilde, 2006; Wulfert, Seufert, et al., 2021).

#### 3.4.1.3 Software Ecosystem

Although we emphasized that business ecosystems can involve virtual participants and conduct transactions via digital media, we introduce the software ecosystem as a subset of digital business ecosystems (Jansen and Cusumano, 2013, p. 17) that is often described in the literature (Manikas and Hansen, 2013, p. 1298). The term software ecosystem was first coined by Messerschmitt and Szyperski (2003). In line with the characterization of business ecosystems (e.g., Adner, 2017; Jacobides et al., 2018; Moore, 1993), Jansen, Finkelstein, et al. (2009, p. 9) emphasized that the software ecosystem's participants pursue a common objective (i.e., increasing the ecosystem value with additional software complements) and have sophisticated relationships among each other. Messerschmitt and Szyperski (2003, p. 13) referred to a software ecosystem[261] as a "collection of software products that have some given

---

ecosystems linking independent participants allowing for collaboration and coopetition, create an overall solution for the (final) customer that no single participant could have achieved on its own. Compared to the individual participants' value creation, the overall ecosystem's value creation already represents an innovation. For this reason, it makes little sense to distinguish artificially between digital business ecosystems and business ecosystems.

[260] In contrast, Tsujimoto et al. (2018, p. 52) noted that there "is no consensus on the definition of a digital [business] ecosystem."

[261] Instead, other authors prefer the term innovation ecosystem emphasizing the common provision of software artifacts and extensions based on a common technological platform as the overarching objective of the ecosystem (Adner, 2006; Autio, 2022). Although there only

degree of symbiotic relationships." Following Selander et al. (2013), Ghazawneh and Henfridsson (2015, p. 200) defined software ecosystems centered around a focal innovation platform as a "collective of organizations having a common interest in the prosperity of a digital platform for leveraging their application development." Bosch (2009, p. 112) defined a software ecosystem as "consisting of the set of software solutions that enable, support, and automate the activities and transactions by the actors in the associated social or business ecosystems and the organizations that provide these solutions."[262] From these definitions, Jansen and Cusumano (2013, p. 17 f.) carved out common themes, namely ecosystem participants, relations among actors, and the software at the core of the ecosystem. These themes are used by Jansen, Finkelstein, et al. (2009, p. 188 f.) to conceptualize software ecosystems, and integrated by Jansen and Cusumano (2013, p. 18) in their definition on software ecosystems. The authors defined software ecosystems as "a set of actors functioning as a unit and interacting with a shared market for software and services, together with the relationships among them. These relationships are frequently underpinned by a common technological platform or market and operate through the exchange of information, resources, and artifacts" (Jansen and Cusumano, 2013, p. 18).[263] In analogy to business ecosystems, the participants have a "positive influence on the economic success" (Kittlaus and Clough, 2009, p. 25) of the overall software ecosystem. The relationships among participants and their extension developments are "frequently underpinned by a common [innovation] platform [...] and operate through the exchange of information, resources, and artifacts" (Jansen, Finkelstein, et al., 2009, p. 18). Manikas and Hansen (2013, p. 1297) stressed that increased ecosystem generativity enabled through a common technological platform "results in a number of software solutions or services."

---

exists a vague delimitation between the terms, Benedict (2018) highlights, that innovation ecosystems do not necessarily evolve around a focal platform. A common focal innovation platform, to which the ecosystem participants are affiliated, is inherent in the software ecosystem. Since we follow a platform-centered perspective in this thesis, we use the term software ecosystem.

262 In addition to the definition, Bosch (2009, p. 111 ff.) differentiated three types of software ecosystem. While operating system-centric software ecosystems emerge around operating systems, application-centric software ecosystems build upon dedicated applications. End-user programming software ecosystems rely on domain-specific programming languages.

263 It should be noted here that Jansen and Cusumano (2013, p. 20) emphasized the role of transaction platforms as extension markets for exchanging third-party developers' extensions in the software ecosystem. Hence, software ecosystems include not only a focal innovation platform as a technological basis upon which third-party complements are developed (Gawer and Cusumano, 2002; Tiwana, Konsynski, and Bush, 2010), but also a transaction platform for the exchange of these complements.

Moreover, software ecosystems often feature a transaction platform for exchanging extensions developed by third-party developers (Dal Bianco et al., 2014; Jansen, Finkelstein, et al., 2009). As described in section 3.2.2, innovation platforms can be the center of larger software ecosystems, and can therefore form the foundation for third-party developers' complements and products (Jansen, 2013; Tiwana, Konsynski, and Bush, 2010). The high degree of generativity in software ecosystems leads to an "unpredictable development of the whole ecosystem for all actors" (Benedict, 2018, p. 459). Despite a focal participant's power over other ecosystem participants, Tiwana (2015a, p. 269) states that developers in a software ecosystem follow an "ecosystem-wide vision." This vision is either mutually co-developed by the majority of participants or defined by the focal participant.

Iansiti and Levien (2004a, p. 81 ff.) described two major roles with related strategies as building blocks of software ecosystems. The hub controls the focal innovation platform in a software ecosystem. The hub can either follow a keystone or a dominator strategy (Iansiti and Levien, 2004a, p. 82 ff.). While a keystone shares value with other ecosystem participants and enables them to co-create value, a dominator "seeks to extract as much value from the ecosystem" (Van den Berk et al., 2010, p. 128) for itself. Besides the hub, the ecosystem is made up of niche players. A niche player "acts to develop or enhance specialized capabilities that differentiate it from other firms in the network, leveraging resources from the network while occupying only a narrow part of the network itself. Niche players individually do not have broad-reaching impacts on other ecosystem participants, but collectively they constitute the bulk of the ecosystem both in terms of total mass as well as variety" (Iansiti and Levien, 2004a, p. 90). Most software ecosystem participants can be described as niche players who do not compete with the hub, but contribute to the overall ecosystem purpose by adding value in a specialized area, with capabilities and solutions (Kittlaus and Clough, 2009, p. 25 ff.). Referring to Hagel, Brown, et al. (2008, p. 5), Van den Berk et al. (2010, p. 128) further differentiated niche participants into influencer, hedger, and disciple. Influencers support and commit early to the focal platform owner's strategy. Hedgers apply a multi-homing strategy developing their extensions for multiple platforms (Idu et al., 2011; Koh and Fichman, 2014). In contrast, disciples adhere to a single focal platform (i.e., single-homing) (Van den Berk et al., 2010, p. 128). We depict the ecosystem participants and the relationships among these participants in software ecosystems, described by Iansiti and Levien (2004a) following Van den Berk et al. (2010, p. 128), in Fig. 3.27.

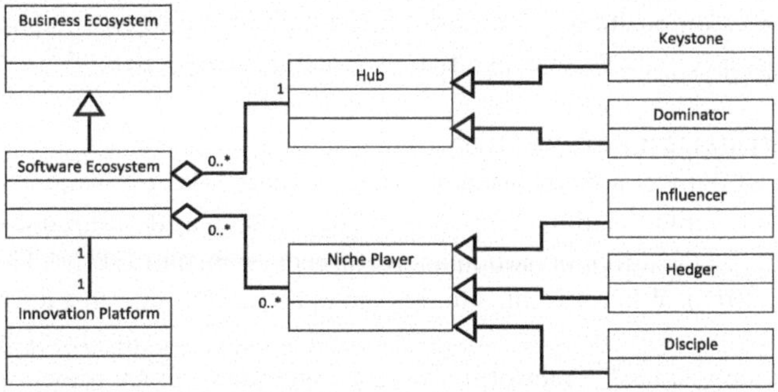


**Fig. 3.27** Relations in Software Ecosystems (van den Berk et al., 2010, p. 128)

In the following section, we focus on e-commerce ecosystems as digital business ecosystems that surround a focal transaction platform in e-commerce. These transaction platforms provide the virtual environment for the ecosystem participants to conduct retail transactions (Wulfert, Woroch, Strobel, Seufert, et al., 2022). These ecosystems are likely to involve software ecosystems fostered by focal innovation platforms that provide the necessary application services for conducting e-commerce transactions (Aulkemeier, Paramartha, et al., 2016).

### 3.4.2 Platform-Centered Ecosystem in E-Commerce

Considering the aforementioned explanations on e-commerce and (digital) business ecosystems,[264] we conceptualize e-commerce ecosystems as digital business ecosystems in the context of e-commerce (Wulfert, 2023; Wulfert and Dennhardt, 2023; Wulfert, Woroch, Strobel, Seufert, et al., 2022; Yi and Ming, 2011).[265] The unfolding of e-commerce ecosystems is structured along Adner's (2017, p. 40

[264] We intentionally wrote the term "digital" in brackets because of the aforementioned blurry distinction between business ecosystems and digital business ecosystems (e.g., Jacobides et al., 2018; Moore, 1993; Moore et al., 2003).

[265] E-commerce ecosystems apply the concept of digital business ecosystems, as transactions are supported by electronic means or even conducted digitally in e-commerce (Choi et al., 1997; Schütte and Vetter, 2016; Zwass, 1996).

f.) perspectives on business ecosystems, differentiating between an ecosystem-as-affiliation perspective and an ecosystem-as-structure perspective.[266]

#### 3.4.2.1 Affiliation Perspective

From an ecosystem-as-affiliation perspective, e-commerce ecosystems can be structured by a "set of roles" (Adner, 2017, p. 42). These roles may describe a standardized relation between the different participants of an ecosystem on an abstract level (Jacobides et al., 2018). While Eisenmann, Parker, et al. (2009) introduced four generic types of participants in business ecosystems—namely participants from the demand sides and supply sides[267], platform providers, and platform sponsors—as generic participants of a business ecosystem, these roles were further specified for the context of retail by Böttcher, Rickling, et al. (2021). Böttcher, Rickling, et al. (2021, p. 5) introduced ten generic roles adopted by participants of business ecosystems in the context of brick-and-mortar retail. The authors described retailers, IT solution providers, service providers, initiatives, producers, data suppliers, researchers, logistics, and consumers (Böttcher, Rickling, et al., 2021, p. 5 f.).[268] E-commerce ecosystems share many participant types with physical retail environments (e.g., manufacturers, customers) but also involve a variety of additional participants (e.g., content providers, software providers, third-party developers) (Böttcher, Rickling, et al., 2021; Schütte and Wulfert, 2022a; Wulfert, Woroch, Strobel, Schoormann, et al., 2023). Focal platforms in e-commerce ecosystems match a number of participants based on customers' article demands and enable retail transactions between them. Focal platforms can also behave competitively toward other ecosystem participants (i.e., dual role) (Wulfert and Schütte, 2022).

In the following, we provide two exemplary ecosystem depictions that apply a layered description of different participants affiliated with an e-commerce ecosystem.[269] Yi and Ming (2011) structured e-commerce ecosystems along a core layer, a support layer, a derivative layer, and an environment (Fig. 3.28). The core

---

[266] While the former perspective highlights the affiliated participants and their isolated activities, the latter perspective emphasizes the overall ecosystem's value proposition and multilateral value co-creation (Adner, 2017, p. 40 f.).

[267] In the context of e-commerce, demand- and supply-side participants resemble the customers and (third-party) sellers (Wulfert, Woroch, Strobel, Seufert, et al., 2022).

[268] Böttcher, Rickling, et al. (2021, p. 5) further differentiated retailers into online retailers, branch retailers, and multi-channel retailers. While online retailers only sell their products online, multi-channel retailers obtain several sales channels involving online and offline activities (Verhoef, Lemon, et al., 2009).

[269] Ecosystem participants affiliate with the focal transaction platform and engage within the ecosystem "purely out of self-interest" (Engert et al., 2022, p. 4).

layer includes buyers, sellers, and the focal transaction platform. The support layer involves additional participants supporting e-commerce transactions such as payment providers and logistics service providers. The derivative layer is concerned with technology and infrastructure services providing the necessary application services.[270] The environment describes the e-commerce ecosystem's political, economic, and social surroundings.

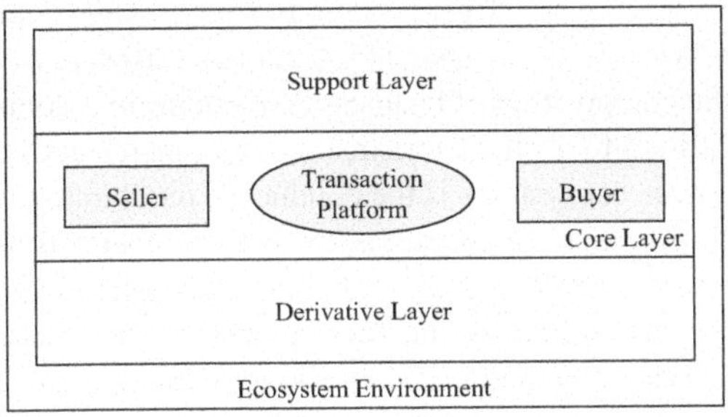


**Fig. 3.28** Composition of E-Commerce Ecosystems (according to Yi and Ming, 2011)

Gui et al. (2018, p. 616) structured e-commerce ecosystems in five layers. These layers are termed the core layer, the key layer, the correlated layer, the extended layer, and the environmental layer. Corresponding to these layers, Gui et al. (2018, p. 616) described the e-commerce ecosystem participants as the core population, the key population, the associated population, and the extended population. While the core layer consists of the focal platform with affiliated customers and third-party sellers, the key layer involves service providers, such as information services, logistic service providers, IT technology services, and marketing services, supporting the conduct of retail transactions.[271] The correlated layer involves ecosystem participants "who have carried out business cooperations" (Gui et al., 2018, p. 616) with participants from the core layer and key layer and helped "improve their service levels and capabilities" (Gui et al., 2018, p. 616). The authors mentioned, among others,

[270] Although Yi and Ming (2011, p. 208) did not further specify this layer and only provide "industry associations, Internet marketing services, technology outsourcing business, e-commerce consulting services," it seems to include innovation platforms with adjacent third-party developers.

[271] Gui et al. (2018, p. 616) further differentiated service providers in the key layer into transaction service providers, supporting service providers, and derivative service providers. However, the authors did not further specify the three types of service providers, and they did they also did not assign exemplary service providers to the three categories

financial institutions, media platforms, and shopping guides as examples of the associated population. The extended layer involves organizations that are not involved in conducting (retail) transactions but that "can affect the growth, maturity, and profitability of the" (Gui et al., 2018, p. 616) e-commerce ecosystem. For instance, the authors mentioned research institutions, industrial associations, and legal departments as part of the extended population. The environmental layer is described as the political, social, and economic environment in which the e-commerce ecosystem participants operate (Gui et al., 2018, p. 617). In Fig. 3.29, we depict the affiliation of e-commerce ecosystems as suggested by Gui et al. (2018, p. 616).

In addition to the composition of business ecosystems in e-commerce, Gui et al. (2018, p. 615) identified four characteristics of e-commerce ecosystems. First, the purpose of e-commerce ecosystems is the conduct of retail transactions and related activities. Second, e-commerce ecosystems involve a number of different participant types. Third, the e-commerce ecosystem relies on "value sharing and co-evolution" (Gui et al., 2018, p. 616). Fourth, the focal platform coordinates and organizes resource sharing, service complementation, and joint innovation.

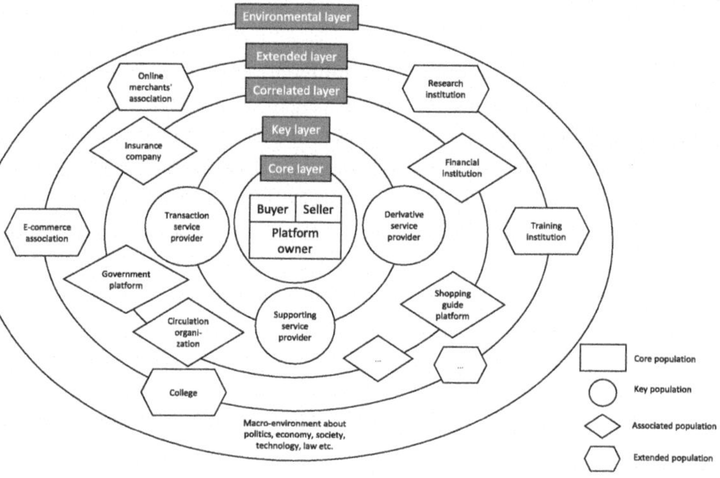


**Fig. 3.29** Composition of E-Commerce Ecosystems (according to Gui et al., 2018, p. 616)

#### 3.4.2.2 Structure Perspective

From an ecosystem-as-structure perspective (Adner, 2017), the participants of e-commerce ecosystems aim at the exchange of physical or digital products via electronic communication media, with of the (final) customer as overarching purpose (Becker and Schütte, 2004; Laudon and Traver, 2019; Wulfert, 2023). The overall e-commerce ecosystem fulfills the necessary functions, bridging the discrepancies in time, space, and quantity from (raw material) suppliers to final customers (Becker and Schütte, 2004; Laudon and Traver, 2019; Schütte, 2017). The ecosystem participants create value for the final customer through the creation of complementarities, through resource sharing, and through exchanging funding, capital, and information benefiting from and aggregating each other's core competencies (Yi and Ming, 2011, p. 207). Each of the ecosystem participants performs specific activities that materialize the ecosystem value proposition when accumulated. E-commerce ecosystems evolve dynamically. This is caused by participants joining and departing, creating new and breaking existing relationships.[272] Existing participants continuously alter their product assortment and service portfolio, further affecting established transaction relationships. Moreover, the market positions of single e-commerce ecosystem participants can evolve dynamically, continuously altering the relationships between participants and the ecosystem's structure (Kawa and Walesiak, 2019; Wulfert, Seufert, et al., 2021).

Since the ecosystem's overarching value proposition for the (final) customer is pivotal for the ecosystem-as-structure perspective, Adner (2017, p. 43) highlighted that the value proposition also needs to "materialize."[273] E-commerce ecosystems are usually formed around a focal transaction platform that provides a virtual environment for the conduct of retail transactions (Kawa and Walesiak, 2019; Turban, Whiteside, et al., 2017; Wulfert, 2023; Wulfert, Woroch, Strobel, Seufert, et al., 2022). The focal transaction platform orchestrates the different types of e-commerce ecosystem participants, such as manufacturers, sellers, logistic services providers, and customers (Böttcher, Rickling, et al., 2021; Wulfert, Woroch, Strobel, Schoormann, et al., 2023). The focal platform owner defines the alignment structure of e-commerce ecosystems by establishing governance rules to control access to the ecosystem (Blaschke and Brosius, 2018; Hein, Schreieck, Riasanow, et al., 2020; Hein, Schreieck, Wiesche, et al., 2016; Ofe and Sandberg, 2023), and has sovereignty

[272] Adner (2017, p. 44) also described these relationships as "links" between the ecosystem participants. For instance, material, information, influence, and funds can be transferred via these relationships.

[273] The materialization of the value proposition is the prerequisite for the platform owner and ecosystem participants involved in a transaction to capture value in the form of fees. The value proposition of ecosystem participants is summarized in Table 3.5.

over the ecosystem via orchestration measures (Maruping and Matook, 2020; Rickmann et al., 2014; Wulfert, Woroch, and Strobel, 2024). The focal platform owner also needs to keep track of the multilateral relations among the e-commerce ecosystem participants to ensure efficient resource utilization and increase value generation.

Following developments in customer convenience and increased competition, e-commerce ecosystem participants in general and retailers in particular augment products with additional services and provide additional non-product-related services (Grewal, Roggeveen, et al., 2017; Muschkiet et al., 2023; Schütte and Vetter, 2016; Schütte, 2017; Turban, Whiteside, et al., 2017). This shift from a goods-dominant logic to a service-dominant logic (Burkhalter et al., 2021; Vargo and Lusch, 2004, 2008) allows for the involvement of multiple ecosystem participants in the value creation process (Adner, 2017; Akana, 2020). Making use of research on the service-dominant logic (e.g., Vargo and Lusch, 2004, 2008; Vargo, Maglio, et al., 2008), it transpires that the (final) customer is (more or less) involved in the process of value creation. In the context of transaction platforms in e-commerce, Zwass (2010, p. 13) defined co-creation as "the participation of consumers along with producers in the creation of value in the marketplace" and differentiated between autonomous co-creation and sponsored co-creation Zwass (2010, p. 21 ff.).[274] Co-created goods are mostly digital, non-rival, and easily excludable (Zwass, 2010, p. 14). Moreover, the platform owner benefits from the customers' integration in the value creation through additional information, such as electronic word of mouth (Hennig-Thurau, Gwinner, et al., 2004, p. 39), ratings, and sentiments (Aulkemeier, Daukuls, et al., 2016; Bundeskartellamt, 2015; Wulfert, Betzing, et al., 2019).[275] For the platform owner, "orchestrating value co-creation implies continuously collecting, processing, filtering, sharing and updating the most relevant aspects of (infinite) digitized information with other actors of the ecosystem" (Burkhalter et al., 2021, p. 264). Grönroos (2012, p. 1520) defined value co-creation "as the joint actions by a customer (or another beneficiary) and a service provider during their direct interactions."[276] Following Burkhalter et al.'s (2021, p. 264) involvement of "other

[274] While autonomous co-creation activities, such as the production of declarative content, trust creation in reputation systems, and word-of-mouth production, are independent of the supplying organization (Zwass, 2010, p. 22 ff.), the customer is integrated into the producers' value chain (e.g., ideation, product co-design, product testing, product promotion) in sponsored co-creation (Zwass, 2010, p. 25 ff.).

[275] Value co-creation requires application services for customers to participate (Zwass, 2010, p. 11).

[276] It is important to emphasize that, compared to Vargo and Lusch (2004, 2008); Vargo, Maglio, et al. (2008), "another beneficiary" Grönroos (2012, p. 1520) can also involve addi-

actors" and Grönroos' (2012, p. 1520) description of other beneficiaries, we take the concept of value co-creation more broadly as referring to the collaboration of multiple ecosystem participants for a joint value creation (Adner, 2006, 2017; Wulfert, Woroch, Strobel, Seufert, et al., 2022).[277] The customer is not necessarily involved in creating the value but experiences the "value in use" (Lim, Kim, et al., 2018, p. 126). The broader understanding of value co-creation in the e-commerce ecosystem is supported by Akaka and Lusch (2012, p. 14 f.), and involves "multiple actors" in the value creation process. In the context of (e-commerce) ecosystems, Burkhalter et al. (2021, p. 264) described value co-creation beyond the focal platform owner's boundaries.

With regard to the ecosystem-as-structure perspective, e-commerce ecosystems are highly dynamic in terms of affiliated participants and the available product and service assortment. This results from participants joining and departing from the ecosystem, and also from product assortments adapted to environmental changes and altering customer demands (Wulfert, Seufert, et al., 2021).[278] Adaptations in the participant structure and service offering may also affect network effects present within the e-commerce ecosystem. Another consequence of the dynamic nature of e-commerce ecosystems is the existence of fuzzy ecosystem boundaries (Gao, Han, et al., 2019, p. 143). The focal platform owner needs to monitor environmental and economic developments (Wulfert, Woroch, and Strobel, 2024) to be able to adapt its own ecosystem and lock in existing participants. This also requires the flexible adjustment of the ecosystem's value proposition to be competitive on an ecosystem-level (Gao, Han, et al., 2019, p. 143).[279]

---

tional and other participants in the multilateral value creation. The customer can be one such participant.

[277] In this regards, Burkhalter et al. (2021, p. 267) described services with reference to Vargo and Lusch (2004) and Lusch and Nambisan (2015) "as the mutual process of applying specialized competences (knowledge and skills) for the benefit of other actors and the actor itself."

[278] The the assortment's continuous adjustment toward environmental developments and changing customer demands in e-commerce environments also leads to alterations of the ecosystem.

[279] For further research on platform and ecosystem-level competition, see also Rietveld and Schilling (2021) and Cennamo (2021).

Based on the two affiliation-based descriptions of e-commerce ecosystem (Gui et al., 2018; Yi and Ming, 2011) and on our own research (e.g., Schütte and Wulfert, 2022a; Wulfert and Karger, 2022b; Wulfert, Woroch, Strobel, Schoormann, et al., 2023), we introduce ten participant types in e-commerce ecosystems, and their major activities, amplifying the value creation. E-commerce ecosystems include the manufacturers and (final) customers, together with intermediating retailers and wholesalers,[280] and also integrate additional participants who are responsible for providing the virtual transaction environment, such as the innovation platform provider, the infrastructure provider, and third-party developers (Böttcher, Rickling, et al., 2021; Schütte and Wulfert, 2022a).[281] The innovation platform provider "owns a core element of the technological system" (Gawer and Henderson, 2007, p. 6). Besides these e-commerce ecosystem participants, Wulfert, Woroch, Strobel, Schoormann, et al. (2023) emphasized the role of additional service providers, such as logistics service providers, content providers, and payment providers, for the execution of transactions in e-commerce. In addition, investors provide the necessary monetary foundation for e-commerce, and government agencies set the legal framework for conducting e-commerce transactions (Wulfert, Woroch, Strobel, Schoormann, et al., 2023). In Table 3.5, we summarize potential participant types in e-commerce ecosystems.

In the context of e-commerce, it is important to note that it is possible and most likely for single ecosystem participants to take on multiple roles (Schütte and Wulfert, 2022a, p. 213). For instance, sellers do not only act as resellers of goods from different manufacturers, but also take on the role of a transaction platform owner who matches and orchestrates retail transactions between different market participants and service providers (Wulfert and Schütte, 2021, p. 604). Another example would be the provision of innovation services in addition to providing the transaction platform in e-commerce ecosystems (e.g., Amazon Web Services, Alibaba Cloud).

---

[280] The participants from the traditional business ecosystem of retailers (Schütte and Wulfert, 2022a) are depicted at the top of Fig. 3.30.

[281] Developers and their respective third-party extensions for transaction platforms as part of a retailer's digital business ecosystem (Schütte and Wulfert, 2022a) are depicted at the bottom of Fig. 3.30.

**Table 3.5** Possible Participants in E-Commerce Ecosystems (e.g., Wulfert, Woroch, and Strobel, 2024; Wulfert, Woroch, Strobel, Schoormann, et al., 2023)

| Participant Type | Description | Source |
|---|---|---|
| Manufacturer | Manufacturers provide an e-commerce ecosystem with products and services. | (Mishra and Tripathi, 2020), (Kaoud et al., 2020), (Balachandran and Mohammadian, 2015) |
| Seller | Sellers resell products purchased from manufacturers to customers. They can employ a variety of business models and sell via digital marketplaces. | (Boudreau and Hagiu, 2009), (Choi et al., 1997), (Timmers, 1998), (Zwass, 1996) |
| Transaction Platform Owner | Transaction platforms orchestrate electronic transactions between sellers and customers and implementing strategic instruments to influence e-commerce ecosystem participants. | (Boudreau and Hagiu, 2009), (Hagiu and Wright, 2015a), (Kawa and Walesiak, 2019), (Wulfert, Seufert, et al., 2021) |
| Social Media Platform | Social media platforms foster interaction of users and sharing of user-generated content. | (Baghdadi, 2016), (Lilien, 2016), (Turban, Whiteside, et al., 2017) |
| Customer | Customers demand and purchase products and services from demand-side participants. | (Lee and Wang, 2019), (Turban, Whiteside, et al., 2017) |
| Service Provider | Service providers support transactions in e-commerce ecosystems. Exemplary service providers offer payment, content, and logistics services. | (Abbu et al, 2021), (Han et al., 2020), (Deufel and Kemper, 2018), (Eickhoff et al., 2017), (Kaoud et al., 2020), (Yang, Ou, et al., 2017) |
| Innovation Platform Owner | Platform providers provide the digital platform including application and infrastructure services for conducting e-commerce transactions. | (Aulkemeier, Paramartha, et al., 2016), (Eisenmann, Parker, et al., 2009), (Tiwana, Konsynski, and Bush, 2010) |
| Developer | Third-party developers develop extensions (e.g., shop themes, interfaces for other digital marketplaces, or feature add-ins) based on boundary resources provided by a focal platform. | (Alami et al., 2015), (Can and Yong-Quan, 2014), (Dowie et al., 2017), (Ghazawneh and Henfridsson, 2013) |
| Investor | Investors provide the necessary monetary foundation such that e-commerce ecosystem participants can engage in production, transactions, and consumption. | (Böttcher, Rickling, et al., 2021), (Kwak, et al., 2019) |
| Government | Government participants define the environment (e.g., antitrust law, taxes, innovation support) for domestic and international transactions in an e-commerce ecosystem. | (Aulkemeier, Iacob, et al., 2017), (Kang, 2021), (Satariano, 2020), (Tian, Vakharia, et al., 2018) |

With regard to the ecosystem-as-a-structure perspective, e-commerce ecosystems face two major directions of integration, extending the value proposition for the final customer and other ecosystem participants. First, e-commerce ecosystems can integrate participants from traditional brick-and-mortar business ecosystems (e.g., manufacturers, retailers, customers) and from virtual e-commerce environments (Schütte and Wulfert, 2022a). The extension of e-commerce ecosystems into the physical brick-and-mortar environments fosters omni-channel scenarios and allows access to an extended service portfolio involving offline services (Verhoef, Lemon, et al., 2009).[282] Even pure online retailers need to cope with other participants from brick-and-mortar environments, such as manufacturers or logistics service providers. In this regard, we consider e-commerce ecosystems as integrated ecosystems consisting of a retailer's traditional business ecosystem (i.e., physical supply chain from manufacturer to (final) customer) and a retailer's digital business ecosystem providing the virtual environment for conducting retail transactions via platforms (Schütte and Wulfert, 2022a). This integrated ecosystem can provide additional value to customers. Second, e-commerce ecosystems can integrate innovation platform services into their focal platform, forming a hybrid core of the ecosystem.[283] Integrating software and infrastructure services and additional development boundary resources, the platform owner can provide ecosystem participants with holistic solutions including application services that extend the platform owner's value proposition (Gawer, 2021b; Wulfert and Karger, 2022b). Hence, we consider hybrid platforms that incorporate the features of innovation platforms and transaction platforms as valuable for e-commerce ecosystems (Schütte and Wulfert, 2022a). Providing application services to other ecosystem participants (Aulkemeier, Paramartha, et al., 2016) and attracting additional third-party developers to the ecosystem (Wulfert, Woroch, Strobel, Seufert, et al., 2022), increase the overall generativity of the ecosystem and propels new business opportunities for ecosystem participants. In Fig. 3.30, we depict the two directions of integration that form an integrated ecosystem for e-commerce ecosystems.

---

[282] Recently, DigitalCommerce360 (2023) reported that e-commerce ecosystems that integrate brick-and-mortar businesses "exhibited strong growth" in 2022.

[283] In section 3.2, we discussed hybrid platforms orchestrating transactions among ecosystem participants and providing innovation services for the development of platform extension.

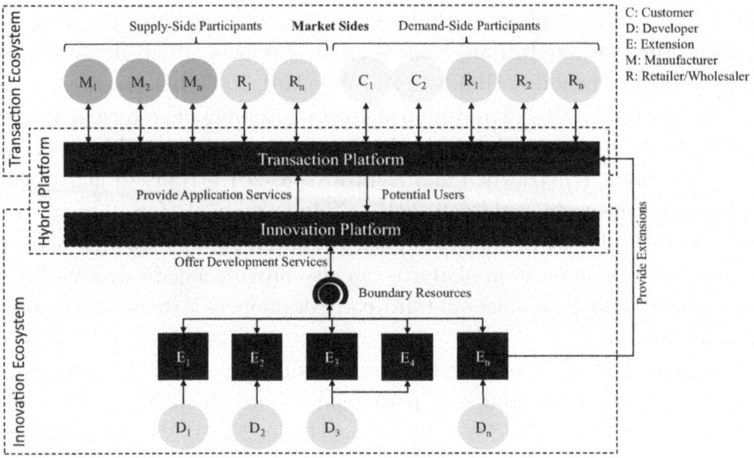


**Fig. 3.30** Integration of Transaction and Innovation Ecosystem

## 3.5 Summary of Core Concepts

In this section, we summarize the description of core terms that form the conceptual foundation of this cumulative dissertation. We define and briefly describe e-commerce, platforms, boundary resources, e-commerce ecosystems, and participants within these ecosystems.

Our research on platforms is focused on the context of e-commerce as the domain of application. Following a discussion on the broad and narrow definition of e-commerce, we focus our research endeavors on electronically-supported commerce transactions of products and services including value-added services via the Internet (i.e., narrow definition) (Choi et al., 1997, p. 33).[284] We focus our investigation on businesses and customers as transaction partners.[285]

In e-commerce, electronically-supported transactions via the Internet require appropriate IS. Innovation platforms can be a possible solution providing the appli-

[284] The broad definition of e-commerce is similar to conceptualizations of electronic business (e.g., Chaffey, 2015; Kalakota and Whinston, 1996; Zwass, 1996). However, we consider e-commerce as a subset of electronic business besides electronic education and electronic entertainment (Wirtz, 2020, p. 15).

[285] This focus involves the combinations possible of these transaction partners (i.e., B2B, B2C, C2B, and C2C).

cation services for multiple possible business models in e-commerce (e.g., transaction platforms) (Alt and Zimmermann, 2019; Timmers, 1998) because of their modularity and extendability (Baldwin and Woodard, 2009; Jansen, 2013; Tiwana, Konsynski, and Bush, 2010). Innovation platforms resemble socio-technical systems upon which software extensions can be developed by (third-party) developers using boundary resources (Ghazawneh and Henfridsson, 2015; Hanseth and Lyytinen, 2010; Tiwana, Konsynski, and Bush, 2010).[286] Innovation platforms are formed by multiple internal software modules designed and developed by the owner of the platform (Fig. 3.8). The innovation platform can also provide a dedicated marketplace for the exchange and deployment of third-party developers' extensions (Jansen and Cusumano, 2013).

Transaction platforms are described as a business model in e-commerce that orchestrates two or more groups of participants more efficiently compared with bilateral exchanges (Hagiu and Wright, 2015a; Hagiu, 2007).[287] Orchestrating multiple market sides, transaction platforms benefit from network effects (Armstrong, 2006; Briscoe, Odlyzko, et al., 2006; Hagiu and Rothman, 2016; Shapiro and Varian, 1998; Voigt and Hinz, 2015). Transaction platforms in e-commerce aggregate the assortment of affiliated third-party sellers and provide it to demand-side participants (i.e., customers) through a single digital channel (Täuscher and Laudien, 2018; Teller and Elms, 2010). The platform owner governs the platforms and the transactions orchestrated including the granting of access and the definition of transaction fees (Fig. 3.11). A platform that provides innovation services and transaction services simultaneously is described as a hybrid platform (Gawer, 2021b; Wulfert and Karger, 2022a).

Boundary resources enable the interaction between the platform and possible participants on technical and social levels (Dal Bianco et al., 2014; Wulfert, 2023). They can be conceptualized as boundary objects (Star, 2010; Star and Griesemer, 1989) in the context of innovation platforms and transaction platforms interfacing with different viewpoints and providing necessary access to a platform's core services for extension development and electronic communication (Ghazawneh, 2012; Wulfert, 2023). Boundary resources are part of a platform's architecture and influenced by governance decisions (Boudreau, 2010; Pon et al., 2014; Tiwana, 2014). While providing access to core services, boundary resources encapsulate the platform's internal modules from direct access and manipulation (Baldwin and Clark,

[286] These boundary resources are provided by the platform owner.

[287] Transaction platforms can be described as socio-technical systems intermediating between different types of participants on a technological basis (Abdelkafi et al., 2019; Haselhoff and Harwardt, 2022; Parker, Van Alstyne, and Choundary, 2016).

2000; Baldwin and Woodard, 2009; Dal Bianco et al., 2014). We differentiate technical boundary resources, such as APIs, SDKs, and IDEs, forming the technological basis for the access of a platform's core modules and supplementary social boundary resources, such as documentation and personal support, complementing and harnessing technical boundary resources (Dal Bianco et al., 2014; Wulfert, 2023).

We introduced the e-commerce ecosystem as a manifestation of platform-centered digital business ecosystems in the context of e-commerce. Following Adner's (2017) ecosystem-as-structure perspective, the objective of e-commerce ecosystems is the exchange of products and services for the purpose of the (final) customers via the Internet (Laudon and Traver, 2019; Wulfert, 2023; Wulfert, Woroch, Strobel, Seufert, et al., 2022). An e-commerce ecosystem consists of a variety of different and independent actors each involved or at most interested in the participation and assistance in commerce transactions.[288] These natural and legal actors can be described as "ecosystem participants."[289] In platform-centered ecosystems, such as e-commerce ecosystems, a participant needs to be affiliated to the focal platform (Hagiu, 2007; Hagiu and Wright, 2015b). The participants are orchestrated by the focal platform owner to enable efficient transactions.

We identified ten participant types in e-commerce ecosystems and a high probability of single participants taking on multiple roles within an ecosystem.[290] The taking on of multiple roles can unfold economic potential, such as benefiting from additional sales data in e-commerce ecosystems, but need to be supported by proper IS.[291] The focal platform owner in e-commerce ecosystems is likely to take on an additional seller role with its own assortment behaving competitively to other third-party sellers. We provide an overview of the major concepts used in this cumulative dissertation in Table 3.6.

---

[288] The set of participants within an e-commerce ecosystem describes the "ecosystem structure" (Adner, 2017, p. 42) that affects the ecosystem's overall value creation.

[289] Although the ecosystem participants are affiliated to a specific ecosystem, we also consider their participation in multiple ecosystems in particular (i.e., multi-homing) and sales to adjacent business partners in general.

[290] We summarize the possible participant types in e-commerce ecosystems in Table 3.5.

[291] In this cumulative dissertation, we especially shed light on a retailer's dual role as platform owner and seller on a transaction platform (Wulfert and Busch, 2022; Wulfert and Schütte, 2022). The retailer can even take on a triple role if providing innovation service in a hybrid platform as the center of the e-commerce ecosystem.

**Table 3.6** Pivotal Concepts and Definitions

| Concept | Definition |
|---|---|
| Electronic Commerce | E-commerce is "the use of electronic means and technologies to conduct commerce, including within-business, business-to-business, and business-to-consumer interactions" (Choi et al., 1997, p. 33). |
| Innovation Platform | Innovation platforms provide an "extensible codebase of a software-based system that provides core functionality shared by the modules that interoperate with it and the interfaces through which they operate (Tiwana, Konsynski, and Bush, 2010, p. 676). |
| Transaction Platform | Transaction platforms orchestrate two (or more) previously distinct market sides and enable (retail) transactions between them (Armstrong, 2006; Hagiu and Wright, 2015a; Rochet and Tirole, 2003). |
| Boundary Resource | Boundary resources are boundary objects that act as interfaces between a focal platform and additional ecosystem participants on technical and social levels (Dal Bianco et al., 2014; Ghazawneh, 2012; Wulfert, 2023). |
| E-Commerce Ecosystem | E-commerce ecosystems are manifestations of digital business ecosystems in the context of e-commerce centered around a focal platform (Wulfert, Woroch, Strobel, Seufert, et al., 2022). |
| Ecosystem Participants | Ecosystem participants are the natural or legal entities affiliated to the focal platform of the ecosystem to conduct transactions among each other or their adjacent affiliates (Böttcher, Rickling, et al., 2021; Corallo, 2007). |
| Participants' Multiple Role | An ecosystem participant can take on multiple roles simultaneously in an e-commerce ecosystem (e.g., retailer's dual role) (Wulfert and Schütte, 2021, 2022). |

# Platform Types in E-Commerce Ecosystems 4

## 4.1 Paper 1: A Bibliometric Analysis of Platform Research in E-Commerce: Past, Present, and Future Research Agenda

**Abstract**

*An important economic sector influenced by the development of platforms is e-commerce.*[1] *The most successful companies in e-commerce employ platform business models and strive to provide other companies with application services. Despite growing economic importance and rising research interest, thus far, no attempts were made to structure existing research into platforms in e-commerce. Hence, a quantitative bibliometric analysis of 7,463 platform-related papers in the context of e-commerce was conducted. The papers were published in major conferences, journals, and books from 1993 to 2021. The authors identified a continuous development of platform research in e-commerce, with the continuous development characterized by three major periods of research. Furthermore, four clusters in platform research are outlined, i.e., business models, social commerce, infrastructure, and socio-technical characteristics. These clusters can serve as a foundation for future*

---

[1] This version of the article has been accepted for publication, after peer review but is not the Version of Record and does not reflect post-acceptance improvements, or any corrections. The Version of Record is referenced as follows: T. Wulfert and E. Karger (2022b). "Shaping Digital Platforms in E-Commerce: Developing an Architecture Framework." In: *Proceedings of the 26th Pacific Asia Conference on Information Systems (PACIS2022)*. Virtual Conference, pp. 1–17

---

**Supplementary Information** The online version contains supplementary material available at https://doi.org/10.1007/978-3-658-45198-1_4.



T. Wulfert, *Selected Perspectives on Platforms in E-Commerce Ecosystems*,
https://doi.org/10.1007/978-3-658-45198-1_4

*research. The conducted bibliometric analysis contributes to scientific research by offering an objective and systematic overview of platform research in e-commerce.*

*This preprint is the comprehensive version of the abbreviated version published. We provide further insights derived from our bibliometric analysis in the Electronic Supplementary Material in Appendix A.4.*

| Fact sheet of publication 1 | |
|---|---|
| Title | A Bibliometric Analysis of Platform Research in E-Commerce: Past, Present, and Future Research Agenda |
| Authors | Tobias Wulfert<br>Erik Karger |
| Publication Type | Journal |
| Publication Outlet | Corporate Ownership & Control |
| VHB-Jourqual3-Ranking | C |
| Status | Published |
| Research Question | How has the literature concerning platforms in e-commerce evolved and what is its bibliometric structure? What are the topics being investigated within research into platforms in e-commerce, and what topical clusters exist within this field? What are possible future research avenues within the area of platforms in e-commerce? |
| Research Approach | Bibliometric Analysis (Donthu, Kumar, Mukherjee, et al., 2021) |
| Full Citation | T. Wulfert and E. Karger (2022a). "A Bibliometric Analysis of Platform Research in E-Commerce: Past, Present, and Future Research Agenda." In: *Corporate Ownership & Control* 20.1, pp. 185–200 |

### 4.1.1 Introduction

Platforms have enriched many industry sectors, for example, accommodation, entertainment, and transportation. The most valuable companies (e.g., Apple, Amazon, and Alibaba) employ platform business models orchestrating multi-sided markets or from the centre of rich innovation ecosystems (Gawer, 2021b; Messerschmitt and Szyperski, 2003; Tiwana, 2014). These platform companies form digital business ecosystems with their surrounding participants (e.g., manufacturers, retailers, third-party developers), which serve (end) customers (Adner, 2017; Wulfert and Karger, 2022b). In extant literature, platforms are conceptualized as organizational capabilities, technology sets, two-sided markets, and extendable systems (McIntyre and Srinivasan, 2017; Sun, Gregor, et al., 2015). Platforms are investigated in a variety of different research areas, for example, in the construction and in the management literature (de Reuver et al., 2017; McIntyre and Srinivasan, 2017).

As an important economic sector, e-commerce boasts an expected global revenue of \$ 7.4 quadrillion and 24.5% of total retail sales (Lebow, 2021). E-commerce involves electronically supported retail transactions without spatial and temporal restrictions (Zwass, 1996). The most successful organizations in e-commerce operate platform business models to benefit from the economic effects of two-sided markets (e.g., matching and subsidizing selected participants) and to execute powerful digital platforms with their application services (Merton, 2021; Parker, Van Alstyne, and Choundary, 2016; Wulfert and Karger, 2022b). In 2020, Amazon, for example, generated more than 60% of its total \$ 340 billion in revenue (from product and service sales) from commission fees of third-party sellers on its transaction platform (Amazon, 2019, 2021). The cumulated revenue of ecosystem participants is even higher. On average, the revenue realized by ecosystem participants is up to three times higher than the revenue generated by the focal platform (Delteil et al., 2020). Platforms form the centre of e-commerce ecosystems, which are surrounded by networks of independent participants (Wulfert, Woroch, Strobel, Seufert, et al., 2022). These focal platforms (e.g., Amazon Marketplace, Walmart Marketplace, and Alibaba.com) actively orchestrate ecosystem participants by executing governance rules (Hein, Schreieck, Riasanow, et al., 2020). They match supply-side and demand-side participants to enable retail transactions involving additional actors, for example, content providers (Böttcher, Rickling, et al., 2021; Wulfert, Seufert, et al., 2021). The matching service and other necessary application services are provided by digital platforms (Wulfert and Karger, 2022b).

In 1993, first investigations of e-commerce platforms appeared with the ignition of e-commerce business models (Zwass, 1996). During recent years, however, research experienced rising interest on platforms in general (Liu, Li, et al., 2021), and on platforms in e-commerce in particular (Wulfert and Karger, 2022b; Wulfert and Schütte, 2022; Wulfert, Woroch, Strobel, Seufert, et al., 2022). The dominant types of platforms in e-commerce research and practice are platform business models (i.e., transaction platforms) and innovation platforms. While transaction platforms orchestrate participants from multiple market sides (Hagiu and Wright, 2015b; Rochet and Tirole, 2003), innovation platforms provide the technological infrastructure for e-commerce (Schütte and Wulfert, 2022a; Wulfert and Karger, 2022b). A combination of both types (i.e., hybrid platform) can be considered as a third relevant platform type in e-commerce (Gawer, 2021b; Schütte and Wulfert, 2022a).

Nowadays, research into e-commerce platforms is difficult to oversee, and spreads across a variety of different disciplines. This is unfortunate, since e-commerce platforms are likely to be of increasing relevance for practitioners and researchers in e-commerce. Thus far, research has only provided a general overview of platform research in the economics and management domain (Liu, Li, et al., 2021). However, to the best of the authors' knowledge, no attempts exist to structure the current body of research into e-commerce platforms as a whole. The authors believe that such a broad analysis of existing research can help consolidate and merge existing research endeavors and integrate the different perspectives on that topic. In view of that, the goal is to extend this view by conducting an interdisciplinary, bibliometric analysis of research into e-commerce platforms as a whole. Against this backdrop, the following research questions are addressed.

**RQ1:** How has the literature concerning platforms in e-commerce evolved and what is its bibliometric structure?

**RQ2:** What are the topics being investigated within research into platforms in e-commerce, and what topical clusters exist within this field?

**RQ3:** What are possible future research avenues within the area of platforms in e-commerce?

In total, 7,463 articles and publications that deal with e-commerce platforms were identified and analyzed. Given the large number of sources, the authors chose a bibliometric approach to find answers for the first two research questions. A bibliometric analysis uses a quantitative approach to analyze and summarize large amounts of literature. A literature review, by comparison, is where relatively less literature is analyzed qualitatively (Donthu, Kumar, Mukherjee, et al., 2021). To answer the third research question, the conducted bibliometric analysis was complemented with a qualitative analysis of relevant and often-cited literature. Hereby, the goal was to obtain a better and detailed understanding of the different research streams.

The remainder of this research paper is structured as follows. First, e-commerce as a research context in general are introduced, and platforms as a construct of interest in particular. Second, the authors elicit the scientific approach to the bibliometric analysis. Third, the results from the bibliometric analysis are present. The authors provide a general overview of the literature corpus analyzed, detail the performance analysis, and describe focus topics of platforms in e-commerce. Fourth, the contributions of this article are discussed and a future research agenda for e-commerce platforms is provided. The article is closed with a brief summary of the results and the limitations of the study are outlined.

### 4.1.2 Theoretical Background

In e-commerce, transactions involving information, products, and services are executed using information and communication technology (Nath et al., 1998). E-commerce can be characterized as a form of electronic business exploiting technologies, such as electronic data interchange and the internet (Choi et al., 1997). Alt (2020) provides an overview of different electronic business types, including platform business models.

An important construct in e-commerce on both infrastructure and business model level are platforms (Schütte and Wulfert, 2022a). Different notions for understanding platforms (e.g., two-sided market, extendable system) and various research areas (e.g., construction or management literature) investigating platforms already exist (McIntyre and Srinivasan, 2017). The technological and economic perspectives (Bughin and van Zeebroeck, 2017) are especially relevant in the context of e-commerce (Hänninen et al., 2018). Abdelkafi et al. (2019) emphasize that transaction platforms (economic perspective), innovation platforms (technological perspective), and hybrid platforms (combination of both perspectives) are especially relevant to research on e-commerce.

Transaction platforms are virtual loci on which participants can conduct retail transactions (Turban, Whiteside, et al., 2017). These transaction platforms match and orchestrate organizations as well as individual participants from various markets and social groups to form a dynamic ecosystem (Gawer, 2021b). Transaction platforms benefit from two-sided market concepts, matching two or more previously independent market sides and generating value from transactions between the participants. Direct and indirect network effects are exploited to further propel the transaction platform (Rochet and Tirole, 2003). Network effects for participants and economies of scale for platform owners are strong drivers for the success of transaction platforms (Briscoe, Odlyzko, et al., 2006). From a customer point of view, transaction platforms "resemble retail agglomerations" (Hänninen et al., 2018), integrating the range of articles of participating suppliers, retailers, and wholesalers through a single digital channel (Wulfert and Schütte, 2022). Furthermore, transaction platforms can offer participants a variety of retail-related services (Wulfert, Seufert, et al., 2021).

Innovation platforms form the technological infrastructure for e-commerce transactions. Baldwin and Woodard (2009) define an innovation platform as "a set of stable components that supports variety and evolvability in a system by constraining the linkages among the other components." Tiwana, Konsynski, and Bush (2010) emphasize that third-party developers can use these stable components as an "extensible codebase [...] that provides core functionality shared by the modules that interoperate with it and the interfaces through which they interoperate." While the stable components form the core of the innovation platform, the third-party modules resemble the periphery (Staykova and Damsgaard, 2015). The platform core and periphery are evolvable, with only the interface specifications that should remain stable over a longer period of time to avoid necessary module adjustments for the developers (Baldwin and Woodard, 2009). Innovation platforms provide the application services necessary for executing the business model-related business routines (Wulfert and Schütte, 2022). Third-party developers implement additional modules (e.g., shop themes, feature add-ins) and provide them via dedicated extension marketplaces (Wulfert, Woroch, Strobel, Seufert, et al., 2022).

### 4.1.3 Scientific Approach

In this study, a bibliometric analysis were conducted to map and analyze the research into platforms in e-commerce. First introduced by Pritchard in (1969), bibliometric studies have increasingly gained relevance and are nowadays a common method to analyze literature. In a bibliometric study, quantitative and statistical methods are used to analyze bibliometric data (Broadus, 1987; Donthu, Kumar, Mukherjee, et al., 2021; Pritchard, 1969). Bibliometric methods can be used to understand a research field's evolution and development (Caviggioli and Ughetto, 2019; Khanra, Dhir, Kaur, et al., 2021; Khanra, Dhir, and Mäntymäki 2020; Tandon et al., 2021) and to organize the current literature on a research topic (Khanra, Dhir, Parida, et al., 2021). A wide range of methods is used in bibliometric studies (Merigó et al., 2015). The objective of bibliometric studies is to measure not only the contributions of research constituents, referred to as performance analysis (Cobo et al., 2011; Ramos-Rodríguez and Ruíz-Navarro, 2004), but also their relationships (Baker, Pandey, et al., 2020; Cobo et al., 2011; Ramos-Rodríguez and Ruíz-Navarro, 2004), called science mapping (Donthu, Kumar, Mukherjee, et al., 2021).

Scholars have used bibliometric studies in a broad variety of different disciplines and domains. Examples of topics that are investigated in bibliometric studies are, among others, social media and networks (Leung, et al., 2017; López-Carril et al., 2020; Shiau et al., 2017), digital marketing (Krishen, et al., 2021), family businesses

(Bekiaris and Papanastasiou, 2021), business model research (Budler et al., 2021), and platforms (Liu, Li, et al., 2021). Bibliometric studies were also used to provide an overview of publications in journals (e.g., Journal of Business Research) (Donthu, Kumar, Pattnaik, 2020; Merigó et al., 2015).

#### 4.1.3.1 Data Collection

The first step consisted of collecting data that can be used for bibliometric analysis. Examples of data that can be used for bibliometric studies are, among others, keywords, titles, journals, as well as information about authors and their institutions (Gutiérrez-Salcedo et al., 2017; Moral-Muñoz et al., 2020). For the collection of bibliometric data, several bibliographic databases exist and they all have different characteristics (Moral-Muñoz et al., 2020). Of all available bibliometric databases, the Web of Science (WoS) and Scopus are often considered to be the two largest ones (Forliano et al., 2021). Scopus is used as the database for data collection because it covers a higher number of journals than WoS (Paul and Criado, 2020). Furthermore, Scopus has been chosen by several other bibliometric studies over the past years (e.g., Caviggioli and Ughetto, 2019; Jagals et al., 2021; Khanra, Dhir, Kaur, et al., 2021; Tandon et al., 2021).

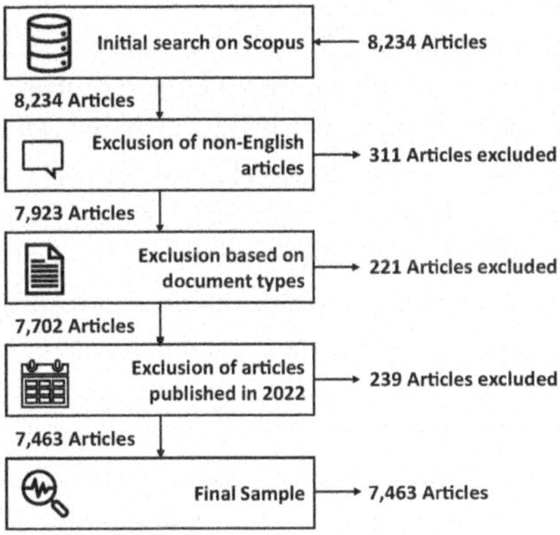


**Fig. 4.1** Process of Data Collection

To collect data from Scopus, the following search term were used: (("platform") AND ("e-commerce" OR "e commerce" OR "electronic commerce")). Title, abstract, and keywords were searched for the occurrence of the search term. Apart from "e-commerce" and "e commerce," the written-out spelling "electronic commerce" were also used to make sure to cover all relevant research. The search on Scopus led to an initial sample of 8,234 results. After that, the authors conducted several exclusion steps. First, all non-English articles were excluded. This led to an elimination of 311 articles. Next, 221 irrelevant document types were excluded. Furthermore, the authors follow the argumentation of Forliano et al. (2021) and Massaro et al. (2016) and believe that "more recent publications had not had the time to receive an adequate number of citations" (Forliano et al., 2021, p. 3). To improve the comparability of the bibliometric results (Forliano et al., 2021), articles published in 2022 were excluded. After the final exclusion step, a final sample of 7,463 articles remained (Fig. 4.1). This sample was exported in a Comma-Separated Values (CSV) file for the further analysis that is outlined in the next subsection.

#### 4.1.3.2 Data Analysis

The analysis of the bibliometric data is the next step after the data collection. For the bibliometric analysis of data, researchers have access to and can use a lot of different software, libraries, and tools (Jagals et al., 2021; Moral-Muñoz et al., 2020). The software R were used as the foundation for the analysis. R is a software environment that covers a lot of functionality like visualization capabilities and statistical and mathematical functions (Derviş, 2019). The authors complemented the usage with the Bibliometrix library, which can be used for comprehensive systematic mapping analysis (Aria and Cuccurullo, 2017; Moral-Muñoz et al., 2020).

Furthermore, VOSviewer complemented the used toolset, and is a tool for the visualization and evaluation of bibliometric data (Khanra, Dhir, Kaur, et al., 2021; van Eck and Waltman, 2014). In this study, VOSviewer was used for the analysis of the keyword co-occurrences (Fahimnia et al., 2015; Khanra, Dhir, Kaur, et al., 2021; van Eck and Waltman, 2014). While VOSviewer proved to be highly suitable for creating a visualization of keyword co-occurrences, Biblioshiny has several options for the statistical analysis of the bibliometric data (Jagals et al., 2021). Moreover, both tools allow the importation of data that is exported from Scopus (Moral-Muñoz et al., 2020). Consequently, it was decided to use both tools in a complementary manner. For the derivation of future research avenues, the set of papers was limited and the most cited articles were investigated qualitatively (Tandon et al., 2021). However, to ensure that the derived future research agenda is up to date, only the most cited articles that were published within the last four years were analyzed.

**Table 4.1** Overview of Author, Document, and General Information

| General information | |
|---|---|
| Timespan | 1993–2021 |
| Sources | 2,783 |
| Documents | 7,463 |
| Average citations per document | 8.767 |
| References | 183,484 |
| **Author and document information** | |
| Different authors | 13,017 |
| Author appearances | 22,558 |
| Single-authored documents | 1,237 |
| Documents per author | 0.573 |
| Authors per document | 1.74 |
| Co-authors per document | 3.02 |
| Collaboration index | 1.93 |
| Different keywords | 15,403 |

### 4.1.4 Findings

#### 4.1.4.1 Main Information

The findings from the bibliometric analysis are presented and structured in a general overview of the final literature sample, a performance analysis, and an elaboration on periods of platform research in e-commerce. A summary of different metrics of the final literature sample is provided in Table 4.1. The final sample consisted of 7,463 scientific publications from 1993 to 2021. The above-mentioned publications were published in 2,783 different sources, among which are conference papers, journals, and books. On average, the documents are cited 8.8 times by other publications. In total, the publications from the final sample have 13,017 different authors who have cited 183,484 different references. Furthermore, the final set included a total of 15,403 different keywords.

In total, 22,558 authors appeared on the publications, with only 1,237 single-authored documents, which is equal to 16.58%. In comparison to other bibliometric studies, this value is the lowest. A comparison of recent bibliometric studies is shown in Table 4.2. The low value could indicate a comparatively high complexity of the subject matter; this makes projects difficult when dealing with single-authored documents and requires larger teams of authors. On average, each author has contributed to 0.537 articles and each article has 1.74 authors. In comparison, the value of 0.537 articles for each author is the second highest among those which were compared. This might also be a hint to the involvement of domain experts authoring a number of articles.

**Table 4.2** Comparison of Bibliometric Studies

| Article | Secinaro, Brescia, et al. (2021) | Secinaro, Dal Mas, et al. (2021) | Uluyol et al. (2021) | (Jagals et al. (2021) | This study |
|---|---|---|---|---|---|
| Topic | Data quality for innovation and accounting management | Blockchain in accounting and auditing | Waqf research | Data governance | E-commerce platforms |
| Documents | 159 | 93 | 527 | 780 | 7,463 |
| Documents per author | 0.305 | 0.443 | 0.599 | 0.367 | 0.537 |
| Collaboration index | 3.6 | 2.83 | 2.53 | 3.26 | 1.93 |
| Single-authored documents | – | 29% | 50% | 22.18% | 16.58% |
| Source dominance | 3.46 | 1.58 | 2.36 | 1.59 | 2.68 |

For the identification of cooperation and collaboration among researchers, the collaboration index (CI) is applied. The CI is calculated by dividing the total number of authors of multi-authored articles by the total number of multi-authored articles (Elango and Rajendran, 2012; Koseoglu, 2016). The CI of the final sample is 1.93, resembling a relatively low collaboration between researchers. Given the small number of single-authored documents, this value is interesting to observe since it shows that very large author teams, as well as contributions that are written in single authorship, are also rare.

The first research article focusing on platforms in the context of e-commerce was published in 1993 (Klein, 1993). Since 1993, the number of publications concentrating on platforms has steadily increased. In the year 2000, 107 documents had already been published. The following years have brought publication peaks, with 183 publications in 2004 and 351 in 2010. After 2010, the number of publications decreases, with only 213 documents in 2015. This decline can be explained with the consequences of the global financial crisis and the European debt crisis. These crises resulted in decreased customer spending and decreased research focusing on risk management and the reinvention of e-commerce (Song, Huang, et al., 2013; Wang, 2013). Since then, the analysis revealed a sharp increase in the number of publications, with 535 in 2018 and as many as 1,226 in 2021. The intensified research into e-commerce platforms highly correlates with the identified third research period that focuses on social commerce involving customer-generated content and electronic word of mouth in the transactions (Rosario et al., 2016; Xiang et al., 2017). This period also focuses on the extended application of existing technologies (e.g., recommender systems) and exploration of new technologies (e.g., internet of things, blockchain) (Lu et al., 2015; Zhang and Wen, 2017). The overview of the number of publications per year on platform research in e-commerce is depicted in Fig. 4.2.[2]

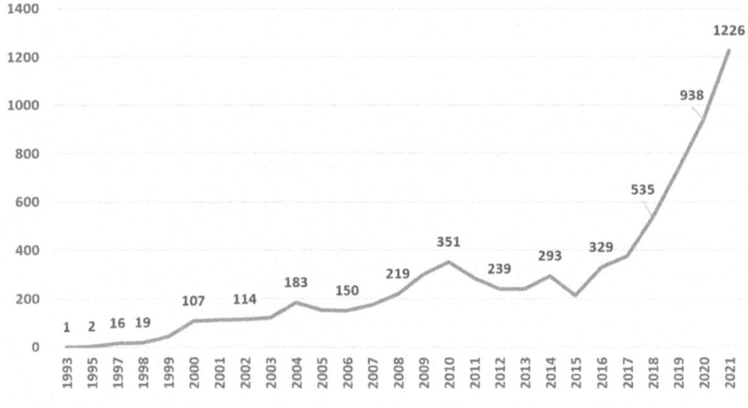


**Fig. 4.2** Number of Publications per Year

In Fig. 4.3, the distribution of publications among different research domains is shown. A publication was assigned to a discipline based on the outlet in which it appeared. For the generation of Fig. 4.3, the data of Scopus were used, which link a journal or a conference to a certain discipline. Since a given source can belong to

[2] In the original publication, the figures are represented in grey shading owing to the specifications of the journal. We included the colored version of the figures within this cumulative dissertation.

more than one discipline, the sum of the publications shown in Fig. 4.3 (12,972) is higher than the total number of publications in this article's final sample (7,463).

If the distribution of research contributions across disciplines is scrutinized, a strong dominance of computer science is found with a total of 5,156 publications, followed by engineering. This is not necessarily surprising, as computer science and engineering are the traditional disciplines that first addressed the topic of platforms from a technical or socio-technical focus. This also involves the investigation of certain technologies in the context of e-commerce platforms. Examples of those technologies are blockchain (Chen, Chen, et al., 2018; Ying et al., 2018), machine learning (Gupta et al., 2014), and big data (Xie et al., 2016; Ye, Wang, et al., 2013). Computer science and engineering as the most dominant disciplines are followed by business and management. While engineering and computer science are mainly concerned with technical topics, business-and-management-related research is mostly concerned with managerial implications of e-commerce platforms. Other disciplines with a high number of contributions are decision sciences, mathematics, social sciences, and economics.

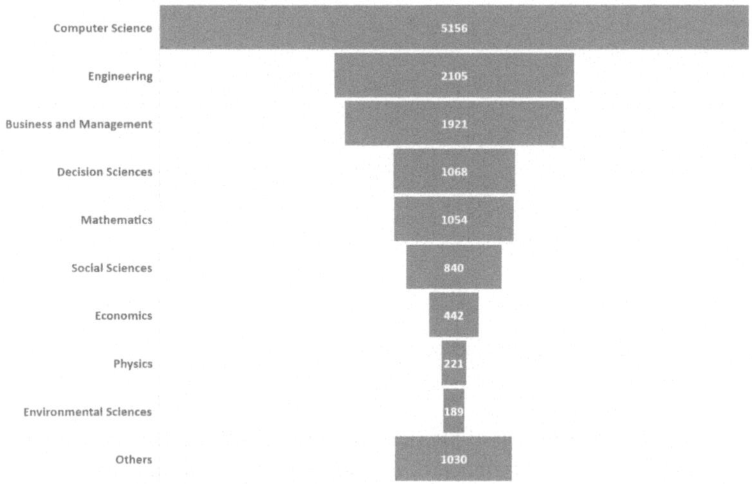


**Fig. 4.3** Number of Publications per Research Domain

Table 4.3 shows the distribution of contributions among different types of publications. The majority of research contributions from the final sample was published in conference proceedings (4,613 papers), followed by 2,468 journal articles. The publication in conference proceedings and the discussion at conferences with usually shorter review cycles compared to journals reflect research progress being made in short cycles in e-commerce. The number of book chapters (156), reviews (134), and books (22) focusing on platform research in the context of e-commerce is significantly smaller.

**Table 4.3** Distribution of Contributions among Different Types of Publications

| Type | No. |
|---|---|
| Conference paper | 4,613 |
| Journal article | 2,468 |
| Book chapter | 156 |
| Review | 134 |
| Book | 22 |
| Others | 70 |

#### 4.1.4.2 Performance Analysis

Performance analysis is a standard practice in reviews to present the research constituent's performance and can be found in the majority of bibliometric studies (Donthu, Kumar, Mukherjee, et al., 2021). Possible units of analysis are, for example, outlets and sources, institutions, articles, or authors (Jagals et al., 2021; Massaro et al., 2016; Thelwall, 2008).

Table 4.4 and Table 4.5 give an overview of the most relevant sources, based on the number of published articles. While Table 4.4 shows the 20 most relevant sources in total, Table 4.5 shows only the most relevant journals. In Table 4.4, the authors notice a clear dominance of conferences, which is not surprising in view of the distribution of contribution types shown in Table 4.3. With a total of 309 articles, the ACM International Conference Proceedings Series is the most frequent source in total, closely followed by the Lecture Notes in Computer Science with 296 publications. This strong dominance of the first two sources is not surprising, as they publish articles from a variety of different conferences.

With a large gap, the Proceedings of the International Conference on Electronic Business follows in third place. It is noteworthy that among the 20 most relevant sources, the names of only three journals appear. These are Electronic Commerce Research and Applications, Sustainability, and Electronic Commerce Research. Likewise, the distribution of disciplines that is depicted in Fig. 4.3 is very well reflected in the outlets, since most conferences and journals belong to engineering and computer science.

**Table 4.4** Distribution of Articles among Conference Proceedings

| Rank | Source | Articles |
|---|---|---|
| 01 | ACM International Conference Proceeding Series | 309 |
| 02 | Lecture Notes in Computer Science (including subseries Lecture Notes In Artificial Intelligence And Lecture Notes In Bioinformatics) | 296 |
| 03 | Proceedings of the International Conference on Electronic Business | 132 |
| 04 | Advances In Intelligent Systems and Computing | 122 |
| 05 | Journal Of Physics: Conference Series | 85 |
| 06 | Communications In Computer and Information Science | 80 |
| 07 | Lecture Notes in Business Information Processing | 75 |
| 08 | IFIP Advances in Information and Communication Technology | 72 |
| 09 | Proceedings Of the International Conference on E Business and E Government | 69 |
| 10 | Electronic Commerce Research and Applications | 67 |
| 11 | Applied Mechanics and Materials | 66 |
| 12 | 2011 2nd International Conference on Artificial Intelligence, Management Science And Electronic Commerce, Aimsec 2011- Proceedings | 51 |
| 13 | CEUR Workshop Proceedings | 51 |
| 14 | Sustainability | 51 |
| 15 | Advanced Materials Research | 42 |
| 16 | Proceedings Of The ACM SIGKDD International Conference On Knowledge Discovery And Data Mining | 41 |
| 17 | International Conference On Information And Knowledge Management, Proceedings | 38 |
| 18 | Lecture Notes In Electrical Engineering | 38 |
| 19 | Electronic Commerce Research | 34 |
| 20 | Decision Support Systems | 32 |

**Table 4.5** Distribution of Articles Among Journals

| Rank | Source | Articles |
|---|---|---|
| 01 | Electronic Commerce Research And Applications | 67 |
| 02 | Sustainability | 51 |
| 03 | Electronic Commerce Research | 34 |
| 04 | Decision Support Systems | 32 |
| 05 | IEEE Access | 30 |
| 06 | Journal Of Retailing And Consumer Services | 23 |
| 07 | Computers & Industrial Engineering | 20 |
| 08 | Information & Management | 20 |
| 09 | International Journal of Production Economics | 20 |
| 10 | Journal Of Theoretical and Applied Electronic Commerce Research | 19 |
| 11 | Future Generation Computer Systems | 18 |
| 12 | International Journal of Information Management | 17 |
| 13 | International Journal of Electronic Commerce | 16 |
| 14 | Complexity | 15 |
| 15 | European Journal Of Operational Research | 15 |
| 16 | Expert Systems with Applications | 15 |
| 17 | Industrial Management and Data Systems | 15 |
| 18 | Journal Of Theoretical and Applied Information Technology | 15 |
| 19 | Management Science | 15 |
| 20 | Boletin Tecnico/Technical Bulletin | 14 |

Table 4.6 shows the 25 most productive affiliations and institutions. This list is strongly dominated by Chinese institutions. Among the 25 affiliations with the most published articles, 21 are located in China and two are from Hong Kong. With the Nanyang Technological University and the National University of Singapore, there are also two affiliations from Singapore within this list. This is rather surprising, since Singapore is only in the 19th place on the list of the most productive countries (see Table 4.7). The two affiliations from Singapore might indicate that those two universities are the most dominant ones in Singapore in terms of platform and e-commerce research since they are responsible for 73 of Singapore's 96 publications.

**Table 4.6** Overview of the most productive affiliations

| Rank | Affiliations | Country | Articles |
|---|---|---|---|
| 01 | Alibaba Group Holding Limited | China | 115 |
| 02 | Tsinghua University | China | 92 |
| 03 | Zhejiang University | China | 89 |
| 04 | Wuhan University | China | 79 |
| 05 | Chinese Academy of Sciences | China | 70 |
| 06 | Beijing University of Posts and Telecommunications | China | 65 |
| 07 | South China University of Technology | China | 48 |
| 08 | City University of Hong Kong | Hong Kong | 47 |
| 09 | Hong Kong Polytechnic University | Hong Kong | 45 |
| 10 | University of Science and Technology of China | China | 45 |
| 11 | Shanghai Jiao Tong University | China | 44 |
| 12 | Renmin University of China | China | 42 |
| 13 | Shanghai University | China | 42 |
| 14 | Tongji University | China | 42 |
| 15 | Ministry of Education of the People's Republic of China | China | 40 |
| 16 | Harbin Institute of Technology | China | 40 |
| 17 | Huazhong University of Science and Technology | China | 39 |
| 18 | Nanyang Technological University | Singapore | 38 |
| 19 | Xiamen University | China | 38 |
| 20 | Beijing Jiaotong University | China | 37 |
| 21 | Zhejiang Gongshang University | China | 37 |
| 22 | National University of Singapore | Singapore | 35 |
| 23 | Sun Yat-Sen University | China | 35 |
| 24 | School of Management Fudan University | China | 35 |
| 25 | Fudan University | China | 35 |

With 3,092 articles, Chinese authors are responsible for around 41% of the sample's 7,463 articles. The United States (US) follows in second place with about one third of the articles. India holds the third place with 500 articles. The most productive European countries are the United Kingdom (321 publications) and Germany (310 articles). An overview of the 20 most productive countries is shown in Table 4.7.

**Table 4.7** Overview of the most productive countries

| Rank | Country | Publications |
|---|---|---|
| 01 | China | 3,092 |
| 02 | United States | 1,010 |
| 03 | India | 500 |
| 04 | United Kingdom | 321 |
| 05 | Germany | 310 |
| 06 | Taiwan | 276 |
| 07 | Australia | 201 |
| 08 | Hong Kong | 175 |
| 09 | Italy | 168 |
| 10 | Canada | 163 |
| 11 | Spain | 156 |
| 12 | South Korea | 152 |
| 13 | Indonesia | 134 |
| 14 | Malaysia | 129 |
| 15 | France | 109 |
| 16 | Greece | 103 |
| 17 | Netherlands | 98 |
| 18 | Japan | 97 |
| 19 | Singapore | 96 |
| 20 | Switzerland | 82 |

The duopoly consisting of China and the US is also well reflected in the collaboration map depicted in Fig. 4.4. In this figure, the collaboration between authors from different countries is shown. The thicker the red line is between two countries, the more collaboration there is between authors of those two nations. Furthermore, the total productivity of a country is shown in blue colors, with a darker blue referring to a higher number of articles. Here it is obvious that the US and China as the most productive countries also enter into the most collaborations with authors from other countries.

**Fig. 4.4** International Collaboration Map. (Created with Biblioshiny)

#### 4.1.4.3 Trending Topics and Content Analysis

Following the performance analysis, the focus is on the keywords applied by the authors of the publications. After providing an overview of the most used keywords in the final sample, three phases of platform research in e-commerce were derived, based on the application of keywords. The first four ranks are occupied by synonyms of the context area, i.e., e-commerce, with a total of 7,810 appearances (i.e., electronic commerce 5,118, e-commerce 1,127, sales 907, commerce 658), followed by internet (543) and world wide web (426). Social networking appeared 405 times on publication in the final literature sample, which reflects the focus on social commerce in recent years of research (Huang and Benyoucef, 2013; Wang and Zhang, 2012). While marketing (397) is an important activity in e-commerce increasing sales, Information Systems (IS) (395) form the necessary infrastructure for conducting retail transactions. From the perspective of this research, IS are likely to be composed of innovation platforms as the "core" of e-commerce ecosystems (Blaschke, Haki, et al., 2019). The 25 most used keywords from the final literature sample are shown in Table 4.8.

The keyword analysis revealed three phases of platform research in e-commerce. The first phase (Platform 1.0) from 1993 to 2004 consists of 716 publications and is concerned with the emergence of the internet facilitating first electronic business models. The platform's focus is on the technological foundation for conducting e-commerce. The second phase (Platform 2.0) ranges from 2005 to 2014 and includes a total of 2,400 publications. This phase is concerned with new business models in e-commerce from a transaction platform perspective and focuses on advanced architectures and services for conducting e-commerce. The second phase also includes first social commerce considerations. The third phase (Platform 3.0) from 2015 to 2021 consists of 4,347 publications. Research in this phase focuses on social commerce, the exploitation of new technologies in e-commerce, and their integration in existing and new platforms.

As a result of the keyword co-occurrence analysis, four clusters of platform-related research in e-commerce were identified (Fig. 4.5). For the identification of a common theme within the four clusters and for the extraction of avenues for future research with regard to each cluster, the authors conducted an in-depth analysis of the ten most cited papers per cluster (Baker, Pandey, et al., 2020; Fahimnia et al., 2015; Xu, Chen, et al., 2018).

**Table 4.8** The 25 most used keywords

| Rank | Keyword | Amount |
|---|---|---|
| 01 | Electronic Commerce | 5,118 |
| 02 | E-commerce | 1,127 |
| 03 | Sales | 907 |
| 04 | Commerce | 658 |
| 05 | Internet | 543 |
| 06 | World Wide Web | 426 |
| 07 | Social Networking | 405 |
| 08 | Marketing | 397 |
| 09 | Information Systems | 395 |
| 10 | E-Commerce | 376 |
| 11 | Data Mining | 349 |
| 12 | Websites | 337 |
| 13 | Information Management | 305 |
| 14 | Web Services | 295 |
| 15 | Big Data | 292 |
| 16 | Decision Making | 282 |
| 17 | Online Shopping | 278 |
| 18 | Recommender Systems | 278 |
| 19 | Competition | 268 |
| 20 | Artificial Intelligence | 263 |
| 21 | Costs | 250 |
| 22 | Information Technology | 243 |
| 23 | Social Media | 227 |
| 24 | Mobile Commerce | 223 |
| 25 | Supply Chains | 216 |

The first cluster involves publications related to platform business models and individual business model components. Besides publications on business process modeling (Karagiannis and Kühn, 2002), general and e-commerce-specific business model components (e.g., revenue structure, value proposition, key partners) are addressed (Osterwalder and Pigneur, 2002; Osterwalder, Pigneur, Bernarda, et al., 2014). While Lin, Zhou, et al. (2020) focus on pricing and value propositions, other researchers address supply chain financing and logistics services required in e-commerce (Qin et al., 2020; Yu, Zhao, et al., 2021). More generally, institutional

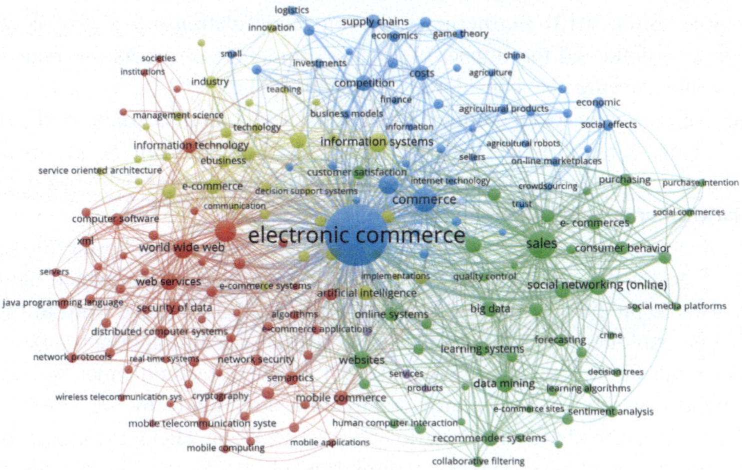


**Fig. 4.5** Keyword Co-Occurences. All keywords >50

enablers (e.g., web experience, extent of coordination) for the usage of e-commerce are identified (Chatterjee et al., 2002). This cluster also investigates how e-commerce and platforms involving different business models impact transaction costs involving different business models (Abhishek et al., 2016; Kauffman and Mohtadi, 2004). Information transparency is seldomly favored in e-commerce ecosystems because of data confidentiality constraints and participants' opportunism (Zhu, 2004). On the contrary, an increased number of intermediaries (e.g., platforms) in e-commerce can reduce transaction costs (e.g., search costs) (Dukes and Liu, 2016).

The second cluster is concerned with platforms in the context of social commerce and related technologies to exploit customer-generated content and word of mouth. Social commerce is based on individuals sharing with others their knowledge, experiences, and information about the products purchased and services consumed, thereby creating electronic word of mouth (Hajli, 2014; Huang and Benyoucef, 2013; Liang and Turban, 2011). Publications in this research cluster are concerned with the user interface design of platforms in e-commerce, involving social commerce constructs (e.g., ratings, recommendations, reviews). Huang and Benyoucef (2013) developed a framework with design principles for social commerce platforms. User centricity is considered important for valuable social commerce transactions (Shin, 2013). Furthermore, the impact of relationship quality on social commerce revenue and the impact of social commerce constructs on repurchase intention and trust are also considered (Hajli, 2015; Hajli, 2014) . The role of trust is widely discussed in social commerce (Hawlitschek, Teubner, Adam, et al., 2016; Wang, Huang, et al., 2018). Another research stream deals with the provision of electronic word of mouth. Electronic word of mouth can be provided via a transaction platform or crawled externally (Rosario et al., 2016). The impact of customer reviews depends on linguistics, semantics, sentiment, and source. Furthermore, the impact of customer reviews differs depending on the platform on which the customers used to publish the reviews (Xiang et al., 2017). Electronic word of mouth can also be exploited by recommender systems that are an important vehicle for increasing sales in e-commerce (Lu et al., 2015). These recommender systems or recommendation services can be provided by innovation platforms and can be integrated in transaction platforms (Rosario et al., 2016; Wang, Huang, et al., 2018). In e-commerce, recommender systems for digital and physical goods involving specific algorithms exist (Lu et al., 2015).

The third cluster is concerned with platforms providing the necessary infrastructure for e-commerce (i.e., innovation platforms). This involves the computing infrastructure (DeCandia et al., 2007; Zeng, Benatallah, et al., 2004) and dedicated (web) services for conducting e-commerce transactions (Bertolino et al., 2009; Bravetti et al., 2004). Dedicated recommender systems can also provide e-commerce-related services for the use of electronic word of mouth (Lu et al., 2015). This cluster also involves research regarding the alignment between business processes, application services, and infrastructure services (Aerts et al., 2004; Kuk and Janssen, 2013). Innovation platforms in e-commerce need to support a variety of e-commerce business models (Yu, Zhao, et al., 2021). These platforms can also incorporate new technologies, such as blockchain and the internet of things (Zhang and Wen, 2017). Since in e-commerce, platforms form the center of larger e-commerce ecosystems of independent participants, application system interfaces are necessary to enable ecosystem-wide data interchange (Reussner et al., 2003). To facilitate this exchange, the focal platform can facilitate ecosystem-wide interface standardization (Wulfert, Woroch, Strobel, Seufert, et al., 2022).

The fourth cluster focuses on the socio-technical characteristics of IS and their architecture in e-commerce. Innovation platforms in e-commerce enable the development and application of third-party extensions to serve a variety of different business models and retailers' requirements (Aulkemeier, Paramartha, et al., 2016). The extension development involves developers as additional participants (Salminen and Teixeira, 2013). Due to varying resource demand, cloud applications are especially suitable for e-commerce contexts, enabling flexible resource allocation provided by external providers (Crabb, 2014). Involving a number of cloud providers that have an on-demand recovery function can also avoid service downtimes (Addo et al., 2014). The research stream focusing on IS also involves security considerations as cross-sectional function in e-commerce architectures (Huang, Zhao, et al., 2011). Another research stream is concerned with the investigation and application of different architecture paradigms for e-commerce platforms. Besides research into traditional multi-tier architectures (Chen and Feng, 2016), the focus of architectural research in e-commerce is on service-oriented architectures involving a variety of service providers (Aulkemeier, Schramm, et al., 2016; Pop and Gîfu, 2018; Yaddarabullah et al., 2019). Furthermore, microservice architectures and decentralized approaches in peer-to-peer or networked contexts are researched (Li, Zhou, et al., 2010; Wu, Ding, et al., 2019; Xia et al., 2010). Table 4.9 summarizes the four research clusters with selected research streams and possible avenues for future research. The topics for future research will be elaborated in the next section.

### 4.1.5 Discussion

The analysis of 28 years of platform research in e-commerce resulted in the identification of three phases of research and four topical clusters of research. Based on these four clusters, the authors derived potential streams for future research and formulated concrete research questions worth pursuing. For this analysis, set of papers was limited to the last four years and the most cited articles were investigated for future research endeavors. Eight avenues for future research are proposed to advance the body of knowledge on e-commerce platforms. These avenues are summarized together with exemplary research questions in Table 4.9.

#### 4.1.5.1 Business Model and Transaction Costs

The bibliometric cluster pertaining to business models and transaction costs of platform research in e-commerce includes two important streams for future research. The first stream is concerned with the description of e-commerce business models and with the digital transformation of affiliated participants in e-commerce ecosystems. Since participants in e-commerce ecosystems are likely to take on multiple roles (Hawlitschek, Teubner, Adam, et al., 2016; Wulfert and Schütte, 2022), future research can investigate the success of different role configurations (e.g., consumer-provider, seller-provider) and their impact on other e-commerce ecosystem participants. Moreover, platform owners' measures for providing transparency with regard to the actions of single participants can be analyzed (Cennamo, 2021). Future research can also examine how cross-channel effects impact the business model choice for ecosystem participants in multi-sided markets. A particular research question to pursue is the impact of the sales channel chosen (e.g., direct sales, platform sales) for product sales (Tian, Vakharia, et al., 2018). In this regard, future research can also develop metrics that support the decision for selecting the appropriate platform to affiliate with.

The focal platform of e-commerce ecosystems may induce a certain degree of digital transformation for affiliated participants. Hence, the focal platforms' impact on the organizational structure and business processes at these ecosystem participants needs to be analyzed (Li, Su, et al., 2018). Specific calls were made for longitudinal and international case studies on platform-induced digital transformation in e-commerce (Li, Su, et al., 2018). Furthermore, it might also be worthwhile to investigate how platforms wield power over affiliated participants with regard to their specific business models (e.g., multi-channel, omni-channel). Another research question can address specific measures the platform owners implement to wield power over affiliated participants.

Moreover, future research can develop new business models and platform configurations that go beyond current developments (e.g., branded product platform, live stream e-commerce) (Reinartz et al., 2019). This also involves the exploitation of new technologies (e.g., blockchain, mixed reality, quantum computing) for value creation (Reinartz et al., 2019). New business opportunities stem from the provision of enhanced customer word-of-mouth evaluations (e.g., reviews, comments) to other ecosystem participants (Ireland and Liu, 2018). Particularly, new text mining and fraud detection algorithms need to be developed to enhance review reliability. In addition, recommendation engines need to be improved to offer real-time comprehensive recommendations to participants with regard to products and transactions. These engines should incorporate a number of additional influencing factors, such as product returns, browsing and clicking behavior, and customers' relationships (Guan et al., 2019). Since omni-channel scenarios are of increasing popularity and are adopted by more and more companies (Verhoef, Kannan, et al., 2015), detailed investigations are necessary to determine the impact of different prices in different channels (Yan et al., 2020).

#### 4.1.5.2 Social Commerce

While analyzing the literature on platforms in the context of social commerce, the authors identified two major streams for future research. First, platforms as intermediaries in social commerce cumulate huge volumes of customer-generated content and electronic word of mouth in the form of customer reviews and recommendations. Since the virtual environment makes product testing quite complicated, reviews of previous customers are a major source for product evaluation, trust in transaction partners, and risk aversion. In social commerce, this involves a variety of roles (e.g., customers, sellers) and different settings (e.g., mobile commerce) (Gibreel et al., 2018; Kong, et al., 2020). Hence, future research not only needs to investigate means to increase the motivation for providing product and business partner reviews but also needs to investigate algorithms to evaluate the reliability of the reviews . Since existing research has already revealed potential intrinsic and extrinsic rewards for reviewers (Khern-am-nuai, et al., 2018), future research needs to verify alternative reward methods in representative settings and investigate how different platform configurations and designs impact the attraction of reviewers and the number of reviews. Moreover, the utility of customer reviews in general for different participant types needs to be analyzed (Chevalier et al., 2018; Wu, Fan, et al., 2018). Another research question worth pursuing would be the investigation of the impact of additional reviewer information (e.g., personality, demographics, financial information, risk perception, emotional state) when analyzing reviews in general and review utility in particular (Peng et al., 2019; Shen et al., 2019; Tran, 2021). The utility analysis also involves the effect of reviewer rewards on product sales and seller revenues. Furthermore, potential moderating factors for the relationship between extrinsic motivation and intention to contribute customer reviews should be considered (Wang, Lin, et al., 2019).

A second major research direction needs to investigate the role and impact of influencers on platforms in social commerce. While influencers have proven their power in marketing, their role in social commerce environments needs to be empirically tested (Mittendorf, 2018). A possible research question might be the possibility of trust transfer between influencer and different ecosystem participants. Moreover, the impact and utility of single influencers can be analyzed. Although existing research has developed metrics for assessments (Arora Bansal, et al., 2019), these metrics can potentially be enhanced by platform-specific metrics and information on the influencers' personality traits and network.

#### 4.1.5.3 Technology and Infrastructure

New or evolving technologies can have a big impact on and a lot of potential for e-commerce and associated platforms. A technology that is increasingly discussed in research and practice is the blockchain. First proposed in 2008 as the underlying technology for an electronic payment system (Nakamoto, 2008), blockchain is nowadays investigated for a lot of different use cases and industries. An idea that is particularly interesting for e-commerce platforms is the concept of blockchain-based marketplaces. Subramanian (2017) was the first to propose blockchain-based marketplaces, which are supposed to be a counter design of conventional marketplaces with one central controlling instance or company (Karger, et al., 2021). Perhaps it might be worthwhile to investigate the potential that blockchain could have as an underlying technology for platforms in e-commerce, or if blockchain-based marketplaces could form a new type of e-commerce platform. Furthermore, recent research believes that blockchain could serve as a technical foundation and system for data quality improvement and assurance (see, e.g., An et al., 2020; Karger, et al., 2021; Wang, Liu, et al., 2018). Future research might address if and how blockchain could lead to benefits when it comes to the management of data that are generated by, or that are part of, e-commerce platforms. Finally, it might be interesting to further analyze potential benefits, risks, and implications of using and accepting cryptocurrencies in e-commerce platforms (Albayati et al., 2020).

Another topic of increasing relevance is the analysis of data. The use cases and benefits of analyzing data in the context of e-commerce is manifold. In existing research, different algorithms and technologies are used for the purpose of data analysis. Examples include graph neural networks (Fan et al., 2019), deep learning (Yang, Li, et al., 2020), and methods of reinforcement learning (Hu, et al., 2018). In previous research, it can be observed that individual algorithms and methods have been used for specific applications or scenarios. According to the authors' understanding, there is a lack of comparative overviews comparing different methods for one use case. There is still little knowledge about the extent to which a particular algorithm or method of analysis is superior to another algorithm or method of analysis for a given use case (Hütsch and Wulfert, 2022a). Future research should therefore focus to investigate what technology is most suitable for a given use case by comparing different results and algorithms.

#### 4.1.5.4 Socio-Technical System

IS research as discipline is concerned with knowledge about the interaction of the technological and social systems. Besides technical application services, IS research also includes organizational and human factors (Martinsons and Chong, 1999; Petter et al., 2012). The authors identified two potential streams for future research into e-commerce platforms: technology acceptance and platform configurations.

An important aspect that is often considered within IS research is technology acceptance. A variety of models were developed to measure how technology is perceived and accepted by its users. Among the most popular ones are different Technology Acceptance Models (TAMs) (e.g., Venkatesh, Viswanath, et al., 2003) and the Unified Theory of Acceptance and Use of Technology (UTAUT) model

(Venkatesh, Thong, et al., 2016). In the context of e-commerce, there are a few examples of attempts that tried to measure acceptance based on the UTAUT (Chen, Rashidin, et al., 2021; Habeeb et al., 2021; Odusanya et al., 2019) or TAM (Panjaitan et al., 2019; Yadav and Mahara, 2019). These studies are, however, most often focused on, for example, specific kinds of enterprises (Habeeb et al., 2021) or certain types of products (Yadav and Mahara, 2019). However, to date, research has not yet investigated acceptance for all products, countries, and industries (Cui et al., 2019). Furthermore, as stated, new types of e-commerce platforms might emerge, for example, blockchain-based marketplaces. These new technical designs could also have implications for acceptance, which still need to be addressed by future research.

E-commerce involves electronic transactions among a variety of different actors. In the virtual environment, these transactions rely on participant information as signals in electronic transactions (Meents and Verhagen, 2018). However, the ways in which the participant's signals impact purchase intention and repurchase intention are still vague in research. E-commerce platforms can standardize and validate participant signals. Those platforms also offer customers a central touchpoint with a unified user interface for a wide assortment even across multiple devices. Beyond standardized user interfaces, future research may investigate how actors can distinguish themselves in standardized and predefined platform environments. Furthermore, the impact of different platform configurations and shop themes on customer experience, satisfaction, and loyalty is worthwhile investigating (Faraoni et al., 2019). Moreover, research on domain-specific (e.g., e-commerce) conceptualizations of platforms and their role in overarching ecosystems is still sparse (de Reuver et al., 2017; Wulfert and Karger, 2022b). With successful platforms evolving toward hybrid configurations providing application services and attracting third-party developers (Gawer, 2021b), the degree and impact of integrating innovation platform services in e-commerce can be analyzed by future research. For hybrid platforms, the integration and harmonization of transaction and innovation services can be further analyzed with reference to business-IT alignment considerations. Future research may also involve longitudinal analyses of successful hybrid platforms. Platform configurations need also to be reflected in current and future enterprise architecture investigations. Future research can also develop architecture models for ecosystem-wide collaboration in general and e-commerce platforms in particular. Table 4.9 summarizes the identified research streams and potential research questions for future research:

**Table 4.9** Summary of research streams and research questions

| Cluster | Research Streams | Potential Research Questions |
|---|---|---|
| Business Model and Transaction Costs (Blue) | Description of business models, digital transformation of ecosystem participants | What are the impacts of specific role configurations in e-commerce ecosystems? How can the appropriate platform in e-commerce be determined for affiliation? How are e-commerce ecosystem participants transformed, from a longitudinal perspective, when affiliating to a specific platform? What (strategic) measures can platform owners implement to wield power over affiliated participants in e-commerce? What are the future platform-based business models in e-commerce? How can new technologies be exploited to generate value in e-commerce? How can recommendation engines be improved and customized using emerging technologies? |
| Social Commerce (Green) | Value of customer-generated content, role of virtual influencers on platforms | How can the motivation for creating reviews in e-commerce be increased? What are successful means of reviewer rewards and platform configurations that would lead to accelerated review generation and provisions? How can the utility of customer reviews be increased for ecosystem participants? How can trust be transferred using influencers in social commerce? How can metrics measuring influencers' impact on participants and transactions in social commerce be improved? |
| Technology and Infrastructure (Red) | Application of blockchain in platforms, advanced data analysis | How could blockchain be used in the context of e-commerce platforms? How can e-commerce platforms be implemented as blockchain-based marketplaces? How can blockchain be used for the management or storage of data in the context of e-commerce? How should a blockchain be designed when it is used in the context of e-commerce platforms? What are implications of accepting cryptocurrencies on e-commerce platforms? What is the best algorithm or technology for analyzing user behavior in e-commerce platforms? |
| Socio-Technical System (Yellow) | Technology acceptance, platform configurations | What are efficient user interface configurations for e-commerce? How can participants differentiate themselves on standardized platforms in e-commerce? How can an efficient user interface in e-commerce be designed and evaluated? What is the impact of new technologies on user acceptance in e-commerce? How can the technology acceptance of e-commerce be measured? What are successful platform configurations in terms of customer experience, satisfaction, and loyalty? How can innovation services propel the success of platforms in e-commerce? |

### 4.1.6 Conclusion

Besides an increase in research interest with 1,226 publications in 2021, a continuous development of the research field is depicted, i.e., platforms in e-commerce. The conducted keyword analysis revealed three phases of platform research in e-commerce. While the first phase depicted the technical foundation for conducting electronic business, the second phase is characterized by research into a variety of business models. The third phase is centered around emerging commerce on social media platforms and the exploitation of customer-created content and word of mouth with new technologies. Overall, the keyword co-occurrence analysis showed four clusters of platform research in e-commerce. The first cluster involved publications related to platform business models in e-commerce. The second cluster was concerned with social commerce platforms. The third cluster focused on innovation platforms as necessary infrastructure for e-commerce. The fourth cluster investigated the socio-technical characteristics of platforms in e-commerce. While the first cluster addressed transaction platforms and the third cluster focused on innovation platforms, the second cluster and the fourth cluster contained research into both platform types. For each cluster, potential research questions for future research were derived.

With this article, the authors make the following contributions. First, this paper aimed to conduct a quantitative bibliometric analysis of research into platforms in e-commerce. The last 289 years of the research field from its first appearance in 1993 involving 7,463 publications was analyzed. Hereby, a structured overview of this research field was given by identifying relevant scientific clusters, the most cited articles, and the most important topics and keywords during different periods. Furthermore, this research aims to establish a first foundation and structure for future research into e-commerce platforms. Based on an in-depth analysis of recent articles, several interesting research streams and questions for future research were derived.

This study has a few limitations and constraints that must be considered. First, given the large number of identified publications, it was not possible for us to conduct a manual analysis of all articles. Therefore, although it is able to give an overview of platform research in e-commerce, a detailed analysis of this topic and its areas of research cannot be given. Instead, the authors limited the qualitative analysis to a limited number of recent articles to identify future research opportunities. Nevertheless, the presented overview should guide future research in selecting relevant research gaps. Furthermore, the data collection for the bibliometric analysis relied on only one scientific database, namely Scopus. Although Scopus is a widely used database for bibliometric studies and covers a large number of relevant sources, other databases might contain publications and articles that were not covered by this study. Moreover, Scopus only contains scientific sources that are published in, for example, journals, books, and conference proceedings. Other documents like grey literature, patents, and papers published on platforms like arXiv were therefore intentionally not considered in the sample. Since the article's scope was to analyze scientific literature, it is therefore possible that additional types of sources might lead to additional insights.

This article identified two major avenues for future research. First, research might investigate in more details the areas of interest the authors identified and answer the potential research questions summarized in Table 4.9. Second, future research can extend the scope of this bibliometric analysis by considering additional scientific databases and more practitioner-oriented grey literature.

## 4.2 Paper 2: Shaping Digital Platforms in E-Commerce—Developing an Architecture Framework

**Abstract**

*In 2024, the global e-commerce revenue is expected to reach $ 6.4 trillion,*[3] *covering 21,8% of overall retail sales.*[4] *Digital platforms*[5] *form the core of e-commerce ecosystems, supporting various e-commerce business models with dedicated application services. In e-commerce, there are a variety of platforms with different services and architectures. Currently, however, there is no framework for digital platforms in the specific context of e-commerce. Hence, we apply a design science approach and develop a framework for digital platforms in e-commerce. We analyzed the architecture of selected digital platforms in e-commerce and conducted expert interviews. As a result, our article is the first to propose a tailored platform architecture for the e-commerce context, which is composed of four layers.*

[3] In the original manuscript, the number was incorrectly given as quadrillion. To maintain consistency of content, we have corrected the error accordingly (EMarketer, 2022).

[4] This version of the article has been accepted for publication, after peer review but is not the Version of Record and does not reflect post-acceptance improvements, or any corrections. The Version of Record is referenced as follows: T. Wulfert and E. Karger (2022b). "Shaping Digital Platforms in E-Commerce: Developing an Architecture Framework." In: *Proceedings of the 26th Pacific Asia Conference on Information Systems (PACIS2022)*. Virtual Conference, pp. 1–17

[5] Throughout this cumulative dissertation, we use the term innovation platform instead of digital platform. In using this terminology, we would emphasize the difference to transaction platforms. Transaction platforms are a possible business model powered by innovation platforms in the architecture framework developed.

*The framework involves different ecosystem participants and a marketplace for the provision of third-party extensions. Furthermore, we outline future research opportunities and give an outlook on the further development of our framework.*

| Fact sheet of publication 2 | |
|---|---|
| Title | Shaping Digital Platforms in E-Commerce: Developing an Architecture Framework |
| Authors | Tobias Wulfert<br>Erik Karger |
| Publication Type | Conference Proceeding |
| Publication Outlet | Pacific Asia Conference on Information Systems |
| VHB-Jourqual3-Ranking | C |
| Status | Published |
| Research Question | How should an architecture framework for platforms be designed for the context of e-commerce? |
| Research Approach | Design Science Research (Hevner, March, et al., 2004) |
| Full Citation | T. Wulfert and E. Karger (2022b). "Shaping Digital Platforms in E-Commerce: Developing an Architecture Framework." In: *Proceedings of the 26th Pacific Asia Conference on Information Systems (PACIS2022)*. Virtual Conference, pp. 1–17 |

### 4.2.1 Introduction

In 2024, the global e-commerce revenue is expected to reach $ 6.4 trillion, covering 21,8% of overall retail sales (eMarketer, 2020). Retail transactions are conducted electronically to a certain extent in e-commerce (Turban, Whiteside, et al., 2017). Brick-and-mortar retailers have faced increasing competitive pressure in recent years, especially from e-commerce. This process has been intensified, in particular by the COVID-19 pandemic (Nicola et al., 2020). As a result, brick-and-mortar retail is increasingly penetrated with information technology such as in-store technology (Grewal, Roggeveen, et al., 2017), shopping companion apps (Wulfert, Betzing, et al., 2019), smart products and additional smart services (Beverungen, Müller, et al., 2019). Electronic transactions and the increasing penetration with technology pose additional requirements for retailers' information systems and offer business opportunities for software providers and third-party developers.

Digital platforms can provide brick-and-mortar and e-commerce retailers with the necessary IT functions for a number of different business models (Tiwana, 2014). In research and practice, the term "platform" has varying connotations. These describe a platform as a digital marketplace enabling economic transactions, or define it as a technical foundation for the development of extensions and innovations (McIntyre and Srinivasan, 2017; Sun, Gregor, et al., 2015; Tiwana, Konsynski, and Bush, 2010). We refer to digital platforms as an "extensible codebase of a software-based system that provides core functionality shared by the modules that interoperate with it and the interfaces through which they operate" (Tiwana, Konsynski, and Bush, 2010, p. 676). The digital platforms can form the "technical core" (Blaschke, Haki, et al., 2019, p. 4) of companies and the surrounding ecosystem. As a technological artifact, digital platforms can be the foundation upon which (third-party) developers can create additional modules (e.g., localizations, themes) based on the digital platform services provided (Baldwin and Woodard, 2009; Tiwana, Konsynski, and Bush, 2010).

In the context of e-commerce, digital platforms need to support a variety of business models, such as e-shops, e-auctions, and digital marketplaces, and new formats, such as live stream shopping and virtual shopping (Laudon and Traver, 2019; Timmers, 1998). For the support of these business models in e-commerce, there are more than 600 different digital platforms (e.g., WooCommerce, Shopify) that compete for retailers and third-party developers (Datanyze, 2021). Third-party developers can extend these platforms with additional modules, providing additional functionality and user interface themes (Eaton et al., 2015). Digital platforms and business ecosystems have already been investigated from different angles and perspectives. For instance, Alami et al., 2015 analyzed three digital platforms fostering business models in e-commerce focusing on ecosystem health. Akatkin et al., 2017 proposed a framework for partitioning industries in semantic core, actors, and offered digital services with the core obtaining platform characteristics that enable the development of additional modules for the expansion of the ecosystem. Henfridsson, Nandhakumar, et al., 2018 described the structure of innovation ecosystems with connections between devices, networks, services, and contents. Evans and Gawer 2016 took a more holistic perspective, differentiating between innovation platforms, transaction platforms, integrated platforms, and investment platforms on a company level.

Although general conceptualizations of digital platforms exist (e.g., Blaschke, Haki, et al., 2019; Derave et al., 2020) and although calls for industry-specific adaptations have been raised (de Reuver et al., 2017), research on digital platforms as technological foundation in e-commerce is sparse. Compared to other industries, e-commerce poses a number of additional requirements for digital platforms. Among others, digital platforms in e-commerce need to deal with a diverse set of ecosystem participants, need to allow a digital representation of the different assortments of the participants, need to deal with a high fluctuation in the number of transactions (e.g., holiday season), need to deal with new shopping trends (e.g., virtual shopping), and need to deal with specifics of retailers' business models (Schütte and Wulfert, 2022a; Teller and Elms, 2010; Turban, Whiteside, et al., 2017; Wulfert and Schütte, 2022). Given the high relevance that digital platforms have in practice for e-commerce, the above-mentioned lack of research is surprising. Due to the variety and heterogeneous structure of available digital platforms in e-commerce, it is difficult for retailers to select an appropriate platform. A partitioning of e-commerce ecosystems can assist retailers in the selection of a digital platform by structuring participants, by providing pathways that link participants and other platforms the participants want to affiliate with, and by highlighting the role of the digital platform (Wulfert, Woroch, Strobel, Seufert, et al., 2022). Accordingly, we consider it necessary to close this research gap. We intend to achieve this by adding an explicit partitioning of e-commerce ecosystems from an architectural perspective to the current body of knowledge and to provide practitioners with a tailored platform architecture framework for e-commerce. Addressing this research gap, and following de Reuver et al.'s (2018) call for industry-specific applications of the digital platform concept, this article addresses the following research question -

> RQ: How should an architecture framework for platforms be designed for the context of e-commerce?

For the identification of a platform design in e-commerce and for the development of an e-commerce platform architecture framework, we apply a design science theory (DSR) approach (Hevner, March, et al., 2004). We develop an enterprise architecture perspective considering single architecture layers as theoretical lens (Niederman and March, 2019) and incorporate knowledge from a literature review on platforms in e-commerce to partition and structure e-commerce ecosystems in general and digital platforms in e-commerce in particular (Wulfert and Busch, 2022; Wulfert, Woroch, Strobel, Seufert, et al., 2022). Leveraging on this perspective, we analyzed three digital platforms in e-commerce in detail and aggregated their architectural patterns for a general e-commerce platform architecture framework (Yin, 2018). We augment this draft with ten interviews conducted with domain experts. Our platform framework for e-commerce provides managers guidance on the selection of digital platforms and participants management. Furthermore, our platform framework also contributes to the scientific literature by providing an explicit partitioning of e-commerce ecosystems to position existing and future research.

The remainder of this research paper proceeds as follows: First, we introduce related literature concerning our theoretical lens on enterprise architecture layers and platform types in e-commerce. Second, we sketch our DSR approach comprising a multiple-case analysis and expert interviews. Third, we present our e-commerce platform architecture framework. Finally, we discuss our findings, provide a research outlook, and briefly summarize our results.

## 4.2.2 Theoretical Background

### 4.2.2.1 Enterprise Architecture Perspective as Theoretical Lens

An enterprise architecture can be described as a coherent whole of principles, methods, and models that are used in the design and realization of an enterprise's organizational structure, business processes, information systems, and infrastructure (Jonkers et al., 2006). The enterprise architecture, therefore, reflects the fundamental organization of a system and its relationships to the environment (Keller, 2017). Since companies are increasingly permeated by technology, enterprise architecture management needs to consider information systems at the application layer and infrastructure layer and their alignment with the corporate strategy and business model (Ahlemann, 2012; Pereira and Sousa, 2004; Schekkerman, 2004; Winter and Fischer, 2006). Organizations and researchers have established several reference frameworks for enterprise architecture management (e.g., The Open Group Architecture Framework (TOGAF), the Zachman Framework, Architektur integrierter Informationssysteme (ARIS)). However, these frameworks need to be instantiated to be applied in concrete organizational contexts (Cloutier et al., 2010).

Different architecture types, such as business, application, and infrastructure architecture, are formed for the individual components of an enterprise architecture that are developed using specific modeling languages like ArchiMate and Multi-Perspective Enterprise Modelling (MEMO). The structure of enterprise architectures usually follows a layered approach that reflects the proximity or distance of business problems to information systems. The following layers are established in enterprise architecture management literature: the strategy layer, the business actor layer, the business process layer, the application layer, and the infrastructure layer (Fig. 2) (Ahlemann et al., 2012; Schekkerman, 2004; Winter and Fischer, 2006). The strategy layer offers a strategic perspective on an enterprise and involves a business model and a company's value propositions. The business actor layer consists of internal and external actors of an organization. The business process layer involves processes to support the business model and the external provision of products and services. The application layer consists of applications services and components to support the business processes. The infrastructure layer contains hardware and software for the execution of application components (Schekkerman, 2004; Winter and Fischer, 2006). Information systems research is mainly concerned with information systems architectures that consider the business layer as well as the organizational structure and process organization of a company. Information systems architectures comprise a high-level sketch of a company's application architecture involving (parts of) the business layer and the infrastructure layer In line with literature on enterprise architecture in general and existing platform conceptualizations in particular (e.g., Blaschke, Haki, et al., 2019; Schekkerman, 2004), we apply a layered approach for the development of our e-commerce platform framework.

The layered approach is applied by general platform frameworks and reference architectures in e-commerce (e.g., Aulkemeier, Schramm, et al., 2016; Blaschke, Haki, et al., 2019). While Aulkemeier, Schramm, et al., 2016 differentiate a business layer, an application layer, and a technology layer for their service-oriented architecture, Blaschke, Haki, et al., 2019 orient their platform taxonomy along a service layer, an ecosystem layer, a core layer, and an infrastructure layer. Huang, Chai, et al., 2019 focus on a participant layer and a platform layer defining several standards for the exchange of information. Staykova and Damsgaard, 2015 also replicate this differentiation in their "core" and "periphery" classification. Hein, Schreieck, Riasanow, et al., 2020 partition digital platform ecosystems into an ecosystem layer, a boundary layer, and a digital platform layer.

To enable cooperation among organizations and interaction of information systems across organizational boundaries and architectural layers (Santos and Eisenhardt, 2005), boundary resources are required to translate and map information between potentially different socio-technical systems. Boundary resources enable different ecosystem participants and information systems to interact without direct coordination Star, 2010; Star and Griesemer, 1989. Furthermore, boundary resources also allow access to modules of an information systems architecture and integrate developers as additional participants in an ecosystem (Eaton et al., 2015). These resources can also be used as a control mechanism for the application layer and the infrastructure layer (Ghazawneh and Henfridsson, 2013).

#### 4.2.2.2 Platform Types in E-Commerce

There are different notions for understanding digital platforms and various research areas investigating platforms (McIntyre and Srinivasan, 2017; Sun, Gregor, et al., 2015). Platforms are investigated in four research streams (Gawer, 2014; Sun, Gregor, et al., 2015; Thomas et al., 2014): (1) as organizational capabilities and business infrastructure in the organizational capabilities literature (Rai et al., 2006), (2) as a technology set on which a product family can be established (Fichman, 2004), (3) as a two-sided or multi-sided market as intermediary in the economics literature (Armstrong, 2006; Rochet and Tirole, 2006; Tan, Lu, et al., 2015), and (4) as a system or architecture for extension development and execution in the operation management literature (Cusumano and Gawer, 2002; Tiwana, Konsynski, and Bush,

2010; West, 2003). The technological view and the economic view (Bughin and van Zeebroeck, 2017; Gawer, 2014; Reillier and Reillier, 2017; Thomas et al., 2014) are especially relevant in the context of e-commerce (Hagberg et al., 2016; Hagiu and Wright, 2015b; Hänninen et al., 2018). From the perspective of our theoretical lens described above, the technological view resembles the application layer and the infrastructure layer, while the economic view is more concerned with the business layer and related business processes. In this research, we focus on the digital platform as the "technical core" (Blaschke, Haki, et al., 2019, p. 4) of an e-commerce ecosystem supporting a focal company's business model (e.g., e-shop, e-auction, digital marketplace) (Asadullah et al., 2018b; Wulfert, Woroch, Strobel, Seufert, et al., 2022).

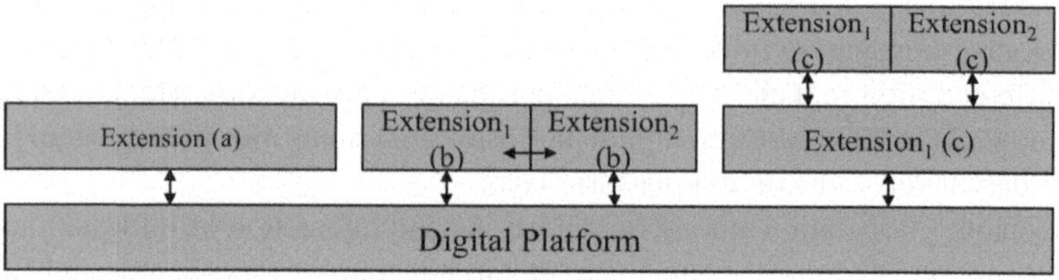


**Fig. 4.6** Types of Extensions on Digital Platforms (Jansen and van Capelleveen, 2013)

Digital platforms form the operating systems for e-shops, digital marketplaces, and other business models, and provide the application services necessary for executing the business processes and environments for developing external modules (Aulkemeier, Paramartha, et al., 2016; Hanseth and Lyytinen, 2010; Tiwana, Konsynski, and Bush, 2010). Baldwin and Woodard (2009, p. 19) define a digital platform "as a set of stable components that supports variety and evolvability in a system by constraining the linkages among the other components." Tiwana et al. (2010, p. 675) emphasize that third-party developers can use these stable components as an "extensible codebase [...] that provides core functionality shared by the modules that interoperate with it and the interfaces through which they interoperate." While the stable components form the core of the digital platform, the third-party modules resemble the periphery (Staykova and Damsgaard, 2015). The platform core and periphery are evolvable; only the interface specifications should remain stable over a longer period of time to avoid necessary module adjustments for the developers (Baldwin and Woodard, 2009). Interfaces correspond to the boundary resource concept introduced in the theoretical lens. A digital platform can also be described as "layered modular technology architectures in business networks" (Kazan, Tan, et al., 2018, p. 186). This perspective fits to the theoretical lens introduced above. While digital platforms usually exploit economies of scale and scope with increased efficiency, product variety through reusability, and reconfiguration of modules or services, they may utilize further economic effects as the center of a broader innovation ecosystem Cusumano and Gawer, 2002; Gawer, 2009b; Rochet and Tirole, 2003. Third-party extensions and innovations based on a digital platform result in the formation of an innovation ecosystem also responsible for disseminating developed extensions (Jansen and Cusumano, 2013; Messerschmitt and Szyperski, 2003). Jansen and van Capelleveen, 2013 have introduced three types of interfaces on digital platforms. These are (a) independent extensions, (b) extensions that can communicate among each other, and (c) interdependent extensions that build upon other extensions (Fig. 4.6).

The ecosystem surrounding a pivotal digital platform can be described along the dimensions technological core, ownership of the platform, availability of an extension market, and affiliation of third-party developers Jansen and Cusumano, 2013. The power of digital platforms rests on their architectural modularity (Baldwin and Clark, 2000; Tiwana, Konsynski, and Bush, 2010), catalyzing the re-configurability of technical and organizational components to accelerate generativity and value creation. The components of single extensions are strongly interconnected with each other but weakly connected to the digital platform through technical boundary resources (Baldwin and Clark, 2000; Tiwana, 2015a). This orchestration of the focal platform and surrounding extensions results in a system "with intentionally minimized interdependences between the single subsystems" (Sanchez and Mahoney, 1996, p. 65). The concept of modularity results in a high degree of generativity allowing for modules and business models initially not intended by the platform owner. The generativity is caused by the shared innovation and contribution of a diversified audience of third-party developers whose actions are only partially controllable (Witte and Zarnekow, 2018).

### 4.2.3 Design Science Research Approach

To pursue the research question of this research paper, we apply a DSR approach and present a first design iteration of a platform framework for e-commerce (Fig. 4.8) (Hevner, March, et al., 2004). This framework can be described as a model, or more particularly as an architecture (March and Smith, 1995; Vaishnavi, Kuechler, et al., 2019). Our framework is designed as a "meta-artifact" (Iivari, 2003) and represents a "general solution concept" (Vaishnavi, Kuechler, et al., 2019) that is applicable to the implementation of platforms powering business models, for instance, shops and marketplaces, in e-commerce. With reference to our theoretical lens, this research paper focuses especially on the application layer and its support of various business models in e-commerce. DSR is a research method aiming to create artifacts that propose new solutions to real-world problems (Vom Brocke and Maedche, 2019). Besides newly designed socio-technical artifacts, design knowledge that provides a fuller understanding of why the artifacts enhance the relevant application contexts can be a goal and a result of a DSR process, too (Gregor and Hevner, 2013; Vom Brocke, Winter, et al., 2020). A DSR approach should involve three different cycles (i.e., the relevance cycle, the design cycle, and the rigor cycle) connecting the three DSR pillars environment, knowledge base, and design (Hevner, 2007). While the rigor cycle grounds our design in scientific literature, the relevance cycle adds knowledge from our application domain (i.e., e-commerce)-gathered through three case studies and ten interviews with domain experts-to the design cycle. In the design cycle, we abstract from the information collected in the relevance cycle and the rigor cycle and aggregate them to an architecture overview of platforms in e-commerce. We present the results of a first design iteration in this research paper. For the rigor cycle, we developed our theoretical lens in an enterprise architecture perspective (Ahlemann et al., 2012; Schekkerman, 2004; Winter and Fischer, 2006). We also leveraged existing research on enterprise architectures and patterns in existing literature (Derave et al., 2020; Hütsch and Wulfert, 2022b; Wulfert and Schütte, 2022). We use existing architectures and ontologies in e-commerce as a basis for our design phase.

#### 4.2.3.1 Multiple-Case Study

For the relevance cycle, we conducted a multiple-case analysis on three digital platforms in e-commerce (i.e., WooCommerce, Shopify, Magento) (Eisenhardt, 1989b; Yin, 2018). These platforms provide the necessary application services for retailers to conduct e-commerce transactions (Wulfert, Woroch, Strobel, Seufert, et al., 2022). The multiple-case study approach allowed us to analyze the "same phenomenon in [...] in a variety of situations" (Yin, 1981, p. 101). The shared phenomenon of our three cases is their success in e-commerce measured by number of supported e-commerce websites (Builtwith, 2022; Datanyze, 2021) and their design as an extensible codebase enabling third-party developers to design third-party extensions (Tiwana, Konsynski, and Bush, 2010). We have selected these platforms to derive knowledge from successful designs of the past and go "beyond a single success story" (Chandra Kruse and Seidel, 2017, p. 186). The extensions to the platforms can significantly alter the scope of the e-commerce platforms and supported business models.

While Eaton et al., 2015 and (Ghazawneh and Henfridsson, 2010) analyze a single digital platform with its surrounding ecosystem (i.e., iOS), Ritala, Agouridas, et al., 2013 comparatively investigate two innovation ecosystems and Hänninen et al., 2018 analyze four competing digital marketplaces (i.e., Alibaba, Amazon, eBay, Rakuten) in a multiple-case study approach. Following the layered architecture perspective for enterprises established in our theoretical lens (Fig. 4.7), our case selection focuses on the application layer providing the necessary service for executing retailers' business models in e-commerce. Introducing a theoretical lens to a multiple-case study is a common vehicle in information systems research in general (Battleson et al., 2016; Cao et al., 2009; Tyworth, 2014) and e-commerce research in particular (Park et al., 2012; Phang et al., 2010). We selected the digital platforms of WooCommerce, Shopify, and Magento Commerce for our case analysis. These three platforms have a high and international reach, are the ones with the most active users, support many business models in e-commerce, and have dedicated extension markets to even extend the functionality with third-party apps. Moreover, these e-commerce platforms are used most by the top websites (website usage) in e-commerce (Builtwith, 2022). For these reasons, we consider the selected platform to be a good cross section of available digital platforms for e-commerce (Cichocki, 2018).Our sample therefore allows for a high generalizability for the first iteration of the platform architecture framework in e-commerce.

For the platform analyses, we leveraged a plethora of primary and secondary data sources to collect architectural information on each case that is publicly available (Hänninen et al., 2018). First, we searched for information on the official websites of the three platforms. In this part of the analysis, we tried to limit ourselves to objective information. We critically analyzed advertising texts or content with the intention of making a profit and did not include them in our analysis if we had any doubts. We also analyzed information and texts from other media that dealt with the three platforms. These were, for example, newspapers, industry journals, and practitioner magazines. The case data is analyzed through the theoretical lens of enterprise architecture, with the objective of identifying architectural elements and orchestrating in an individual architecture for each platform (Niederman and March, 2019). Consequently, we aggregated the individual architectures to an overall e-commerce architecture framework and augmented it with theoretical findings from our literature analysis. Table 4.10 summarizes meta-information on the selected digital platforms and particular companies.

**Table 4.10** Selected Digital Platforms in E-Commerce

| | WooCommerce | Shopify | Magento Commerce |
|---|---|---|---|
| Responsible Organization | Automattic | Shopify | Adobe |
| Headquarters | San Francisco, USA | Ottawa, Canada | San Jose, USA |
| Website usage | 29% | 19% | 8% |
| Founded | 2008 | 2006 | 2005 |
| DP Launch | 2011 | 2006 | 2008 |
| Recent Version | 5.9.0 | 2021–10 | 2.4.3-p1 |
| Number of e-commerce websites | 3.3 Mio. | 3.6 Mio. | 186,000 |

WooCommerce is an extension to the free content management system, WordPress, adding e-shop functions. WooCommerce's code relies to a large degree on PHP and MySQL. As an open-source project, it can be extended and adapted by additional extensions (Patel et al., 2011). WooCommerce was initially published as a digital platform in e-commerce by WooThemes in 2011. WooCommerce executes more than 3.3 million e-commerce websites. The WooCommerce ecosystem is based on WordPress as a software platform and is coordinated by a privately owned entity (Automattic) with multiple extension markets to which participants can submit extensions for free. WooCommerce features a headless architecture decoupling the user interface on various devices from the backend processes.

Shopify is a digital platform that allows customers, such as retailers and manufacturers, to open an e-shop and sell products online. In 2004, the development of the platform started after the developers were unable to find a satisfactory software platform for an online store. The platform was made available for the public in 2006. Shopify executes more than 3.6 million e-commerce websites. Furthermore, Shopify enables retailers to create and design e-commerce websites, handle payment processes, manage orders, and track orders. The Shopify ecosystem is based on a software platform and is coordinated by a privately owned entity (i.e., Shopify Inc.) with an extension market to which participants can submit extensions after a screening. Shopify features a headless architecture to support an increasing amount of different customer touchpoints on different devices decoupling the user interface and backend.

Magento Commerce (Magento) is a digital platform that provides application services for e-commerce websites of varying business models. The Magento platform was initially launched in 2008. Several versions have been released since then. Magento executes more than 186,000 e-commerce websites. The source code of Magento is open source and Adobe offers a free, open-source version of its product for small companies and developers. This basis product is bundled with additional features and extended services in Magento Commerce Pro. The Magento Commerce ecosystem is based on a software platform and is coordinated by a privately owned entity (i.e., Adobe) with an extension market to which participants can submit extensions after making a payment.

#### 4.2.3.2 Expert Interviews

In addition to the case analyses, we conducted a total of ten interviews. The interviews were with enterprise architects and domain experts of leading e-commerce companies (e-business) and digital platforms, and were about their overall e-commerce architecture and platform interfaces. Initially, we approached more than 50 successful platforms and e-commerce companies. To allow for a maximum generalizability of the results, we approached experts from e-businesses in different e-commerce ecosystems and platforms supporting a variety of business models. The interviewees were asked about the structure of their enterprise architecture and if reference architectures (e.g., TOGAF) or a standardized enterprise architecture management approach is applied. Furthermore, the interviews included questions regarding each architectural layer and specific tools used to enable the particular business models within the application layer. As requested by the interviewees, we did not mention company names. None of the interviewees works for a case company mentioned above. We depict an overview of the ten interviews in Table 4.11. The interviews were conducted via electronic communication media in a semi-structured manner between December 2021 and March 2022. The interview structure in the form of an interview guide was sent to the interviewees in advance. The interviews were conducted either in English or in German depending on the interviewees' preferences, and were subsequently transcribed (Kuckartz, 2010). For the analysis of the transcripts, we apply a qualitative content analysis. Particularly, we applied the content-structuring approach with the architectural layers' initial set of codes (Aulkemeier, Schramm, et al., 2016; Mayring, 2014). We reached a stable set of codes already after the first five interviews. We chose a single sentence as the coding unit, while a paragraph of an interview can be considered as a context unit (Mayring, 2014). Multiple codes can also be applied to a single sentence if the sentence has several meanings. We used MAXQDA for the coding and the subsequent construction of a quantitative graphical representation (Kuckartz and Rädiker, 2022). We include exemplary quotes from the interviews in the results section. This additional information augments the architecture framework developed by analyzing the platforms. Data saturation with regard to our framework was achieved after the eighth interview (Guest et al., 2006).

**Table 4.11** Overview of Expert Interviews

| # | Interviewee's role | Number of employees | Revenue (in $) | Business type | Duration |
|---|---|---|---|---|---|
| 1 | Head of Product | ~200 | ~11 m | E-Business | ~35 mins |
| 2 | Account Manager | ~49,000 | ~16 bn | E-Business | ~55 mins |
| 3 | Team leader IT development | ~200 | ~1 m | E-Business | ~50 mins |
| 4 | Head of Engineering | ~4,700 | ~800 m | Platform | ~46 mins |
| 5 | Chief Technology Officer | ~25 | ~5 m | Platform | ~20 mins |
| 6 | Chief Digital Officer | ~500,000 | ~140 bn | E-Business | ~60 mins |
| 7 | Enterprise Architect | ~45 | ~11 m | Platform | ~20 mins |
| 8 | Enterprise Architect | ~7,000 | ~3 bn | E-Business | ~46 mins |
| 9 | VP Customer Success | ~270 | ~28 m | Platform | ~58 mins |
| 10 | Enterprise Architect | ~40 | n.a. | Platform | ~47 mins |

### 4.2.4 Platform Framework for E-Commerce

Based on our case analysis, we develop a general e-commerce platform architecture in analogy to Blaschke, Haki, et al., 2019 conceptualization of the general digital platform. However, we focused on e-commerce as requested by Reuver et al., 2017 in their call for industry-specific applications of the platform concept. The synthesized e-commerce platform architecture is depicted in Fig. 4.7. The framework consists of four architectural layers. A generalization of the multiple cases augmented by exemplary quotes from our interviews is used to describe the business actor layer, the business layer, the application layer, and the infrastructure layer

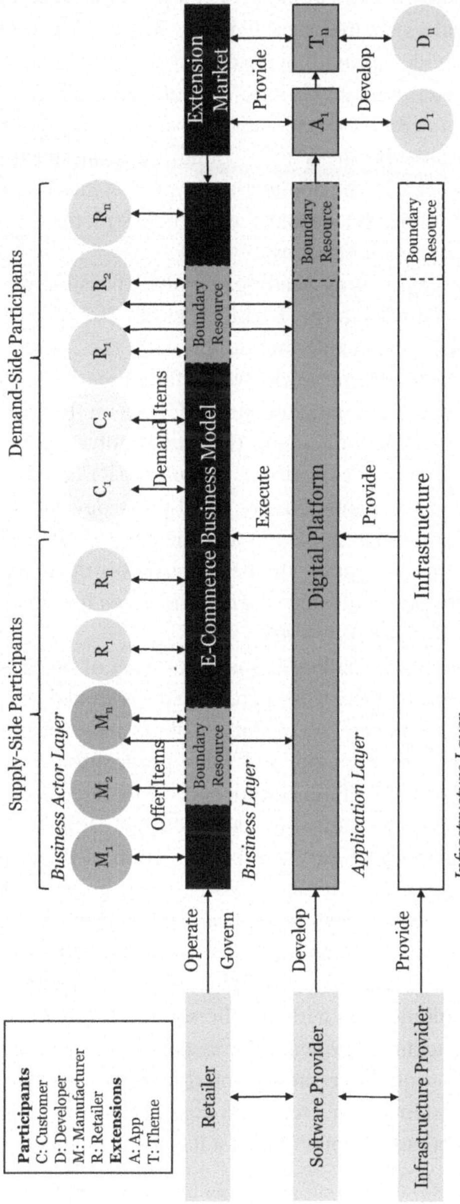


**Fig. 4.7** Platform Architecture Framework for E-Commerce

(top-down). We have identified customers, developers, manufacturers, and retailers (also includes wholesalers) as major participants in e-commerce ecosystems. Furthermore, software providers providing the digital platform to execute the necessary business services for e-commerce business models are required and infrastructure providers offering the necessary computation power are also required (Schütte and Wulfert, 2022a). While the business layer contains relevant business processes for e-commerce business models, the application layer provides necessary application services. These application services make use of infrastructure services. These four layers are outlined in more detail below.

The *business actor layer* (gray shading) involves possible participant types in an e-commerce ecosystem (Böttcher, Rickling, et al., 2021). We include demand-side (i.e., downstream) participants and supply-side (i.e., upstream) participants exchanging articles. In this framework, we differentiate between three different user groups and stakeholders. First, customers (C) are on the demand-side procuring merchandise, whereas manufacturers (M) are supply-side participants selling merchandise on an e-commerce platform. Retailers (R) can act as both supply-side and demand-side participants. Finally, third-party developers (D) provide the data ecosystem with additional apps (A) and themes (T). Retailers can integrate the developed extensions (i.e., apps, themes) to support the e-commerce business model chosen ("We provide dedicated coding guidelines for developers to ensure a minimum level of quality," interview #9).

The *business layer* (black shading) consists of a set of business processes necessary for the execution of the e-commerce business model that the retailer had selected. Although the concrete specification of business processes depends on the retailer's selected business model and value proposition, Becker and Schütte, 2004 introduce retail-specific reference processes. The three digital platform cases natively support the operation of e-shops as electronic storefront. The integration of additional extensions is necessary for the execution of different business models such as digital marketplaces of e-auctions (e.g., Shopify, WooCommerce). The interviewees have emphasized that retailers can either configure their own business processes in a number of platforms (e.g., interview #7) or they need to adhere to entrenched standards (e.g., interview #4). The business layer also involves boundary resources for the communication with suppliers and customers on a personal level and on an information technology level. For instance, suppliers can update product descriptions via Application Programming Interfaces (APIs) connecting a supplier's product information systems with a retailer's digital platform ("Of course, we offer several APIs to connect our customer's and adjacent companies' systems with our platform," interview #3).

The *application layer* (blue shading) consists of the digital platform and additional third-party extensions. The digital platform is developed by a software provider (e.g., Shopify, Magento) or a community of developers (e.g., WooCommerce) that provide the necessary application services to execute selected e-commerce business models ("We did not apply a given reference architecture. But as we are a large company, we have a core team of architects defining group-wide standards that each single e-commerce business needs to adhere to," interview #7). In the case of WooCommerce, an additional digital platform (i.e., WordPress) is required to execute the application services (i.e., nested platform). For the development of platform extensions, developers make use of the development services provided by the digital platform. Software providers intentionally implement technical boundary resources (e.g., APIs and Software Development Kits (SDKs) and provide social boundary resources (e.g., documentation and technical support) to allow for a diversity of third-party modules enriching the set of features provided by the digital platform (Eaton et al., 2015; Ghazawneh and Henfridsson, 2013). Third-party extensions can be apps (A) as feature add-ins (e.g., interfaces to merchandise management systems or other digital platforms, additional e-commerce functions, enhanced analytics) and shop themes (T). However, most of the interviewees from the application layer stated that they apply a headless approach, leaving the theme design to the retailer and focusing on the business processes ("We propagate an API-first approach, leaving the user interface design for diverse sets of devices to the retailer. We believe it is essential for them to differentiate from competitors," interview #9). These extensions make use of the boundary resources provided by the digital platform for third-party developers. Our case analysis revealed that these different types of extensions are often provided via dedicated extension markets. These extension markets can be provided by the platform owner (e.g., Shopify, Magento), the owner of a nested platform (e.g., WooCommerce), or by a third party (e.g., WordPress).

The *infrastructure layer* (white shading) provides the necessary infrastructure services, such as computing power and storage for the digital platforms, and hence the support of an e-commerce business model. WooCommerce allows for an on-premise installation on dedicated and virtualized servers. The digital infrastructure and related services can, however, also be provided as cloud services (e.g., Shopify, Magento) (Constantinides et al., 2018).

### 4.2.5 Discussion and Outlook

The developed platform architecture framework for e-commerce is a systematic representation of digital platforms from the perspective of enterprise architecture. Furthermore, it is a conceptualization of digital platforms in the specific context of e-commerce as requested by de Reuver et al., 2017. For practitioners, we propose prescriptive design knowledge for the selection of a digital platform to successfully establish an online sales channel and develop an e-commerce ecosystem with additional participants, utilizing our platform architecture framework for e-commerce. Moreover, third-party developers may use our architecture to position themselves adequately in such ecosystems and discover new business opportunities (Chandra Kruse, Seidel, and vom Brocke, 2019). We contribute to research on digital platforms by providing an explicit partitioning of e-commerce ecosystems from an architectural perspective. The architecture framework is also intended to support researchers in addressing specific points of interest in future research, such as digital platform architecture and the use of boundary resources in e-commerce ecosystems. The theoretical lens applied, together with the enterprise architecture perspective, is a useful vehicle for partitioning and structuring e-commerce ecosystems. This perspective led to a layered model of the ecosystem with a differentiation between the infrastructure layer, the digital platform layer, the business layer, and the business actor layer. Furthermore, this perspective is also aligned with the understanding of digital platforms as layered modular technology architectures (Kazan, Tan, et al., 2018). The following discussion is organized along the architectural layers introduced with the enterprise architecture lens in the background section and depicted in Fig. 4.7, followed by an elicitation on limitations and an outlook to future research on digital platforms in e-commerce ecosystems.

Considering the *business actor layer*, it is important to note that a single actor may also take on several roles within a single e-commerce ecosystem (Eisenmann, Parker, et al., 2009). A supplier who demands merchandise from an e-shop and a service provider who provides infrastructure services can therefore be the same actor For instance, in addition to the digital platform, WordPress also offers a hosting service for the infrastructure the forms the basis of WooCommerce. Adobe and Shopify usually bundle their digital platform with infrastructure services in cloud prescription models (McCombs and Banh, 2009; Neufeld, 2018). Digital platforms are an important vehicle to digitally connect participants from different markets and united societies (Hein, Schreieck, Riasanow, et al., 2020; Rochet and Tirole, 2003; Rochet and Tirole, 2006). Digital platforms in e-commerce establish new paths of value creation for retailers, software providers, and third-party developers (Hedman and Kalling, 2003). Retailers increasingly engage in software development, opening their self-developed digital platforms and providing them to other competitors. An example from Europe is the REWE Group with its spinoff commercetools and its subsidiary fulfillmenttools offering digital platform and fulfillment services for competing retailers. Besides its e-commerce business with e-shop and digital marketplace, Amazon provides infrastructure services via Amazon Web Services and also offers its digital platform for e-commerce to other retailers (Amazon, 2021). The increasing penetration of retail with information technology and the success of e-commerce offer new opportunities for software providers in an industry that previously neglected a necessary digital transformation (Schütte and Vetter, 2016). Developers can enhance the focal platform's scope of functions with extensions and generate revenue through the provision of such extensions (Ghazawneh and Henfridsson, 2013; Tiwana, Konsynski, and Bush, 2010). Fostering a high degree of generativity among developers and a diversity of extensions (e.g., feature add-ins, themes, connectors to other platforms and information systems), software providers can significantly enhance the value of their surrounding ecosystem and allow for unintended but successful developments (Aulkemeier, Paramartha, et al., 2016; Blaschke and Brosius, 2018). Across industries, the value generated by the surrounding ecosystem participants is up to three times higher than the revenue generated by a focal platform such as WooCommerce or Shopify (Delteil et al., 2020).

The *business layer* can profit from new business models enabled through the application of digital platforms in retail in general and e-commerce in particular (Grewal and Levy, 2007). Digital platforms, for instance, also foster sustainable development initiatives simplifying the setup for marketplaces selling refurbished merchandise and developing a circular economy (Schroeder et al., 2019). The increasing sales of smart products will lead to additional sales of smart services and retailers transforming from selling products to providing services (Beverungen, Müller, et al., 2019). This will also alter the business model from one-time sales to continuous service delivery and service fees. Besides digital platforms facilitating e-commerce business models and direct customer contact, more specialized platforms supporting selected tasks in e-commerce are established. An example of such a platform is fulfillmenttools offering logistics and fulfillment services to a network of brick-and-mortar stores acting as fulfillment centers. This transformation is intended to enhance customers' omni-channel experience with shortened delivery times and closer interaction with merchants.

From an *application layer* perspective, the selection of an appropriate digital platform is an important decision for retailers. Since there could be more than one retailer in a specific e-commerce sector, the selection of an appropriate digital platform supporting the retailer's specific business model is a crucial endeavor

(Senyo et al., 2019). Especially in sectors in which digital marketplaces dominate and winner-takes-all tendencies with strong network effects of incumbents limit the financial scope of new entrants, an appropriate digital platform can be a competitive advantage (Eisenmann, 2008; McIntyre and Srinivasan, 2017). Boundary resources are implemented to connect upstream participants with the supply chain to ensure a proper supply of articles for the ecosystem and a maximum level of article availability for downstream participants (e.g., customers). It might be useful to extend and customize the digital platform with extensions and themes available from third-party developers via dedicated extension marketplaces (Mantin et al., 2014). Our multiple-case study revealed that extension marketplaces usually accompany digital platforms in e-commerce. These marketplaces serve as a central repository for the enrichment of the digital platform with additional features. The extension of a digital platform might be necessary to support a retailer's business model in e-commerce (e.g., WooCommerce). The conducted business model should be considered when selecting a digital platform, as there are also platforms available that natively support other business models in e-commerce besides e-shops. For instance, commercetools natively supports digital marketplaces in e-commerce without the need for installing third-party extensions. Multihoming, the provision of extensions to a variety of digital platforms, would increase the market size for developers (Koh and Fichman, 2014). However, this approach is mostly only possible within the ecosystem of a specific company (Idu et al., 2011). Multihoming across digital platforms is usually restricted by platform-specific boundary resources that require an adaption of extensions for each digital platform. Since multihoming allows developers to easily switch between supported platforms, owners of digital platforms intentionally design proprietary boundary resources to lock in developers (Cennamo, 2021).

From an *infrastructure layer* perspective, standardization is expected for cloud services, such as storage and computing power (Blind and Mangelsdorf, 2016). Computational power is considered a commodity (Carr, 2003). Hence, retailers should focus on user interface design and accordingly procure digital platform services from software providers and infrastructure services (interview #7). Our multiple-case study revealed that the three digital platforms analyzed can be executed in a cloud with Shopify and Magento, being available as software as a service. Consequently, retailers procure services that enable their particular business model without a need for infrastructure or application management ("Procuring infrastructure services from cloud providers enable us to scale our e-commerce business flexibly, especially in periods with high demands such as holiday seasons," interview #6). The current state of this research also has its limitations. The downside of applying a theoretical lens in general is intentionally hiding information from reality (Niederman and March, 2019). As a result, we depict a possible second iteration of our e-commerce platform framework in Fig. 4.8. Future research may also analyze digital platforms in e-commerce from a different lens, focusing on additional but different aspects of the concept. Since digital platforms are part of ecosystems as socio-technical systems, the perspective of actor-network theory could lead to a better understanding of the interplay between ecosystem actors and information technology used (Kotusev and Kurnia, 2020; Tatnall, 2005). Structuration theory can also be applied to ecosystems in e-commerce for an analysis of digital platforms in their social environment with an action area, a structure, and capacity (Giddens, 1991). Structuration theory can also be used to analyze the role of digital platforms in societies that are increasingly penetrated with information technology, and is a perspective that is often applied in information systems research (Jones and Karsten, 2008).

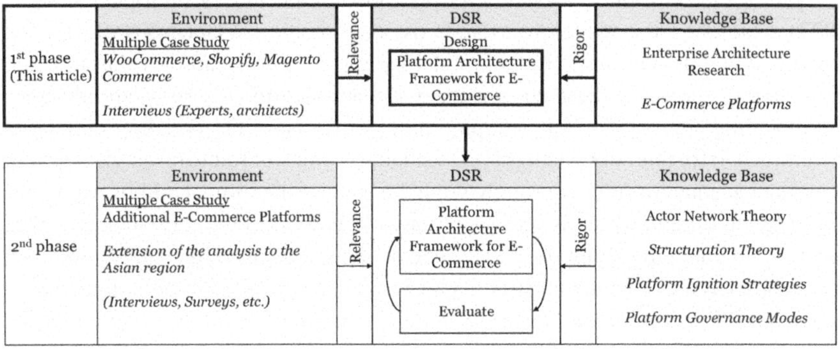


**Fig. 4.8** Design Science Research Approach of 1st and 2nd Design Iteration

Moreover, we did not further analyze different governance modes applied by the owners of digital platforms, and we disregarded strategies for the establishment of the digital platforms attracting a huge number of retailers and developers (Hein, Schreieck, Wiesche, et al., 2016; Parker and Van Alstyne, 2016). As part of its platform governance, Shopify reserves the right to close online stores that violate governance rules. Customers, who believe that an online store violates any of the governance rules, are encouraged to report this to Shopify such that Shopify can remove the online store, if necessary. Platform partitioning and structure of the e-commerce ecosystem can also be influenced by ignition strategies employed by software providers to establish their digital platforms in the market (Parker and Van Alstyne, 2016). To be successful, it is important for platforms to reach a threshold of retailers that execute their business models by using the digital platform and third-party developers that provide a variety of extensions (Alami et al., 2015). The attraction of developers can be amplified by implementing sophisticated boundary resources (Eaton et al., 2015; Ghazawneh and Henfridsson, 2013) and by opening the platform such that developers have access to additional platform services (Hein, Schreieck, Riasanow, et al., 2020).

Furthermore, we only focus on abstract elements of a digital platform in e-commerce and its surrounding ecosystem. The level of abstraction can be reduced by, for instance, discussing the developed framework with additional interviewees. Future research may analyze the identified elements of digital platforms in e-commerce in further detail. More specifically, architecture and process descriptions for business model configuration that consider the peculiarities of the e-commerce environment can be developed. Architectural patterns can therefore be developed for specific elements of the e-commerce platform architecture (Wulfert and Schütte, 2022). Extending the modeling to an ecosystem requires adaptations of existing enterprise architecture and process modeling approaches because of the dynamic and complex nature of these ecosystems (Anwar and Gill, 2019). Adapting existing modeling approaches or developing new domain-specific modeling languages (Frank, 2013) for this context might be a valuable avenue for future research. Another important avenue for future research would be the investigation of additional digital platforms in e-commerce, as our set of cases is limited to the three popular platforms WooCommerce, Shopify, and Magento. Our platform architecture framework for e-commerce can be demonstrated and instantiated with additional digital platforms (e.g., BigCommerce, salesforce commerce, ShopWired). Most importantly, additional platforms from the Asian-Pacific region should be incorporated in the second design iteration (Fig. 4.8). Furthermore, it may be interesting to consider other forms of digital platforms and their role in e-commerce ecosystems, such as platforms supporting fulfillment in omni-channel scenarios (e.g., fulfillmenttools). We also intend analyzing the role of boundary resources in platform-centered e-commerce ecosystems.

### 4.2.6 Conclusion

Based on a DSR approach, we have developed a platform architecture framework for e-commerce as an industry-specific conceptualization of business ecosystems. The framework is grounded in existing literature on architectures and layers in e-commerce, taking into account an in-depth analysis of the architecture of three digital platforms in e-commerce (i.e., WooCommerce, Shopify, Magento). Furthermore, we enriched our data collection by conducting ten interviews with domain experts. Here, we aimed to gain additional insights about the particular architecture the experts had developed and the specific boundary resources they had implemented for the interconnection with other ecosystem participants. Our framework partitions e-commerce ecosystems from an enterprise architecture perspective (i.e., layered approach) and positions digital platforms on the application layer providing the necessary application services for the execution of business processes. Digital platforms establish a new path for value creation for retailers, software providers, and third-party developers in e-commerce ecosystems. These ecosystems have the potential to connect societies. For the development of the e-commerce-specific conceptualization, we took an architectural perspective as theoretical lens and subdivided the architecture in an infrastructure layer, an application layer, a business layer, and a business actor layer. The digital platform is situated in the application layer, it is developed by a software provider, it exploits infrastructure services mostly provided in cloud environments, and it provides the necessary application services not only for executing e-commerce business models but also for developing third-party platform extensions. Digital platforms in e-commerce usually require additional feature extensions to support business models other than e-shops. Besides feature add-ins, other modules such as themes and enhanced analytics are possible. These extensions require sophisticated boundary resources to allow for a high degree of generativity among developers. The extensions are provided via dedicated extension marketplaces as part of the digital platform. The architectural perspective involves manufacturers, retailers, wholesalers, customers, developers, software providers, and infrastructure providers as participant types.

## 4.3 Paper 3: Developing Multi-Sided Markets in Dynamic Electronic Commerce Ecosystems—Towards a Taxonomy of Digital Marketplaces

**Abstract**

*Multi-sided marketplaces*[6] *have proven to be a successful business model in the dynamic electronic commerce environment.*

---

[6] Throughout this cumulative dissertation, we use the term transaction platform instead of multi-sided marketplaces that is used within this research paper as a synonym for the term digital marketplace and electronic marketplace. In using this terminology, we would emphasize the difference to innovation platforms. Transaction platforms are able to orchestrate multiple market sides consisting of a variety of different participant types.

*footnoteThis version of the article has been accepted for publication, after peer review but is not the Version of Record and does not reflect post-acceptance improvements, or any corrections. The Version of Record is referenced as follows: T. Wulfert, S. Seufert, et al. (2021). "Developing Multi-Sided Markets in Dynamic Electronic Commerce Ecosystems—Towards a Taxonomy of Digital Marketplaces." In:* Proceedings of the 54th Hawaii International Conference on System Sciences (HICSS 2021). *Maui, Hawaii, pp. 6133–6142 There exists a variety of multi-sided marketplaces differing in their provided features and services for their participants. Existing taxonomies often focus on value creation and business-to-business transactions. We apply Nickerson et al.'s taxonomy development approach. We especially incorporated aspects of orchestrating the distinct market sides and governance dimensions. The developed taxonomy for multi-sided marketplaces consists of 21 dimensions and 99 characteristics in total. We have applied our taxonomy to 44 multi-sided marketplaces and identified asymmetries between the market sides concerning Monetization, Network Effect Amplifiers and Provided Services. We emphasize that the taxonomy is not only an artifact for classifying the current situation of an multi-sided marketplace but can also be used by multi-sided marketplace owners to derive directions for the future development. We illustrated how these developments can be conducted by examples for five dimensions of our taxonomy.*

| Fact sheet of publication 3 | |
|---|---|
| Title | Developing Multi-Sided Markets in Dynamic Electronic Commerce Ecosystems—Towards a Taxonomy of Digital Marketplaces |
| Authors | Tobias Wulfert<br>Sarah Seufert<br>Christian Leyens |
| Publication Type | Conference Proceeding |
| Publication Outlet | Hawaii International Conference on System Sciences |
| VHB-Jourqual3-Ranking | C |
| Status | Published |
| Research Objective | Develop a taxonomy for transaction platforms for B2B, B2C and C2C transactions exchanging (physical) merchandise in e-commerce. |
| Research Approach | Taxonomy Development in IS (Nickerson, Varshney, et al., 2013) |
| Full Citation | T. Wulfert, S. Seufert, et al. (2021). "Developing Multi-Sided Markets in Dynamic Electronic Commerce Ecosystems—Towards a Taxonomy of Digital Marketplaces." In: *Proceedings of the 54thHawaii International Conference on System Sciences (HICSS 2021)*. Maui, Hawaii, pp. 6133–6142 |

### 4.3.1 Introduction

As Electronic Commerce (e-commerce) [7] is accountable for 16.1 percent of total retail sales in the US in the second quarter of 2020 (U.S. Department of Commcerce, 2020), it seems to be a profitable sales channel for trading companies. Despite the continuously increasing total sales, the e-commerce itself represents a highly competitive environment (Grewal and Levy, 2007). To gain competitive advantage, trading companies establish digital multi-sided marketplaces (M (Mantin et al., 2014) in which the owner acts as an intermediary between two or multiple distinct market sides enabling trading transactions between them making use of the economic effects of multi-sided marketplaces, mainly network effects (Armstrong, 2006; Parker, Van Alstyne, and Choundary, 2016; Rochet and Tirole 2003; Rochet and Tirole, 2006). Companies operating multi-sided marketplace business models matching suppliers and consumers in e-commerce have up to four times higher valuations than traditional businesses (Libert et al., 2014). For example, Amazon's revenue generated with its multi-sided marketplace is already accountable for nearly 60 percent of Amazon's total e-commerce revenue (Amazon, 2019). Exploiting network effects (Shapiro and Varian, 1998), these multi-sided marketplace form the core of emerging digital business ecosystems with networks of independent participants (McIntyre and Srinivasan, 2017) who may profit from the success and dissemination of the multi-sided marketplace among demand-side participants (Briscoe, Odlyzko, et al., 2006). The value of the multi-sided marketplace-mediated business ecosystem for each participant increases with more participants within the network (Briscoe, Odlyzko, et al., 2006). These digital business ecosystems are subject to continuous and dynamic change caused by participants joining and departing (Corallo, 2007). Thus, decisions regarding the future development of an existing multi-sided marketplace or the positioning of a new multi-sided marketplace within these ecosystems are crucial to reach and exceed a critical mass of participants in the dynamic e-commerce networks (Grewal, Roggeveen, et al., 2017; Täuscher and Laudien, 2018). With many retailers and wholesalers developing from reseller mode to multi-sided marketplace (Hagiu and Wright, 2015b; Tian, Vakharia, et al., 2018) and winner-take-all dynamics in platform ecosystems (Eisenmann, 2008), the positioning of a multi-sided marketplace in an available niche and future development gets even more complicated. Thus, the owner's governance decisions should aim at locking in participants and disabling multihoming while offering a variety of boundary resources and standardized processes (Täuscher and Laudien, 2018).

---

[7] To comply with the spelling used within this cumulative dissertation, we added a hyphen to the e-commerce abbreviation.

Additionally, multi-sided marketplace owners need to decide on which further additional services to offer to participants to differentiate from competing multi-sided marketplace-mediated ecosystems (Reillier and Reillier, 2017). Previous literature on e-commerce mainly focuses on Business-to-Business (B2B) marketplaces when deriving common characteristics and developing taxonomies (Dai and Kauffman, 2001; Giaglis et al., 2002; Wang and Archer, 2007). Täuscher and Laudien (2018) develop a taxonomy for general digital marketplaces that takes a business model perspective focusing on value generation, delivery and capture excluding markets on which the owners appear as competitors. We will address this literature gap by conceptually and empirically developing a taxonomy for multi-sided marketplace for B2B, Business-to-Customer (B2C), and Customer-to-Customer (C2C) transactions exchanging (physical) merchandise in e-commerce. It can be used for analyzing the current state of an multi-sided marketplace and suggest steps for designing and refining the business model. We draw on the economic literature on multi-sided marketplaces and network effects focusing on the matchmaking between and orchestration of the distinct market sides (Armstrong, 2006; Rochet and Tirole, 2003) added by aspects of platform governance as major building blocks of multi-sided marketplaces (Boudreau and Hagiu, 2009; Hein, Schreieck, Wiesche, et al., 2016). From a research perspective we serve the description goal (Gregor, 2006) by providing a classification scheme for digital marketplaces in IS research. Following a design science approach with the taxonomy as an artifact (Hevner, March, et al., 2004) we support multi-sided marketplace owners with an overview on potential governance decisions to foster network effects for an increase in the number of participants. For researchers, we propose a classification for further analysis of multi-sided marketplace.

The remainder of this research is structured as follows: firstly, we outline related literature on multi-sided marketplaces. Secondly, we introduce our research methodology for developing taxonomies in IS research according to Nickerson, Varshney, et al. (2013). Next, our developed taxonomy for classifying multi-sided marketplaces is described, followed by an elaboration on its application by multi-sided marketplace owners for designing the future operation of the multi-sided marketplace. Finally, we discuss and summarize our findings.

### 4.3.2 Related Literature

Since the ignition of the internet several business models for the e-commerce have been developed (Timmers, 1998; Turban, Whiteside, et al., 2017; Urbaczewski et al., 2002). In e-commerce "electronic means and technologies [are used] to conduct commerce" (Choi et al., 1997). The degree to which the transaction is carried out digitally varies in the literature on a continuum between a completely digital transaction (Laudon and Traver, 2019) and only a small part of the procurement process (Timmers, 1999). Introducing multi-sided marketplaces in e-commerce, we draw on the concept of multi-sided marketplaces (Armstrong, 2006; Evans, 2003; Hagiu and Wright, 2015b; Rochet and Tirole, 2003; Rochet and Tirole, 2006). multi-sided marketplaces emphasize the economic effects inherent to electronic marketplaces (Hänninen et al., 2018). They differ from the traditional value chain of (offline) retailers and electronic shops insofar, as that multi-sided marketplaces match manufacturers on the supply side with end customers on the demand side. Other retailers or wholesalers may also interact with a multi-sided marketplace as a supplier or may also demand goods from the multi-sided marketplace that is controlled by the multi-sided marketplace owner. The multi-sided marketplace owner can be either one (e.g., Walmart Marketplace) or a conglomerate of the aforementioned parties (e.g., Opodo) or even an independent third-party (e.g., eBay). multi-sided marketplace subsume procurement-, sales-side and additional participants acting as intermediaries between these sides without acquiring property over the traded articles (Turban, Whiteside, et al., 2017). Although we introduce multi-sided marketplace in e-commerce, we focus on the two dominant market sides in e-commerce namely suppliers and customers. The core value proposition of the multi-sided marketplace owner thus is the facilitation of the matchmaking between participants from demand- and supply-side and the enabling of (retail) transactions between them (Hagiu and Wright, 2015b). Hagiu and Wright (2015b) argue that participants of a multi-sided marketplace require some affiliation with the multi-sided marketplace. However, the way in which participants must affiliate with the multi-sided marketplace is not further defined and can be interpreted differently (e.g., contract, registration, cookies). Although the concept of multi-sided marketplaces is also present in stationary trading with shopping malls or some variants of trading such as agency trade (Abhishek et al., 2016) and commission business (Müller-Hagedorn et al., 2012), network effects for participants (lower transaction costs for search and initiation) and economies of scale for multi-sided marketplace owners (marginal costs for adding another supplier or article are almost zero) are even stronger in the context of the internet. multi-sided marketplaces offer a digital representation of the diverse assortment of articles offered by supply-side participants. We focus on multi-sided marketplaces transacting (physical) merchandise excluding the exchange of services. The assortment can be described as the periphery of the multi-sided marketplace while the core is the multi-sided marketplace itself offering merchandise related services to supply- and demand-side participants as described analogously in the platform literature (Baldwin and Woodard, 2009; Staykova and Damsgaard,

2015). From a customer's point-of-view multi-sided marketplaces "resemble retail agglomerations" Hänninen et al., 2018, p. 155 integrating the range of articles of a network of participating suppliers, retailers and wholesalers through a single digital channel (Teller and Elms, 2010). multi-sided marketplaces match two or more previously distinct markets (gatekeeper) (Rochet and Tirole, 2003) and form the center of a digital business ecosystem connecting dynamic networks of independent suppliers and customers that share a common interest that is the success of the ecosystem (Corallo, 2007; McIntyre and Srinivasan, 2017). Thus, they are able to exploit direct and indirect network effects to further propel the ecosystem to reach and exceed critical mass (Rochet and Tirole, 2006; Tiwana, 2014) making the ecosystem more valuable for each participant (Briscoe, Odlyzko, et al., 2006). Applying graph theory, network effects can be described as triadic closures (focal or membership) in social affiliation networks on one side or on multiple sides involving the multi-sided marketplace at the core (Easley and Kleinberg, 2010). The ecosystem dynamically evolves as actors join or depart while creating new or interrupting prior stable triadic closures. The participants in digital business ecosystems are interdependent. They both cooperate to achieve common objectives while competing for scarce resources (Corallo, 2007). With direct (or same-side) network effects the value of a multi-sided marketplace for a (demand- or supply-side) participant increases with the size of the network on the same side. This is due to the higher potential of exchange between these participants (e.g., of product review, of knowledge) (Eisenmann, Parker, et al., 2006; Shapiro and Varian, 1998). An indirect (or cross-side) network effectin multi-sided marketplaces arises if the benefit to participants from supply side depends on the number of participants on the demand side and vice versa (e.g., supply side of the multi-sided marketplace subsidizing the demand side) (Armstrong and Wright, 2007; Rochet and Tirole, 2003). Indirect network effects require the presence of the cross-group side network effects in both directions (Hagiu and Wright, 2015b; Shapiro and Varian, 1998).

### 4.3.3 Scientific Approach

A taxonomy can be both an artifact for classification and the process of developing the artifact (Bailey, 1994). Taxonomies as an artifact are multidimensional classification systems in which objects are categorized into complete and disjoint groups by applying decision rules (Doty and Glick, 1994). The benefit of developing a taxonomy is that it creates an artifact for comparing and contrasting the objects of a given domain so that common properties of the objects can be discovered (Gregor, 2006). Furthermore, taxonomies are considered helpful for the analysis of complex research areas (Nickerson, Varshney, et al., 2013). For the taxonomy development we apply the methodology as proposed by Nickerson, Varshney, et al. (2013) using a hybrid approach including both conceptually and empirically derived dimensions (Fig. 4.9). Thus, we can include already established dimensions from literature while deriving dimensions from our sample of 44 multi-sided marketplaces. This method is based on the "Three-Level Measurement Model" developed by (Bailey, 1984). According to the definition of Nickerson et al. (Nickerson, Varshney, et al., 2013), a taxonomy consists of a series of dimensions consisting of mutually exclusive and exhaustive characteristics. The characteristics within each dimension must be complete and disjunct.

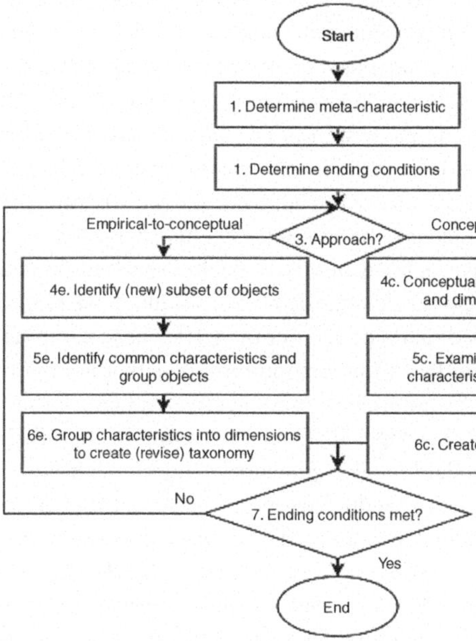


**Fig. 4.9** Method for Taxonomy Development (Nickerson, Varshney, et al., 2013)

The development of the taxonomy begins with the definition of the meta-characteristic (*1*) (Nickerson, Varshney, et al., 2013). The meta-characteristic is superior to the other characteristics and all other characteristics are derived from it. The purpose of the meta-characteristic is to ensure that the characteristics presented in the taxonomy are related to each other and have not been loosely collected. Thus, the meta-characteristic must be chosen in such a way that it can be derived from the purpose of the taxonomy. In this way it should be ensured that the taxonomy follows its purpose and benefits the chosen target group (Nickerson, Varshney, et al., 2013). From a research perspective we serve the description goal (Gregor, 2006) by providing transparency about multi-sided marketplaces in e-commerce. We aim at supporting owners and developers of multi-sided marketplace in making decisions in the development process regarding business model and measures to increase the number of participants. The taxonomy should enable researchers to classify multi-sided marketplaces in e-commerce from a multi-sided marketplace and governance perspective. Next, the ending conditions are determined (*2*). This step is necessary due to the iterative development process (Nickerson, Varshney, et al., 2013). The ending conditions are divided into subjective and objective ending conditions. As soon as each of the defined ending conditions applies, the development of the taxonomy is completed (Nickerson, Varshney, et al., 2013). We apply the ten objective ending conditions as proposed by Nickerson, Varshney, et al. (2013). These are all relevant objects analyzed, no merge or split of an object, each characteristic of the taxonomy must be assigned at least once, no dimension or characteristic added, no dimension merged or split, every dimension has to be unique with unique characteristics, and the characteristics have to be mutually exclusive as well as collectively exhaustive. Although we have defined mutually exclusive as an objective ending condition, it is possible that several characteristics are simultaneously fulfilled in the dimensions by one multi-sided marketplace in Fig. 4.10. These combinations form an additional characteristic in the formal taxonomy tuples that are not represented for reasons of graphical simplification. For the subjective ending conditions we also follow Nickerson et al.'s (2013) proposal. Thus, our taxonomy should be concise, robust, comprehensive, extendable and explanatory. Before each iteration, in the third step (*3*) we choose between an inductive empirical-to-conceptual (represented with (e) in Fig. 4.10) and a deductive conceptual-to-empirical (represented with (c) in Fig. 4.10) approach for the development of the taxonomy (Bailey, 1984). In the empirical-to-conceptual approach, a selection of (new) objects is determined (*4e*) followed by the identification of common characteristics of these objects and classification of the objects according to the identified characteristics (*5e*). Then, the characteristics can then be assigned to dimensions that are added to the taxonomy (*6e*). If the conceptual-to-empirical approach is chosen, we set up the characteristics and dimensions of the objects based on a literature research and existing knowledge (*4c*). Next, the objects are examined according to these characteristics and dimensions (*5c*). The characteristics and dimensions are added to the taxonomy if necessary (*6c*). After each iteration the taxonomy is checked for the fulfillment of the defined ending conditions (*7*). Further iterations are performed and steps three to seven are repeated until all ending conditions are met (Nickerson, Varshney, et al., 2013). Our taxonomy fulfilled all ending conditions after two iterations of each approach. Once the development of the taxonomy is complete, it must be evaluated to ensure that it is useful for the target group (Nickerson, Varshney, et al., 2013). To evaluate the usefulness of our taxonomy we classified 44 multi-sided marketplaces mainly operating in the US and Europe using publicly available data from articles and web pages. The numbers in brackets in Fig. 4.10 represent the number of classified multi-sided marketplaces for each characteristic. Unlike the formal taxonomy tuples, the characteristics are not always mutually exclusive in this graphical representation. Thus, they may exceed the column sum of 44.

### 4.3.4 Taxonomy

Our taxonomy for classifying digital marketplaces in e-commerce consists of 21 dimensions with 99 characteristics in total (Fig. 4.10). Overall, we have grouped the dimensions following the multi-sided markets approach (Armstrong, 2006; Rochet and Tirole, 2003) to the participants they relate to (supply- and demand-side) and general marketplace dimensions (Eisenmann, 2008). We will briefly introduce each dimension with its related characteristics and mention examples from our classification for each characteristic.

The **Article Type** dimension relates to the aggregated assortment of articles offered on the multi-sided marketplace. It describes the articles that are either *Physical Merchandise* (e.g., Wish) or *Digital Merchandise* (e.g., Bonanza) (Choi et al., 1997; Kawa and Walesiak, 2019). Digital articles consist of data, information and knowledge without any form of physical representation (e.g., music downloads) (Loebbecke, 1999). Digital articles are special in terms of indestructibility, transmutability and reproducibility and pose different requirements for storage and distribution on the multi-sided marketplace (Choi et al., 1997).

The name of the second dimension is **Business Orientation**. This dimension aims at the choice of the mode of e-commerce and describes the type of participants on the multi-sided marketplace (Kawa and Walesiak, 2019). This results in the characteristics *B2B* (e.g., Alibaba), *B2C* (e.g., Cratejoy) and *C2C* (e.g., Facebook Marketplace), all of which can be used in combination with each other (e.g., Amazon as B2B and B2C, eBay as B2C and C2C).

The dimension **Industry Scope** distinguishes multi-sided marketplaces according to their market orientation. A distinction is made between *horizontal* multi-sided marketplaces (e.g., Walmart Marketplace) and *vertical* multi-sided marketplaces (e.g., Discogs, Zalando). While horizontal multi-sided marketplaces cover a variety

| | Dimension | Characteristics | | | | | | |
|---|---|---|---|---|---|---|---|---|
| Marketplace | **Article Type (c)** | Physical Merchandise (44) | Digital Merchandise (9) | | | | | |
| | **Business Orientation (c)** | B2B (16) | B2C (31) | C2C (9) | | | | |
| | **Industry Scope (c)** | Vertical (14) | Horizontal (30) | | | | | |
| | **Centralization (c)** | Active (31) | Passive (13) | | | | | |
| | **Owner Behavior (c)** | Neutral (35) | Competitive (9) | | | | | |
| | **Price Discovery (c)** | Fixed Prices (42) | Auction (4) | Negotiation (14) | | | | |
| | **Geographic Focus (c)** | Local (2) | Domestic (8) | Multinational (34) | | | | |
| | **Subsidy-Side (e)** | None (3) | Demand-Side (38) | Supply-Side (3) | | | | |
| Demand-Side | **Market Access (c)** | Open (39) | Closed (5) | | | | | |
| | **Fee Structure (e)** | Free (41) | Fee-Based (5) | | | | | |
| | **Affiliation Type (e)** | Direct Interaction (9) | Registration/Contract (42) | | | | | |
| | **Review System (e)** | None (16) | Supply-Side Review (22) | Article Review (25) | | | | |
| | **Network Effect Amplifiers (e)** | None (17) | Blog / Magazine (20) | Customer Q & A / Forum (12) | Social Events (4) | | | |
| | **Provided Services (e)** | None (18) | Information Services (24) | Loyalty Services (11) | | | | |
| Supply-Side | **Market Access (c)** | Open (18) | Closed (26) | | | | | |
| | **Sales Channels (c)** | Brick-and-mortar Commerce (41) | Electronic Commerce (41) | | | | | |
| | **Fee Structure (e)** | Free (16) | Commission Fee (26) | Listing Fee (5) | Access Fee (8) | Subscription Fee (22) | Cost per Click (4) | |
| | **Review System (e)** | None (23) | Demand-side review (21) | | | | | |
| | **Boundary Resources (e)** | Website (44) | Personal Support (31) | File Upload (16) | API (19) | | | |
| | **Network Effect Amplifiers (e)** | None (34) | Comparison with Competitors (6) | Demand-Side Newsletter (10) | | | | |
| | **Provided Services (e)** | None (9) | Sales Processing (23) | Payment Handling (13) | Fulfillment Services (12) | Marketing and Analytical Services (22) | Infrastructure Services (4) | Training Services (4) |

**Fig. 4.10** Taxonomy for Digital Marketplaces

of product groups from different industries, vertical multi-sided marketplaces focus on a single domain or a few industries (Täuscher and Laudien, 2018).

The degree of **Centralization** describes the elements of the purchase process that are executed or supported by the multi-sided marketplace (Wang and Archer, 2007). This dimension is part of the governance structure dimension (Hein, Schreieck, Wiesche, et al., 2016). *Passive* multi-sided marketplaces focus on the matching between demand- and supply-side participants and providing article information (e.g., Craigslist, Kijiji). In contrast, *Active* multi-sided marketplaces offer additional merchandise-related services for the participants and depict nearly the whole purchasing process including clearing, purchase processing etc. (e.g., Amazon, Walmart Marketplace).

The characteristics of the **Owner Behavior** dimension distinguish between *Competitive* owners who sell their own articles on the multi-sided marketplace and *Neutral* owners who do not offer their own articles or compete in single product groups. While the former are often multi-sided marketplaces that developed from online shops that opened-up themselves to include external supply-side participants (e.g., Amazon, Otto Market), the latter are multi-sided marketplaces on which the owner does not sell own products and solely acts as an independent intermediary (e.g., Alibaba, Etsy).

(Täuscher and Laudien, 2018) have identified three characteristics regarding the **Price Discovery** on the multi-sided marketplaces. Either the supply-side participant may set *Fixed Prices* for the articles (e.g., Zalando), the price is determined during an *Auction* process (e.g., eBay), or the discovery is a *Negotiation* between supply- and demand-side participants (e.g., mercateo.com). The negotiation characteristic also includes price on request in B2B marketplaces as prices can be set individually for each customer.

The **Geographic Focus** dimension describes the geographic reach of a multi-sided marketplace and also gives a hint on the heterogeneity of participants (Berendes et al., 2020; Kawa and Walesiak, 2019). *Local* multi-sided marketplaces focus a very small region such as a single city center (e.g., mercato). *Domestic* multi-sided marketplaces are only available in one country and/or a single language (Kawa and Walesiak, 2019) (e.g., Best Buy). *Multinational* multi-sided marketplaces reach multiple countries and also match participants globally (e.g., Amazon).

Multi-sided marketplaces usually implement asymmetric pricing policies with market sides that are being subsidized by the other(s) (Evans, 2003; Rochet and Tirole, 2003). This dimension relates to the pricing policy governance dimension (**Subsidy-Side**) of multi-sided marketplaces (Armstrong, 2006) and distinguishes multi-sided marketplaces that either subsidize *Demand-Side* (e.g., Yatego), *Supply-*

*Side* (e.g., JOOR), or try to equally charge (*None*) the market sides (e.g., Craigslist, Facebook Marketplace).

The following dimensions will be described for demand- and supply-side participants individually as the governance mechanisms are usually asymmetric (Boudreau and Hagiu, 2009; Hein, Schreieck, Wiesche, et al., 2016). With the **Market Access Demand-Side** dimension, the access prerequisites for demand-side participants are analyzed. If a multi-sided marketplace is assigned to the characteristic *Open*, participants on the demand-side can join a market freely or after an initial registration (e.g., Amazon, Facebook Marketplace). For a multi-sided marketplace with the characteristic *Closed*, a participant must pass an application process to join the marketplace (e.g., Walmart Marketplace).

The dimension **Demand-Side Fee Structure** refers to the pricing policy used for the demand-side. We have identified multi-sided marketplaces that are *Free to Use* for users on the demand-side (e.g., Facebook Marketplace, Walmart Marketplace) and multi-sided marketplaces that are *Fee-Based* (e.g., Restposten.de). Furthermore, we identified multi-sided marketplaces that use both free to use options and fee-based options on the demand-side (e.g., Amazon, eBay).

As participants have some **Type of Affiliation** with the multi-sided marketplace (Hagiu and Wright, 2015b), we have identified *Direct Interaction* and *Registration/Contract* as possible affiliation types for demand-side participants. For the direct interaction characteristic information is exchanged between demand-side participants and multi-sided marketplace (e.g., product description or price). This is especially true for price comparison websites (e.g., idealo). More formal instances of the affiliation are registration with an agreement to the terms and conditions (e.g., Atalanda) that result in preliminary contracts between the multi-sided marketplace and demand-side participants.

Trust is an important governance-dimension in multi-sided marketplaces (Boudreau and Hagiu, 2009; Hein, Schreieck, Wiesche, et al., 2016). For the implementation of a trust-building review system (**Review System Demand-Side**) (Täuscher and Laudien, 2018), we identified multi-sided marketplaces that either have *None* review system (e.g., Stylight), demand-side may review supply-side participants (e.g., eBay) or demand-side participants review the merchandise traded (e.g., Amazon).

As network effects are inherent to multi-sided marketplace (Hagiu and Wright, 2015b), we have derived characteristics implemented by the multi-sided marketplace owner to increase network effects for demand-side participants under the **Demand-Side Network Effect Amplifiers** dimension. We have derived *None* (e.g., Rakuten), *Blog/Magazine* (e.g., Reverb), *Customer Q&A / Forum* (e.g., Best Buy), and *Social Events* as characteristics from our sample of multi-sided marketplaces. The former foster the interaction among demand-side participants and make it more valuable for them to join or remain on the multi-sided marketplace. With social events (e.g., virtual shopping show) the multi-sided marketplace tries to enhance the matching (e.g., Alibaba).

The **Demand-Side Provided Services** dimension includes further services provided by the multi-sided marketplaces or its owner for the demand-side participants. These services go beyond the pure matchmaking (Hänninen et al., 2018). For this dimension we have identified *None*, *Information Services*, and *Loyalty Services* as characteristics. If no additional services besides the matching are offered the multi-sided marketplace focuses on its core value proposition (e.g., Hood.de). With information services further and more detailed article information is provided that extends those provided by the supply-side participants (e.g., Alibaba.com). Loyalty services offer (paid) bonus programs for parts of the demand-side participants (e.g., Wayfair).

The **Supply-side Market Access** dimension examines multi-sided marketplaces from the perspective of how supply-side participants can enter the market. The characteristics of this dimension are *Open* (e.g., Etsy) and *Closed* (e.g., Wish) analogous to the demand side.

According to (Kawa and Walesiak, 2019) the **Sales Channels** of the supply-side participants aggregated on the multi-sided marketplace can be *Brick-and-Mortar* (e.g., gearbest.com) and *Electronic* (e.g., Discogs). A combination of both characteristics is also possible with physical stores operating an online shop as an additional sales channel (e.g., Atalanda).

The dimension **Supply-Side Fee Structure** distinguishes marketplaces according to the structure for pricing the supply-side participants. We identified the following pricing methods as characteristics of this dimension: *Free*, *Commission Fee*, *Listing Fee*, *Access Fee*, *Subscription Fee* and *Cost-per-Click*. These characteristics also appear in combination with each other. The characteristic Free includes all multi-sided marketplace that offer a free option to use the marketplace on the supply-side (e.g., Craigslist). Multi-sided marketplaces using commission fees deduct a certain percentage from the turnover of a transaction (e.g., Amazon, eBay). Listing fees are paid once by the supply-side participant when initializing the offer (e.g., eBay, Etsy). Access fees are paid once upon registering a membership for accessing the multi-sided marketplace (e.g., Rakuten, Yatego). A subscription fee is an amount that is paid monthly or annually (e.g., Alibaba, Bonanza, Cratejoy). Cost-per-Click means the payment of a fixed amount each time a link to the supplier-side user's shop is called by a demand-side user (e.g., idealo, everysize).

We have examined *None* (e.g., eworldtrade.com) and *Demand-Side Review* (e.g., wucato.de) as characteristics for the trust-building **Review System Supply-Side** mechanism.

**Boundary Resources** enhance the interaction with the multi-sided marketplace by providing technical and social interfaces (Eaton et al., 2015; Ghazawneh and Henfridsson, 2013). Besides, we have derived *Website*, *Personal Support*, *File Upload*, and *API* as characteristics from our sample of multi-sided marketplaces. Every multi-sided marketplace from our sample offers a dedicated back-end on its website for supply-side participants for listing articles and managing transactions. Personal support is also offered via a number of channels ranging from product-related to technical inquiries (e.g., mercato.com). For the interaction between the IS of the multi-sided marketplace owner and supply-side participants simple file uploads (e.g., everysize) or sophisticated APIs (e.g., Amazon.com) are possible. The latter boundary resources also include a technical documentation as file or API specification.

**Supply-Side Network Effect Amplifiers** subsume artifacts implemented by the multi-sided marketplace owner to propel network effects for supply-side participants. They focus either on the interaction between supply-side participants (direct network effects) or the improvement of the matching with demand-side participants. If *None* further artifacts are implemented, the multi-sided marketplace focuses on reaching critical mass (e.g., Yatego). Offering a *Comparison with Competitors* for supply-side participants, they are able to adjust prices and assortments based on the market environment (e.g., Zalando Connected Retail). *Demand-side Newsletter* is a customized newsletter for demand-side participants, including merchandise offered by supply-side participants (e.g., Etsy).

The **Supply-Side Provided Services** dimension includes further services provided by the multi-sided marketplace or its owner for the supply-side participants. These services go beyond the pure matchmaking (Hänninen et al., 2018). For the provided services for the supply-side we have derived seven characteristics from our sample of multi-sided marketplaces. Besides the matchmaking *None* additional services are offered by some multi-sided marketplaces (e.g., Craigslist). *Sales Processing* are executed by other multi-sided marketplaces to support supply-side participants in the communication with the demand-side participants (delivery dates, invoicing) (e.g., Walmart Marketplace). With *Payment Handling* the multi-sided marketplace coordinates the whole payments process or even offers a marketplace specific payment method (e.g., eBay). Offering *Fulfillment Services* the multi-sided marketplace takes on the logistics for the supply-side participants at least for part of their assortment (warehousing, order picking, shipping) (e.g., eBay) (Semeijn et al., 2005). This may also include goods return processes (e.g., Amazon). *Marketing and*

*Analytical Services* offer supply-side participants product highlighting on the multi-sided marketplace or further advertising campaigns. Additionally, supply-side participants are provided with analysis of sales, prices and fraud alerts (e.g., Newegg). With additional *Infrastructure Services* the multi-sided marketplace owner offers innovation platform services for the marketplace participants (e.g., Amazon) not associated with the core trading business (compute power, storage or development environments) (Tiwana, Konsynski, and Bush, 2010). *Training Service* offers introduction to the functionalities of the multi-sided marketplace, to enable supply-side participants to sell their assortment through the multi-sided marketplace (e.g., XOM Materials).

### 4.3.5 Evaluation and Application

For the selection of the sample of multi-sided marketplaces, special attention was paid to ensuring a wide range of multi-sided marketplaces and a high degree of diversity. From a maturity model perspective (Reillier and Reillier, 2017; Tiwana, 2014), we have selected multi-sided marketplaces from the ignition stage (e.g., Otto Market is expected to launch 2020) to the mature stage (e.g., Amazon Marketplace was launched in 2000). We take into account huge multi-sided marketplaces with regards to annual sales revenue (e.g., Alibaba) and smaller ones (e.g., Bonanza). Our sample includes pure online players (e.g., eBay, Zalando) and former brick-and-mortar retailers (e.g., Walmart) starting their own multi-sided marketplace. The multi-sided marketplaces offer physical (and partially digital) merchandise. However, multi-sided marketplaces trading services are out-of-scope of our analysis. Although the developed taxonomy resembles an artifact for analyzing a static point in time, it can be used by multi-sided marketplace owners to plan the future development of the multi-sided marketplace. Thereby, multi-sided marketplace owners ought to focus on governance decisions aiming at increasing the number of participants on both sides of the multi-sided marketplace as a key driver for increased revenues (Armstrong, 2006; McIntyre and Srinivasan, 2017; Shapiro and Varian, 1998). In the following we emphasize the importance of five exemplary dimensions for fostering competitive advantage of multi-sided marketplaces through increasing network effects.

Firstly, the classification of our multi-sided marketplace sample revealed that **Network Effect Amplifiers** are not implemented by 39% of multi-sided marketplaces for demand-side participants. 77% of the analyzed multi-sided marketplaces have not implemented additional amplifiers for supply-side participants. As network effects in general and their strength in particular are key to the success of multi-sided marketplaces (McIntyre and Srinivasan, 2017), amplifiers ought to be implemented to propel triadic closures and thus increase network effects resulting in a higher number of participants. While for the demand side additional loyalty services can be offered, for the supply side customized newsletters for demand-side participants that integrate selected supply-side content can be implemented.

Secondly, the role of the multi-sided marketplace as a trustworthy intermediary can be augmented by additional **Review Systems** rating either demand-, supply-side participants, or the merchandise sold. Additional mechanisms for creating trust are useful especially when the brand of the multi-sided marketplace does not per se create enough traction. However, 36% of multi-sided marketplaces under investigation do not implement review systems in which demand-side participants can engage. 52% of multi-sided marketplaces do not use review mechanism for supply-side participants rating consumers. Hence, adding the possibility for reviewing the opposite side or merchandise bought increases the level of trust on the multi-sided marketplace and for the transaction partner. Including this participant generated content will also foster (direct and indirect) network effects.

Next, to acquire additional supply-side participants a multi-sided marketplace owner can extend the range of **Provided Services**. 79% of our multi-sided marketplace sample already provide additional services. These extra services can simplify the retail transaction for supply-side participants by taking on responsibility for retail functions such as transaction processing or payment handling. This reduces barriers for entry for supply-side participants as these functions do not have to be implemented by themselves. Acquiring further supply-side participants will positively affect the demand side because of indirect network effects and enabling focal closures via the multi-sided marketplace. While multi-sided marketplaces mature they typically provide additional services to participants on both sides to improve their competitive positioning and establish a further source of revenues. The multi-sided marketplaces take on (parts of) the trading functions formerly executed by retailers (Hagiu and Wright, 2015b). eBay has introduced fulfillment services in response to Amazon and is also planning to offer return handling.

Fourthly, all multi-sided marketplaces within our sample provide a dedicated website as **Boundary Resource** for supply-side participants on which the merchandise information and orders can be managed. However, only 36% of the analyzed multi-sided marketplaces offer the possibility of file upload and 43% provide an API to create and update this information. Granting further levels of multi-sided marketplace openness to supply-side participants significantly increases number of complementors and additional merchandise on the multi-sided marketplace (Boudreau, 2010). Exploiting indirect network effects, this will further increase the number of demand-side participants and thus the number of transactions between the market sides. After opening the multi-sided marketplace, it may take some time for participants to adopt the changes in the boundary resources as technological shifts in multi-sided marketplace may cause difficulties (Anderson and Tushman, 1986; Henderson and Clark, 1990). Otto Market offers an additional REST-API for order processing and merchandise master data management that is currently available in version three.

Lastly, a multi-sided marketplace owner can open or simplify the **Market Access** for supply-side participants by decreasing barriers for entry to foster participation on the supply-side. The application procedure required by 59% of multi-sided marketplace under investigation can be shortened or eliminated. Hence, more supply-side participants are attracted that increase competition among supply-side participants, widen the range of merchandise and can also increase the number of demand-side participants applying indirect network effects. On the down-side, this may cause more potential fraud merchandise and may result in negative network effects for demand-side participants. Thus, multi-sided marketplace owners can use our taxonomy as an artifact to augment their base of supply- and demand-side participants to increase network effects and (potentially) achieve competitive advantage.

### 4.3.6 Discussion

Besides the proposed suggestions for multi-sided marketplace owners, the application of our taxonomy on the sample of 44 exemplary multi-sided marketplaces revealed further implications for researchers and practitioners. We propose a taxonomy for analyzing and describing multi-sided marketplaces in e-commerce considering economic concepts of multi-sided marketplaces, network effects and governance. Partitioning the taxonomy in marketplace, supply- and demand-side centered dimensions, we identified several asymmetries regarding the market sides. In line with (Rochet and Tirole, 2003) 86% of the analyzed multi-sided marketplaces monetize the supply side while subsidizing the demand side. Beyond the value capture the asymmetric design of the multi-sided marketplaces privileges supply-side participants regarding the dimensions fee structure, boundary resources, network effects facilitators and provided services. The asymmetry results in a total of 47 characteristics for the supply side and only 23 for the demand side. Moreover, e-commerce is introduced as an highly competitive environment with dynamically changing ecosystems (Corallo, 2007; Grewal and Levy, 2007). We identified winner-takes-all-dynamics and strong network effects of incumbent multi-sided marketplaces which requires a thorough positioning of new entrants in an available niche (McIntyre and Srinivasan, 2017). Jet.com, a formerly independent multi-sided marketplace that became a subsidiary of Walmart, was shut down during our analysis by Walmart that wants to pool activities and participants on its own multi-sided marketplace. Rakuten recently announced to shut down its multi-sided marketplace in Germany. Another example of the dynamics in e-commerce ecosystems are multi-sided marketplaces (e.g., Zalando, Atalanda) that introduced reduced commission fees in the peak period of COVID-19. Hence, these multi-sided marketplaces not only attract more supply-side participants and increase the number of transactions but also accelerate network effects that may have positive effects on the competitive positioning.

Our taxonomy is designed both as an artifact for analyzing the current situation of a multi-sided marketplace and as guidance for multi-sided marketplace owners providing directions for the future development. It can be applied in several stages of multi-sided marketplace maturity (Reillier and Reillier, 2017; Tiwana, 2014). In the ignition stage it can be used for designing or positioning a new multi-sided marketplace while it can be applied in the maturity stage for evolving and refining the positioning of an already existing multi-sided marketplace. Implementing further characteristics of a dimension can potentially attract additional participants. This increases the number of triadic closures as well as network effects and will result in additional transaction that will enhance the number of transaction with increasing revenues. The identified dimensions in general and the specific manifestation of their characteristics pose additional requirements for the IS of the multi-sided marketplace. A major example is the introduction of additional services for supply- or demand-side participants. While pure multi-sided marketplaces merely focus on the digital matching between the market sides (Hänninen et al., 2018), a majority (80%) of multi-sided marketplaces within our sample already provided additional services. When introducing these additional services, further IS components are required by the multi-sided marketplaces (e.g., logistics modules for fulfillment services). The services provided by a multi-sided marketplace are key to the success and may form a competitive advantage. Thus, additional services are provided by multi-sided marketplaces as they mature to compete in the dynamic e-commerce environment (Wang, Zheng, et al., 2008). Besides the provided services, multi-sided marketplaces as intermediary between supply- and demand-side participants have an important role as trusted third-party in e-commerce (Hagiu and Wright, 2015b). Especially because merchandise that is purchased digital cannot undergo a sensorial screening a priori to the transaction by customers. Multi-sided marketplaces implement review systems for the participants or the traded merchandise to offset this drawback and further propel the role of a trustee. In addition to these review systems, the reputation of the multi-sided marketplace and the trust in the brand of the merchandise are further factors for establishing trust (Dellarocas, 2003). Review systems are especially important for multi-sided marketplaces in the ignition stage for building reputation and for products with a low brand recognition. As the brand of the supply-side participants are often masked (intentionally or unintentionally) by the multi-sided marketplace, closing the access to the multi-sided marketplace by implementing a more sophisticated application process (Market Access dimension) and increasing product quality will potentially increase the reputation of the multi-sided marketplace (McAfee and Brynjolfsson, 2017).

This research work also has its limitations. Although our sample of 44 analyzed multi-sided marketplace covers a variety of different multi-sided marketplace, we neither claim to have a representative sample nor to analyze all multi-sided marketplace available. We have randomly selected the analyzed multi-sided marketplaces with focus on multi-sided marketplaces operating in Europe and the US. Because of the existing number of incumbents and newly established multi-sided marketplaces, a complete analysis was not possible for this research. The classification of the 44 objects represent a snapshot as the multi-sided marketplaces continuously evolve (Staykova and Damsgaard, 2015). Our taxonomy is extendable to cope with the dynamics of e-commerce ecosystems as proposed by (Nickerson, Varshney, et al., 2013). In line with existing literature we considered network effects as black box (McIntyre and Srinivasan, 2017) and do not analyze their specifics and strength dependent on the mentioned amplifiers. Future research may investigate the strength of network effects exploited by a multi-sided marketplace. We also consider the availability of participants a "construct that could be strategically manipulated" (McIntyre and Subramaniam, 2009). Following literature on ecosystems, their participants are considered as independent (Corallo et al., 2007). Nevertheless, participants deciding to sell and/or purchase through a multi-sided marketplace as the center of such an ecosystem adhere to the terms and conditions set by the multi-sided marketplace owner(s).

### 4.3.7 Conclusion

Our taxonomy for multi-sided marketplaces consists of 21 dimensions with 99 characteristics in total grouped according to the multi-sided marketplace participants. While ten of the dimensions were derived conceptual-to-empirical, the remaining eleven were derived empirical-to-conceptual. Our taxonomy for multi-sided marketplaces contributes to the body of literature by integrating the concepts of multi-sided marketplaces and network effects with platform governance dimensions in the context of e-commerce. We developed our taxonomy based on an already proofed taxonomy development method in the IS discipline (Nickerson, Varshney, et al., 2013) and include aspects of matchmaking between and orchestration of the distinct market sides. The classification of 44 multi-sided marketplaces using the developed taxonomy revealed directions for future development of multi-sided marketplaces that can be used by practitioners (i.e. multi-sided marketplace owners) to attract additional participants and further propel their marketplace (as illustrated by five examples).

Future research may evaluate additional multi-sided marketplaces using our taxonomy and may enhance the existing taxonomy with additional dimensions and characteristics reflecting future developments in e-commerce (Nickerson, Varshney, et al., 2013). We also suggest future research to take the strength of network effects into account and also make an in-depth analysis of the influence of network effect amplifiers on the strength of both direct and indirect network effects. Another important avenue for future research is to analyze the implications of the decision by the multi-sided marketplace owner to include further features and services into the multi-sided marketplace and thus take on tasks previously executed by other participants. Especially with regards to additional requirements for IS caused by the addition of further services. We will interview software architects from selected multi-sided marketplaces and analyze the role and composition of IS to derive a software architecture for multi-sided marketplaces. As the architecture may vary across multi-sided marketplaces, we will derive archetypes based on our classification and developed architectural patterns for types of multi-sided marketplaces.

# Architecture of Platforms in E-Commerce 5

## 5.1 Paper 4: Retailer's Dual Role in Digital Marketplaces—Reference Architectures for Retail Information Systems


**Abstract**

*Digital marketplaces*[1] *have entered the retail sector and have proven to be a successful business model compared to traditional retailing.*[2] *Established retailers are increasingly launching digital marketplaces as well as participating in marketplaces of pure online companies. Retailers transforming to digital marketplaces orchestrate formerly independent markets and enable retail transactions between participants while simultaneously selling articles from their own assortment to customers in the digital marketplace (dual role). A retailer's dual role must be supported*


[1] Throughout this cumulative dissertation, we use the term transaction platform instead of the digital marketplace.

[2] This version of the article has been accepted for publication, after peer review but is not the Version of Record and does not reflect post-acceptance improvements, or any corrections. The Version of Record is referenced as follows: T. Wulfert and R. Schütte (2021). "Retailer's Dual Role in DigitalMarketplaces—Reference Architecture for Retail Information Systems." In: *SN Computer Science* 3.206, pp. 1–17


**Supplementary Information** The online version contains supplementary material available at https://doi.org/10.1007/978-3-658-45198-1_5.







T. Wulfert, *Selected Perspectives on Platforms in E-Commerce Ecosystems*, https://doi.org/10.1007/978-3-658-45198-1_5

*by retail information systems.*[3] *However, this support is not explicitly represented in existing reference architectures for retail information systems. Thus, we propose to develop a reference architecture for retail information systems that facilitates the orchestration of supply- and demand-side participants, selling their own articles, and providing innovation platform services. We apply a design science research approach and present nine architectural requirements that a reference architecture for a multi-sided market business model in retail needs to fulfill (dual role, additional participants, affiliation, matchmaking, variety of services, innovation services, smart services, aggregated assortment, and boundary resources) from the rigor cycle. From the first design iteration, we propose four exemplary, conceptual architectural patterns as a solution for the requirements (matchmaking for participants, innovation platform services, boundary resources, and aggregated assortment). These patterns can form a conceptual reference architecture that guides the design and implementation of information systems.*

| Fact sheet of publication 4 | |
|---|---|
| Title | Retailer's Dual Role in Digital Marketplaces—Reference Architecture for Retail Information Systems |
| Authors | Tobias |
| | Wulfert |
| | Reinhard Schütte |
| Publication Type | Journal Article |
| Publication Outlet | SN Computer Science |
| VHB-Jourqual3-Ranking | n.a. |
| Status | Published |
| Research Objective | Derive architectural requirements for a retail information system caused by the transformation of a retailer in reseller mode to a marketplace owner and develop architectural patterns that address these requirements. |
| Research Approach | Design Science Research Approach (Peffers et al., 2007) |
| Full Citation | T. Wulfert and R. Schütte (2022). "Retailer's Dual Role in DigitalMarketplaces—Reference Architecture for Retail Information Systems." In: *SN Computer Science* 3.206, pp. 1–17 |

[3] The majority of architectural requirements identified in this research paper need to be fulfilled by innovation platforms powering transaction platforms.

### 5.1.1 Introduction

Catalyzed by the implications of the Covid-19 pandemic, electronic commerce (e-commerce) revenue has increased worldwide and across sectors by 11 percent annually since 2019 and is expected to reach $ 3,300 billion by 2024 (Rotar, 2020). In Europe, e-commerce accounted for EUR 374 billion in total in 2019. As customers are willing to shop at brick-and-mortar retailers only if they feel safe, they increasingly tend to shop for groceries, apparel, jewelry, etc., online and aim for digital end-to-end customer journeys (Bhatti et al., 2020; Dietz et al., 2020; McKinsey, 2020). Additional government mandates that closed brick-and-mortar shops forced retailers to (hastily) establish additional online sales channels and transform their value proposition in favor of digital marketplaces (Dietz et al., 2020; Nicola et al., 2020). Sales via the marketplace accounted for more than 60 percent of Amazon's revenue in the fiscal year 2019 (Wulfert and Schütte, 2021). The pandemic has even amplified digital transformation in retail and wholesale that previously neglected necessary digitalization endeavors (Schütte and Vetter, 2016).

In addition to the possibility of establishing electronic shops, retailers and wholesalers may participate in existing or create their own digital marketplaces (Hagberg et al., 2016; Staykova and Damsgaard, 2015; Van Alstyne et al., 2016). While (Electronic) shops act as resellers in a single market, but digital marketplaces connect previously independent markets, match individual participants from the multiple market sides, and enable (retail) transactions between the parties (Hagiu and Wright, 2015a). Digital marketplaces orchestrate multiple markets and simplify interactions with suppliers, logistic service providers, market researchers, and other actors (Böttcher, Rickling, et al., 2021; Hänninen et al., 2018). These marketplaces focus on monetization of the matchmaking instead of selling articles with higher margins (Choudary, 2015; Evans and Schmalensee, 2016; Ivarsson and Svahn, 2020b). The orchestration causes (merely indirect) network effects for market participants (Shapiro and Varian, 1998), and marketplace owners implement asymmetric pricing mechanisms to monetize the matchmaking (Armstrong, 2006; Rochet and Tirole, 2003). The revenue generated by a digital marketplace and its surrounding ecosystem is up to three times higher compared to that of other business models (Delteil et al., 2020). Although traditional retail companies act as market intermediaries that offer manufacturers' products to customers (reseller mode), the development of digital marketplaces for the implementation of multi-sided markets (as a generalization of two-sided markets) has not been driven by retail, but usually by technology companies. Examples of the tremendous success of digital marketplaces include Amazon, Alibaba, and eBay, which orchestrate multiple market sides (Schütte, 2018). It seems unusual that although retailers are a significant factor in any economy and have traditionally linked markets (e.g., producer and consumer markets) for a long time (Levy, Weitz, et al., 2019), the expansion to the orchestration of multiple market sides with digital marketplaces was not driven by retailers. Thus far, only a few large retail companies have established their own digital marketplaces (e.g., Walmart and REWE). As digital marketplaces often form the preferred touchpoint for many customers, and marketplace owners exclude manufacturers from customers, retailers need to establish their own marketplaces (McKinsey, 2020; Rotar, 2020).

If the owner of a digital marketplace behaves neutrally (Kawa and Walesiak, 2019; Wang and Archer, 2007), the company does not gain ownership of the traded articles (in contrast to e-shops) at any point during the transaction (Hagiu, 2007). A digital marketplace facilitates retail transactions between ecosystem participants by providing interfaces for the interaction (Williamson, 1985). In contrast, competitive marketplace owners possess a dual role and offer their own articles in the digital marketplace (Kawa and Walesiak, 2019; Täuscher and Laudien, 2018). They may compete with other supply-side participants offering similar articles. This dual role as marketplace owner and competitor selling their own articles in the digital marketplace creates additional requirements for a retail information system (Wulfert and Schütte, 2021). Although the importance of multi-sided market business models has seemed to grow, existing literature focuses only on the adaptation of business models and the respective tools for modeling them (Evans and Schmalensee, 2016; Reillier and Reillier, 2017). The consequences of the retailer's dual role for the underlying retail information system-the digital infrastructure for e-commerce-supporting the specifics of the e-commerce and intermediary business model of multi-sided markets are rarely considered (Aulkemeier, Schramm, et al., 2016; Wulfert, Seufert, et al., 2021). A reference architecture may help to decrease setup time for the digital infrastructure supporting digital marketplaces and standardize processes and interfaces (Angelov, Grefen, et al., 2012). Reference architectures also facilitate the introduction of new processes and technologies in e-commerce and allow for better value co-creation within the surrounding ecosystem because of well-documented boundary resources (Eaton et al., 2015; Recker et al., 2021). This standardization may ease participation in the digital marketplace and increase network effects (Eaton et al., 2015; Shapiro and Varian, 1998). Although domain-specific reference architectures for the retail sector, such as the h-model (Becker and Schütte, 2004) and the ARTS model (APQC, 2019; OMG, 2019), do exist, literature dealing with domain-specific reference architectures for e-commerce in general (Aulkemeier, Schramm, et al., 2016) and a retailer's dual role in digital marketplaces in particular is sparse according to our research. Thus, we address this research gap by deriving architectural requirements for a retail information system caused by the transformation of a retailer in reseller mode to a marketplace owner and present focal architectural patterns that address these requirements. For the analysis of architectural requirements for digital marketplaces and the development of architectural patterns, the focus is on the combination of three aspects: the orchestration of formerly independent markets in the sense of multi-sided markets (Armstrong, 2006; Caillaud and Jullien, 2003; Haucap and Heimeshoff, 2018; Rochet and Tirole, 2003), competitive marketplace owners selling their own articles (Hagiu, 2007; Kawa and Walesiak, 2019), and the establishment of digital platforms from a technical software perspective (Gawer and Henderson, 2007; Tiwana, Konsynski, and Bush, 2010). The architectural patterns include aspects of technology platforms that center the marketplace owner's retail information system as a key technological artifact upon which further modules can be developed (Gawer and Henderson, 2007; West, 2003). To derive architectural requirements, we follow a design science research approach (Peffers et al., 2007) with the architectural patterns as artifacts (March and Smith, 1995). We use ArchiMate as the language for modeling enterprise architectures to formally present the architectural patterns in a "unified, unambiguous, and widely understood domain terminology" (Nakagawa et al., 2011). For the presentation of the architectural patterns, we focus on the business and application layers.

This research paper is an extended version of a paper presented at the 23rd International Conference on Enterprise Information Systems. For this special issue of the *SN Computer science* we have modified the introduction, added three architectural requirements and an architectural pattern, analyzed reference architectures in literature for the fulfillment of the requirements and patterns, improved the discussion section, and streamlined the concepts used throughout the text. The remainder of the paper is structured as follows. First, we introduce related literature concerning architectural patterns in Information Systems (IS) architecture and digital marketplaces in e-commerce and present nine architectural requirements. Second, we outline our research approach for deriving architectural patterns. Third, we present three exemplary architectural patterns and elicit additional architectural considerations for digital marketplaces. Finally, we discuss the architectural patterns, present findings from a preliminary reference architecture analysis, and summarize the results.

### 5.1.2 Related Literature

#### 5.1.2.1 Information Systems Architecture and Patterns

Retail information systems include all application systems that are used to support operational tasks in retail. The retail information system supports the execution of the main trading functions and related tasks that bridge the discrepancies in the streams between manufacturers and customers in real goods (goods, services; returns), nominal goods (money, credits), and information across space, time, quantity, and quality (Barth et al., 2015; Levy, Weitz, et al., 2019). Indeed, Retail information system support operational-dispositive, business-administrative, controlling, and corporate planning tasks (Becker and Schütte, 2004). In addition to merchandise management (merchandise planning, logistics, and settlement processes), retail information systems also support business intelligence and necessary corporate-administrative tasks in an integrative manner (Schütte, 2017). Facilitated by the ongoing digitalization of the retail sector, the bridging functions increasingly address digital product and price information and adaptations in payment, logistics, and distribution processes (Becker and Schütte, 2004; Schütte and Vetter, 2016). In e-commerce, transactions are carried out digitally to some degree (Laudon and Traver, 2019). Thus, they build digital infrastructures for executing the trading function in online and offline environments. Reference architectures are reference models (scripts) that serve as architectural templates for the design of a specific architecture in the context of a specific company (Bass et al., 2003; Recker et al., 2021). They are developed by using a specific grammar and guided by an implementation method (i.e., a procedure model) (Recker et al., 2021). An information systems architecture "is a set of high-level models which complements the business plan in IT-related matters and serves as a tool for information systems planning and a blueprint for information systems plan implementation" (Willcocks et al., 1997). Information systems architectures comprise a high-level sketch of the system and application architecture of a specific company and part of its application architecture (Heinrich and Stelzer, 2009).

Concrete information systems architectures deal with one particular company. However, reference architectures abstract from the company's peculiarities, therefore enabling the reuse of architecture components, providing an agreed-upon set of concepts and architectural patterns, and communicating fixed viewpoints (Giachetti, 2010). The development of a reference architecture is often inspired by concrete architectures or other artifacts and thus has a "descriptive nature" (Galster and Avgeriou, 2011). Developing a reference architecture based solely on existing research in a prescriptive manner allows the creation of "a futuristic view of a class of systems" (Galster and Avgeriou, 2011). Reference architectures focus on clarification of innovative patterns and aim to convince domain architects of the architecture qualities. Consequently, concrete systems can be developed according to this research-based architecture (Angelov, Trienekens, et al., 2008). Reference architectures are applied either to standardize existing systems to ensure interoperability or to facilitate the design and improve the quality of a concrete architecture with architectural guidelines (Angelov, Grefen, et al., 2009; Angelov, Trienekens, et al., 2008). They can be used as a starting point for company-specific models to reduce the effort of creating them through reuse of established artifacts and constructs (Winter and Fischer, 2006). A domain-specific reference architecture is a reference model at a high level of abstraction that provides a view of the essential areas of a domain (e.g., AUTOSAR) without having to consist of complete process and data models (Galster and Avgeriou, 2011; Nakagawa et al., 2011). A reference architecture is the mapping of process and data models' functionality onto system modules (Galster, 2015). Domain-specific information systems reference architectures for the retail sector offer a high-level view on architecture components and business functions.

A reference architecture consists of several architectural patterns (Cloutier et al., 2010; Shaw and Garlan, 1996). These patterns are defined as a "named collection of architectural design decisions that are applicable to a recurring design problem" (Taylor et al., 2010). The patterns are reusable solutions to common architectural problems within a given domain (Shaw and Garlan, 1996). Additionally, architectural patterns are often parameterized so that they can be applied to specific issues in different organizational contexts (Taylor et al., 2010). The relation between a reference model and architectural patterns as building blocks of the reference architecture and its manifestation in a concrete architecture for a company are illustrated in Fig. 5.1.

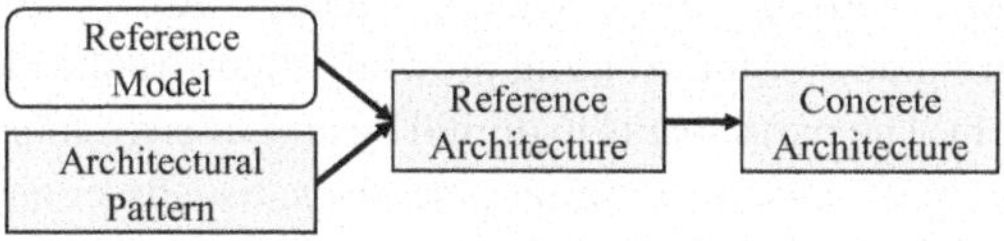


**Fig. 5.1** Relation between Architecture Types (Wulfert and Schütte, 2021)

#### 5.1.2.2 Digital Marketplaces

Introducing the idea of digital marketplaces in e-commerce, we draw on the concept of two-sided markets (Armstrong, 2006; Hagiu and Wright, 2015a). Digital marketplaces match two or more previously distinct markets and exploit direct and indirect network effects to further propel the digital marketplace (e.g., one side of the market subsidizing the other (Armstrong and Wright, 2007; Rochet and Tirole, 2003)). Digital marketplaces have predominantly operated in the Business-to-Customer (B2C) and Customer-to-Customer (C2C) contexts, but they are starting to be used for Business-to-Business (B2B) transactions more frequently (Li and Penard, 2014). Although the concept of marketplaces is also present in brick-and-mortar retail with shopping malls or variants of trading such as agency trade (Abhishek et al.,

2016) and commission business (Müller-Hagedorn et al., 2012), the network effects for participants (lower transaction costs for search and initiation) and economies of scale for market owners (the marginal costs for adding another supplier or article are almost zero) are even stronger in e-commerce. Digital marketplaces form the center of digital business ecosystems in e-commerce (Schütte and Wulfert, 2022a).

With direct (or same-side) network effects, the value of a digital business ecosystem for a demand- or a supply-side participant increases with the size of the network on the same side. This is due to the higher potential of exchange (e.g., of a product review or knowledge) between these actors (Eisenmann, Parker, et al., 2006; McIntyre and Srinivasan, 2017). An indirect (or cross-side) network effect in a digital business ecosystem arises if the benefit to actors on the supply side depends on the number of participants on the demand side and vice versa (e.g., the supply side of the digital business ecosystem subsidizing the demand side) (Armstrong and Wright, 2007; Rochet and Tirole, 2003). A prerequisite for indirect network effects is the presence of cross-group-side network effects in both directions (Hagiu and Wright, 2015a; Shapiro and Varian, 1998). Applying graph theory, network effects can be described as triadic closures (focal or membership) in social affiliation networks on one (direct, black circles) or multiple sides (indirect, dashed circles) in a digital business ecosystem with a focal digital marketplace (Fig. 5.2). The digital marketplace is the foundation for orchestrating multiple market sides (Easley and Kleinberg, 2010). The ecosystem dynamically evolves as actors join or depart, while creating new or interrupting previous stable triadic closures (Wulfert, Seufert, et al., 2021). The participants in digital business ecosystems are interdependent. They cooperate to achieve common objectives, usually competing for scarce

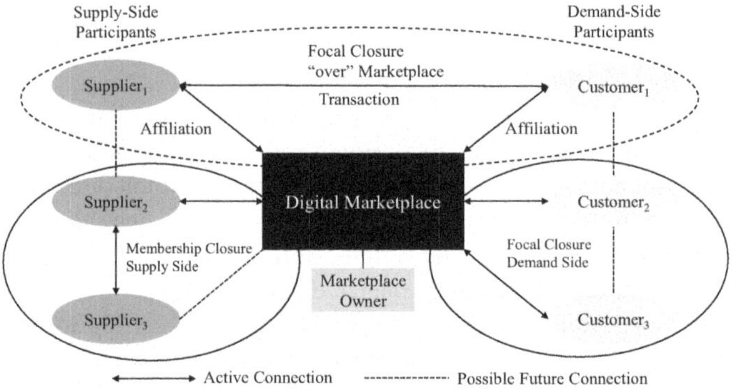


**Fig. 5.2** Triadic Closure in Marketplace-Centered Digital Business Ecosystems

resources at the same time (Corallo, 2007). Supply-side participants offering similar merchandise compete for customers from the demand side.

Establishing a digital marketplace poses additional requirements for a retailer's IS. The further elicitation of digital marketplaces is used to derive and is structured by nine architectural requirements resulting from a retailer's dual role that are summarized in Table 5.1.

**Table 5.1** Architectural Requirements of a Retailer's Dual Role in Digital Marketplaces

| # | Architectural Requirement | Description |
|---|---|---|
| 1 | Retailer's dual role | A retailer as a marketplace owner can also behave competitively with participants (Hänninen et al., 2018; Kawa and Walesiak, 2019). |
| 2 | Additional types of participants | The different participants need to be represented adequately in terms of master data, and matching records are required (Böttcher, Rickling, et al., 2021). |
| 3 | Affiliation with the marketplace | Participants require an affiliation to the marketplace to conduct transactions (Hagiu and Wright, 2015b). |
| 4 | Matching as a core value proposition | Matching between individual participants of the formerly independent market sides must be enabled (Moazed and Johnson, 2016). |
| 5 | Diversity of services | Services offered by the marketplace owner differ in type, scope, and coverage (modularity) and can be added as the digital marketplace evolves (pluggability) (Wulfert, Seufert, et al., 2021). |
| 6 | Innovation platform services | Technical capabilities that enable the creation of innovative solutions by developers (Tiwana, Konsynski, and Bush, 2010). |
| 7 | Smart service provision | Support the management of smart products and provision of smart services (Beverungen, Müller, et al., 2019). |
| 8 | Aggregated assortment | Digitally aggregate the assortment of several supply-side participants (Teller and Elms, 2010). |
| 9 | Sophisticated boundary resources | Provide boundary resources for different participants to connect to the marketplace (Dal Bianco et al., 2014; Eaton et al., 2015). |

**Architectural Requirement 1: Retailer's dual role**
In addition to taking a neutral role by merely facilitating the matchmaking, the marketplace owner can behave competitively with supply-side participants offering its own articles to demand-side participants (Wulfert, Seufert, et al., 2021). Hänninen et al. distinguish pure multi-sided digital platform business models that merely facilitate matches between the supply and demand sides (e.g., eBay, Alibaba, and Rakuten) and multi-sided market business models that extend their own range of articles with independent suppliers and offer further services to them (e.g., Amazon) (Hänninen et al., 2018; Li, 2019). The focus of this research paper is on the retailer's dual role as simultaneously a marketplace owner and a reseller that competes with other supply-side participants (Fig. 5.3).

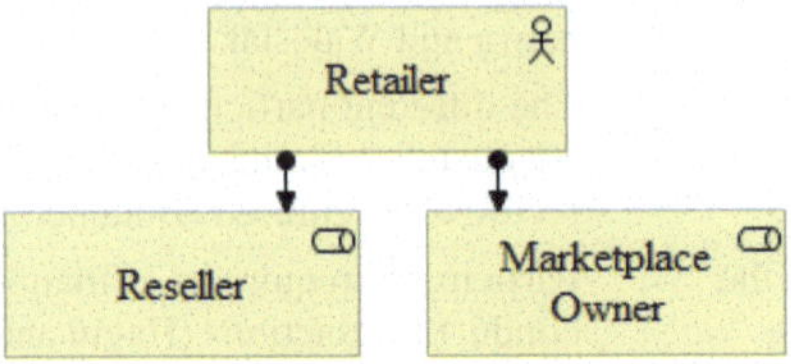


**Fig. 5.3** Retailer's Dual Role (Wulfert and Schütte, 2021)

Additionally, marketplaces can be established based existing brick-and-mortar stores or electronic shops as additional sales or procurement channel (Kawa and Walesiak, 2019). The marketplace owner can be either one (e.g., Walmart Marketplace) or a conglomerate of the participants (e.g., Opodo) or even an independent third party (e.g., eBay) (Wang and Archer, 2007). *Requirement:* The reference architecture should support the orchestration of the market sides (Reillier and Reillier, 2017) and traditional bridging functions with related tasks (Levy, Weitz, et al., 2019; Müller-Hagedorn et al., 2012).

**Architectural Requirement 2: Additional types of participants**
For the development of architectural patterns, we focus on the two most important market sides, suppliers (manufacturers, wholesalers, and retailers) and customers (end customers and retailers). In general, we describe a two-sided market as a specific manifestation of a multi-sided market in e-commerce (Hagiu and Wright, 2015a). Moreover, we also integrate third-party developers and infrastructure providers to support the innovation platform perspective (Gawer and Henderson, 2007; Tiwana, Konsynski, and Bush, 2010). Possible additional market sides are, among others,

advertising partners, logistics service providers, or opinion research agencies. Digital marketplaces differ from the traditional value chain of (offline) retailers and electronic shops in that digital marketplaces match manufacturers on the supply side with end customers on the demand side. Retailers and wholesalers may interact with a digital marketplace as a supplier or may demand articles from the digital marketplace that is controlled by the marketplace owner (Fig. 5.4). The digital marketplace is modeled as a location at which the matching and (parts of) the transaction are executed (Grieger, 2003; Turban, Whiteside, et al., 2017). *Requirement:* The different participants within a digital marketplace should be represented adequately in terms of master data, and records need to ensure that transactions between the participants can be tracked to optimize future matchmaking.

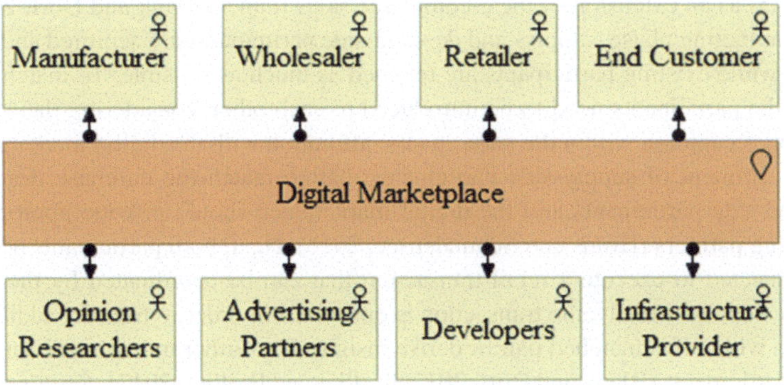


**Fig. 5.4** Digital Marketplace Participants (Wulfert and Schütte, 2021)

**Architectural Requirement 3: Affiliation with the marketplace**

In the offline environment, retailers try to establish relations with their customers by offering, for example, optional loyalty cards or apps (Hanke et al., 2018; Rudolph et al., 2015). Hagiu and Wright (2015b) contend that digital marketplace participants always require some affiliation with the marketplace. However, the way in which the digital marketplace participants must be affiliated is not further defined and can be interpreted differently (e.g., contract, membership, and cookies) (Hagiu and Wright, 2015b). The affiliation is important to improve the likelihood and quality of the matching, as the affiliation requires information about the participants (Evans and Gawer, 2016; Reillier and Reillier, 2017). *Requirement:* The affiliations of the different participants and multiple market sides should be represented and linked to participant profiles to support and improve the matchmaking.

**Architectural Requirement 4: Matching as a core value proposition**
As stated, the orchestration of formerly independent market sides is the core value proposition of a digital marketplace (Armstrong, 2006; Evans and Schmalensee, 2016; Rochet and Tirole, 2003; Rochet and Tirole, 2006). This involves the matching of individual participants in the market sides (Moazed and Johnson, 2016). The matching can be described according to Reillier and Reillier (Reillier and Reillier, 2017) as a process of attracting, matching, and connecting participants to enable (retail) transactions between them. The transaction process and matching are optimized afterward (Reillier and Reillier, 2017). The matching between supply- and demand-side participants can be illustrated in a schematic two-sided sales funnel (Fig. 5.5) as an extension of the e-commerce sales funnel (Blank and Dorf, 2012). In the attracting phase, supply- and demand-side participants are acquired and activated, while existing participants are retained as much as possible. To match both sides, the participants need to be introduced to each other, considering their characteristics captured within the participants' affiliation with the digital marketplace. The assortment of supply-side participants, has to match the purchase desire of demand-side participants and the digital marketplace should provide appropriate matching partners (Evans and Schmalensee, 2016). Next, both participants need to be connected to execute a retail transaction that can be coordinated by the marketplace owner. Finally, the transaction is optimized in order to transact additional articles within this matched pair or derive insights for further transactions between other participants (Blank and Dorf, 2012; Reillier and Reillier, 2017). *Requirement:* The matching process needs to be supported by, business and application services.

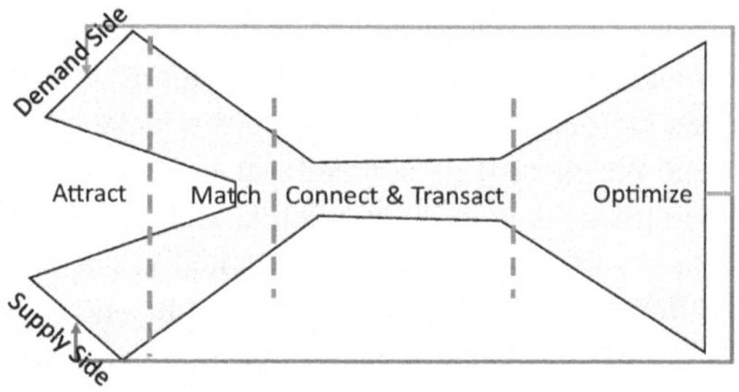


**Fig. 5.5** Matching Market Sides in E-Commerce (Wulfert and Schütte, 2021)

**Architectural Requirement 5: Diversity of services**
Digital marketplaces differ in terms of the type, scope, and coverage of services offered by the marketplace owner (Wulfert, Seufert, et al., 2021). When marketplace owners add additional services, they enhance their core value proposition. The degree of additional services offered by a digital marketplace varies on a continuum from passive matching (e.g., eBay classifieds) to full service offerings with sales processing, fulfillment services, and training services (e.g., Amazon) (Wang and Archer, 2007; Wulfert, Seufert, et al., 2021). Regarding the main bridging function (Levy, Weitz, et al., 2019), a substantial amount may be performed by other ecosystem participants surrounding the focal marketplace or the marketplace owner depending on the degree of centralization of the ecosystem (Hein, Schreieck, Wiesche, et al., 2016; Wang and Archer, 2007). As digital marketplaces typically mature by offering additional services (e.g., Amazon and eBay) (Reillier and Reillier, 2017), the retail information system should be flexible to support the integration of additional services performed by the digital marketplace. A modular design also supports the service continuum of digital marketplaces ranging from pure matchmaking to innovative marketplaces. *Requirement:* The reference architecture should be defined in a flexible and modular way, so it supports the development and integration of services that are not yet part of the business model but are likely to be integrated in the continuing evolution of the digital marketplace.

**Architectural Requirement 6: Innovation platform services**
In addition to trading-related services, digital marketplaces can offer innovation platform services for marketplace participants such as access to sales data or smart product-related data or remote services not associated with the core trading business (Tiwana, Konsynski, and Bush, 2010). These services can also be compute power, storage, or development environments similar to the one that Amazon provides with its Web Services that originated from the variability of demand for computing resources in the e-commerce business (Wittig et al., 2016). Innovation services are the technical capabilities that enable the creation of new solutions (services or software modules) by participating third-party developers (Asadullah et al., 2018a). Integrating transaction and innovation services, the digital marketplace resembles an integrated platform (Evans and Gawer, 2016). The power of innovation platforms rests on their architectural modularity (Baldwin and Clark, 2000; Tiwana, Konsynski, and Bush, 2010), catalyzing the re-configurability of technical and organizational components to accelerate generativity and value creation. The components of single modules are strongly interconnected but weakly connected with the central platform through technical boundary resources (Baldwin and Clark, 2000; Tiwana, 2013a). External modules make use of technical boundary resources

provided by the innovation platform (Eaton et al., 2015; Ghazawneh and Henfridsson, 2013). *Requirement:* The reference architecture should include these innovation platform services and respective boundary resources to enable developers to exploit the offered services.

**Architectural Requirement 7: Smart service provision**
Digital marketplaces not only need to deal with physical and virtual products from various third-party sellers but also need to handle data generated by the continuous hybridization of physical products (Schütte and Vetter, 2016). The augmentation of physical products with sensors, actuators, and software allows previously passive things to interact with their environment in Internet of Things scenarios (Knackstedt, 2006; Porter and Heppelmann, 2014). Data generated by these smart products can be exploited by marketplace owners to offer additional smart services to ecosystem participants that provide value to the customer and the service provider (Beverungen, Müller, et al., 2019). Thus, the digital marketplace needs to manage a variety of smart products; collect, aggregate, and analyze the data generated; and enable the creation and provision of smart services that exploit the data generated to create value for a customer. *Requirement:* The reference architecture should support the management of smart products, data aggregation and analysis, and provision of smart services.

**Architectural Requirement 8: Aggregated assortment**
Digital marketplaces aggregate a digital representation of the diverse assortment of articles offered by supply-side participants. The assortment can be described as the periphery of the digital marketplace, while the core is the digital marketplace itself offering services to supply- and demand-side participants as described analogously in the platform literature (Baldwin and Woodard, 2009; Staykova and Damsgaard, 2015). From the customer perspective, digital marketplaces "resemble retail agglomerations" (Hänninen et al., 2018) integrating the range of articles of participating suppliers, retailers, and wholesalers through a single digital channel (Teller and Elms, 2010). Thus, digital marketplaces further reduce transaction costs (Williamson, 1985) for participants as a variety of articles can be sold or purchased via a single touchpoint with a consistent user experience. Digital marketplaces also reduce the number of intermediaries that participate in a single customer journey and provide a consistent user interface. (Eaton et al., 2015; Ghazawneh and Henfridsson, 2013). E-Commerce in general and the aggregation of the individual assortments of various supply-side participants require a digital representation of the articles

within the assortment (Turban, Whiteside, et al., 2017). *Requirement:* The reference architecture should include a flexible model of the article master data to allow the aggregation of the assortment.

**Architectural Requirement 9: Sophisticated boundary resources**

The establishment of a digital marketplace involves the provision of dedicated boundary resources so that different types of participants from different markets can connect (Wulfert, Woroch, Strobel, Seufert, et al., 2022). Designing boundary resources requires considering a variety of different applications. While Supply-side participants need an interface to upload their assortment to the marketplace, and service providers need interfaces to offer additional (product-related) services. External application developers use a development environment, standard system architecture, or interface descriptions (Dal Bianco et al., 2014). The marketplace owner needs to open its retail information systems to other participants (Eisenmann, Parker, et al., 2009). Although boundary resources allow access to core modules of the marketplace, they also act as a control mechanism, allowing marketplace owners to manage the infrastructure based on the strategy pursued, which increases the chances of achieving market leadership (Eaton et al., 2015; Ghazawneh and Henfridsson, 2013). Boundary resources represent a dimension of governance, defining the boundaries between the marketplace owner and the community of external participants, thus facilitating the realization of strategically relevant decisions about ownership, entry into new markets, or community building (Dal Bianco et al., 2014; Foerderer, Kude, Schuetz, et al., 2019; Hein, Schreieck, Riasanow, et al., 2020). Dal Bianco et al. (2014) differentiate application, development, and social boundary resources (Fig. 5.6). Social boundary resources are used for knowledge transfer, development boundary resources for supporting application development, and application boundary resources for enabling interaction with a focal platform. Application boundary resources (Application Programming Interfaces (APIs), libraries, etc.) are defined as the minimum required for a software ecosystem to be viable. In contrast, development and social boundary resources increase the attractiveness of the ecosystem from the developers' perspective (Dal Bianco et al., 2014). *Requirement:* The reference architecture should propose sophisticated boundary resources to connect all relevant participants with a focal retailer's information system on the business and application layers (Wulfert, Woroch, Strobel, Seufert, et al., 2022).

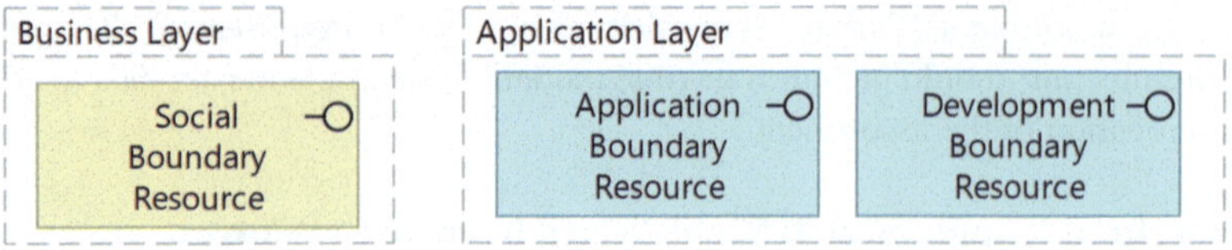


**Fig. 5.6** Overview of Types of Boundary Resources for Digital Marketplaces (Dal Bianco et al., 2014)

### 5.1.3 Research Method

This publication aims to develop architectural patterns for retail information systems supporting the orchestration of multiple market sides, different types of participants in e-commerce, and innovation platform services. We applied a design science research approach as proposed by Peffers et al. (2007) with an emphasis on problem identification, objectives, and solution design, presented in Fig. 5.7. The objective-centered process starts with the problem (1) stated in the introduction. The objectives of the artifact (2) to be developed are the architectural requirements already derived in Sect. 5.1.2.2. Exemplary architectural patterns addressing the objectives (3) are designed and demonstrated (4) as an extension the shortcomings of existing reference architectures. They (5) are evaluated based on informed arguments in the discussion section (Hevner, March, et al., 2004).

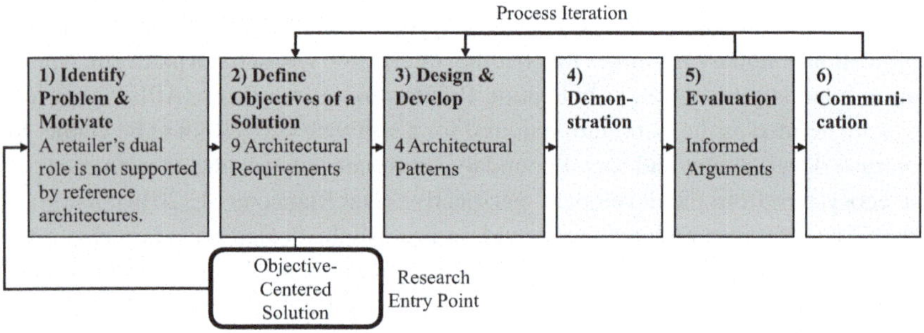


**Fig. 5.7** Design Science Research Approach (Peffers et al., 2007)

The outcome of the artifact design is a model (March and Smith, 1995) or more particular (parts of) an architecture (Vaishnavi, Kuechler, et al., 2019). Developed as "meta-artifacts" (Iivari, 2003), the architectural patterns represent a "general solution concept" (Van Aken, 2004) that is applicable to a class of problems when instantiated in the context of electronic retail (March and Smith, 1995). The architectural patterns (artifacts) are a new solution to a known problem and thus resemble an improvement (Gregor and Hevner, 2013). These patterns can describe major tasks of a digital marketplace and present them in a formal and understandable manner, applying a highly regarded enterprise architecture modeling language (i.e., ArchiMate). ArchiMate is part of The Open Group Architecture Framework (TOGAF) for the development, planning, implementation, and maintenance of enterprise architectures (Wulfert and Schütte, 2021). ArchiMate follows the service-oriented paradigm and is structured along the business, application, and infrastructure layers (Wulfert and Schütte, 2021). For this paper, we focus on modeling the business layer with related actors and processes and the application layer with its functions and services. A design science research project should go through three cycles (Hevner and Chatterjee, 2010). The focus of our overarching research project is the development and evaluation of a reference architecture supporting digital marketplaces in e-commerce with transaction and innovation functions (Evans and Schmalensee, 2016); this design cycle presents architectural patterns pivotal to marketplace business models in e-commerce. Our research approach can be summarized as follows: First, we derive architectural requirements based on a previous literature analysis as presented in sect. 5.1.2.2 (the rigor cycle). Second, we develop conceptual architectural patterns as general solution concepts for these requirements (Iivari, 2015). Thus, we develop domain-specific architectural patterns as building blocks of an overarching reference architecture based on literature (Angelov, Trienekens, et al., 2008; Galster and Avgeriou, 2011). This design cycle focuses on deriving architectural patterns from the rigor cycle and modeling them in ArchiMate. For a future relevance cycle, we will conduct interviews with IT architects and responsible IT staff architecting (parts of) an organization's (information systems) architecture. The retailer's dual role and additional innovation services (Gawer and Henderson, 2007; Tiwana, Konsynski, and Bush, 2010) create additional requirements for retail information systems that need to be reflected in reference architectures for e-commerce. In the following section, we depict four pivotal architectural patterns for the fulfillment of the introduced architectural requirements.

### 5.1.4 Architectural Patterns

#### 5.1.4.1 Pattern 1: Matching of Participants

The first exemplary architectural pattern addressing AR5 is concerned with the matchmaking between participants from different market sides as the core value proposition of a digital marketplace (Armstrong, 2006; Choudary, 2015). The matching sequence, as illustrated in Fig. 5.8 is executed by the retailer or wholesaler in its role as marketplace owner. The matching process is embedded in the matching sequence as proposed by Reillier and Reillier (2017) and introduced in sect. 5.1.2.2. After supply- and demand-side participants are attracted,, participants from independent markets need to be matched. In this context, a customer's desire usually leads to a product search either via search query or category search (Kotler and Keller, 2016). Based on the customer's preferences stored in the customer data, the matching engine situated in the customer and supplier relationship management systems calculates the order of the listed assortment. Thus, the relevance regarding the search term is not the only factor when articles are listed as a result of the customer's query; the preferred supplier may also be considered. Matching participants from the supply and demand sides of the digital marketplace requires an interface between these independent systems to exchange supplier and customer data relevant for the matching. The listing of the assortment is an important internal driver for retailers to increase revenues in e-commerce (Chen, Chiu, et al., 2014). In digital marketplaces, the product listing is even further complicated by the retailer's dual role, causing the question whether to emphasize products from the owner's own assortment or from another ecosystem participant's assortment. A higher priority for the owner's assortment cannot be implemented because of antitrust law considerations (Wulfert, Seufert, et al., 2021). The matching can be initiated proactively to stimulate a customer's desire (e.g., customized newsletters, social media marketing or search engine advertising). After a match has been created successfully, the retail transaction can be executed. To optimize the matching sequence, the supplier and customer data are enriched with information derived from a previously executed transaction, and other demand-side participants in the same cluster may be notified about the previous transaction.

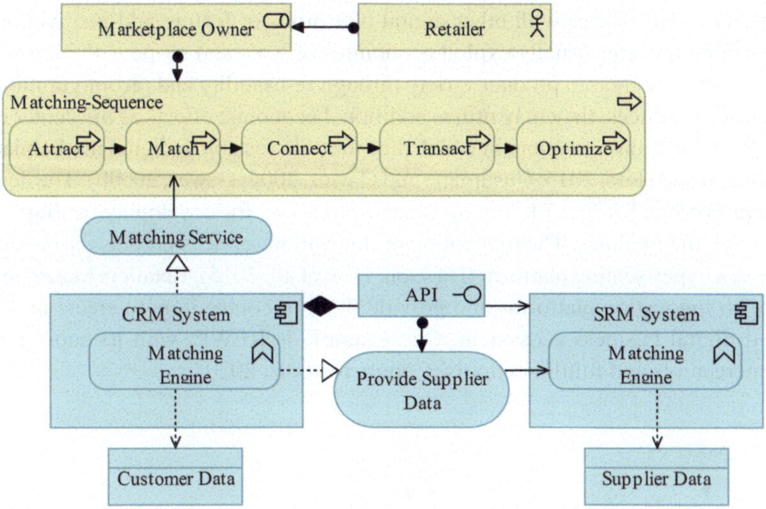


**Fig. 5.8** Matching as a Core Value Proposition (Wulfert and Schütte, 2021)

### 5.1.4.2 Pattern 2: Innovation Services

The second exemplary pattern addresses AR6. This conceptual pattern emphasizes the integration of innovation platform services (Fig. 5.9) in a digital marketplace (Evans and Schmalensee, 2016; Tiwana, Konsynski, and Bush, 2010). Offering innovation services (Tiwana, 2015a) focusing on the development of additional modules or apps requires opening the retail information system and supporting infrastructure for third-party developers by implementing applications (e.g., APIs and Software Development Kits (SDKs) and providing social (e.g., documentation and technical support) boundary resources (Eaton et al., 2015; Ghazawneh and Henfridsson, 2013). External modules are developed using technical boundary resources provided by the innovation platform in the form of APIs services (Eaton et al., 2015; Ghazawneh and Henfridsson, 2013).

The openness of an innovation platform defines which platform services and components from the application and infrastructure layers can be used, modified, and extended by third-party developers (Witte and Zarnekow, 2018). Openness is usually defined by the scope and richness of the interfaces offered by the platform owner (Eisenmann, 2008). The development environment can also be operated by the marketplace owner depending on the degree of openness and the boundary resources provided. Third-party developers implement additional modules such as shop themes, interfaces with other digital platforms, or feature add-ins. Although innovation platforms usually exploit economies of scale and scope with increasing efficiency and increased product variety through re-usability and reconfiguration of modules or services, they may utilize additional economic effects as the center of a broader innovation ecosystem in which they may also establish digital marketplaces (Buxmann and Hess, 2015; Cusumano and Gawer, 2002; Gawer, 2009b). The development process consists of, among others, processes for developing, testing, and deploying the modules. The integration of innovation services propels the development to a hyper-scaling platform (Dawson, Hirt, et al., 2016). Retailers increasingly establish innovation platforms and provide them to competitors to create an integrated digital business ecosystem. One example is REWE, with its subsidiaries commercetools and fulfillmenttools (Commercetools, 2020).

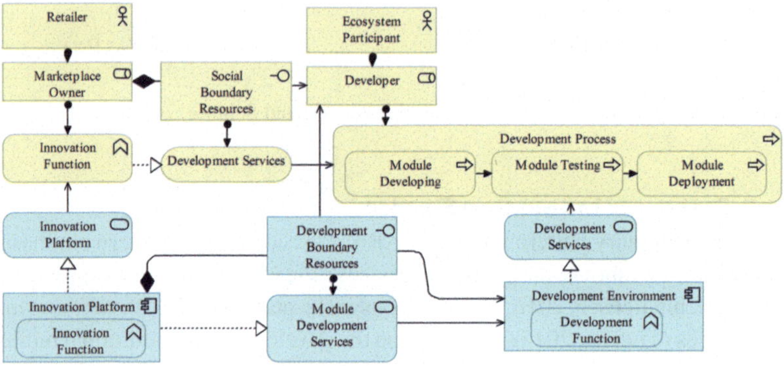


**Fig. 5.9** Innovation Platform (Wulfert and Schütte, 2021)

#### 5.1.4.3 Pattern 3: Boundary Resources

The development boundary resources have been introduced in the context of the innovation platform for the development of external modules by third-party developers (Fig. 5.9). In this subsection, we are specifically concerned with application boundaries, opening the focal marketplace to a variety of participants (Fig. 5.10). These boundary resources resemble interfaces for the participants' information systems to connect with the digital marketplace to exchange information on articles, delivery status, or customer inquiries. The ecosystem participants may involve, among others, manufacturers, wholesalers, retailers, logistic service providers, opinion researchers, content providers, or advertising partners (Böttcher, Rickling, et al., 2021; Schütte and Wulfert, 2022a). The various participants have diverging requirements regarding the interfaces offered. While a wholesaler sells merchandise via the marketplace and needs to exchange article- and customer-related data, a logistic service provider sends delivery updates and requires parcel information. Additionally, the timeliness of the information varies: The description of an article is updated rarely by a content service provider, but delivery updates and payment information are sent more frequently. The application boundary resources are offered by a retailer in its role as the marketplace owner in the transaction platform. They realize APIs that can be used by an ecosystem participant's information systems. Each application boundary resource is augmented by dedicated social boundary resources, such as documentation, training, or guidelines (Dal Bianco et al., 2014). This should guarantee the correct use of APIs by external application systems. There can also be social boundary resources such as general support that are not directly linked to a specific API.

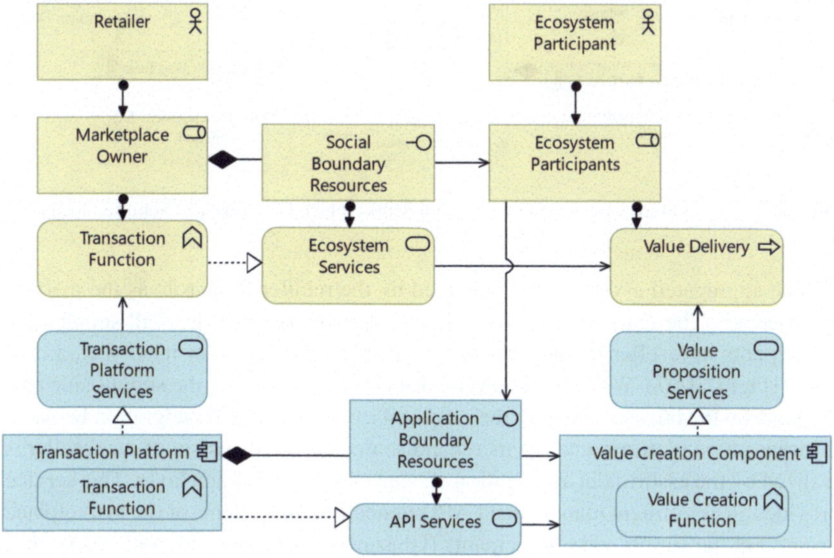


**Fig. 5.10** Boundary Resources of a Digital Marketplace

#### 5.1.4.4 Pattern 4: Aggregated Assortment

A conceptual solution for AR8 is presented in the third exemplary architectural pattern (Fig. 5.11). This conceptual pattern deals with the aggregation of the individual assortments of the different supply-side participants (Teller and Elms, 2010).

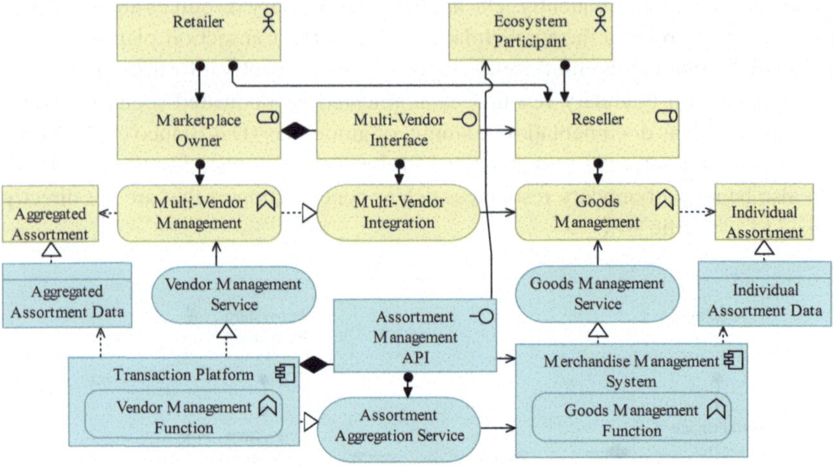


**Fig. 5.11** Aggregated Assortment of a Digital Marketplace (Wulfert and Schütte, 2021)

The aggregated assortment is managed by the retailer in its role as the marketplace owner. The marketplace owner aggregates the assortment of all supply-side participants in reseller mode. This may also include the retailer itself (i.e., a dual role) (Hagiu, 2007). While a multi-vendor integration connects the supply-side participants on the business layer to the marketplace, the technical integration between the transaction platform and the merchandise management system of the reseller is realized by the assortment aggregation service on the application layer. This service provides an assortment management API connecting the systems of the marketplace owner and the supply-side participants (Ghazawneh and Henfridsson, 2013). Following the introduced concept of boundary resources in platform-centered ecosystems, the multi-vendor interface is depicted as a social boundary resource on the business layer, while the assortment management API is modeled as an application boundary resource that acts as an intermediary between the transaction platform and a third-party retailer's merchandise management system on the application layer. For a retailer with a dual role, this means the listing of its own assortment in the marketplace. The assortment of each reseller is stored in an individual assortment business object accounting for possible additional sales channels of the reseller. The aggregated assortment is also captured by the marketplace owner in a business object realized by a data object. The data objects are managed by the transaction platform (for the marketplace owner) and the merchandise management system (for the reseller).

### 5.1.5 Discussion

In this research paper, we present four exemplary architectural patterns that are pivotal for a retailer's dual role in digital marketplaces. They are developed based on nine architectural requirements derived from the literature on e-commerce in general and digital marketplaces in particular. Although there could be more than one digital marketplace for a specific e-commerce sector (Senyo et al., 2019), winner-takes-all tendencies and strong network effects of incumbents limit the existing number of digital marketplaces in a particular domain to only one or a few (Eisenmann, 2008; McIntyre and Srinivasan, 2017). A reference architecture or multiple architectural patterns supporting a retailer's dual role in digital marketplaces simplify the assessment of the retail information system before the transition to a digital marketplace. Thus, the capabilities of the information systems for the marketplace transformation can be assessed more accurately and may decrease the possibility of failure during the establishment of the digital marketplace. Several digital marketplaces failed to establish successful digital marketplaces (e.g., zapatos, jet.com, and Rakuten) either because of the number of participants was small or because of information systems problems (Wulfert, Seufert, et al., 2021). The domain-specific reference architecture can also be used to identify potential gaps within the retailer's retail information system and reduce the overall time to market (Angelov, Grefen, et al., 2012). Moreover, a reference architecture should be augmented by a dedicated procedure for implementing the reference architecture in a specific company (Schütte and Wulfert, 2022b). As reference architectures are designed to be reusable in various companies, they contain abstract descriptions of processes or functions and often provide a predefined process or component alternatives. Thus, the architectures need to be instantiated for the context of a particular company (Schütte and Wulfert, 2022b). These instantiations can result in significant differences across companies (e.g., the matching process at Amazon and Otto Marketplace). An example of such a procedure is the Architecture Development Method as a major building block of TOGAF (Wulfert and Schütte, 2021). A dedicated implementation procedure can be an additional requirement for the support of a retailer's dual role in digital marketplaces in particular and for reference architectures in general.

A retailer's or wholesaler's dual role in a digital marketplace results in several advantages compared to other ecosystem participants (Fig. 5.4). Despite possible antitrust law considerations (Wulfert, Seufert, et al., 2021), the marketplace owner will be eager to prefer the offering of its own assortment to increase reseller revenues. As the marketplace owner controls the touchpoint to the customer, the owner also has information about fast-selling and profitable articles. This information can be used to adjust the assortment of the reseller role to mainly sell profitable articles and leave the long tail (McAfee and Brynjolfsson, 2017) of articles that sell slowly to other ecosystem participants. The concentration on fast-selling articles may also release storage capacity that can be offered as additional, retail-related services to ecosystem participants (Wulfert, Seufert, et al., 2021). With additional sales information, the marketplace owner can calculate articles for which the monetization of the matchmaking (e.g., commission fees) is more profitable than selling these articles in its reseller role. Matching as the core value proposition of a digital marketplace (Armstrong, 2006; Haucap and Wenzel, 2011) relies on correct data concerning customers, suppliers, and articles. Thus, the data needs to be stored accessibly for the matching engines to provide customers with the desired products. The actual article is the customer's focus in the e-commerce environment (Hagberg et al., 2016), while suppliers are excluded by the marketplace owners (McKinsey, 2020). Nevertheless, we propose to include supplier information in the matchmaking process, as customer preferences can be matched to suppliers' properties. Customers who care about their environment may, for example, be likely to buy articles from a supplier that can prove sustainability. Integrating additional innovation services attracts additional participants to the digital marketplace and adds more value propositions. The range and scope of modules developed by third-party developers depend on the openness allowed by the marketplace owner and the provided development boundary resources (Eaton et al., 2015; Ghazawneh and Henfridsson, 2013). These modules can be related to the bridging functions that enhance the retail transactions between supply- and demand-side participants (e.g., shop themes, and vendor management) or go beyond retail-centered purposes. Amazon is a major example of the wide range of external modules with its Web Services stemming from the usage of unused computational power from retail activities (Wittig et al., 2016). Thus, the modeled innovation platform and development environment components are generic enough to cope with the whole continuum of external modules. The innovation pattern should be instantiated according to the intention of a specific marketplace owner. Additional types of participants and markets are attracted to a digital marketplace by opening the marketplace and offering a set of sophisticated boundary resources for each type. The openness or the degree to which a marketplace should be opened to ecosystem participants compared to proprietary parts is a topic often discussed in literature (Benlian et al., 2015; Eisenmann, Parker, et al., 2009; Ondrus et al., 2015; Witte and Zarnekow, 2018; Wulfert, Seufert, et al., 2021). A marketplace owner needs to balance the degree of openness with the risk of losing control over its own marketplace and the surrounding ecosystem (Croitor and Adam, 2020; Hein, Schreieck, Wiesche, et al., 2016). Participants are likely to join an ecosystem via a marketplace creating additional focal closures and propelling network effects only if participation is valuable for them (Shapiro and Varian, 1998). Thus, the matchmaking also needs to take into account these additional types of participants. Digital marketplaces aggregate the assortment of several supply-side participants and require a data model for the articles capable of storing much unstructured data (image, video, exploded-view drawings, spare parts with historical data, etc.). The data model must be designed to be flexible so that it is suitable for different product categories agglomerating the diverse assortments of a number of participants from independent markets (Evans, 2011; Kollmann, 2019). This may lead to a decoupling of the master data storage of a transaction processing system (e.g., enterprise resource planning) from the transaction platform and the system for product information management. This is mainly because not all articles or services of a digital marketplace can be kept on a transaction platform with all available data. While the transaction platform requires high-resolution images for digital representation in the e-commerce environment, the enterprise resource planning system is mainly concerned with financial and inventory data. However, the degree to which article data is stored on the transaction platform and provided by an additional product information system depends on the specific environment of the retailer. The product information system is not modeled in pattern 3 for reasons of graphical simplification. These architectural patterns from the first design iteration should be further evaluated with practitioner insights and aggregated to an overall reference architecture for the information system supporting a retailer's dual role in digital marketplaces.

Aggregating transaction and innovation services, the digital marketplace forms an integrated platform (Evans and Gawer, 2016). The transaction platform is included in pattern 4, while the innovation platform is part of pattern 2. An integrated platform is likely to become a hyper-scaling platform that quickly achieves critical mass and shaping industries (Dawson, Hirt, et al., 2016). Based on our research-based requirements and patterns, we analyzed an initial sample of existing reference architectures from the e-commerce context. The preliminary sample consists of seven reference architectures-the electronic market architecture GEMS (Albers et al., 2004), ECOMOD (Ecomod, 2006), e-ZOCO architecture (Castro-Schez et al., 2010), E-Commerce Service Composition Platform (EC-SCP) (Chi et al., 2010), e-commerce reference architecture (Aulkemeier, Schramm, et al., 2016), integrated architecture for e-commerce (Vetter and Morasch, 2019), and Next-Generation E-Commerce Platform (NGECP) (Huang, Chai, et al., 2019)-for this analysis. The initial analysis is summarized in Table 5.2 in chronological order of the newest reference architecture release (architectural requirement fulfilled: X; architectural requirement partially fulfilled: (X); architectural requirement not fulfilled: -).

**Table 5.2** Preliminary Reference Architecture Analysis

| Reference Architecture | 1 | 2 | 3 | 4 | 5 | 6 | 7 | 8 | 9 |
|---|---|---|---|---|---|---|---|---|---|
| GEMS (Albers et al., 2004) | – | – | x | x | – | – | – | – | – |
| ECOMOD (Ecomod, 2006) | – | – | (x) | – | (x) | – | – | – | – |
| e-ZOCO Architecture (Castro-Schez et al., 2010) | – | – | (x) | (x) | x | – | (x) | x | (x) |
| EC-SCP (Chi et al., 2010) | – | x | x | x | – | – | – | – | – |
| E-commerce Reference Architecture (Aulkemeier, Schramm, et al., 2016) | – | – | (x) | – | x | – | – | – | – |
| Integrated Architecture (Vetter and Morasch, 2019) | – | x | (x) | – | x | x | (x) | x | – |
| NGECP (Huang, Chai, et al., 2019) | – | x | x | x | x | – | – | x | – |

Applying our architectural requirements and developed architectural patterns to selected reference architectures revealed that none of the analyzed reference architectures for e-commerce fully supports the requirements of digital marketplaces and a retailer's dual role. The NGECP fulfills five of the nine architectural requirements resulting from a retailer's dual role (Huang, Chai, et al., 2019). A retailer's dual role was not addressed by any of the analyzed reference architectures. The matching process requirement, the innovation platform services requirement, and the aggregated assortment requirement were also underrepresented in the literature sample (Table 5.2). The peculiarities of e-commerce and digital marketplaces require adaptations and enhancements of existing reference architectures. De Reuver et al. (de Reuver et al., 2017) encourage researchers to conceptualize platforms in different industries such as e-commerce. Therefore, we call for the development of reference architectures supporting digital marketplaces in general and in retail in particular.

This research paper also has its limitations. The developed architectural requirements are neither comprehensive nor complete. Architectural requirements should also be derived from practitioner sources (e.g., interviews and architecture documents) in an additional relevance cycle for a more sophisticated analysis. We plan to derive additional architectural requirements and evaluate the seven architectural requirements as well as the three architectural patterns developed by conducting interviews with IT architects at retailers that operate digital marketplaces. Additionally, the developed architectural patterns must be demonstrated in specific IS implementations in the digital marketplace context. Therefore, an avenue for future research might be the implementation and evaluation in e-commerce contexts and the discussion of these patterns with practitioners (e.g., enterprise architects). Although we claimed that the architectural requirements are derived from our literature analysis and make up the sole foundation for the developed architectural patterns, we need to acknowledge that the architectural requirements are biased from our own understanding of the meta-problem. We incorporated our understanding of the retail-specific problem and the retailer's dual role in digital marketplaces that leads to interpretations regarding the architectural requirements and architectural patterns (Iivari, 2015). Moreover, the chosen enterprise architecture modeling language for developing the architectural patterns leads to an implicit decision for a service-oriented architecture design, as this is inherent to this language connecting actors as well as business, application, and infrastructure layers that use the services (Wulfert and Schütte, 2021). Although service orientation is a well-regarded paradigm, it can at least be questioned whether it is the best approach for modeling the information systems architecture of a digital marketplace with the focus on the retailer's dual role.

### 5.1.6 Conclusion

The main contribution of this research paper is the determination of a retailer's possible dual role in digital marketplaces. We derive nine architectural requirements resulting from a retailer's dual role (dual role, additional participants, affiliation, matchmaking, variety of services, innovation services, smart services, aggregated assortment, and boundary resources) for a retail information system. These requirements resemble a class of problems relevant for digital marketplaces in e-commerce. Additionally, we propose four architectural patterns (matchmaking for participants, innovation platform services, boundary resources, and aggregated assortment) as a conceptional solution to selected requirements. These architectural patterns were developed based on the literature and can be applied to analyze existing reference architecture toward the fulfillment of these requirements. The patterns resemble buildings blocks of a meta-model as a reference architecture for the retail domain. A preliminary analysis of existing reference architectures for e-commerce showed that a retailer's dual role in digital marketplaces is not fully supported. Future research can analyze additional (scientific and practice) concrete architectures and reference architectures for the fulfillment of the requirements and patterns. The architectural patterns may also be improved by consolidating domain knowledge such as company-specific architectures and conducting interviews with information systems architects. Another important avenue for future research may be extending the range of architectural patterns and orchestrating them to a complete reference architecture that includes additional architecture layers. To demonstrate and evaluate the presented and additional patterns, they can be implemented in a concrete or experimental system as proof of concept.

**Acknowledgments** We would like to thank the reviewers of the 23rd International Conference on Enterprise Information Systems and the discussants during the conference for their fruitful comments. These comments helped us to improve the line of argumentation of the paper.

## 5.2 Paper 5: Reference Architectures Facilitating a Retailer's Dual Role in Digital Marketplaces—A Literature Review

**Abstract**
*Electronic commerce and digital marketplaces*[4] *have proven to be successful business models compared with traditional brick-and-mortar retailing.*

*footnote This version of the article has been accepted for publication, after peer review but is not the Version of Record and does not reflect post-acceptance improvements, or any corrections. The Version of Record is referenced as follows: T. Wulfert and J. Busch (2022). "Reference Architectures Facilitating a Retailer's Dual Role on Digital Marketplaces: A Literature Review." In:* Proceedings of the 24th International Conference on Enterprise Information Systems (ICEIS2022) Volume 2. *Online sales and the simultaneous orchestration of participants from independent market sides in digital marketplaces (a dual role) pose additional requirements for information systems. Reference architectures can be used as blueprints for the implementation of information systems for digital marketplaces supporting a retailer's dual role. However, reference architectures in retail were mostly developed for brick-and-mortar environments. The peculiarities of electronic commerce and digital marketplaces require adaptations and enhancements. Thus, we conducted a literature*

[4] Throughout this cumulative dissertation, we use the term transaction platform instead of digital marketplace.

*review following Vom Brocke, Simons, et al. (2009) of 1,357 research papers to identify reference architectures supporting a retailer's dual role in digital marketplaces. We identified seven digital marketplace-specific architecture requirements and analyzed the reference architectures identified according to Angelov, Grefen, et al. (2012). The analysis revealed 13 reference architectures with limited support for a retailer's dual role in digital marketplaces.*

*This research article was slightly corrected in terms of language usage and grammar and extended compared to the version published for ICEIS 2022.*

| Fact sheet of publication 5 | |
|---|---|
| Title | Reference Architectures Facilitating a Retailer's Dual Role on Digital Marketplaces: A Literature Review |
| Authors | Tobias Wulfert<br>Jan Busch |
| Publication Type | Conference Proceeding |
| Publication Outlet | International Conference on Enterprise Information Systems |
| CORE—Ranking | C |
| Status | Published |
| Research Question | What reference architecture in e-commerce support a retailer's dual role on digital marketplaces? |
| Research Approach | Literature Review (Vom Brocke, Simons, et al., 2009) |
| Full Citation | T. Wulfert and J. Busch (2022). "Reference Architectures Facilitating a Retailer's Dual Role on DigitalMarketplaces: A Literature Review." In: *Proceedings of the 24th International Conference on Enterprise Information Systems (ICEIS2022) Volume 2*. Online Event, pp. 494–505 |

## 5.2.1 Introduction

Platform-based business models are a successful way of conducting business, with up to four times higher valuations than traditional companies and three times higher revenue at the ecosystem level that includes participating companies (Delteil et al., 2020; Libert et al., 2014). In addition to industries such as accommodation (e.g. AirBnB and Couchsurfing), mobility (e.g., Uber, Blablacar) or the music (e.g. Spotify, napster) (Evans and Gawer, 2016), platform business models have conquered the retail sector with tremendous success (e.g., Amazon, eBay) (Wulfert, Seufert, et al., 2021). As the largest platform company in electronic commerce (e-commerce), Amazon generated $ 340 billion in revenue from product and service sales in 2020, with more than 60 percent of revenue resulting from commission fees for third-party sellers in its digital marketplace (Amazon, 2021). Digital marketplaces employ a multi-sided market business model and monetize the orchestration of multiple market sides without gaining ownership of traded products (Hagiu, 2007). While electronic shops act as resellers in a single market, digital marketplaces connect previously independent markets with independent participants and enable transactions between them (Armstrong, 2006). Digital marketplaces significantly reduce transaction costs in e-commerce environments for involved participants such as suppliers, logistic service providers and market researchers, by reducing the number of connections through intermediation (Böttcher, Rickling, et al., 2021; Hagiu and Wright, 2015b). Digital marketplaces focus on monetizing the matching between distinct participants instead of selling products with higher margins (Evans and Schmalensee, 2016). The orchestration of digital marketplace participants results in indirect network effects that increase the attractiveness of a digital marketplace for participants (Shapiro and Varian, 1998). As digital marketplaces often form the first touchpoint for many customers, and marketplace owners encapsulate manufacturers from customers, retailers need to establish their own digital marketplaces (McKinsey, 2020). Although the importance of platforms in retail has seemed to grow until today, the literature has focused on the adaption of business models and respective tools for modeling them (Evans and Schmalensee, 2016; Reillier and Reillier, 2017; Wulfert, Seufert, et al., 2021). Consequences for the underlying IS as infrastructure for e-commerce supporting the specifics of e-commerce and the intermediary business model of digital marketplaces are rarely considered (Aulkemeier, Schramm, et al., 2016; Wulfert, Seufert, et al., 2021). A retailer that transformed into a marketplace owner while competing with ecosystem participants selling their own products in the digital marketplace has a dual role that results in additionalIS requirements (Wulfert and Schütte, 2021).

A reference architecture for digital marketplaces may help to identify gaps in a retailer's existingIS and capabilities to be closed. (Kotusev and Kurnia, 2020). Moreover, a reference architecture can decrease the setup time for theIS that support digital marketplaces and help to standardize processes and interfaces (Angelov, Grefen, et al., 2012). Reference architectures serve as architectural templates for the design of company-specific architectures and are prescriptive character for futureIS implementations (Bass et al., 2003; Recker et al., 2021). The blueprints may also facilitate the introduction of new processes and technologies in e-commerce and ease participation in a digital marketplace, resulting in increased networks effects (Eaton et al., 2015; Shapiro and Varian, 1998). Although reference architectures for e-commerce exist (Aulkemeier, Schramm, et al., 2016; Ecomod, 2006; Vetter and Morasch, 2019), literature focusing on the specific requirements of a retailer's dual role in digital marketplaces is sparse. These requirements involve bridging functions for e-commerce (Levy, Weitz, et al., 2019), a dedicated process for matching participants from multiple market sides (Reillier and Reillier, 2017), and additional innovation platform services coping with the increasing hybridization of products (Tiwana, Konsynski, and Bush, 2010). Against this backdrop, our research question is the following:

> **RQ:** What reference architectures in e-commerce support a retailer's dual role in digital marketplaces?

To answer this research question, we conduct a literature review of (reference) architectures on theIS level in e-commerce (involving B2B and B2C business models) (Vom Brocke, Simons, et al., 2009). As the architecture should have a reference characteristic, it needs to be reusable in several companies, should be technology- and vendor-independent, and cover (parts of) theIS architecture (Angelov, Grefen, et al., 2012). Moreover, we analyze identified reference architectures concerning the fulfillment of additional requirements stemming from a retailer's dual role in digital marketplaces in a conceptual matrix (Webster and Watson, 2002).

The remainder of this article proceeds as follows. First, we derive characteristics of reference architectures and introduce architectural requirements for digital marketplaces and for supporting a retailer's dual role. Second, we sketch our research approach for the literature review. Third, we analyze 13 domain-specific reference architectures for the facilitation of a retailer's dual role on digital marketplaces. Fourth, we discuss the results and provide a brief conclusion.

### 5.2.2 Reference Architectures

Reference architectures are reference models that serve as architectural templates for the design of company-specific architectures (Bass et al., 2003). The subject of reference architectures depends on the purpose for which different architectural approaches are used. An enterprise architecture reflects the fundamental organization of a system and its relationships to the environment (Keller, 2017). Reference architectures aim to serve as templates for several companies that need to be instantiated by each company (Winter and Gericke, 2006). Reference architectures can be distinguished primarily according to whether templates are provided for the entire company or for sub-architectures such asIS. Different architecture types, such as business, application, and infrastructure architectures are formed for the individual components of the reference architecture, which are developed using specific modeling languages. The structure of the architectures follows a layered approach, which usually reflects the proximity or distance from business problems toIS (the business layer vs. the infrastructure layer). As companies are increasingly permeated by technology, enterprise architecture management needs to considerIS at the application and infrastructure layers and their alignment with the company's corporate strategy and business model (Ahlemann et al., 2012; Pereira and Sousa, 2004; Winter and Fischer, 2006). acIS research is mainly concerned withIS architectures, which consider the business architecture, organizational structure, and process organization of a company. Furthermore, they comprise a "high-level sketch" of the system and the application architecture of a company (Heinrich and Stelzer, 2009). In addition toIS architectures, there are more specific architectures for software, data, and the technical infrastructure that deal with the more technical details ofIS. These sub-architectures can also be included inIS architectures (Heinrich and Stelzer, 2009). The different architecture types within an enterprise architecture can be structured using architectural meta-models (Bass et al., 2003). The purpose of reference architectures for companies is to make the interaction of business requirements and information technology (business-IT alignment) transparent (Niemann, 2006). Reference architectures are mostly developed by an independent institution (e.g., a research institution or a standardization institute), such as TOGAF (Angelov, Grefen, et al., 2012; Open Group, 2016). Thus, reference architectures are blueprints that form proven templates for companies to economically describe and design the increasingly complex business and IT architectures. Reference architectures abstract from the peculiarities of a company and thus enable reuse of the architecture.

To determine the reusability of reference architectures in e-commerce, in Sect. 5.2.5, we applied six dimensions describing reference architectures (Angelov, Grefen, et al., 2012; Barbosa et al., 2019; Giachetti, 2010). We describe reference architectures according to their specificity, layers, coverage, level of detail, degree of formalization, and purpose. **Specificity** determines the applicability and reusability of a reference architecture and serves to distinguish architectures for a particular context from more holistic ones (Angelov, Grefen, et al., 2012). Industry-specific architectures (e.g., Y-CIM model and the H-model) can be directly applied but are limited to one industry, such as manufacturing or retail (Becker and Schütte, 2004). Technology-specific reference architectures focus on a particular technology or technology area. Vendor-specific architectures focus on system components of a software manufacturer (Giachetti, 2010). Unspecific architectures can be used as domain-independent architectures in different industries (Giachetti, 2010). They abstract from the specifics of individual software manufacturers and are usually oriented toward the processes within a company or industry. Unspecific architectures represent templates that must be specified further for concrete applications in companies. These architectures include Architektur integrierter Informationssysteme (ARIS), the Zachmann framework, and TOGAF (Heinrich and Stelzer, 2009). The following architectural **layers** are usually distinguished based on their proximity to information technology (Ahlemann et al., 2012; Barbosa et al., 2019; Heinrich and Stelzer, 2009; Winter and Fischer, 2006): the motivation, business, application, and infrastructure layers. These layers are also considered in widely used architectural modeling approaches (e.g., ArchiMate) (Open Group, 2019). Following Angelov et al. (Angelov, Grefen, et al., 2009), architecture **coverage** describes the content of the architecture. Components of concrete software modules, and interfaces describe the connections between the individual modules. Protocols define the communication formats between the components. Guidelines are closely connected to procedure models and contain recommendations for action for the implementation of the reference architecture. With regard to the **level of detail**, a distinction is made among detailed, semi-detailed, and aggregated specifications (Angelov, Grefen, et al., 2012; Greefhorst et al., 2006). The level of detail is differentiated by the number of elements and connections within the architecture and the number of aggregation levels. According to Angelov, Grefen, et al. (2012), architectures with many elements and more than two levels of aggregation have a detailed specification, whereas architectures with few elements and no levels of aggregation have an aggregated specification.

The **degree of formalization** describes the type of representation of the reference architecture (Barbosa et al., 2019). Informal architectures are in textual form. Semi-formal architectures use non-standardized notations to describe the architecture. Formal architectures are implemented in standardized notations (e.g., Unified Modeling Language (UML), ArchiMate) (Angelov, Grefen, et al., 2012; Open Group, 2019).

The **purpose** of a reference architecture is either to abstract existing systems to standardizeIS (standardization) or to set an ex-ante requirement facilitating newIS (facilitation) (Greefhorst et al., 2006). Standardization architectures are based on (parts of)IS that have been tested in practice, but facilitation architectures are developed as templates (Angelov, Grefen, et al., 2012). Standardization architectures have a descriptive character (Galster and Avgeriou, 2011). Developing a facilitation architecture based on existing research has a prescriptive character and allows for "a futuristic view of a class of systems" Galster and Avgeriou, 2011, p. 154. Facilitation architectures integrate innovative architectural patterns and aim to convince software and enterprise architectures of the qualities and benefits of the new reference architecture (Angelov, Trienekens, et al., 2008).

### 5.2.3 Digital Marketplaces and Architectural Requirements

Retail transactions are executed using electronic means and digital technologies in e-commerce (Laudon and Traver, 2019). The degree to which the transaction of the physical or digital product is executed digitally is not further specified and can vary from electronic information retrieval in the pre-purchase phase to a completely digital transaction involving a digital product (e.g., music download) (Laudon and Traver, 2019; Timmers, 1998). In this regard, digital marketplaces establish a virtual environment for buyers and sellers to conduct transactions (Turban, Whiteside, et al., 2017), exploit network effects (Park, Mezias, et al., 2004), and implement asymmetric pricing mechanisms (Chen, Chiu, et al., 2014; Rochet and Tirole, 2003). Digital marketplaces apply the concept of two-sided markets (Armstrong, 2006; Hagiu and Wright, 2015b; Rochet and Tirole, 2003). They match two or more previously distinct markets and exploit direct and indirect network effects (Armstrong, 2006). These effects are leveraged to further propel the digital marketplace (e.g., subsidization of a particular participant type) (Armstrong and Wright, 2007; Rochet and Tirole, 2003). Digital marketplaces offer a digital representation of the diverse assortment of products offered by sellers (Teller and Elms, 2010). The digital marketplace can be represented as a virtual location in an architecture model (e.g., ArchiMate) (Wulfert and Schütte, 2021). Using the typology by Staykova and Damsgaard (Staykova and Damsgaard, 2015) the participants with their assortment can be described as the periphery of the digital marketplace while the core is the digital marketplace itself offering services to supply- and demand-side participants. From the customer point-of-view, digital marketplaces "resemble retail agglomerations" Hänninen et al., 2018, p. 155, integrating the range of products of participating suppliers, retailers, and wholesalers through a single digital channel (Teller and Elms, 2010). Thus, digital marketplaces further reduce transaction costs for participants as a variety of products can be sold or purchased via a single touchpoint with a consistent user experience (Wulfert, Seufert, et al., 2021). Digital marketplaces also reduce the number of intermediaries and provide uniform boundary resources (Eaton et al., 2015). In addition to taking a neutral role by facilitating the matchmaking between participants, the marketplace owner can behave competitively with supply-side participants, offering its own products to demand-side participants (Wulfert and Schütte, 2021; Wulfert, Seufert, et al., 2021). The focus of this research paper is on the retailer's dual role as a simultaneous marketplace owner and reseller behaving competitive with other supply-side participants, as introduced by Wulfert and Schütte (Wulfert and Schütte, 2021). digital marketplaces can be established based on an existing offline business in brick-and-mortar stores or electronic shops as additional sales or procurement channels (Kawa and

Walesiak, 2019). Thus, a retailer'sIS need to facilitate the intermediation and orchestration of previously independent sides and traditional retail functions with related tasks (Levy, Weitz, et al., 2019; Reillier and Reillier, 2017). Establishing a digital marketplace and taking on a dual role poses additional requirements forIS (Schmid, 1997; Smolander and Rossi, 2008; Wulfert and Schütte, 2021). For the analysis of the identified reference architectures in section 5.2.5, we refer to the criteria introduced by Wulfert and Schütte (Wulfert and Schütte, 2021). These criteria are augmented by an additional architectural requirement on boundary resources (AR7).

Suppliers and customers as essential participants in e-commerce and brick-and-mortar environments are augmented by advertising partners, logistical service providers, opinion research firms, and others as additional market sides connected by digital marketplaces (Böttcher, Rickling, et al., 2021). To support the innovation platform perspective, we consider third-party developers and infrastructure providers as potential participants in digital marketplaces (Tiwana, Konsynski, and Bush, 2010). Thus, a digital marketplace usually orchestrates multiple market sides (Hagiu, 2009). Retailers and suppliers can either interact with a digital marketplace as suppliers or demand products from other supply-side participants (Wulfert, Seufert, et al., 2021). As a result, the various digital marketplace participants must be effectively represented in terms of master data, and records must be maintained to guarantee that interactions between them can be traced to improve future matchmaking (**AR1**).

According to Hagiu and Wright (Hagiu and Wright, 2015b), participants in a digital marketplace must constantly have some sort of affiliation with it. In the offline environment, brick-and-mortar retailers aim to establish relationships with their customers by providing loyalty cards or shopping companion apps (Wulfert, Betzing, et al., 2019). The manner in which digital marketplace participants affiliate is not clearly defined and can be interpreted in various ways (e.g., contracts, membership or cookies) (Wulfert, Seufert, et al., 2021). Because it requires information about participants, affiliation is necessary to increase the likelihood and quality of the matching among participants (**AR2**) (Evans and Schmalensee, 2016; Reillier and Reillier, 2017).

The key value proposition of a digital marketplace is the orchestration of previously independent market sides (Armstrong, 2006; Evans and Schmalensee, 2016; Rochet and Tirole, 2003). This entails matching single participants from the various sides of the digital marketplace (Moazed and Johnson, 2016). Reillier and Reillier (Reillier and Reillier, 2017) described matching as a process of attracting, matching, and connecting digital marketplace participants in order to facilitate (retail) transactions (**AR3**).

The type, extent, and coverage of services offered by the digital marketplace owner varies depending on the maturity and business model specification (Wulfert, Seufert, et al., 2021). The degree of additional services provided by a digital marketplace ranges from passive matching (e.g., eBay classifieds) to full-service offerings (e.g., Amazon) that include sales processing, logistic services, and training (Wang and Archer, 2007; Wulfert and Schütte, 2021; Wulfert, Seufert, et al., 2021). Depending on the degree of centralization of the digital marketplace (Wang and Archer, 2007), a significant portion of the bridging activities may be performed by other digital marketplace participants or the marketplace owner (Levy, Weitz, et al., 2019; Wulfert and Schütte, 2021). Because digital marketplaces often mature by adding new offerings for participants, such as payment or logistics services, (Reillier and Reillier, 2017), a retailer'sIS should be adaptable to accommodate the integration of new services provided by the digital marketplace (**AR4**) (Wulfert and Schütte, 2021). This requires a flexible and modular architecture with the possibility of integrating new functions and services (pluggability) (Aulkemeier, Schramm, et al., 2016; Wulfert, Seufert, et al., 2021).

Apart from retail-related services, digital marketplaces can provide participants with innovation platform services such as access to sales or smart product-related data, as well as development services unrelated to the core retail business (**AR5**) (Tiwana, Konsynski, and Bush, 2010). Innovation services are technical capabilities that enable participating developers to create new solutions (services or software extensions) and foster a digital marketplaces generativity (Asadullah et al., 2018a; Wulfert and Schütte, 2021). The potential of innovation platforms to catalyze reconfigurability of technical and organizational components to accelerate generativity and value creation is based on their architectural modularity (Baldwin and Clark, 2000; Tiwana, Konsynski, and Bush, 2010).

Digital marketplaces aggregate digital representations of suppliers' assortments (Wulfert and Schütte, 2021). E-commerce in general, as well as the aggregation of distinct supply-side participants' individual assortments, necessitates a digital representation of the products (**AR6**) (Turban, Whiteside, et al., 2017). Products and services can take many different forms, each with its own set of attributes that must be reflected in the product's master data. The assortment supplied by external participants can be described as the periphery of the digital marketplace, whereas the core is the digital marketplace itself, which provides (core) services to its participants (Staykova and Damsgaard, 2015). As a result, digital marketplaces lower transaction costs for participants by allowing a variety of products to be sold or acquired through a single touchpoint with a consistent user experience (Wulfert and Schütte, 2021).

The establishment of a digital marketplace involves the provision of dedicated boundary resources so that participants from different market sides can connect (**AR7**). Designing boundary resources requires considering a variety of different applications. Supply-side participants need an interface to upload their assortment to the marketplace, service providers need interfaces to offer additional (product-related) services. External application developers use a development environment, standard system architecture, or interface descriptions (Dal Bianco et al., 2014). Thus, the marketplace owner needs to open itsIS to other participants (Eisenmann, Parker, et al., 2009). Although boundary resources allow access to core modules of the digital marketplace, they also act as a control mechanism, allowing marketplace owners to manage the infrastructure based on the strategy pursued, which increases the chances of achieving market leadership (Eaton et al., 2015; Ghazawneh and Henfridsson, 2013). Boundary resources represent the governance dimension, defining the boundaries between the marketplace owner and the community of external participants, thus facilitating the realization of strategically relevant decisions about ownership, entry into new markets, or community building (Dal Bianco et al., 2014; Foerderer, Kude, Schuetz, et al., 2019; Hein, Schreieck, Riasanow, et al., 2020). Dal Bianco et al. (Dal Bianco et al., 2014) differentiated application, development, and social boundary resources. Social boundary resources are used to transfer knowledge, development boundary resources to support application development, and application boundary resources to enable interaction with platforms. For this research, we focus on technical boundary resources (i.e., application and development boundary resources) (Dal Bianco et al., 2014). While the former (APIs, libraries, etc.) are defined as the minimum required for a platform ecosystem to be viable, the latter increase the attractiveness of the ecosystem from the developers' perspective (Dal Bianco et al., 2014).

### 5.2.4 Scientific Approach

To adequately address our research question, existing reference architectures and conceptual models in e-commerce are identified and analyzed for the support of digital marketplaces in a scoping approach (Paré et al., 2015). For this purpose, we conducted a literature review following vom Brocke et al. (Vom Brocke, Simons, et al., 2009) in combination with Webster and Watson (Webster and Watson, 2002) to identify reference architectures that facilitate a retailer's dual role in digital marketplaces (Wulfert and Schütte, 2021). A literature review based on a clearly defined process and additional quality criteria ensures traceability, systematicity, and reproducibility of the results (Cram et al., 2020). As an initial narrow search for reference architectures in digital marketplaces did not provide sufficient results, we broadened our literature search to the e-commerce context. According to Cooper (Cooper, 1988), our literature review can be described as follows: We searched for reference architectures in e-commerce as research outcomes. Our goal was to synthesize the literature on reference architectures that facilite a retailer's dual role in digital marketplaces. We neutrally represented our findings from our exhaustive literature review. The organization of our literature review is conceptual. The foundation of our research is the conceptualization of reference architectures as objects of consideration in e-commerce in general and for digital marketplaces in particular. The generalized search query, which was adapted to the syntax of each search engine, is depicted on the left of Figure 1. A search across relevant journals and conference papers was conducted in the scientific databases SCOPUS, Web of Science (WoS), IEEE Xplore (IEEE), AIS electronic library (AISeL), and ACM on 2021-09-28. We used reference architecture and e-commerce as keywords with related synonyms, which resulted in a set of 1,357 preliminary articles.

To ensure an appropriate level of quality, we focused on scientific literature and added additional quality criteria to the search (Randolph, 2009). We excluded non-English and non-German articles, panels, and commentaries. Moreover, the papers had to include the e-commerce context. We excluded architectures that did not cover the entire shopping process in e-commerce, were vendor- or technology-specific, and were concerned only with actors in the business layer (Angelov, Grefen, et al., 2012). In line with our search for reference architectures inIS consisting of information about the business, application, and infrastructure layers, we also excluded articles that focused on the infrastructure layer (i.e., articles on infrastructure and application aspects without applying them in the e-commerce context) and articles with a focus on the business layer only (e.g., articles on e-commerce or sub-types without the inclusion of additional technical layers). After we applied these criteria to terms in the title, abstract, and full text of the sample, we identified 53 potentially relevant publications, leaving 39 after the exclusion of duplicates (Bandara, et al., 2015).

Then, we applied the method for reference architecture analysis by Angelov et al. (Angelov, Grefen, et al., 2012). As analysis criteria, we considered the reference architecture dimensions (Sect. 5.2.2) and architectural requirements (Sect. 5.2.3). The application of reference architecture dimensions and architectural requirements was conducted by two researchers independently. Diverging criteria were discussed in online meetings after each phase. The analysis of the reference architecture dimensions (i.e., the reference architecture check) assessed the reusability of architectures in different organizational contexts. Applying these criteria to the initial set left eight papers. Then, we conducted a one-way back- and forward search and identified five additional papers (Fig. 5.12). We determined architectural requirement fulfillment based on a qualitative analysis of the final set of papers and analyzed the textual descriptions and graphical representations of the architectural requirements. While fulfillment means that they match our textual description, partial fulfillment is indicated when not all aspects are fulfilled (e.g., platform boundary was included but not a development boundary). As the final set of 13 papers must support the retailer's role in brick-and-mortar and e-commerce with its related bridging functions (Levy, Weitz, et al., 2019), we applied the seven architectural requirements. The orchestration of the formerly independent markets sides (Rochet and Tirole, 2003) and additional innovation services (Tiwana, Konsynski, and Bush, 2010) pose additional requirements for a retailer'sIS that must to be reflected in reference architectures for e-commerce (Wulfert and Schütte, 2021). We further analyzed them regarding the fulfillment of the architectural requirements posed by the marketplace operator role and the provision of innovation services (i.e., dual role) (Armstrong, 2006; Tiwana, Konsynski, and Bush, 2010). In the following section, we present the analysis of the 13 reference architectures in a conceptual matrix (Table 5.3).

| Research areas and search string | |
|---|---|
| Reference architecture | Electronic commerce |
| ("Reference Architecture" OR "IT-Architecture" OR "Software Architecture" OR "Information System Architecture") AND ("Electronic Commerce" OR "Electronic Trade" OR "Online Shopping" OR "Online Commerce" OR "E-Commerce" OR "Ecommerce" OR "Electronic Marketplace" OR "Transaction Platform") | |

| Literature screening and selection | | | | |
|---|---|---|---|---|
| **Databases** | **Scopus** | **WoS** | **IEEE** | **AISeL** |
| **Preliminary set** | 393 | 215 | 81 | 668 |
| **Title check** | 121 | 79 | 41 | 11 |
| **Abstract check** | 46 | 27 | 28 | 4 |
| **Full-text check** | 23 | 16 | 12 | 2 |
| **Duplicate check** | 38 | | | |
| **RA check** | 8 | | | |
| **For-/Backward** | 5 | | | |
| **Final set** | 13 | | | |

**Fig. 5.12** Search String and Literature Review Results

## 5.2.5 Reference Architectures for Retail Information Systems

The industry-specific reference architectures forIS in the e-commerce context offer a high-level view of architecture components and business functions (Bass et al., 2003; Becker and Schütte, 2004). The 13 reference architectures identified cover a period from 2004 until 2019. The reference architecture analysis is summarized in Table 5.3 in chronological order of the reference architecture release in its latest version (architectural requirement fulfilled: X; architectural requirement partially fulfilled: (X); architectural requirement not fulfilled: ( ).

**Table 5.3** Conceptual Matrix of Reference Architectures Supporting a Retailer's Dual Role in Digital Marketplaces

| Reference Architecture (Reference) \| Architectural Requirement | 1 | 2 | 3 | 4 | 5 | 6 | 7 |
|---|---|---|---|---|---|---|---|
| GEMS (Albers et al., 2004) | – | x | x | – | – | – | – |
| H-Model (Becker and Schütte, 2004) | – | (x) | – | x | – | – | – |
| ECOMOD (Ecomod, 2006) | – | (x) | – | (x) | – | – | – |
| Platform Architecture (Lan, et al., 2008) | – | (x) | – | (x) | (x) | – | x |
| Model of e-Marketplaces (Matook and Vessey, 2008) | – | – | x | x | – | x | (x) |
| e-ZOCO Architecture (Castro-Schez et al., 2010) | – | (x) | (x) | x | – | x | (x) |
| EC-SCP (Chi et al., 2010) | x | x | x | – | – | – | – |
| Shell model (Schütte, 2011) | – | x | – | x | – | – | (x) |
| E-Commerce Reference Architecture (Aulkemeier, Schramm, et al., 2016) | – | (x) | – | x | – | – | – |
| Pluggable Service Platform (Aulkemeier, Paramartha, et al., 2016) | x | (x) | – | x | x | – | x |
| ARTS (OMG, 2019) | – | (x) | – | x | – | – | – |
| Integrated Architecture (Vetter and Morasch, 2019) | x | (x) | – | x | x | x | – |
| NGECP (Huang, Chai, et al., 2019) | x | x | (x) | x | – | x | – |

The **Global Electronic Market Stands (GEMS)** is a reference architecture for e-commerce that incorporates the manufacturer, retailer, and consumer perspectives on the business layer (Albers et al., 2004). The reference architecture provides an aggregated specification in textual form and was developed to facilitate interactions in digital marketplaces. The architecture describes the affiliation of diverse actors with the digital marketplace required for the entrance control module and defines matching as a core task. The authors described the data records needed for sufficient matching of the supply and demand sides (Albers et al., 2004).

The **H-model** was developed for brick-and-mortar retail and wholesale and consists of 11 main trading functions and detailed process models (Becker and Schütte, 2004). The process models were formally described with event-driven process chains, and the data models were depicted in entity-relationship models. Additionally, the architecture specifies components, interfaces, protocols, and guidelines for architecture and system implementation. As the H-model is derived from several system implementations and the specification of the industry solution for SAP's enterprise resource planning system, it is an attempt to standardize retail processes and supporting information technology. The H-model focuses on the reseller mode, purchasing products from manufacturers and selling them to customers (Hagiu and Wright, 2015b). The affiliation of manufacturers and customers with the retailer is partially modeled in business partner master data. It covers retail-related services (e.g., logistics, procurement). As this reference architecture has a brick-and-mortar focus, it does not consider innovative platform services.

**ECOMOD** was presented in 2006 and consists of reference processes and strategies for e-commerce subsumed under a structuring reference architecture (Ecomod, 2006). The reference processes mainly describe business layer components, and the descriptions are considered semi-detailed. The authors developed an individual notation for the depiction of the reference architecture and related processes. As the research project was based on the input from various companies, the architecture was introduced to standardize the processes and information technology in e-commerce. The reference architecture focuses on purchasing and selling activities with suppliers and customers under consideration. Affiliations with the intermediary are considered only for the consumer side. The orchestration of market sides is not reflected within this reference architecture. Retail-specific services are integrated only partially (Frank, 2004). Products are represented digitally and enriched with unstructured data (e.g., photos, videos), but the manufacturers' and retailers' assortments are not meant to be agglomerated (Frank, 2001).

Lan, et al. (2008) presented a **platform architecture** with e-commerce patterns that follow the service-oriented paradigm. Existing standards, such as service data objects or service component architecture, were adapted for use in e-commerce. The architecture formally presents components in UML and Extensible Markup Language (XML) notation. While the overall transactions were modeled in an aggregated manner in use case diagrams, a purchase order was exemplary detailed using XML. The reference architecture was developed to facilitate transactions in the e-commerce environment. The affiliation of the actors with the platform was implicitly mentioned in the authentication portal. The authors defined several documents and signals (e.g., order and acknowledge signal) as boundary objects for information exchange between actors. In addition to transaction-related services, the authors integrated additional offerings by adding additional (micro) services. They can be developed using development tools from the development tool layer. However, the authors did not intend for them to be used by external developers.

The domain-specific **Model of e-Marketplaces** covers the business, application, and infrastructure layers with components and interfaces in an aggregated manner (Matook and Vessey, 2008). It was developed to support different types of digital marketplaces and standardize their information systems with semi-formal and additional textual descriptions. Matook and Vessey (2008) considered a retailer's dual role as the owner of a digital marketplace and the simultaneous seller of products in their textual description. However, the authors mentioned customers and sellers only as potential participants and did not describe affiliation vehicles. The matching of participants was described as a value-adding activity, and marketplaces can also offer additional retail services such as contract, financial, or logistics services besides an aggregated representation of the various assortments. The authors did not consider any innovation platform service or related development boundary resources. Application boundary resources were described as part of the infrastructure layer.

The **e-ZOCO** architecture includes aggregated specifications for components and guidelines for the business, application, and infrastructure layers (Castro-Schez et al., 2010). The architecture was presented in a semi-formal way with non-standardized illustrations supplemented by descriptions of the texts. It was developed to facilitate digital marketplaces by overcoming "the limitations of centralized architecture" (Castro-Schez et al., 2010, p. 272). The architecture requires affiliation with the digital marketplace to authenticate participants. The matching process was addressed by sophisticated recommendation and search engines that were described in detail. e-ZOCO comprises a product catalog as a pivotal component for managing the aggregated assortment of manufacturers and retailers. Additional services to facilitate the orchestration of the different actors were considered, such as messaging and forum subsystems. The authors integrated middleware for the communications of the different (internal) subsystems but did not mention any specific boundary objects for communication with other actors.

The **Generic EC-SCP** was introduced to facilitate various business models in e-commerce (Chi et al., 2010). It is concerned with components and interfaces on the business and application layers. The aggregated specification was presented semi-formally with non-standardized forms and supplemented with textual descriptions. The core of the architecture is formed by a matchmaking component and an orchestration plan generator. The authors introduced service providers as additional participants besides manufacturers and consumers. Affiliations are represented in the access component managing participant identities and access rights.

The **shell model** as proposed by Schütte (2011) is a task-oriented redesign of the H-model. It consists of four architectures covering the entire value chain across the manufacturer, wholesaler, retailer, and consumer and implicitly mentions data forwarding along the value chain. The retail architecture consists of five shells for the main retail tasks (master data, technical tasks, economic operative tasks, administrative tasks, and decision-oriented tasks) of each actor and the retailer in particular (Schütte, 2017). Each shell consists of a series of tasks that form the components of the architecture. The aggregated specification of the architecture is presented semi-formally with non-standardized circles and rings supplemented with textual descriptions. As it is based on the H-model and integratingIS along a retail value chain, the shell model is a means for standardization. The affiliation is highlighted by specific architectures for the participants. It considers the digital representation of the products, but not the aggregation of diverse assortments in a single digital marketplace.

Aulkemeier et al. (Aulkemeier, Paramartha, et al., 2016) developed a **reference architecture for electronic commerce** and offered an additional view on the application and technological architectural layers besides the business processes. The architecture presents components, interfaces, and guidelines for the business, application, and infrastructure layers. The aggregated specification is depicted formally using ArchiMate. It was derived from a literature review and facilitates various business models in e-commerce. However, the reference architecture focuses on the reselling of products in electronic shops. Because of the reseller focus, the affiliation is integrated only for customers. With this focus, retail-specific services were considered but not additional innovation platform services.

Furthermore, Aulkemeier, Schramm, et al. (2016) suggested an additional reference architecture for a **pluggable service platform** focused on the development and integration of retail-related and external services. The architecture provides an aggregated specification of components, interfaces, and guidelines depicted formally in ArchiMate. In contrast to the previous architecture, the authors focused on the business and application layers. The architecture should facilitate the integration of additional services and enables the development of (innovative) services by third-party developers. It introduces service providers, which provide IT or business services to the retailer, and a platform provider responsible for the inter-mediation between the retailer and service providers or developers as additional participants. The affiliation of these participants is indicated by connecting arrows. A retailer can integrate various external services using defined boundary resources for files, applications, and web services.

The **Association for Retail Technology Standards (ARTS)** introduced an reference architecture for its business process retail models for the brick-and-mortar environment with a focus on the business layer components (OMG, 2019). ARTS is updated by the Object Management Group, which provides open standards for several industries and ARTS is the reference architecture for retail. The components can be characterized as an aggregated specification, and the architecture is modeled with non-standardized notation. The components are detailed in the Business Process Model and Notation language. It focuses on consumer and suppliers as participants but partially integrates affiliations with retailers. The orchestration of independent markets is not addressed i.e. by matching. Retail-specific services are provided as the main focus. As this reference architecture introduces online ordering and pick-up in store, it must include a digital representation of the assortment with unstructured data but no aggregation of the assortment of diverse actors.

Vetter and Morasch (2019) developed a reference architecture involving the components of an **integrated platform** for the brick-and-mortar and e-commerce environment. In addition, the authors provided a mapping for their architecture components to specific SAP software modules that were not considered for further analysis. The reference architecture includes components in the business and application layers. The authors used non-standardized notation to depict their aggregated specification of the architecture. As the reference architecture is based on on the SAP enterprise resource planning system, it is an attempt to standardize the architecture in retail. The authors implicitly mentioned additional participants, such as employees. The affiliation is also only monitored for the customer side, and only retail-specific services are included. Although the authors did not mention external access to their application and infrastructure layers, the components map to the introduced innovative platform services.

Huang, Chai, et al. (2019) introduced an **architecture for a portal-based NGECP**. The NGECP provides an aggregated specification for the actor, business, and application layer components and interfaces. The platform was described in a non-standardized notation supplemented with textual descriptions. The architecture was developed to facilitate a portal-based e-commerce solution. The authors described several additional participants, such as logistics service providers or reference raw material suppliers, and represented their affiliation as matchmaking between the participants, which is seen as the core value proposition. Nevertheless, the authors did not further specify the data records required for proper matching. Moreover, retail-specific services are offered, and the dual role is supported. However, innovation platform services were not mentioned.

### 5.2.6 Discussion and Future Research

We identified 13 architectures for e-commerce that are characterized as reference architectures and analyzed them for the fulfillment of seven additional architectural requirements stemming from the marketplace operator role and simultaneous sales in the digital marketplace in the reseller role (i.e., a dual role). The analysis showed that none of the 13 reference architectures fully supports the requirements resulting from a retailer's dual role in digital marketplaces. Specifically, architectures dealing with digital marketplaces in particular (Albers et al., 2004; Castro-Schez et al., 2010; Huang, Chai, et al., 2019) neither mentioned the marketplace owner as an additional ecosystem participant providing its own products nor refer to additional innovation platform services enabling integration of third-party developers. Not explicitly mentioning a retailer's dual role in an reference architecture may not be mentioned because an actor can take on several roles within a single ecosystem (Corallo, 2007. However, a dual role as a reseller in the digital marketplace and marketplace owner involves information advantages, such as concentration on fast-moving products and more profitable customers (Wulfert and Schütte, 2021). These advantages have led to antitrust law consideration in the past involving recommendation of own products to customers or adjusting prices and commission fees accordingly (KOM, 2020). The implementation of sophisticated boundary resources and the attraction of third-party developers can positively influence the attractiveness of a digital marketplace to other ecosystem participants. Third-party developers can implement additional modules for the digital marketplace, such as shop themes, interfaces with other digital platforms, or feature add-ins. The best fit for the the seven architectural requirements is the reference architecture presented by Huang, Chai, et al. (2019) that focuses on the orchestration of different types of participants (Table 5.3). As none of the analyzed reference architectures fulfills the requirements resulting from a retailer's dual role, future research may develop an reference architecture that supports a retailer's dual role in digital marketplaces. The reseller mode has been fully covered by many brick-and-mortar (Becker and Schütte, 2004; Schütte, 2011) and e-commerce (Aulkemeier, Paramartha, et al., 2016; Chi et al., 2010; Ecomod, 2006) architectures under consideration. Research on conceptual modeling from the last 50 years led to a focus on existing business models and creatin domain-specific description languages (Recker et al., 2021). The brick-and-mortar and e-commerce reference architectures as well as domain-specific description languages may form the foundation for an architecture supporting a retailer's dual role. Innovation platform services are also considered in some reference architectures (Aulkemeier, Schramm, et al., 2016; Lan, et al., 2008; Vetter and Morasch, 2019). Seven of the reference architectures were developed by researchers and research institutes to facilitate the development of newIS and the integration of new concepts and technologies. StandardizingIS in e-commerce was the focus of five reference architectures. In this regard, generic architectures (e.g., TOGAF and the Zachman Framework) might also be used for standardization in e-commerce. However, we did not include these architectures in our literature sample, because they are too generic and abstract to be directly applied in e-commerce, and therefore, will not discuss or support a retailer's dual role. Most of the reference architectures applicable in e-commerce are inspired by a brick-and-mortar background or were initially developed for an offline environment. Although the reference architectures are vendor- and technology-independent and were designed for longevity, as well as reusability, they do not always represent state-of-the-art processes or retail tasks developed 15 years ago. DespiteIS researchers propagated reference models early as crucial vehicles forIS development and implementation, in the era of agile development, they have not prevailed in many industries, including retail, and their value for developers, companies, and researchers has been questioned (Saghafi and Wand, 2014). Nevertheless, recent research claimed that reference models remain important (Recker et al., 2021). Although we focused on reference architectures for e-commerce, the back- and forward search led to the integration of architectures originating in the brick-and-mortar context. These architectures support a retailer's bridging functions that are also relevant in e-commerce (Schütte, 2017). However, they do not consider the specifics of the retailer's dual role, such as innovation platform services (AR5), a digitally aggregated assortment (AR6), or boundary resources for all types of ecosystem participants (AR7) (Table 5.3). Additionally, this review addressed scientific databases, potentially excluding company-specific reference architectures and those proposed by industrial associations. Although the former would be considered too specific during the analysis, the latter can be reused and potentially facilitate a retailer's dual role. Thus, future research may also analyze practice-based reference architectures for the fulfillment of the introduced architectural requirements. Moreover, our set of applied architectural requirements was limited to seven (Wulfert and Schütte, 2021). Although this set is neither complete nor comprehensive, it includes the main requirements resulting from a retailer's dual role on digital marketplaces. The architectural requirements address the retailer's dual role with the marketplace owner as an ecosystem participant behaving competitively with other resellers (Kawa and Walesiak, 2019), with the major requirements matching process as the core value proposition (AR3) (Reillier and Reillier, 2017) and additional innovation platform services (AR5) (Tiwana, Konsynski, and Bush, 2010). Thus, future researchers can derive additional architectural requirements from practitioner interviews and company documentation. Another important avenue for future research is the enhancement of an reference architecture to support the specifics of a retailer's dual role in digital marketplaces (i.e., matching and innovation-related services). The enhanced reference architecture should especially support the necessary activities for orchestrating participants (matching functions) and integrate innovation-related services to develop new services and modules (the infrastructure layer).

### 5.2.7 Conclusion

A retailer operating a digital marketplace while simultaneously selling products in this marketplace in a reseller role takes on a dual role in this digital marketplace ecosystem. This dual role involves the bridging functions from the reseller mode, the matching processes relevant for the marketplace, and additional innovation platform services for attracting third-party developers to the marketplace. Our analysis revealed 38 architectures in the context of e-commerce that were described by six reference architecture-related dimensions and analyzed using seven digital marketplace-specific architectural requirements. We identified 13 reference architectures that address some of the architectural requirements resulting from a retailer's dual role in digital marketplaces. However, none fully supports the requirements of digital marketplaces. The NGECP fulfills five of the architectural requirements resulting from a retailer's dual role in digital marketplaces (Huang, Chai, et al., 2019). A retailer's dual role was not fully addressed by any of the analyzed reference architectures. The matching process requirement (five architectures), the innovation platform services requirement (three architectures), and the aggregated assortment requirement (four architectures) were also underrepresented in the literature sample. Therefore, we call for initiatives to develop dedicated reference architectures supporting a retailer's dual role in digital marketplaces.

# Boundary Resources in E-Commerce Ecosystems

# 6

## 6.1 Paper 6: Developing Design Principles to Standardize E-Commerce Ecosystems: A Systematic Literature Review and Multi-Case Study of Boundary Resources

**Abstract**
*Platform ecosystems have captured a variety of markets, enabling coordination, transactions, and value co-creation between independent actors.*[1] *A focal platform constitutes the central nexus of e-commerce ecosystems and fosters the interaction among ecosystem participants through their boundary resources. Standardizing these interfaces simplifies ecosystem entry for developers and increases the number of participants propelling the network effects, and thus the overall value of the ecosystem. Currently, there is a lack of prescriptive design knowledge guiding platform owners in designing successful e-commerce ecosystems. Addressing this issue, we followed a dual approach, reporting on a systematic literature review in which we identified design requirements and complemented these with a multiple-case study on selected e-commerce ecosystems. Aggregating the requirements resulted*

[1] This version of the article has been accepted for publication, after peer review but is not the Version of Record and does not reflect post-acceptance improvements, or any corrections. The Version of Record is referenced as follows: T. Wulfert, R. Woroch, G. Strobel, S. Seufert, et al. (2022). "Developing Design Principles to Standardize E-Commerce Ecosystems: A Systematic Literature Review and Multi-Case Study of Boundary Resources." In: *Electronic Markets* 32, pp. 1813–1842

**Supplementary Information** The online version contains supplementary material available at https://doi.org/10.1007/978-3-658-45198-1_6.



T. Wulfert, *Selected Perspectives on Platforms in E-Commerce Ecosystems*,
https://doi.org/10.1007/978-3-658-45198-1_6

*in six meta-requirements and 19 design principles that foster the standardization of focal e-commerce platforms. Our design principles simplify the development of complements and enable multi-homing for developers due to possible standardization across ecosystems.*

| Fact sheet of publication 6 | |
|---|---|
| Title | Developing Design Principles to Standardize E-Commerce Ecosystems: A Systematic Literature Review and Multi-Case Study of Boundary Resources |
| Authors | Tobias Wulfert |
| | Robert Woroch |
| | Gero Strobel |
| | Sarah Seufert |
| | Frederik Möller |
| Publication Type | JournalArticle |
| Publication Outlet | Electronic Markets |
| VHB-Jourqual3-Ranking | B |
| Status | Published |
| Research Question | How should the standardization of boundary resources in e-commerce ecosystems be guided to foster network effects? |
| Research Approach | Literature Review (Vom Brocke, Simons, et al., 2009)<br>Multi-Case Study (Yin, 2018) |
| Full Citation | T. Wulfert, R. Woroch, G. Strobel, S. Seufert, et al. (2022). "Developing Design Principles to Standardize E-Commerce Ecosystems: A Systematic Literature Review and Multi-Case Study of Boundary Resources." In: *Electronic Markets* 32, pp. 1813–1842 |

### 6.1.1 Introduction

Platform ecosystems have already captured markets such as accommodation, transportation, music, and retail, with popular focal platforms such as Airbnb, Uber, Spotify, and Amazon (Choudary, 2015; Evans and Schmalensee, 2016; Parker, Van Alstyne, and Choundary, 2016). As the largest platform in electronic commerce (e-commerce), Amazon generated \$ 340 billion in revenue from product and service sales in the fiscal year 2020, with more than 60% of revenues resulting from commission fees from third-party sellers on its marketplace (Amazon, 2019, 2021). Across industries, the revenue realized by ecosystem participants is up to three times higher than the revenue generated by the focal platform (Delteil et al., 2020). Business ecosystems facilitate the interconnectivity of participants within and across previously independent markets. These autonomous participants maintain their relationships with each other using information technology in e-commerce ecosystems. They emerge around focal platforms employing various business models (e.g., marketplaces, auctions) with additional types of participants, such as manufacturers, retailers, and content providers. Choosing a platform strategy is currently a common approach to commercialize digital products and has been adopted by numerous companies (Foerderer, Kude, Schuetz, et al., 2019). The digital channel simplifies the interactions among ecosystem participants and reduces transaction costs (Nicola et al., 2020). Therefore, the success of e-commerce ecosystems depends on exploiting network effects and their strength strongly correlates with the size of the ecosystem (Briscoe, Odlyzko, et al., 2006). Thus, a platform provider must increase the number of actors from different markets by attracting external developers as participants to foster generativity and maximize ecosystem growth (Parker, Van Alstyne, and Jiang, 2017).

Establishing consensus among independent ecosystem participants, standards amplify e-commerce ecosystems by ensuring a proper value proposition (Adner, 2017; Papachristos and Van De Kaa, 2021; Staudenmayer et al., 2005). Technical standards are defined as "the characteristics of a product as well as its level of performance, safety, or quality" (Viardot et al., 2021, p. 11). Standards can be defined generally and within e-commerce ecosystems by standard-setting organizations or competitive market forces that most likely are the platform owners (de facto standard) or enforced by law (de jure standard). Within e-commerce ecosystems, standardization is applied to achieve a "winner-takes-all" strategy, often involving quasi-monopolies (e.g., Amazon) that can only be broken by law (Shapiro and Varian, 1999; Viardot et al., 2021). In recent years, many standards have been introduced to increase interoperability in e-commerce ecosystems that accelerate collaboration in product identification (e.g., GTIN), logistic units (e.g., SSCC), or global data models (e.g., GDM) (GS1, 2021; Guilloux et al., 2013; Narayanan et al., 2009).

In an e-commerce ecosystem, participants can be attracted by reducing barriers to entry on various levels (Porter, 1980), employing governance decisions to open the focal platform (Boudreau, 2010; Hein, Schreieck, Riasanow, et al., 2020), and implementing sophisticated boundary resources for participants to plug in (Aulkemeier, Paramartha, et al., 2016; Eaton et al., 2015; Ghazawneh and Henfridsson, 2013). The provision of platform boundary resources, such as Application Programming Interfaces (APIs) or Software Development Kits (SDKs), involves external application developers focusing on a product, standard system architecture, or communication protocol (Dal Bianco et al., 2014). Boundary resources form an indispensable part of a focal platform's architecture, facilitating the participant's interrelation with the periphery as a stable part (Baldwin and Woodard, 2009; Staykova and Damsgaard, 2015).

Ghazawneh (2012) has taken the concept of boundary objects (i.e., repositories, ideal types, coincident boundaries, and standardized forms) from social science and applied it to (software) ecosystems (Ghazawneh, 2012; Ghazawneh and Henfridsson, 2013; Star and Griesemer, 1989). Based on this elicitation, Dal Bianco et al. (2014) distinguished among application (e.g., APIs), development (e.g., SDKs), and social boundary resources (e.g., documentation, developer portals). Acting as a keystone, the platform owner can actively control the boundary resources as part of the platform governance (Hein, Schreieck, Riasanow, et al., 2020; Hein, Schreieck, Wiesche, et al., 2016; Iansiti and Levien, 2004a). Defining the boundaries between the platform owner and the community of third-party developers and other participants facilitates the realization of strategically relevant decisions about ownership, entry into new markets, and community building (Dal Bianco et al., 2014; Foerderer, Kude, Schuetz, et al., 2019; Hein, Schreieck, Riasanow, et al., 2020). Although approaches to strategically develop boundary resources have been proposed in recent research studies (Hein, Weking, et al., 2019; Star, 2010), standardization of these resources within and across ecosystems has not been considered. Standardized boundary resources are likely to attract additional participants, allow multi-homing, increase the size of the network, and propel network effects (Eaton et al., 2015). Such boundary resources also allow for increased interoperability, modularization, and specialization within an ecosystem and guarantee that complements use boundary resources appropriately (Teece, 2018). The success of platform ecosystems (e.g., Apple iOS, SAP ERP) relies on the provision of sophisticated boundaries for third-party participants that provide additional modules and co-create value for end customers (Eaton et al., 2015; Ghazawneh and Henfridsson, 2010; Hütsch and Wulfert, 2022b).

However, for developers providing additional modules to ecosystems, standards do not exist for the application layer in e-commerce ecosystems other than technological, industry-independent, de facto standards such as programming languages or development tools. Thus, the setup time for developers entering a new e-commerce ecosystem is evident, and multi-homing is only possible with additional efforts. In research, the topic of ecosystems, in general, is of increasing interest (Bogers et al., 2019), but little attention has been paid to the technological level and boundary resources, in particular those enabling the interconnection of autonomous participants (Aulkemeier, Schramm, et al., 2016). Existing design principles in e-commerce focus on User Interface (UI) design (Billewar et al., 2021; Foglie et al., 2010; Resnick and Sanchez, 2004; Zollet, 2014), specific technical boundary resources (Sarkar et al., 2007; Verborgh and Dumontier, 2018), or platforms in service networks (Blaschke, Riss, et al., 2019) and omit the ecosystem perspective. We explicitly focus on the standardization of boundary resources for developers to ease developers' work in e-commerce ecosystems, increase the number of developers and complements, and thus propel network effects. We address the problem of proprietary boundary resources in e-commerce ecosystems and propose a set of standardized boundary resources to platform owners to ease the participation of developers and increase the ecosystems' generativity. Hence, our research question is as follows:

> **RQ:** How should the standardization of boundary resources in e-commerce ecosystems be guided to foster network effects?

To answer our research question, we conducted a systematic literature review and report on multiple case studies extracting prescriptive design knowledge in the form of design principles. Developing prescriptive design knowledge in multiple subsequent case studies that enable reflection of successful designs is an established strategy in Information Systems (IS) research (Gregor and Jones, 2007; Van Aken, 2004). Consequently, design principles offer a suitable mechanism to learn from successful designs of the past (i.e., those cases we have selected) and go "beyond a single success story" to "inform other design endeavors" (Chandra Kruse and Seidel, 2017, p. 186). This strategy is paramount to finding cross-case patterns for successful designs and reflections of design principles as part of the theory for design and action (Eisenhardt, 1989b; Gregor, 2006; Van Aken, 2004). We regard the formulation of design principles as highly relevant, given that we can codify design knowledge about ecosystem design in e-commerce and make it reusable, so that it can be used by others at a different time and reduce the subsequent iterations needed to achieve a successful design (McAdams, 2003). We propose design principles to achieve the standardization of boundary resources in e-commerce ecosystems to reduce barriers to entry for developers and propel network effects. As de Reuver et al. (2017) have called for domain-specific conceptualizations of the focal platform, we focus on boundary resources as the transition zones between the core and the periphery of an e-commerce ecosystem (Staykova and Damsgaard, 2015). Consequently, we develop our design principles focusing on the specific requirements of developers within e-commerce ecosystems that extend the focal platform with additional modules (e.g., shop themes, plug-ins, and integration components).

The remainder of this article proceeds as follows: first, we unfold related literature on e-commerce ecosystems as a specialization of digital business ecosystems and the concept of boundary resources in digital platforms. Second, we present our scientific approach of developing design principles to increase the standardization of boundary resources. Third, we elicit our six meta-requirements with 19 related design principles. Fourth, we discuss our results, and fifth, we conclude with a short summary and research outlook.

## 6.1.2 Related Literature

### 6.1.2.1 E-Commerce Ecosystems

The ecosystem concept was introduced into the business literature by Moore (1993, 1996) as an analogy to biology, in which individual organisms form a population and interact in a habitat. A business ecosystem is formed by independent (or even competing) actors (natural or legal entities), who share a common interest, such as the success of the ecosystem (Corallo et al., 2007; McIntyre and Srinivasan, 2017; Wareham et al., 2014), and who depend on each other (Iansiti and Levien, 2004a) and even co-create value (Blaschke and Brosius, 2018). The perspective of business ecosystems is grounded in organizational boundary theory (Santos and Eisenhardt, 2005; Teece, 2007; Tsujimoto et al., 2018). An ecosystem dynamically evolves and is subject to continuous change, as individual participants join or depart while creating new or interrupting existing relationships (Corallo, 2007; Wulfert, Seufert, et al., 2021).

The relationships and affiliations among the participants in digital business ecosystems are maintained via information technology IT, and value co-creation is realized digitally (Tsujimoto et al., 2018). The participants in digital business ecosystems are interdependent. They cooperate to achieve common objectives, usually at the same time competing for scarce resources (Corallo et al., 2007). Thus, a digital business ecosystem is a complex network of platform-mediated, actor-to-actor interactions, becoming increasingly accessible to end users through third parties' platform complements (Corallo et al., 2007; Wareham et al., 2014). An e-commerce ecosystem is a manifestation of a digital business ecosystem in the context of e-commerce. The value of an e-commerce ecosystem increases for every actor with the addition of each actor in the network (Briscoe, Odlyzko, et al., 2006; Shapiro and Varian, 1998). Although an increasing number of ecosystem participants amplifies inter-ecosystem competition and diminishes exclusive offers, the overall ecosystem value for each participant on the same and different connected markets is increasing (Zhao, von Delft, et al. 2020). For instance, manufacturers offering similar products within an e-commerce ecosystem compete for similar end customers. The additional ecosystem value is driven by direct and indirect network effects[2], increasing the value of participation in a network by actors of both the same and different types (Hinz et al., 2020; Shapiro and Varian, 1998). While this affilia-

[2] With direct and indirect network effects, the value of an e-commerce ecosystem for ecosystem participants increases with the number of similar participants (direct network effects) and the overall size of the network (indirect network effects). This is due to the higher potential of exchange (e.g., products, product reviews, knowledge) between these participants (Eisenmann, Parker, et al., 2006; McIntyre and Srinivasan, 2017; Rochet and Tirole, 2003). A pre-

tion perspective is mainly concerned with the relationships between the individual actors in an e-commerce ecosystem, the structural view of an ecosystem focuses on the overarching purpose of the ecosystem and its value proposition for a focal company's end customers (Adner, 2017). The overarching purpose of e-commerce ecosystems is to exchange physical or digital products via electronic communication media (Adner, 2017; Becker and Schütte, 2004; Laudon and Traver, 2019).

E-commerce ecosystems can be formed around a focal platform that can serve several purposes (i.e., innovation and transaction activities), depending on its type (Evans and Schmalensee, 2016; Evans and Gawer, 2016; Gawer, 2014). The platform supports the focal company's business model and can be characterized as the "technical core" of the ecosystem (Blaschke, Haki, et al., 2019, p. 4). Innovation platforms form the operating systems for online shops, electronic marketplaces, and other business models and provide the application services necessary for executing the business processes and environments for developing external modules (Aulkemeier, Schramm, et al., 2016; Hanseth and Lyytinen, 2010; Tiwana, Konsynski, and Bush, 2010). While platforms usually exploit economies of scale and scope with increasing efficiency and increased product variety through reusability and reconfiguration of modules or services, they may utilize further economic effects as the center of a broader innovation ecosystem (Cusumano and Gawer, 2002; Gawer, 2009b; Rochet and Tirole, 2003). Platform extensions and independent modules developed by third-party developers result in the formation of an innovation ecosystem (Manikas and Hansen, 2013; Messerschmitt and Szyperski, 2003). Innovation ecosystems often include dedicated extension marketplaces responsible for providing developed modules (Jansen and Cusumano, 2013). Digital business ecosystems can be structured by a "set of roles" (Adner, 2017, p. 42), describing a standardized relation between the different participants on an abstract level (Jacobides et al., 2018). Eisenmann, Parker, et al. (2009) have defined four archetypal roles[3] in an e-commerce ecosystem: users from the demand and supply sides, platform providers, and platform sponsors. The platform owner (i.e., the provider and sponsor) implements governance mechanisms for the focal platform to control the autonomous ecosystem participants and facilitates value creation (Hein, Schreieck, Riasanow, et al., 2020). We consider platform owners as architects of the surrounding ecosystem, defining an interaction structure with a set of boundary resources (Helfat and Raubitschek, 2018).

requisite for indirect network effects is the presence of bidirectional cross-group-side network effects (Hagiu and Wright, 2015a; Shapiro and Varian, 1998).

[3] A single ecosystem participant may also take on several roles within a single ecosystem.

The major actors in e-commerce ecosystems are the various manufacturers and end customers with intermediating retailers and wholesalers (Böttcher, Rickling, et al., 2021). The original participants mentioned above are augmented by further participants, such as data suppliers, content providers, and advertising partners in e-commerce ecosystems that provide additional information and media in a digital environment (Wulfert and Schütte, 2022). Software providers implement the necessary IS orchestrating the diverse actors and offer development environments for external developers (Tiwana, Konsynski, and Bush, 2010). Third-party developers may implement additional modules, such as shop themes, interfaces with other digital platforms, or feature add-ins (Wulfert and Schütte, 2021). The participants are connected to the focal platform using a set of boundary resources (Fig. 6.1). The joint value creation effort of the participating actors in e-commerce ecosystems results in the provision of products and services to end customers (Adner, 2017). An e-commerce ecosystem fulfills the necessary functions, bridging the discrepancies in time, space, and quantity within a supply ecosystem from (raw material) suppliers to end customers (Becker and Schütte, 2004; Laudon and Traver, 2019; Schütte, 2017).

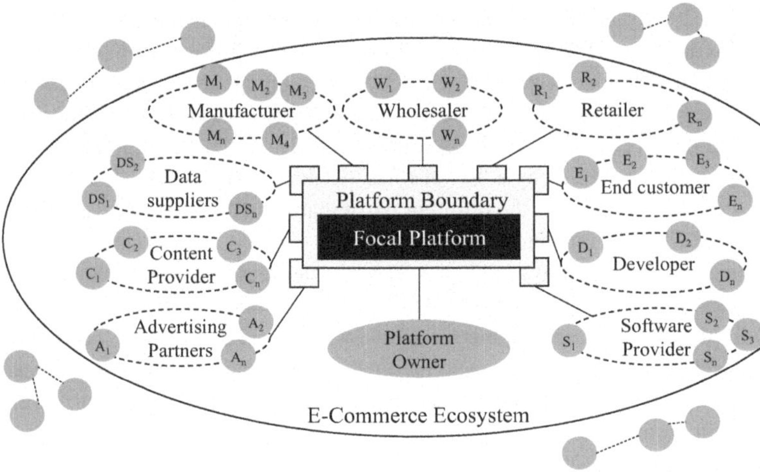


**Fig. 6.1** Schematic View of a Platform Ecosystem in E-Commerce with Archetypal Participants

### 6.1.2.2 Boundary Resources as Theoretical Lens

Initially conceptualized in the context of a case study of collaboration among actors with different perspectives but primarily the same objective, boundary objects explain successful collaboration through standardized methods, as well as the development of a set of objects that enable different groups to work together without direct coordination (Star, 2010; Star and Griesemer, 1989). Boundary objects are "plastic enough to adapt to local needs and the constraints of the several parties employing them, yet robust enough to maintain a common identity across sites" (Star and Griesemer, 1989, p. 393). The initial concept from the field of science and research, based on a case study from the Museum of Vertebrate Zoology in California, included four types of boundary resources (repositories, ideal types, coincidence boundaries, and standardized forms), which did not claim exclusivity and, thus, can be adapted or extended (Star, 2010; Star and Griesemer, 1989). The provision of platform boundary resources enables building a digital business ecosystem involving external application developers around a product, standard system architecture, or communication protocol (Dal Bianco et al., 2014). Even though boundary resources allow access to core modules and integrate developers as participants in the ecosystem, they also act as a control mechanism allowing platform owners to manage the infrastructure based on the strategy pursued, which increases the chances of achieving market leadership (Eaton et al., 2015; Ghazawneh and Henfridsson, 2013).

Designing boundary resources requires considering a variety of different applications. Therefore, it is necessary to find a form that, on the one hand, supports easy software development due to its slenderness and, on the other hand, leaves open creative space so that the innovative ideas of the developer communities can be included in the cultivation of the digital business ecosystem (Dal Bianco et al., 2014; Ghazawneh and Henfridsson, 2013). Boundary resources, as mentioned in the introduction, represent a dimension of platform governance, defining the boundaries between the platform owner and the community of third-party developers, thus facilitating the realization of strategically relevant decisions about ownership, entry into new markets, or community building (Dal Bianco et al., 2014; Foerderer, Kude, Schuetz, et al., 2019; Hein, Schreieck, Riasanow, et al., 2020). In this way, platform boundary resources encompass more than the provision of merely technically relevant resources. Within this research, different types of platform boundary resources that involve both technological and organizational interaction between platform owners and ecosystem participants (e.g., developers) are applied as a theoretical lens to shape e-commerce ecosystem design principles (Dal Bianco et al., 2014; Foerderer, Kude, Schuetz, et al., 2019; Ghazawneh, 2012). To this end, we draw on theoretical work on platform boundary resources that build on each other (Fig. 6.2).

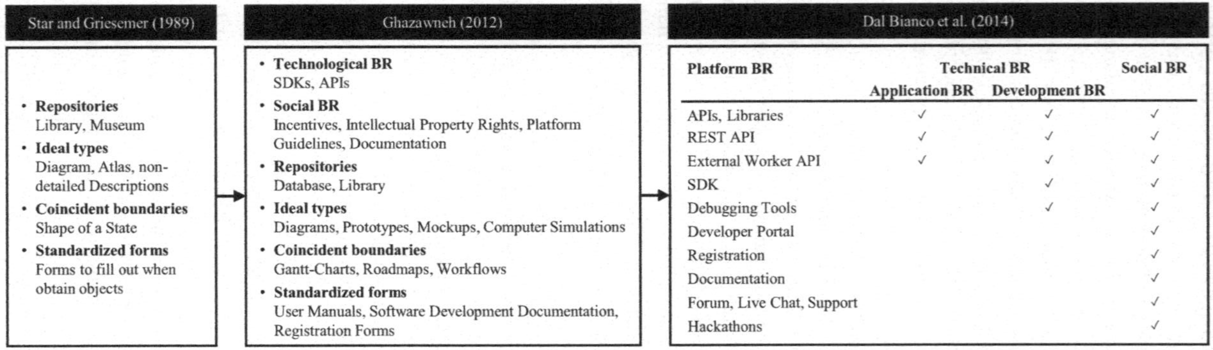


**Fig. 6.2** Development of Platform Boundary Resource Types (Dal Bianco et al., 2014; Ghazawneh, 2012; Star and Griesemer, 1989)

### 6.1.3 Methodology

This paper aims to provide a theoretically grounded and empirically validated set of design principles for the standardization of e-commerce ecosystems based on their boundary resources. Design principles are meta-artifacts that belong to the theory for design and action, according to Gregor (2006) taxonomy of theory types. Consequently, they are not artifacts per se, but meta-artifacts that enable the codification of design knowledge and make it reusable in different application scenarios (Chandra et al., 2015; Iivari, 2003). Since design principles are nascent theories, we devised a strategy to generate design principles from two data sources suitable to develop theory (Gregor and Hevner, 2013; Möller, Guggenberger, et al., 2020). First, we report on an in-depth systematic literature review, which we used to identify knowledge engraved in the literature corpus and extracted in the shape of design requirements (Schryen et al., 2015). Second, we complement these findings with a multiple-case study using comparative analysis to identify relevant design knowledge suitable for abstraction and codification into design principles (Yin, 2018). The dual approach (Fig. 6.3) that we propose is expressively suitable to generate theory, as the systematic literature review ensures the inclusion of existing knowledge in theory building (Webster and Watson, 2002), and the multi-case study complements it through its high "likelihood of generating novel theory" (Eisenhardt, 1989b, p. 546).

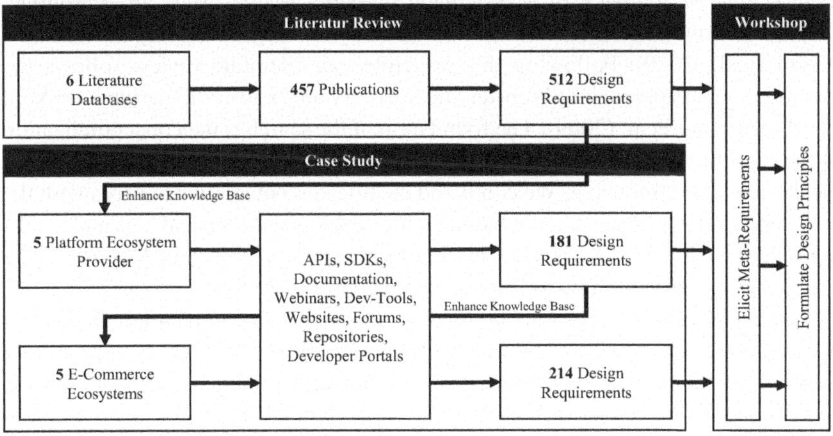


**Fig. 6.3** Overall Research Process

Since our view on e-commerce ecosystems is focused on platform boundary resources, the case data and the literature review (i.e., the coding of the papers) are viewed through a theoretical lens (Niederman and March, 2019) that we developed during the elaboration of the theoretical foundation. The theoretical lens is a product of an iterative analysis of the literature conceptualizing platform boundary resources (Appendix A.5.1 in the Electronic Supplementary Material). Generally, we started with the generic classification of boundary resources into social boundary resources, platform boundary resources, and development boundary resources, following Dal Bianco et al. (2014). By entangling these concepts with more nuanced ones, we constructed a hierarchy of concepts, with the final leaf being the theoretical lens fragment that guided our coding process. For example, we specified platform boundary resources to APIs and coincident boundary (Star and Griesemer, 1989). We detailed the latter more to accommodate our inductive findings, from which we judged that further detailing of the concept was required and sensible (Appendix A.5.2 in the Electronic Supplementary Material). In the following sections, we describe in more detail how both methods (i.e., the systematic literature review and the multi-case study) intertwine conceptually to formulate design principles.

#### 6.1.3.1 Systematic Literature Review

The first phase aims to analyze the existing literature and the associated deduction of initial design requirements. Based on process and quality criteria, the methodological approach of a systematic literature search offers an appropriate framework to ensure the traceability, systematicity, and reproducibility of the results (Cram et al., 2020). Following this principle, our literature review follows the methodological approach of Webster and Watson (2002), in combination with Vom Brocke, Simons, et al. (2009). The foundation of the search is the conceptualization of the object under consideration (i.e., boundary resources) and its integration into the domain of e-commerce. We combined the relevant boundary resources with the conceptual variety of the domain based on the developed theoretical lens and resulting search terms (Appendix A.5.2 in the Electronic Supplementary Material). The boundary resources search terms are aggregated in Figure A.4.

In total, 84 search strings were performed in six databases, with initially 8,994 hits in April 2021. To ensure an appropriate level of quality, additional quality criteria were added to the search. Excluded were non-English language articles; panels and commentaries; purely technical articles (e.g., articles that focus exclusively on technological aspects without applying them to an e-business platform ecosystem context); and articles with a pure e-business focus (e.g., articles that focus exclusively on e-business or subtypes without adopting a corresponding e-commerce ecosystem perspective). Based on the title and abstract considering the quality criteria, and

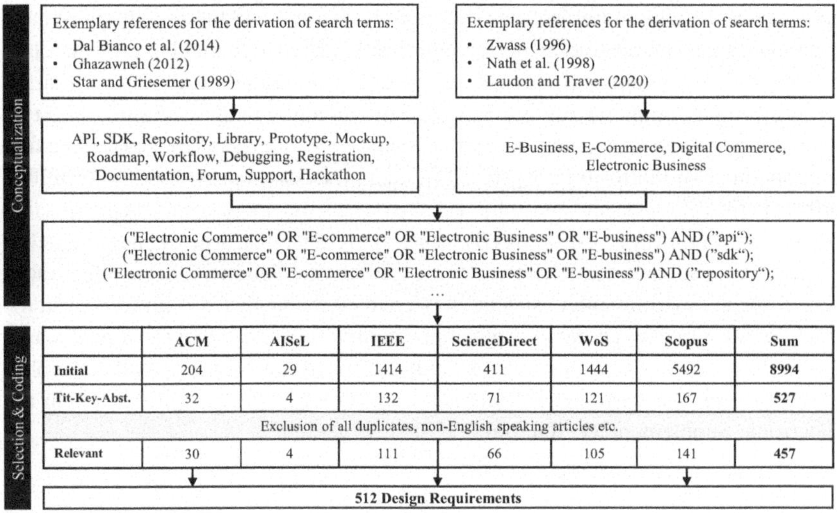


**Fig. 6.4** Literature Review and Coding-Process

following the approach of Bandara et al. (2015), 527 relevant publications could be identified within the initial set. After excluding duplicates, 457 publications were defined as the final literature sample. The articles within the sample were independently analyzed using full-text screening, and the relevant text passages were coded to extract the requirements. The 14 conceptualized boundary resources served as the basis for coding, which was inductively adapted as needed. Based on the coding scheme, 512 requirements were extracted in four coding iterations (Fig. 6.4).

#### 6.1.3.2 Multi-Case Study Analysis

To develop design principles that are grounded in empiricism, we report on a multiple-case study (Eisenhardt, 1989b; Yin, 2018). Explicitly, we chose a multiple-case study since it enabled us to collect design knowledge more broadly from a series of cases that might differ in environmental aspects but share a common phenomenon Yin (2018). Selecting multiple cases is paramount in identifying cross-case patterns and as the basis for comparative analysis (Eisenhardt, 1989b; Van Aken, 2004). For example, Avdiji et al. (2020) have demonstrated the power of cross-case analysis and the abstraction of design knowledge from that process. In our research, we identified ten cases as the basis to derive design knowledge about thriving e-commerce ecosystems. Having multiple cases is necessary to transcend the specificity of each individual case and abstract generally applicable knowledge that benefits the class of the artifact (Gregor and Hevner, 2013; Lee, Pries-Heje, et al., 2011), that is, standardized e-commerce ecosystems. We structured the multi-case study into two thematic categories to accommodate their shared environment (Yin, 2018). The first set contains five Platform Ecosystem Providers (A), cases providing the technical infrastructure of successful e-commerce ecosystems, and the second set contains five E-Business Platform Ecosystems (B), cases applying a platform business model in e-commerce (Schütte and Wulfert, 2022a; Wulfert and Schütte, 2021). While these sets differ in environmental aspects, they enabled the analysis of how ecosystems work in e-commerce, giving us ample opportunity to generate theory from all cases (Table 6.1). Further details on cases analyzed regarding business model, number of developers, and participant orientation can be found in Appendix A.5.1 in the Electronic Supplementary Material.

**Table 6.1** Key Performance Indicators of Case Platforms

| Platform Ecosystem Providers (A) | Shopware | Magento | SAP Commerce Cloud | Salesforce Commerce Cloud | OXID eShop |
|---|---|---|---|---|---|
| Establishment | 2004 | 2008 | 2012* | 1999 | 2003** |
| Country of origin | Germany | USA | Germany | USA | Germany |
| Number of employees | 300 | 375 | – | 49.000 | 51–200** |
| Anual revenue (billion euro) | 0,02 | 84,7 | – | 14,5 | – |
| E-Business Platform Ecosystems (B) | Amazon | Walmart | Ebay | Etsy | Otto |
| Establishment | 1994 | 1962 | 1995 | 2005 | 1949 |
| Country of origin | USA | USA | USA | USA | Germany |
| Number of employees | 1.298.000 | 2.200.000 | 13.300 | 1.240 | 49.895 |
| Anual revenue (billion euro) | 326,9 | 473,5 | 9,1 | 1,4 | 15,6 |

* = Year of publication of the SAP Cloud Platform ** = Values refer to the OXID eSales AG

Since we wanted to produce prescriptive design knowledge that enables designers to design e-commerce ecosystems more efficiently and learn from past success stories, we collected design knowledge from highly successful ecosystems in e-commerce (Chandra Kruse, Seidel, and vom Brocke, 2019). Consequently, the environment of each case indicated a high degree of success in establishing highly successful ecosystems in e-commerce. Correspondingly, the shared phenomenon that we investigated is how these cases leverage the utility and potential of standardized platform boundary resources. In selecting the cases, we explicitly targeted "large" platform owners with substantial ecosystems. We also tried to include prominent and successful cases from China, such as Alibaba and JD. However, the language barrier proved to be an obstacle too difficult to overcome without a significant loss of comprehension of the underlying data, which is why we opted to select only cases with data available in the German or English languages. In addition, we opted to include only successful cases to codify "successful" knowledge and make it applicable to others (Chandra et al., 2015; Gregor, Chandra Kruse, et al., 2020). Subsequently, we excluded Rakuten, given its transition from a marketplace to a cash-back platform and its decrease in global availability. In summary, we selected focal platforms of successful e-commerce ecosystems on the business and infrastructure layer, in terms of size and identity (Cennamo, 2021), that leverage the utility and potential of standardized boundary resources. Ultimately, we identified the following ten cases that we used to reflect and abstract our findings (Gregor, 2009). While category A consists of the cases of Salesforce Commerce Cloud, Magento, SAP Commerce Cloud, Shopware, and OXID eShop, category B contains Amazon marketplace, Walmart marketplace, eBay, Etsy, and Otto (Appendix A.5.1 in the Electronic Supplementary Material). We leveraged a plethora of publicly available data sources to collect information on each case. Specifically, we analyzed the case data through the theoretical lens developed above, ensuring that we framed it distinctively in the context of platform boundary resources (Niederman and March, 2019). Naturally, we screened only official information on each case. We did not limit ourselves to the homepages of the companies alone and therefore screened all company sources on different platforms (e.g., GitHub, YouTube). Since we analyzed the data through the lens of platform boundary resources, we included technical documentation on APIs and SDKs as primary data sources. In addition, we looked through available information, codes, documentation, and meta-data on GitHub for each case. Moreover, we analyzed corresponding development tools offered by platform providers for development or prototyping to fully understand the necessary steps to utilize the platform from the ecosystem participants' perspective. Appendix A.5.1 in the

Electronic Supplementary Material contains an extract of sources leveraged for each case company selected and includes a matrix providing transparency on the impact of the case companies on our design principles (Table A.5.1.3). In the dual approach, we extracted detailed design requirements that stayed on a very narrow conceptual level and close to the wording and content of the underlying literature and/or case data (Fig. 6.4). The procedure resulted in an accumulated set of 900 design requirements from the literature and case analysis (Fig. 6.3). Since design principles require addressing a class of artifacts rather than an instance, we correspondingly elevated the set of design requirements to a higher order of meta-requirements (MRs) (Walls et al., 1992). In addition, developing design principles based on meta-requirements ensures value grounding, meaning that no design principles exist without fulfilling at least one requirement (Goldkuhl, 2004). We did this by, first, excluding duplicates from the design requirements as well as those we identified as not being crucial. In a second step, following the recommendations of Koppenhagen, et al. (2012), we used logical content aggregation in a series of workshops by the author team to distill the most relevant requirements of the class of artifact, that is, meta-requirements (Thoring et al., 2020). Based on that aggregation process, we generated six meta-requirements for designing boundary resources in e-commerce ecosystems to guide our formulation of design principles (Möller, Guggenberger, et al., 2020).

### 6.1.4 Formulation of Design Principles

This section reports on the meta-requirements and design principles for standardized boundary resources in e-commerce ecosystems that we developed based on the literature corpus and the multi-case study. In the following, we introduce the six aggregated and condensed meta-requirements, alongside a short explanation (Table 6.2). The first meta-requirement (MR1) addresses the need to tailor boundary resources to accommodate the particular needs of different ecosystem participant types in terms of access to the ecosystems and communication levels. MR2 addresses the need for openness for the ecosystem, expressed in an open architecture or the possibility to contribute third-party applications. MR3 addresses issues of trust and risk, demanding that boundary resources in e-commerce ecosystems implement mechanisms for security when selecting, for example, third-party modules or supplies on marketplaces to lower the chance of fraud. MR4 requires that these ecosystems provide a minimal shared understanding of institutionalized rules and standards to enhance the quality of components and streamline integration into the ecosystem.

MR5 (development environment) requires the provision of streamlined workflows for developers and standardized mechanisms to contribute third-party modules (e.g., integrated development environments or demos). Last, MR6 requires the curation of the ecosystem by a platform owner through providing documentation and good practices in early communication of changes to the codebase to enhance reliability for developers. Table 6.2 summarizes the meta-requirements with a corresponding short title and using the modal verb "should" (Offermann et al., 2010).

**Table 6.2** Meta-Requirements for Designing Boundary Resources in E-Commerce Ecosystems

| # | Short Title | Meta-Requirement |
|---|---|---|
| MR1 | Tailored Boundary Resources | Boundary resources in e-commerce ecosystems should be tailored to the requirements of ecosystem participants. |
| MR2 | Openness of the Ecosystem | Boundary resources in e-commerce ecosystems should foster the openness of the ecosystem and use an open architecture. |
| MR3 | Trust and Risk | Boundary resources in e-commerce ecosystems should implement mechanisms to enhance trust and security when selecting third-party modules. |
| MR4 | Rules and Standards | Boundary resources in e-commerce ecosystems should institutionalize a minimal common understanding of rules and standards for participants. |
| MR5 | Development Environment | Boundary resources in e-commerce ecosystems should provide developers with streamlined workflows to contribute third-party modules. |
| MR6 | Ecosystem Curation | Boundary resources in e-commerce ecosystems should actively curate the ecosystem through mechanisms such as documentation and communication. |

Figure 6.5 links the six meta-requirements above with 19 design principles that we explain in detail below. The mapping diagram depicts the fulfillment of requirements by the design principles, with links between them, and provides each design principle with a short textual description (Möller, Guggenberger, et al., 2020).

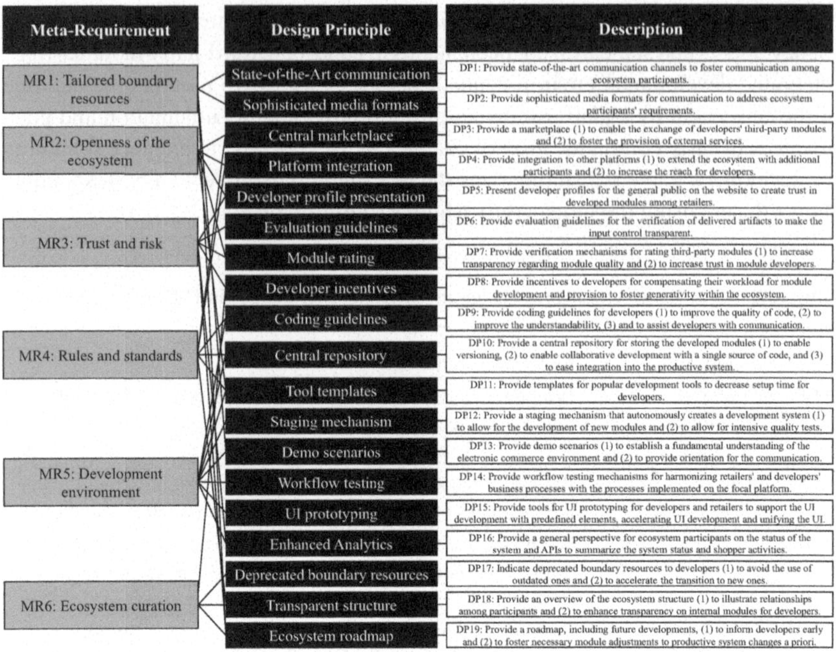


**Fig. 6.5** Meta-Requirements and Design Principles for Boundary Resources in E-Commerce Ecosystems

### 6.1.4.1 Tailored Boundary Resources

> **DP1:** Provide **state-of-the-art communication channels** to foster communication among ecosystem participants.

**Rationale:** The establishment and maintenance of relationships among ecosystem participants rest on synchronous and asynchronous communication. This communication can be fostered by state-of-the-art communication channels, such as developer forums to exchange ideas and discuss problems (Magento, 2021d; SAP, 2021a), blogs to inform developers about future developments and provide developers with a platform to present their new projects and solutions (Liang, Ho, et al., 2011; Otto SE, 2021c), and an online code editor with syntax highlighting for collaboration (Salesforce, 2021g). In addition to these, the platform owner can implement a dedicated support system (Ahmed, 2020), offer a live chat bot that answers questions

(Elmorshidy, 2011; Li, Qiu, et al., 2017), and present frequently asked questions and solutions for already-recognized issues (Büttner et al., 2019). These channels should create and foster social interactions to satisfy developers' social needs for belonging and support (Liang, Ho, et al., 2011). Social interaction can also be addressed by hosting hackathons to create new solutions for the e-commerce ecosystems (Otto SE, 2021a).

**DP2:** Provide **sophisticated media formats** for communication to address ecosystem participants' requirements.

**Rationale:** As developers are habituated to sophisticated media formats because of prior personal and professional experiences, boundary resources using sophisticated media formats should be presented by the platform owner. This will increase the overall communication quality with ecosystem participants in general and developers in particular. The platform owner can offer online collaboration with synchronized editing, voice messaging, web conferencing, messaging, and video streaming to communicate boundary resources to developers (Sokiyna and Agel, 2020). Our case study analysis revealed that interactive webinars with gamification elements and micro learning are often offered to engage developers in training (Salesforce, 2021f; SAP, 2021b). Furthermore, platform providers organize online keynotes and virtual face-to-face meetings with developers (Salesforce, 2021b). In addition to the application of media formats, all platform cases analyzed provide extensive textual descriptions. This is especially relevant for documentation on technical boundary resources provided to developers (Etsy, 2021a; GitHub, 2021; Otto SE, 2021b).

### 6.1.4.2 Openness of the Ecosystem

**DP3:** Provide a **marketplace** (1) to enable the exchange of developers' third-party modules and (2) to foster the provision of external services.

**Rationale:** A central marketplace for providing third-party modules and external services (e.g., address validation, logistics services, content provision), integrated into the focal platform and provided by the platform owner, can encourage the intention to use these modules and services among ecosystem participants (Hajli, 2013). The platform owner as an intermediary increases trust in external modules and services (Verhagen et al., 2006). The analysis of our case studies demonstrated that there is no consensus established to offer a dedicated store for modules to be installed and a separate one for services or a store including both types. While Salesforce provides a store for external modules (i.e., app exchange) and a store for services provided by partners (i.e., partner marketplace) (Salesforce, 2021c,e), SAP integrates both types within a single store, Magento offers additional modules, and Walmart only provides self-developed, ready-to-use modules without a dedicated store (Magento, 2021e; SAP, 2021e; Walmart, 2021b).

> **DP4:** Provide **integration to other platforms** (1) to extend the ecosystem with additional participants and (2) to increase the reach for developers.

**Rationale:** An existing e-commerce ecosystem can be extended by connecting the focal platform with its provided services and products to other platforms and their surrounding (e-commerce) ecosystems. This increases the dissemination of the focal platform and enables developers to reach additional participants of other ecosystems. Piggybacking on an existing ecosystem is a well-known strategy for establishing and disseminating platforms (e.g., PayPal using eBay to attract participants) in electronic business (Evans and Schmalensee, 2016; Parker, Van Alstyne, and Choundary, 2016). To establish a relation and implement a connection between two or more platforms, integrations services, connectors, protocol conversion, or message transformation are required. This integration ensures a seamless flow between the platforms without any interruption for developers or customers. Moreover, the focal platform may optimize its pluggability for other platforms to increase the generativity and variety of the ecosystem and to attract additional participants, increasing network effects (Aulkemeier, Paramartha, et al., 2016). Integration services to other electronic marketplaces, shops, or payment platforms are often deployed via stores providing external modules (Magento, 2021h; Oxid AG, 2021b). Software vendors such as Salesforce or SAP integrate their shopping systems with their portfolio of applications.

#### 6.1.4.3 Trust and Risk

> **DP5:** Present **developer profiles** for the general public on the website to create trust in developed modules among retailers.

**Rationale:** The developer profiles help clarify which independent or certified developer has contributed to the e-commerce ecosystem. This transparency increases trust in external modules (Teubner and Dann, 2018). Providing detailed profiles, including portraits of the developers, creates relationships on a social level (Hajli, 2013). The publication via the focal platform, as an intermediary between developers and users of the modules, increases the established trust (Tan, Lu, et al., 2015). At Magento, for example, the number of created and accepted contributions of a developer is published. There is also a scoring system to determine the contributors of the month, the quarter, and the year (Magento, 2021c).

**DP6:** Provide **evaluation guidelines** for the verification of delivered artifacts to make the input control transparent.

**Rationale:** It should be transparent which evaluation guidelines have to be followed by the developers to comply with the guidelines determined by the platform owner and the requirements of the platform (Hein, Schreieck, Wiesche, et al., 2016; Tiwana, Konsynski, and Bush, 2010). In general, these guidelines improve the quality of external modules and make quality ratings transparent for developers and users of the modules. These guidelines can even be transferred to other platforms using platform-independent criteria (Hesse, Teubner, et al., 2020; Teubner and Dann, 2018). At Etsy, approval review criteria have to be reviewed before the artifact is released to the public (e.g., applications must not sidestep the API to retrieve or post Etsy data) (Etsy, 2021a).

**DP7:** Provide **verification mechanisms for rating third-party** modules (1) to increase transparency regarding module quality and (2) to increase trust in module developers.

**Rationale:** To further strengthen the trust in third-party modules, module ratings can be used. The module ratings represent the trustworthiness and usefulness from the user perspective. This helps the developer get feedback from the users and improve the quality of their modules (Dalvi et al., 2016). In addition, the third-party modules can be awarded gold, silver, and bronze ratings for their quality and support from the perspective of the platform provider. Metrics can include how many times a module has been used, how many ratings have been given, how good the ratings were, how many updates there have been to the module, how long the developer has been registered on the platform, and how many total extensions the developer has created (Shopware, 2021b). Another aspect is that the developers must be registered and often authorized to publish their modules (e.g., at the Amazon Marketplace developers receive their own IDs).

**DP8:** Provide **incentives to developers** for compensating their workload for module development and provision to foster generativity within the ecosystem.

**Rationale:** While the relationships between the focal platform and its customers play a significant role in social commerce, this can also be applied to the relationship between the platform and the developer (Liang, Ho, et al., 2011). Rewards and incentives should be introduced to stimulate developers' generativity and engage them to contribute to e-commerce ecosystems with additional modules. In general, incentives can increase the loyalty of developers and are often offered on a monetary basis. However, incentive plans should not only be influenced by the number of modules provided and their reach among ecosystem participants, but also by their overall quality. Receiving a reward can also be seen as a social incentive (Magento, 2021c).

#### 6.1.4.4 Rules and Standards

**DP9:** Provide **coding guidelines** for developers (1) to improve the quality of code, (2) to improve the understandability, (3) and to assist developers with communication.

**Rationale:** The use of coding guidelines has already been proven in open source projects (Szyjewski, 2019). It becomes a prerequisite in an ecosystem where participants with different perspectives and no direct communication need to coordinate. This might include guidelines for programming languages, stylesheet languages, scripting languages, markup languages, or the interplay between languages at different architectural levels. Compliance with coding guidelines should be supported by offering utilities that automatically adjust the code or offer live support to developers within the development environment. The coding guidelines can be based on established standards, such as the PSR family for developing extensions in PHP, which provide a syntactic and semantic framework for code design (Szanto, 2021). Tools such as ESLint or Standard JS can be used in JavaScript to automatically check compliance with coding guidelines and adjust them if needed.

**DP10:** Provide a **central repository** for storing the developed modules (1) to enable versioning, (2) to enable collaborative development with a single source of code, and (3) to ease integration into the productive system.

**Rationale:** Collaboration among ecosystem participants should be supported by an easily accessible, central repository containing the status of available e-commerce ecosystem modules. This enables coordination of projects as well as suggestions for features or bug fixes from the community. Using a central repository further increases the activity level of individual ecosystem projects and enables goals to be achieved faster (Alshomali et al., 2017). In this context, the version control system Git, enabling distributed and transparent tracking of code changes, emerged as a de facto standard (Govil et al., 2020). Cloud-based services combine version control with free hosting space, as well as social features, making them ideal for collaboration among different participants in an e-commerce ecosystem (Alshomali et al., 2017). By offering their boundary resources, they also enable deployment processes to be established, positively affecting system integration in the long term Wei et al. (2020).

**DP11**: Provide **templates** for popular **development tools** to decrease setup time for developers.

**Rationale:** Establishing the conditions for easy and rapid development of new services around the focal platform further increases the opportunity for more third-party developers appreciating and expanding the e-commerce ecosystem. The setup effort for potential third-party developers can be reduced by providing templates and instructions for the optimal configuration of development tools. This includes the provision of pre-sets for popular integrated development environments, settings for utilities, and the configuration of local development web servers (Magento, 2021f; Oxid AG, 2021a; Shopware, 2022c). Furthermore, customized plug-ins and extensions for integrated development environments or browsers can noticeably shorten the initial training period (Salesforce, 2021d).

#### 6.1.4.5 Development Environment

**DP12:** Provide a **staging mechanism** that autonomously creates a development system (1) to allow for the development of new modules and (2) to allow for intensive quality tests.

**Rationale:** The development environment and the productive environment should be very similar but separate from each other. Before new modules are released in the productive environment, they should be extensively tested for their functionalities in a development system that contains sample data. Sandbox environments can be used for this purpose. In the sandbox, test users can be created who can act as sellers or customers (eBay, 2021b). Orders, shipments, and returns can be simulated to test API connections with the focal platform (Otto SE, 2021b; Walmart, 2021a). After intensive testing, the new module should be smoothly transferred to the productive environment of the focal platform.

**DP13:** Provide **demo scenarios** (1) to establish a fundamental understanding of the electronic commerce environment and (2) to provide orientation for the communication.

**Rationale:** As the possibilities in e-commerce are manifold, the developers should familiarize themselves with implementation opportunities. Demo scenarios can be a helpful starting point because they can also identify unforeseen opportunities and obstacles (Svátek et al., 2015). Various demos have been offered at Magento. Business-to-Business (B2B) demos, front-end demos, back-end demos, and administration demos are available. The SAP Customer Experience Lab offers numerous showcases for the SAP Commerce Cloud (SAP, 2021c). In this way, communication with the provider or between developers can be supported.

> **DP14:** Provide **workflow testing mechanisms** for harmonizing retailers' and developers' business processes with the processes implemented on the focal platform.

**Rationale:** The business workflow must run correctly in e-commerce ecosystems; if it does not, significant financial damage can occur. It is still a considerable challenge for developers to design, analyze, control, and diagnose a business workflow due to the complexity of the e-commerce environment (Wu and Lin, 2011). To support this, the workflow should be visualized transparently. Templates can be used to represent different situations and to help the user understand the meaning of the business process more quickly (Wu and Lin, 2011). Appropriate modeling languages should then be used accordingly. For example, commonly used workflow models are based on active networks, on event-driven process chains, on speech act theory, on Petri nets, and on Unified Modeling Language (Cai, 2012). In addition to understanding and representation, the possibility should be provided for the workflow to be tested. It is imperative to verify and validate the functionalities, especially if new functionalities have been implemented (Walunj and Sadafale, 2013).

> **DP15:** Provide tools for **UI prototyping** for developers and retailers to support the UI development with predefined elements, accelerating UI development and unifying the UI.

**Rationale:** The development of a prototype generally indicates the feasibility of a business process (Aulkemeier, Paramartha, et al., 2016). In the context of UI, the prototype can also be employed to assess the usability of the e-commerce website (Aulkemeier, Iacob, et al., 2019). Users can be observed and questioned afterward, for example on their impressions, complaints, and suggestions (Mushthofa et al., 2018). To avoid every third-party developer having to observe the user again, the platform provider should provide predefined and evaluated elements. It also helps to ensure that the elements are unifying. Magento, for example, even presents the UI components in detail in the documentation for the developers.

**DP16:** Provide a general perspective for ecosystem participants on the **status of the system and APIs** to summarize the system status and shopper activities.

**Rationale:** In e-commerce, it is becoming increasingly important to address customers individually and provide them with suitable and customized offers (Da Silveira et al., 2001). For this, tools must be available that customers understand in real time, for example, during navigation (Marcondes et al., 2017). Customers can in future also use big data to find an e-commerce website that suits them individually (Malhotra and Rishi, 2021). One step below is the analysis for errors in the system, for example, in the case of interface errors. Further steps can be triggered automatically (Marcondes et al., 2017). In addition, different ecosystem participants need the analysis to be accessible to non-experts (Diamantini et al., 2015). Therefore, enhanced analytics are recommended. On this basis, further steps can then be derived to influence shopper activities positively in economic terms.

#### 6.1.4.6 Ecosystem Curation

**DP17:** Indicate **deprecated boundary resources** to developers (1) to avoid the use of outdated ones and (2) to accelerate the transition to new ones.

**Rationale:** An e-commerce ecosystem must evolve to be sustainable because of strategic decisions and technological developments (Wareham et al., 2014). This involves the evolution of the focal platform as the core of the ecosystem and external modules (periphery), while the boundary resources must be relatively stable (Baldwin and Woodard, 2009; Staykova and Damsgaard, 2015). Nevertheless, the boundary resources must also be updated over time, creating new versions while abandoning deprecated ones. Hence, the principle of deprecated boundary resources should ensure that deprecated resources are appropriately handled by the platform owner when new resources are designed, and that proper versioning of boundary resources is established. Deprecated boundary resources should be announced a priori with an appropriate lead time, and transition paths should be demonstrated (Ghanam et al., 2012; Magento, 2021b). While our case study analysis demonstrated that most of the platforms indicate deprecated boundary resources (eBay, 2021a; Etsy, 2021b; Shopware, 2021a), transition plans are seldom employed, often leaving developers with the task of analyzing changelogs on their own. In addition, the stability of boundary resources should be guaranteed, as exemplified by Salesforce's three-year plan from their first launch (Salesforce, 2021a).

**DP18:** Provide an overview of the **ecosystem structure** (1) to illustrate relationships among participants and (2) to enhance transparency on internal modules for developers.

**Rationale:** Focal platforms in e-commerce ecosystems comprise several interdependent and interacting components (Aulkemeier, Iacob, et al., 2019; Yu and Ni, 2013). Especially in agile e-commerce projects documentation quality is often neglected, resulting in insufficient transparency concerning the overall system architecture (Sensuse et al., 2020). However, the expansion of core e-commerce functionalities leads to challenges in integrating new services into the core system (Aulkemeier, Paramartha, et al., 2016). Therefore, platform providers need to provide appropriate social boundary resources that document a platform's present architecture and clarify the interrelationships of internal services with application boundary resources. The documentation should also explain key underlying architectural principles (e.g., command query responsibility segregation) (Magento, 2021a).

**DP19:** Provide a **system roadmap**, including future developments, (1) to inform developers early and (2) to foster necessary module adjustments to productive system changes a priori.

**Rationale:** The evolution of the focal platform envelops additional features and addresses further markets, as the core of an e-commerce ecosystem affects third-party developers if previously stable boundary resources are adapted (Eisenmann, Parker, et al., 2011; Hermes, Kaufmann-Ludwig, et al., 2020). Platform owners can present roadmaps for the communication of additional platform features and forthcoming boundary resource adjustments so that developers can leverage new features within their existing and new modules. Roadmaps enable developers to plan necessary adjustments and keep modules operational (Ghanam et al., 2012). Our case study analysis demonstrates that platform ecosystem providers communicate significantly more future platform adjustments than e-business platform owners. While Magento releases new software versions quarterly with a priori announcements and SAP offers a detailed web-based roadmap explorer and management summary (Magento, 2021g; SAP, 2021d), Etsy only provides an abstract roadmap (Etsy, 2021a). Walmart does not provide exact dates for developments and only randomly announces future developments via the "what's new" section of the developer blog (Walmart, 2021c).

### 6.1.5 Discussion

This section presents a preliminary evaluation, followed by the paper's scientific and practical contribution, and the limitations of this research. Focal platforms form the core of e-commerce ecosystems. While standards already exist on several levels (e.g., process harmonization across a supply chain and interface standards such as EDIFACT) for e-commerce ecosystems, the relation between the focal platform and external developers lacks standardization, in addition to de facto standards for software providers and implementation guidelines. Our 19 design principles were developed following the second strategy proposed by Iivari (2015), generalizing from ten specific platform ecosystems in e-commerce, supported by an extensive literature review. As prescriptive statements, design principles are reusable for a multitude of instances of a system class (Kuechler and Vaishnavi, 2008; Venable, 2006). Thus, we follow the framework for "light reusability evaluation of design principles" (Iivari, Rotvit Perlt Hansen, et al., 2021, p. 291), along with the criteria accessibility, importance, novelty and insightfulness, actability and guidance, and effectiveness, as suggested by Möller, Guggenberger, et al. (2020), for the evaluation of our design principles. Our preliminary evaluation was executed analytically (Jenkins, 1985; Väyrynen and Iivari, 2015). As design principles for a class of systems form a unit of prescriptive knowledge, they should be evaluated as a comprehensive set (Iivari, Rotvit Perlt Hansen, et al., 2021).

Our design principles can be applied by platform owners (i.e., software providers, e-commerce companies) in e-commerce ecosystems. The class of systems under consideration (i.e., focal platforms) is apparent to practitioners. The design principles are accessible, as we apply a language known to domain experts. We utilized the framework suggested by Chandra et al. (2015), resulting in a structured design principle formulation and, based on this, we provide a detailed set of 19 design principles for a holistic standardization of the focal platform, focusing on developers. Our design principles are important to practitioners, as the significance of platforms in dynamic e-commerce ecosystems increases and platforms penetrate almost every sector (Choudary, 2015; Evans and Schmalensee, 2016; Wulfert, Seufert, et al., 2021). In addition, platforms are introduced to augment physical products, with additional smart services fostering a hybridization of products. A comprehensive set of 19 design principles focusing on the standardization of e-commerce ecosystems for developers is a novel approach. Our theoretical lens of boundary objects can provide new insights for ecosystem control, curation, and development. Our design principles provide immediate utility, as they are actable and directly implementable for platform owners, with exemplary design instances (Kuechler and Vaishnavi, 2011). Although we provide guidance, including design instances for several design principles, platform owners can be flexible regarding the specific implementation of the prescriptions. Platform owners can decide on the selection of boundary resources and add others to best support their specific e-commerce ecosystems. Thus, we achieved a reasonable balance between guidance and flexibility (Chandra Kruse, Seidel, and Purao, 2016). The design principles can be effectively reused by the owner of a focal platform in an e-commerce ecosystem on different (architectural) levels. We tailored the design principles for actor, business, and application level, using our theoretical lens. The design principles (positively) affect the development of third-party modules by accelerating the overall degree of standardization within and across ecosystems.

As scientific contribution, we provide a novel set of 19 design principles for a class of IS (i.e., focal platforms in e-commerce ecosystems). These design principles guide the relation between platform owners and third-party developers by standardizing boundary resources across e-commerce ecosystems (Hein, Schreieck, Riasanow, et al., 2020). Design principles for boundary resources in e-commerce ecosystems specialize the general research on boundary resources by Star and Griesemer (1989) and Ghazawneh (2012). We emphasized the role of boundary resources for the proliferation of e-commerce ecosystems and their role in propelling network effects to reach a minimum threshold of ecosystem participants (Eisenmann, Parker, et al., 2009; Parker, Van Alstyne, and Choundary, 2016). We also proposed a differentiation of focal platforms in e-commerce ecosystems (i.e., platform ecosystem providers, e-business platform ecosystems) in line with Gawer (2021b).

Although we propose design principles for the standardization of e-commerce ecosystems with a focus on boundary resources, for the practical contribution it needs to be discussed which ecosystem actors have the power to set (technical) standards within and across e-commerce ecosystems (Iansiti and Levien, 2004a). The owners of a focal platform can promote ecosystem-wide standards as part of their governance function, but only have a minor interest in standardization across e-commerce ecosystems (Cennamo, 2021), as cross-ecosystem standardization may enable multi-homing of participants. Cennamo (2021) has argued that platforms with an established base of participants are likely to increase barriers to entry (e.g., by proprietary boundary resources) to lock in ecosystem participants and increase the quality of extensions developed externally, requiring registration fees, extensive certifications, and quality checks for modules. With its focal platform, Amazon is already setting proprietary standards for product stocking and shipping requirements, differentiating itself from other ecosystems, to lock in ecosystem participants (Jacobides et al., 2018). Independent software vendors that offer a software platform for e-commerce (e.g., Magento, Shopware) can also act as the standard-setting organizations, as defined by Viardot et al. (2021).

The derivation of design principles for boundary resources in e-commerce ecosystems indicated a tendency for software vendors to focus on the development of boundary resources for third-party developers, while e-business platforms are more concerned with platform boundary resources linking supply-side participants. Social boundary resources are employed equally by the case companies selected. In contrast to platform owners, developers significantly profit from the standardization of ecosystems, enabling multi-homing and increasing the dissemination of their modules. The increased reach of developers can also be desirable for other ecosystem participants, such as content providers and suppliers (GitHub, 2021). These participants require a different set of design principles for standardization, as they are less interested in developing modules and more interested in business processes and social boundary resources. Based on our multi-case study results, we also took an internal platform perspective, investigating possible responsible agents for implementing or benefitting from the design principles. We allocated our 19 design principles to responsible agents at focal platforms in e-commerce ecosystems (i.e., Business Process Owner (BPO), User Interface Designer (UID), Internal Platform Developer (IPD), and infrastructure provision). The roles were derived based on the layers of digital platforms, as proposed by Zutshi and Grilo (2019). The mapping in Table 6.3 indicates the potential relevance of each design principle for the four roles. Overall, the mapping indicates that internal platform development (3), responsible for implementing APIs and SDKs for third-party developers, is concerned with most of the design principles. While business process owners (1) are responsible for integrating third-party development at the business level, user interface designers (2) should ensure a standardized and uniform user interface across modules. The infrastructure provision (4) ensures flexible computing power for the focal platform and is also involved in ecosystem orchestration.

**Table 6.3** Association of Design Principles to Roles in the Focal Platform

| Role in the Platform | Design Principle |
|---|---|
| (1) Business Process Owner | DP1, DP2, DP3, DP4, DP5, DP13, DP14, DP16, DP18, DP19 |
| hline (2) User Interface Designer | DP1, DP2, DP4, DP9, DP10, DP11, DP13, DP15 |
| hline (3) Internal Platform Developer | DP1, DP2, DP3, DP5, DP8, DP9, DP10, DP11, DP12, DP13, DP14, DP16, DP17, DP18, DP19 |
| hline (4) Infrastructure Provision | DP5, DP6, DP7, DP8, DP9, DP12, DP13, DP16, DP17, DP18, DP19 |

When applying our design principles, platform owners need to carefully balance the standardization of boundary resources without limiting developers' flexibility and freedom in pursuing their ideas. Excessive standardization may contradict developers' generativity and, thus, reduce the overall value proposition for customers (Adner, 2017). Standardization of boundary resources in e-commerce ecosystems to attract additional participants is a particular trade-off in favor of the generativity of developers. We argue that a higher degree of standardization of boundary resources will enhance network effects and, as a result, the overall value of the ecosystem for participants. Opening the focal platform, offering access to additional sections of the platform core, and integrating with other platforms may attract additional developers; however, the pursuit of openness can reduce a focal platform provider's control over the e-commerce ecosystem (Hein, Schreieck, Riasanow, et al., 2020).

Star (2010) and Hein, Weking, et al., (2019) have emphasized that the definition of standards is never complete. Rather, it is a continuous process deriving residual categories, developing boundary objects, and updating or creating new standards. We have analyzed a snapshot of the boundary resources of our ten cases for one point in time. As e-commerce ecosystems and their focal platforms, including boundary resources for participants, are subject to dynamic evolution (Wulfert, Seufert, et al., 2021), our design principles may also require an update in the future. Therefore, a process for standardizing boundary resources can be established in analogy to Hein, Weking, et al., (2019). The management of external developers and boundary resources is a crucial activity for an e-commerce ecosystem's success (Cennamo, 2021), as we reflect in MR6 on the curation of the ecosystem and related design principles. Decisions on boundary resources are structural for the platform, affect its core architecture, constrain or propel platform evolution, and are therefore difficult to imitate by competing platforms (Blaschke, Haki, et al., 2019; Cennamo, 2018).

This research also has certain limitations. We could not differentiate between boundary resources provided by platform ecosystem providers and e-business platform ecosystems, as the boundaries between these two categories are increasingly blurred, with retailers and wholesalers in e-commerce also offering software and cloud services (e.g., Amazon, REWE). Hence, we considered five leading examples from both types for our case studies. For focal platforms integrated into enterprise system portfolios (e.g., SAP, Salesforce), it was difficult to determine the boundary resources specific to e-commerce. We only considered software components directly related to the shopping modules. The ten global and European platforms were selected based on their overall dissemination within e-commerce and total revenues. In line with the literature on design principles (Chandra Kruse, Seidel, and vom Brocke, 2019; Gregor, Chandra Kruse, et al., 2020), we strived to codify prescriptive knowledge from successful cases. The underlying shared phenomenon focuses on the exploitation of standardized platform boundary resources in successful e-commerce ecosystems. In addition, we excluded social commerce platforms from our case study analysis, although it is a field of increasing importance for e-commerce. The specifics of social commerce, compared to B2B and Business-to-Customer (B2C) commerce, require adjusted design principles for standardization; for example, trust that is fostered by the focal platform is even more important in Customer-to-Customer (C2C) transactions. Moreover, we did not explicitly consider different maturity stages when selecting our cases (Muzellec et al., 2015; Parker, Van Alstyne, and Choundary, 2016; Reillier and Reillier, 2017). An ecosystem in the launch stage might require less sophisticated boundary resources, as the number of services provided for participants is often limited (Wulfert, Seufert, et al., 2021). Beyond that, we focused on one research stream (i.e., our interactions on the technical and organizational level) when developing our theoretical lens (Foerderer, Kude, Schuetz, et al., 2019).

### 6.1.6 Conclusion and Outlook

Based on the theoretical lens of boundary object theory, we developed 19 theoretically grounded and empirically validated design principles for the standardization of focal platforms in e-commerce ecosystems. The design principles address the standardization of the meta-requirements of tailored boundary resources, the openness of the ecosystem, trust and risk, rules and standards, the development environment, and ecosystem curation. In our argumentation, we stressed the importance of developers for e-commerce ecosystems and their requirement of improved standardization. Our design principles simplify the development of external modules and possibly enable multi-homing for developers due to standardization across ecosystems. Extending our preliminary research, we will conduct interviews with domain experts and knowledgeable senior managers (e.g., consultants, software developers, and platform owners) following the guidelines of Iivari, Rotvit Perlt Hansen, et al. (2021) to further evaluate our design principles in future research. These interviews can also be used to evaluate the relevance of our 19 design principles for different roles at focal platforms and to verify the appropriateness of our design principles for different stages of platform and ecosystem maturity. Future research might also derive patterns of design principles for types of focal platforms in e-commerce ecosystems (e.g., platform ecosystem providers and e-business platform ecosystems) and specific boundary resource archetypes. It might also be interesting for future research to consider the role of boundary resources in unsuccessful cases. As we explicitly focused on third party developers when deriving design principles for the standardization of digital business ecosystems, future research might address the requirements of different ecosystem participants and summarize prescriptive knowledge. An interesting avenue for future research would be the aggregation of several sets of design principles for different ecosystem participants for a holistic design theory integrating the perspectives of all relevant participants. In the same vein, multi-dimensional modeling of all participants' perspectives is necessary to improve IS' quality. Applying actor-network theory can lead to a deeper understanding of the relations between participants in an ecosystem. As our research is concerned with B2B and B2C commerce, our design principles can be evolved by future research to cope with the specifics of social commerce. Future research may also employ other theoretical lenses when developing design principles (e.g., a design and engineering perspective, platform governance).

## 6.2 Paper 7: Let's Join Forces: Boundary Resources as Enablers of Value Cocreation in E-Commerce Ecosystems

**Abstract**
*Collaboration and value cocreation*[4] *are important drivers of the continuous growth of e-commerce, which is expected to reach US $ 6.4 trillion in 2024 despite current global crises.*[5] *Only a few transaction platforms currently dominate e-commerce (e.g., Amazon, Walmart), but other participants are likely to join these platforms' ecosystems. Third-party developers can provide extensions to these ecosystems to enhance the platforms' functionality, but third-party developers' role in e-commerce ecosystems' success and generativity remains underexamined in academia. In this paper, we investigate boundary resources' ability to attract and manage third-party developers in e-commerce ecosystems. An analysis of 14 interviews with domain experts yielded design knowledge summarized in seven design principles that can help owners of innovation and transaction platforms in e-commerce implement*

[4] In this research paper, the term "co-creation" is written without a linking hyphen ("cocreation"). Though, we prefer to use the more common spelling of "co-creation" with a linking hyphen in line with leading publications on value co-creation (e.g., Grönroos, 2012; Vargo and Lusch, 2004, 2008).

[5] This version of the article has been accepted for publication, after peer review but is not the Version of Record and does not reflect post-acceptance improvements, or any corrections. The Version of Record is referenced as follows: T. Wulfert, G. Strobel, et al. (2024). "Let's Join Forces: Boundary Resource as Enablers of Value Cocreation in E-Commerce Ecosystems." In: *Electronic Commerce Research* Forthcoming

*boundary resources to attract third-party developers, thereby amplifying collaboration and value cocreation, as well as the e-commerce ecosystem's overall generativity and success.*

| Fact sheet of publication 7 | |
|---|---|
| Title | Let's Join Forces: Boundary Resources as Enablers of Value Cocreation in E-Commerce Ecosystems |
| Authors | Tobias Wulfert |
| | Gero Strobel |
| | Hiep Hoang |
| Publication Type | Journal Article |
| Publication Outlet | Electronic Commerce Research |
| VHB-Jourqual3-Ranking | C |
| Status | Under Review—1st round |
| Research Question | How can the platform boundary be designed to orchestrate participants in e-commerce ecosystems? |
| Research Approach | Semi-Structured Interviews (Misoch, 2015)<br>Qualitative Content Analysis (Mayring, 2014) |
| Full Citation | T. Wulfert, G. Strobel, et al. (2024). "Let's Join Forces: Boundary Resource as Enablers of Value Cocreation in E-Commerce Ecosystems." In: *Electronic Commerce Research Forthcoming.* |

### 6.2.1 Introduction

Digital technologies, as enablers and catalysts of digital transformation, propel inter-company collaboration and value cocreation in platform-centered ecosystems (Fitzgerald et al., 2013; Ghazawneh and Henfridsson, 2013; Huber, Hurni, et al., 2022; Vargo and Lusch, 2008). Platform-centered ecosystems require a set of boundary resources to allow for orchestrated value cocreation (Ghazawneh and Henfridsson, 2013; Huber, Hurni, et al., 2022). The focal platform's owner provides these resources as part of platform governance to decide which third parties have access to which boundary resources and how they can be used for value cocreation to satisfy customers (Boudreau, 2010; Hein, Schreieck, Riasanow, et al., 2020; Huber, Kude, et al., 2017; Wang, Deng, et al., 2022). A prominent example of such value cocreation is e-commerce, with an expected accumulated global revenue of US $ 6.4 trillion in 2024 (eMarketer, 2020). Several successful transaction platforms (i.e., digital marketplaces) currently dominate e-commerce—e.g., Amazon, Alibaba, eBay, Flipkart, Mercado Libre, and Rakuten–through their surrounding ecosystems (Wulfert, Seufert, et al., 2021). Amazon generated US $340 billion in revenue from product and service sales in fiscal year 2020, with more than 60 percent of its revenue coming from commission fees from third-party sellers in its marketplace (Amazon, 2021). The revenue generated within these ecosystems is up to three times higher than that of the focal platform (Delteil et al., 2020). These transaction platforms are powered by e-commerce-specific innovation platforms—e.g., Magento Commerce, Shopify, and WooCommerce (Wulfert and Karger, 2022b)—that provide the infrastructure and application services needed to conduct e-commerce transactions (Aulkemeier, Schramm, et al., 2016; Wulfert and Karger, 2022b).

E-commerce ecosystems comprise various independent participants–e.g., manufacturers, sellers, customers, and service providers–that are (digitally) connected (Böttcher, Rickling, et al., 2021). These participants compete for scarce resources (e.g., products, supplies) while pursuing the common target of fulfilling customer demand (Adner, 2017). As e-commerce transactions occur via electronic means (Zwass, 1996), the platform owner must implement sophisticated boundary resources to manage the various ecosystem participants (Ghazawneh and Henfridsson, 2013; Wulfert, Woroch, Strobel, Seufert, et al., 2022). Ghazawneh (2012) transferred the concept of boundary objects (e.g., repositories, ideal types, coincident boundaries, and standardized forms) from social science to (software) ecosystems (Ghazawneh and Henfridsson, 2013; Star and Griesemer, 1989). Dal Bianco et al. (2014) used this description to distinguish between three layers of boundary resources in platform-centered ecosystems: application boundary resources (e.g., APIs); development boundary resources (e.g., SDKs); and social boundary resources (e.g., documentation, developer portals). As an "ecosystem leader" (Moore, 1996), a platform's owner in an e-commerce ecosystem can define boundary resources while setting and enforcing governance rules (Adner, 2017; Hein, Schreieck, Wiesche, et al., 2016). Defining the boundaries and demarcation points between a focal platform and ecosystem participants (Santos and Eisenhardt, 2005) facilitates the execution of strategically relevant decisions about ownership, entry into new markets, and community building (Dal Bianco et al., 2014; Hein, Schreieck, Riasanow, et al., 2020).

Although many standards for information exchange have been introduced in e-commerce (GS1, 2021; Guilloux et al., 2013; Narayanan et al., 2009), extant research on how standardized boundary resources are defined remains scarce (e.g., Engert et al., 2022; Wulfert, Woroch, Strobel, Seufert, et al., 2022). Most extant literature either focuses the analysis on individual aspects of boundary resource design, e.g., the user interface (e.g.,Billewar et al., 2021; Zollet, 2014) or refers to dedicated scenarios (Petrik, 2022). Against this backdrop, our research question is as follows:

> **RQ:** How can the platform boundary be designed to orchestrate participants in e-commerce ecosystems?

To answer the research question, we provide design principles for the shaping of boundary resources in e-commerce ecosystems based on 14 expert interviews and a subsequent qualitative content analysis (Mayring, 2014). Furthermore, we proposed design guidance in the form of a boundary resource model comprising a set of design rules for platforms in e-commerce ecosystems to reduce barriers to entry for participants and to propel network effects (Cennamo, 2021) based on a content-structuring approach and our elicitation of focal platform types (Gawer, 2021b). The remainder of this article proceeds as follows. First, we introduce the concepts of e-commerce ecosystems and platform boundary resources. Second, we present our design science research approach involving 14 expert interviews. Third, we introduce a boundary resource model for e-commerce ecosystems. Finally, we summarize and discuss our results and propose future research directions.

## 6.2.2 Related Literature

E-commerce ecosystems emerge around focal platforms, orchestrating ecosystem participants and enabling transactions among them (Wulfert and Karger, 2022b; Wulfert, Woroch, Strobel, Seufert, et al., 2022). These platform owners can define and enforce governance rules for interactions with focal platforms (Boudreau, 2010; Hein, Schreieck, Wiesche, et al., 2016; Tiwana, 2014). The focal platform's owners design and implement boundary resources for interactions with other ecosystem participants as a building block of their platform governance (Eaton et al., 2015; Ghazawneh and Henfridsson, 2013; Tiwana, Konsynski, and Bush, 2010).

### 6.2.2.1 E-Commerce Ecosystems and Governance

Introduced to the economics literature by Moore (1993, 1996), digital business ecosystems are complex networks of platform-mediated actor-to-actor interactions involving digital technologies (Wulfert, Woroch, Strobel, Seufert, et al., 2022). Independent participants are linked by a common goal-namely the overall ecosystem's success (Adner, 2017). Thus, an e-commerce ecosystem is a manifestation of a digital business ecosystem in the e-commerce context, with participants conducting digital transactions (Wulfert, Woroch, Strobel, Seufert, et al., 2022; Zwass, 1996). An e-commerce ecosystem's value for participants increases with the addition of each participant to the network, resulting in direct and indirect network effects (Shapiro and Varian, 1998). E-commerce ecosystems typically evolve around focal transaction platforms that function as virtual loci through which participants conduct retail transactions (Tsujimoto et al., 2018; Turban, Whiteside, et al., 2017).

Innovation platforms power transaction platforms, providing necessary application services (Schütte and Wulfert, 2022a; Wulfert and Karger, 2022b).

Transaction platforms (e.g., Amazon Marketplace, Walmart Marketplace, Alibaba.com) match and orchestrate organizations and individual participants from various markets and social groups to form dynamic ecosystems (Corallo, 2007; Gawer, 2021b). The major supply-side participants on transaction platforms are manufacturers and sellers (Eisenmann, Parker, et al., 2009; Hein, Schreieck, Riasanow, et al., 2020). Transaction platforms offer various retail-related services for participants, e.g., payment or fulfillment services (Wulfert, Woroch, Strobel, Seufert, et al., 2022). E-commerce platforms also increasingly provide innovation services (e.g., APIs, computing power), enabling development of third-party extensions and attracting external developers as additional ecosystem participants (Wulfert, Woroch, Strobel, Seufert, et al., 2022). Innovation platforms provide the technological infrastructure and necessary application services to conduct e-commerce transactions (Aulkemeier, Schramm, et al., 2016). Innovation platforms form the "technical core" of e-commerce ecosystems (Blaschke, Haki, et al., 2019), supplying sophisticated boundary resources to enable development of extensions (e.g., shop themes, feature add-ins) by third-party developers as major supply-side participant type (Hein, Schreieck, Riasanow, et al., 2020; Wulfert, Woroch, Strobel, Seufert, et al., 2022). These extensions expand innovation platforms' application services and their generativity (Blaschke and Brosius, 2018). Hybrid platforms incorporate the two aforementioned platform types' characteristics and provide transaction and innovation services (Gawer, 2021b), thereby increasing platform owners' reach and their ecosystems' value to incumbents and new participants (Schütte and Wulfert, 2022a).

As e-commerce ecosystems comprise several independent participants, these participants' interactions, collaborations, and value cocreation require a certain degree of governance (Schreieck et al., 2016; Tiwana, 2014; Tiwana, Konsynski, and Bush, 2010), which has been identified as an important determinant of an ecosystem's success (Alves et al., 2017; Tiwana, 2014). As Tiwana (2014, p. 39) described it, "[g]overnance broadly refers to who decides what in a platform's ecosystem". Ecosystem governance comprises the set of rules, norms, and policies to which the platform adheres to build its ecosystem (Reillier and Reillier, 2017). Ecosystem governance involves direct and indirect measures to control the ecosystem and participant interactions within an ecosystem (Tiwana, 2014). The focal platform's owner in a platform-centered ecosystem "sets, and often enforces, the governance rules, determines timing" (Adner, 2017, p. 48), and reaps the lion's share of ecosystem revenues. Extant literature has structured and aggregated governance rules and decisions within several governance dimensions (Hein, Schreieck, Riasanow, et al.,

2020; Hein, Schreieck, Wiesche, et al., 2016; Reillier and Reillier, 2017; Tiwana, 2014). Among others, ecosystem governance involves decisions on access to the focal platform; implementation of boundary resources for access to the focal platform; the division of value among ecosystem participants, including the ecosystem's pricing structure; and conflict resolution with divergent objectives among participants (Evans and Schmalensee, 2016; Evans and Gawer, 2016; Reillier and Reillier, 2017; Wulfert, Woroch, Strobel, Seufert, et al., 2022). Aside from pricing mechanisms, Boudreau and Hagiu (2009, p. 187) proved that "platform regulation including contractual, technological, and information design" that governs platform openness is key in establishing ecosystems. The authors differentiated between the regulation of access and interactions among participants already on the multi-sided platform. Hein, Schreieck, Wiesche, et al. (2016) proposed governance structures, resources and documentation, accessibility and control, trust and perceived risk, pricing, and external relationships as dimensions of ecosystem governance, which supports decisions on the focal platform's architecture and the overall ecosystem's composition. In platform-centered ecosystems (e.g., e-commerce ecosystems), the focal platform's architecture augments governance rules (Tiwana, 2014).

#### 6.2.2.2 Boundary Resources

When boundary resources first emerged in social science, they were referred to as "boundary objects", functioning as interfaces between different social worlds (Star and Griesemer, 1989) and enabling different actors to work cooperatively without the need for direct management (Star, 2010). Boundary resources accomplish this by being easily adaptable to individual use cases and recognizable to all participants (Star and Griesemer, 1989). Considering that boundary resources function as interfaces between different participants (i.e., firms), they move within companies' organizational boundaries (Santos and Eisenhardt, 2005). Furthermore, the theory of organizational boundaries provides the base context of ecosystems (Santos and Eisenhardt, 2005; Tsujimoto et al., 2018).

Adopting this concept in e-commerce ecosystems supports the design of third-party applications (Ghazawneh and Henfridsson, 2013), thereby lowering entrance barriers to the ecosystem for new participants (Hein, Schreieck, Riasanow, et al., 2020). Boundary resources facilitate access to platforms' core resources (Eaton et al., 2015). In providing boundary resources, a platform owner also defines the platform's governance rules (Adner, 2017; Hein, Schreieck, Wiesche, et al., 2016). In particular, boundary resources influence a platform's openness (Ghazawneh and Henfridsson, 2010; Parker, Van Alstyne, and Choundary, 2016). By providing these resources, a platform can attract more participants (Ghazawneh and Henfridsson, 2013. These resources also can be used to secure the platform and establish control over the platform and its participants (Ghazawneh and Henfridsson, 2013). Boundary resources are designed as part of governance rules and can be implemented as part of a platform's architecture (e.g., APIs) (Eaton et al., 2015; Ghazawneh and Henfridsson, 2013; Tiwana, Konsynski, and Bush, 2010).

Dal Bianco et al. (2014) identified three layers of boundary resources. The first, application boundary resources, refers to resources that provide access to the platform for a third-party application, e.g., an API that the ecosystem provides. Second, development boundary resources help third parties develop such applications. Finally, social boundary resources facilitate communication between an ecosystem's provider and third parties. A development boundary resource could be a SDK, while a social boundary resource could be a blog. These three layers are not explicit, i.e., one boundary resource can belong to different layers simultaneously (Dal Bianco et al., 2014). For example, a developer forum simultaneously can help develop an application and facilitate communication among ecosystem participants.

### 6.2.3 Research Approach

Following vom Brocke et al.'s (2020) guidance, this publication is part of a multi-paper design science research project (e.g., Schütte and Wulfert, 2022a; Wulfert and Karger, 2022b; Wulfert, Woroch, Strobel, Seufert, et al., 2022). We developed a design framework for e-commerce ecosystems and reported on the development of evaluated design principles for the design of boundary resources based on Wulfert, Woroch, Strobel, Seufert, et al. (2022). The research drew from an interview study of 14 domain experts and utilized the framework for minimum reliability evaluation that Iivari, Rotvit Perlt Hansen, et al. (2021) proposed as a theoretical lens to contribute applicable design knowledge and address these design principles' usability by matching them with the relevant platform types, mainly innovation and transaction platforms, in the e-commerce environment. To formulate the design principles, we applied the structure that Chandra et al. (2015) proposed, including material property and possible actions. Based on the anatomy of design principles (Gregor, Chandra Kruse, et al., 2020), the platform owner can implement our derived principles to attract third-party developers to e-commerce ecosystems. Implementing these principles should ease participation into the ecosystems and propel generativity.

#### 6.2.3.1 Data Collection

This study's objective was to transform implicit design knowledge into explicit, prescriptive knowledge, thereby enabling its utilization (Chandra Kruse, Seidel, and Purao, 2016). Therefore, we conducted semi-structured interviews to address knowledge holders directly within the domain. To create a broad and diverse knowledge base to make the results generalizable (Iivari, Rotvit Perlt Hansen, et al., 2021), experts from all e-commerce ecosystem areas (e.g., platform operators, retailers) and all platform types (e.g., innovation, transaction) were integrated into the study. We mapped the interviewees' company-internal roles to the platform layers that Zutshi and Grilo (2019) identified and related roles (i.e., business process owner, user interface designer, internal platform developer) on the basis of purposeful sampling (Guest et al., 2006; Wulfert, Woroch, Strobel, Seufert, et al., 2022). During the data collection process, we contacted 45 innovation (e.g., Shopify, Big Commerce, SAP) and transaction (e.g., Amazon, eBay, Mercado Libre) platforms in e-commerce from Europe and the US through publicly available addresses and connected with selected employees via professional social media in a purposeful sampling approach (Guest et al., 2006). Altogether, we conducted 14 interviews over a six-month period (Table 6.4).

Our semi-structured interviews were based on a previously known guide to allow for the greatest possible flexibility in examining explicit knowledge (Alsaawi, 2014; Misoch, 2015). Following the research question and objectives, the interview guide was structured thematically using the 19 design principles and extended thematically with five quality criteria (effectiveness, accessibility, importance, novelty and insightfulness, and actability and appropriate guidance) based on Iivari et al.'s (2021) evaluation framework to combine conceptual design knowledge and implicit expert knowledge to provide explicit design knowledge in the form of design principles for implementing ecosystem boundary resources in the e-commerce environment.

During the interviews, the study's purpose, individual design principles, and quality criteria were explained to the experts in detail and presented with corresponding examples (Alsaawi, 2014). Based on this, the experts were invited to evaluate the design principles using quality criteria. Special emphasis was placed on the evaluation of the design principles' importance and novelty to mirror the insights gained in theory with practice and to classify them accordingly. At least two researchers conducted all the interviews, and the interviewer's role was changed continuously to ensure the highest degree of objectivity. Interviewees 5, 7, and 12 submitted written responses to the interview guide, so we clarified uncertainties and inconsistencies within shorter interviews. The interviews were conducted and recorded electronically because of COVID-19 restrictions and social distancing practices.

**Table 6.4** Overview of the Interview Study Participants

| # | Date | Role | Length | Type | Staff | Revenue |
|---|---|---|---|---|---|---|
| 1 | 12-6-21 | Business Process Owner | 35 min. | Transaction | 200 | 11 M |
| 2 | 12-13-21 | Business Process Owner | 55 min. | Transaction | 49K | 16 B |
| 3 | 12-20-21 | Internal Platform Developer | 50 min. | Transaction | 200 | 1 M |
| 4 | 1-11-22 | User Interface Designer | 46 min. | Innovation | 4.7K | 800 M |
| 5 | 1-19-22 | User Interface Designer/ Internal Platform Developer | 20 min | Innovation | 2.5K | 5 M |
| 6 | 1-10-22 | Business Process Owner | 60 min | Transaction | 500K | 140 B |
| 7 | 1-26-22 | Internal Platform Developer | 20 min | Innovation | 49K | 17.1 B |
| 8 | 2-11-22 | Internal Platform Developer | 46 min | Transaction | 7K | 3 B |
| 9 | 2-23-22 | Business Process Owner | 58 min | Innovation | 270 | 28 M |
| 10 | 4-22-22 | Business Process Owner | 42 min | Innovation | 107K | 28 B |
| 11 | 4-29-22 | Business Process Owner | 42 min | Innovation | 107K | 28 B |
| 12 | 5-16-22 | Internal Platform Developer | 33 min | Innovation | 40 | n. a. |
| 13 | 6-3-22 | Internal Platform Developer | 34 min | Innovation | 107K | 28 B |
| 14 | 6-7-22 | User Interface Designer | 45 min | Innovation | 107K | 28 B |

#### 6.2.3.2 Data Analysis

To prepare for the coding and associated analysis of the interviews, they were transcribed completely verbatim (Nunkoosing, 2005). Based on the transcripts, a qualitative content analysis was conducted, which facilitated a systematic analysis of the object of communication regardless of its form (Schreier, 2014). The coding and associated content structuring were conducted in the form of deductive category assignment, with the objective to extract information based on predefined categories (Mayring, 2014). For this purpose, a two-level category system based on the design principles as first-order structuring categories and the five central quality criteria from Iivari et al.'s (2021) evaluation framework as second-order categories was constructed to conduct a detailed analysis (Fig. 6.6). Furthermore, relevant practical examples or best practices were coded based on the research question. The study's authors collected relevant text passages that could not be categorized and aggregated them after the respective coding rounds.

The study's researchers conducted the coding iteratively and independently. Following Mayring (2014), a single sentence was defined within the coding scheme as the smallest unit to be coded, and a paragraph as the corresponding context unit, to ensure the highest possible level of detail. The coding scheme was adapted inductively after each coding round and extended based on the insights gained.

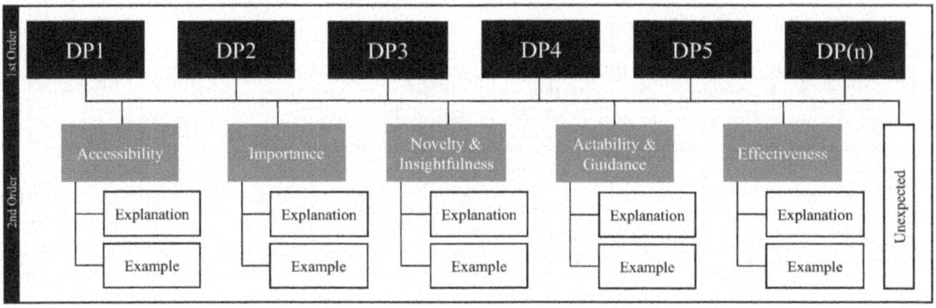


**Fig. 6.6** Coding Scheme for the Content Structuring Approach

## 6.2.4 Design Principles for Boundary Resources

Next, we presented the updated design principles for boundary resources in e-commerce ecosystems. The model managed the internal platform roles and ecosystem participants using a set of seven design principles (DP 1–7) for boundary resources (Fig. 6.7). As for platform roles, we identified three—Business Process Owner (BPO), User Interface Designer (UID), and Internal Platform Developer (IPD) (Zutshi and Grilo, 2019)—that described the interviewees' roles. Platform owners can implement boundary resources to facilitate communication among ecosystem participants and attract third-party developers (Cennamo, 2021; Wulfert, Woroch, Strobel, Seufert, et al., 2022). Summarizing design principles provides a structure for developers and platform owners, and enhances the accumulated design knowledge's tangibility. This utilization of abstract design knowledge was the central practical requirement, according to the interviewees: "What I think would help comprehension is if there would be some overarching structure among the design principles" (#5).[6] This enabled us to derive seven design principles-ecosystem conversation, extension marketplace, ecosystem extension, standards and guidelines, quality assurance, analysis and monitoring, and boundary resource evolution-that enhance the boundary resources design framework with insights from practices.

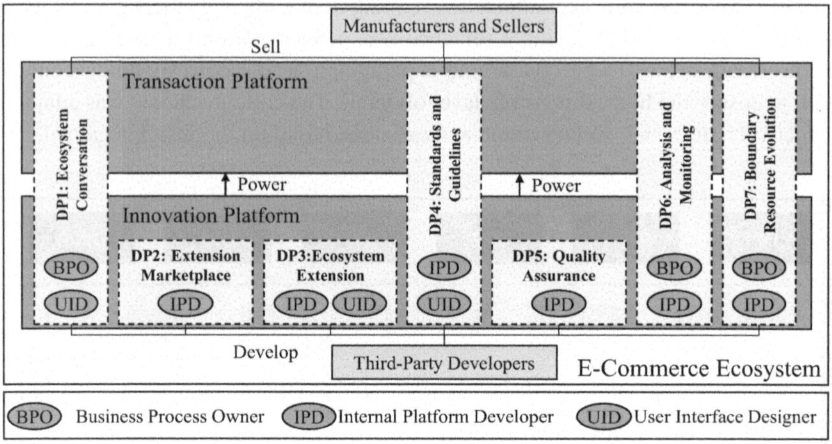


**Fig. 6.7** Boundary Resources for Platforms in E-Commerce Ecosystems

### 6.2.4.1 Ecosystem Conversation

Analogous to social boundary resources (Dal Bianco et al., 2014), the platform boundaries in e-commerce ecosystems provide mechanisms for communication among participants. Conversation mechanisms address the business actor layer (Aulkemeier, Schramm, et al., 2016) and involve dedicated channels, multiple media sources, and diverse demo scenarios (Wulfert, Woroch, Strobel, Seufert, et al., 2022). Through conversation channels, participants can contact the platform and easily exchange information about their projects and products (Petrik and Herzwurm, 2020). They resemble an interface to propel human-to-human communication among ecosystem participants and with the focal platform. A developer forum, blog, e-learning suite, or instant messaging application can be implemented

[6] The numbers in brackets refer to the overview of interviews in Table 6.4 and indicate a specific interviewee.

as a communication channel (e.g., Slack or Discord) (#4, #5, #7). Conversation channels should fit ecosystem participants' individual requirements: "Communication channels and media need a fit for the target audience" (#5). For instance, most IPDs deemed a written API description to be appropriate (#5, #6, #11). Media formats describe the form of the information provided via conversation channels (Wulfert, Woroch, Strobel, Seufert, et al., 2022). The information and media exchanged via these channels should address each participant's needs: "It's not the most important to be sophisticated in the media format, but that it's usable and helpful for the audience" (#5). However, the number of channels and media shared must be chosen carefully to avoid overwhelming participants. It is also necessary to be transparent about which information is communicated via which channel to avoid confusion (#10). As a specific media format, demo scenarios showcase important processes or features of the platform for ecosystem participants (Svátek et al., 2015). They also provide a starting point in big e-commerce ecosystems with many different possible processes. Interviewees viewed demo scenarios as offering potential shortcuts for otherwise lengthy coordination processes (#8, #10): "Demo scenarios are quite essential because if you don't have them, you're going to manually answer these questions" (#4). Therefore, we proposed DP1: Provide the platform with mechanisms for communication so that ecosystem participants can exchange information transparently and easily.

The selection of appropriate conversation channels and media formats is particularly relevant for transaction platforms orchestrating a variety different of ecosystem participants, e.g., sellers, customers, and service providers (Böttcher, Rickling, et al., 2021). This may involve several different social boundary resources accompanying application boundary resources. For innovation platforms that mainly attract external developers, the communication focus is on development boundary resources (#6, #11). Communication with developers can be customized to a high degree and limited to only a few media formats on diverse channels: "For the support of extension development, written documentation is mostly sophisticated" (#7).

#### 6.2.4.2 Extension Marketplace

Platform extensions enhance a platform's features and functionalities (#8), and retailers and other ecosystem participants can use them (Jansen, 2013). These extensions can be either software artifacts (Jansen, 2013) or service-oriented (#3). While third-party developers implement software extensions, service providers can offer additional services (#14). Different types of extensions usually are provided via central extension marketplaces in e-commerce ecosystems (Wulfert and Karger, 2022b): "Implementing a marketplace and fostering a network of partners with their own modules and a marketplace [are] a key development for a growing tech company. So, we are planning to do this in the future to keep being attractive on the market and to get a unique selling proposition" (#12). Thus, we proposed DP2: Provide the platform with an extension marketplace so that ecosystem participants can provide and use complementary extensions.

An extension marketplace offers business opportunities for third-party developers and service providers (Jansen, 2013) who both can be incentivized to provide additional modules and services (#5). Incentives for developers can take the form of money and other non-monetary rewards (e.g., developer awards, special extension placement) (#11, #8). Monetary and non-monetary incentives can increase third-party developers' loyalty to the ecosystem (Liang, Ho, et al., 2011). If the platform owner does not provide a central solution, developers might collaborate elsewhere without exploiting their full generativity (#12). Thus, the platform owner can attract several developers to extend the platform's features with a central marketplace: "If we introduce incentives, we get more developers working on extensions in the ecosystem and, overall, the offer of extensions becomes broader and better" (#7). The extension marketplace can implement evaluation guidelines as internal rules of the platform for rating third-party extensions. These offer input control before extensions are provided in the marketplace (Hein, Schreieck, Wiesche, et al., 2016). The platform owner can make the evaluation guidelines transparent for third-party developers so they can optimize their extensions and make them more compatible (Jansen, 2013). This transparency can be achieved by offering written guidelines or automatic evaluation tools (#13): "You must have certain criteria that you can evaluate your modules against" (#4). Aside from these internal guidelines, other ecosystem participants can conduct module rating, which can take the form of external control (Hein, Schreieck, Wiesche, et al., 2016) or a star rating system with optional comments (#2, #7). Thus, ecosystem participants can share their opinions (i.e., word of mouth) with other potential users (Wulfert, Betzing, et al., 2019). However, reaching a threshold of opinions and filtering relevant reviews are crucial issues for platforms (#9): "It is crucial who is allowed to rate and who can verify a rating or opinion" (#3). The ratings must contain useful information for the developers (#4). Developer profiles present developers' professional information, skills, and rewards (Hajli, 2013). These profiles are linked to platform extensions provided via an extension marketplace and allow for assessing each extension with reference to the developer or provider (Kude, Huber, et al., 2019). For extensions that a group of developers or a dedicated software provider develop, the company's overall reputation can be displayed, including reference projects (#10): "We are planning to include extended developer profiles in our marketplace, as we agree this is an important factor in building trust" (#5).

An extension marketplace for software extensions is particularly relevant for innovation platforms (Jansen, 2013; Wulfert and Karger, 2022b), as they provide the necessary application services for business models in e-commerce (Aulkemeier, Paramartha, et al., 2016) that can be extended through third-party extensions (#12). Additional services—e.g., fulfillment services, credit assessments, and address validations—also can be provided on transaction platforms in dedicated marketplaces.

#### 6.2.4.3 Ecosystem Extension

An ecosystem's reach can be enlarged not only through an extension marketplace, but also an extension of the ecosystem itself (Parker and Van Alstyne, 2016). Connecting or integrating an additional platform to the focal platform allows for more ecosystem participants in general and developers in particular to access the ecosystem: "This is also very effective for customers who rely on having as many partners or marketplaces [in the ecosystem] as possible" (#10). This also may enable access to new markets: "The fact is that we are trying to expand our ecosystem so that we can connect the other markets" (#8). Extending the ecosystem allows more participants to get involved, which may lead to more collaborative work on individual projects: "Very few major things are developed by one person alone these days; you always have teams" (#11). Thus, we proposed DP3: Provide the platform with boundary resources to connect to an additional platform to allow more developers to access the platform and enhance collaboration.

Extending the ecosystem by integrating additional platforms is particularly relevant for innovation platforms, allowing external developers to extend their reach, as well as for potential users of their extensions. Connecting with additional platforms is handled by IPDs that implement the necessary boundary resources and possible abstraction layers: "The biggest one for me on platform integrations is that no two platforms will have the same data model underpinning them, so there's a bit of abstraction that you have to place in there" (#4). Collaboration among the extended ecosystem can be fostered by a central repository that supports effective value cocreation (#4, #8, #11). Providing a central repository (e.g., Git) for versioning facilitates collaborative coding and integration. This can be either a central repository that the internal platform developers manage, or community-specific repositories (#14). Extending the ecosystem by integrating additional platforms is particularly relevant for innovation platforms, allowing external developers to extend their reach, as well as for potential users of their extensions. Connecting with additional platforms is handled by IPDs that implement the necessary boundary resources and possible abstraction layers: "The biggest one for me on platform integrations is that no two platforms will have the same data model underpinning them, so there's a bit of abstraction that you have to place in there" (#4). Collaboration among the extended ecosystem can be fostered by a central repository that supports effective value cocreation (#4, #8, #11). Providing a central repository (e.g., Git) for versioning facilitates collaborative coding and integration. This can be either a central repository that the internal platform developers manage, or community-specific repositories (#14).

#### 6.2.4.4 Standards and Guidelines

Standards and guidelines are important when working collaboratively on software projects (Aulkemeier, Iacob, et al., 2019; Huber, Kude, et al., 2017), and they become inevitable when working in an e-commerce ecosystem with many participants who have different perspectives: "So, standards are valuable when you have to work with a lot of developers" (#9). In particular, the use of coding guidelines can improve code quality, understandability, and communication: "So, coding guidelines are useful because they support structuring the code, the maintainability, and the readability" (#4). Using coding guidelines is a matter of scalability, i.e., greater numbers of participants tend to be more effective, while with smaller projects, participants' efficacy is limited: "For me, this is a question of scaling: With a small team of five engineers, writing coding guidelines is rather idle" (#11). Furthermore, using templates for major development tools helps implement new services around a platform. Tool templates reduce barriers to entry for third-party developers and accelerate overall implementation: "If people can work immediately, it is definitely a very enormous push" (#8). These templates also ease the onboarding of new third-party developers: "This lowers the entry barrier for new developers who may not be familiar with the tool" (#12). Furthermore, providing predefined and evaluated user interface prototypes can facilitate the development of an e-commerce website's user interface. These prototypes help evaluate websites' designs before too much time is invested in their development: "We also show these drafts selectively to existing and future users in order to incorporate feedback into the development before implementation" (#12). Generally, defining the user interface prototype beforehand accelerates implementation of a unified user interface. Such prototypes nullify all new discussions about an artifact's design: "One of the biggest discussions is always: How do I design my application? And because we predefine it, they basically already have a lot of that discussion behind them" (#8). UIDs are involved in designing (visual) boundary guidelines. Therefore, we proposed DP4: Provide the platform with standards and guidelines so that users can propel inter-ecosystem collaboration to accelerate development, reduce developers' barriers to entry, and harmonize design.

This design principle is relevant for innovation and transaction platforms. While the former implements coding guidelines and tool templates to enhance distributed extension development among developers (#7, #11), the latter provides UI guidelines for retailers for harmonized shop designs to create a continuous experience for customers (#3, #6): "These guidelines will guide us on how we can better achieve the good, consistent user experience across all of our platforms" (#1).

#### 6.2.4.5 Quality Assurance

The quality of third-party extensions developed for the ecosystem must be ensured (Wulfert, Woroch, Strobel, Seufert, et al., 2022), so when developing an extension, sophisticated testing and verification methods should be provided (#11, #12). Establishing a development system similar to the production system allows for conducting quality tests during the development's early stages (Wulfert, Woroch, Strobel, Seufert, et al., 2022). Sandbox environments, in which all available functionalities can be tested with test data, are common (#14). However, it is not possible to rely exclusively on test data, which do not correspond in scope and quality to production data: "You can't replicate production data into a staging system. This is what makes staging mechanisms quite difficult" (#4). Furthermore, quality gates should be implemented to ensure proper development when extensions are transported between the test and productive environments: "Important for quality management is to integrate quality gates when transporting to the next stage. In those, the artifact is evaluated against predefined quality criteria in analogy to evaluation guidelines (e.g., functionality, security, performance)" (#11). An implementation pipeline also needs to be built to ensure the extension's correct functionality and integration, which may vary depending on the developer and user's needs (#10, #11). Aside from extension development, another key challenge for developers concerns the design, analysis, and monitoring of business workflows on the basis of their complexity in the e-commerce environment (Wu and Lin, 2011). Templates can be used to represent different situations and help the user understand the meaning of the business process quickly (Wu and Lin, 2011). Furthermore, a testing mechanism for the whole workflow to harmonize the retailer and developer's business processes can increase quality. However, individual workflows' complexity and length make this quite challenging: "I find it difficult to offer testing of a workflow because they can be enormously long, complex, and diverse" (#2). Ideally, the workflow tests are performed automatically, but this presents a huge challenge for the developers: "It is a lot of upfront labor to implement automated workflow testing, but you need it, especially for scaling your business with numerous ecosystem participants" (#4). Nevertheless, the benefit of implementing such mechanisms is huge, and it can minimize hard labor (#4, #8, #10), resulting in increased generativity, as the developers need not test the workflow themselves: "Through automation, you release cognitive potential within a project because people can then simply occupy themselves with other topics" (#10).Therefore, we proposed DP5: Provide the platform with testing and verification mechanisms so that developers can conduct extensive quality tests for the developed extension, as well as test their functionality.

Innovation platform owners need to implement this design principle to ensure third-party extensions' compatibility with platform services (Aulkemeier, Paramartha, et al., 2016) and to provide users of the extensions with acceptable quality (#8, #10).

#### 6.2.4.6 Analysis and Monitoring

Business analytics provide ecosystem participants with key performance indicators and enable sophisticated analyses of past and current transactions, including analysis of big data generated in e-commerce (Huber, Kude, et al., 2017). Owing to the variety of possible ecosystem participants, business analysis should be accessible to non-experts and filterable for different tasks and roles (Diamantini et al., 2015). The degree of information must be appropriate for each participant so that they can cope with the volume of information (#10). This is particularly relevant to BPOs: "We must show detailed information to our users, like how many views they got, how many searches, particular keywords, and how many clicks, and those types of things. The more data that you can provide on that level, the better it is in terms of building a relationship" (#4). Providing tools to analyze this data and predefined reports can provide ecosystem participants with business insights and start potential business actions, e.g., promotions (#8). Furthermore, overall system performance in general and the status of boundary resources in particular must be monitored (#3, #13). This requires the platform owner and developers to be informed about the system's status in real time. This information can be used to fix errors manually or automatically (Marcondes et al., 2017). Developers also are informed about possible system and boundary resource vulnerabilities (#11). Third-party and internal developers rely on boundary resource monitoring for their extensions' proper functioning (#5): "Supporting admin and developers by providing an easy system status overview and an informative error log is a key to success" (#7). Therefore, we proposed DP6: Provide the platform with analysis and monitoring tools so that users can perform sophisticated business analyses, as well as fix errors that occur as quickly as possible.

While transaction platforms create and provide analyses of current and past transactions in e-commerce ecosystems for a variety of ecosystem participants, innovation platforms provide information on the overall system status in general and boundary resources in particular to their customers and external developers.

#### 6.2.4.7 Boundary Resource Evolution

Aside from boundary resources' current status, their evolution at a strategic level is relevant to internal and external developers. This strategic evolution reflects a platform's business strategy and involves the boundary structure, deprecation, and roadmap. The platform owner should provide a transparent structure for the provided features and boundary resources. As contemporary software artifacts tend to be complex, making their structure transparent is key to developing extensions (#2). Third-party developers need to know which interfaces are mandatory and how they interact with others (Aulkemeier, Paramartha, et al., 2016). This structure also includes responsibilities for boundary resources and exchanged messages (#9). As the platform evolves, the structure must be updated regularly (#11). However, boundary resources are likely to hide the platform's internal implementation, and they are visible only to ecosystem participants. Only a few platforms provide structural diagrams and information on internal modules (#1, #9): "There's a point at which internal transparency is intentionally not desired" (#4). Although the literature on platforms recommends keeping boundary resources relatively stable (Baldwin and Woodard, 2009), the evolution of platforms necessitates creating, updating, and abandoning boundary resources. Thus, it is critical to differentiate between boundary resource versions and prescribe migration paths for ecosystem participants (#7, #12). To a certain degree, new boundary resources that are backward-compatible also can be implemented (#6). It is also important to remind companies and developers that they are using outdated interfaces (#3): "Before I tell our customers and developers, 'but the change was in the release notes', I prefer to go and talk to them. Notes do not matter; conversation is key" (#9). Unlike past-oriented deprecation, a boundary resource roadmap describes new platform developments and future boundary resource evolutions. New boundary resources can be introduced, while old ones are abandoned during an initial beta phase before the public rollout (#9). Some platforms even provide regular roadmaps on a monthly or quarterly basis (#4, #9, #13). Moreover, it is important to train developers and other participants in new features, if applicable (#7). The roadmap can include technical and feature aspects (#11). However, interviewees warned that a fixed roadmap creates expectations among participants and may result in path dependency (#3): "We usually have a phase over process, where we start deprecating the old module and move it into the new module" (#1). Against this backdrop, we proposed DP7: Provide the platform with transparency regarding its structure and boundary resource updates so that users can use appropriate boundary resources and adapt to upcoming changes.

The platform structure's evolution and updates for boundary resources are provided by innovation and transaction platforms in e-commerce. While innovation platforms focus on development boundary resources (#5, #9, #10), transaction platforms provide information for third-party sellers on application boundary resources (#4, #6, #8). The deprecation and roadmap of boundary resources should be announced with an appropriate lead time to third-party developers so they can adapt to future changes (#1). Nevertheless, according to the interviewees, deprecation is far more important than communicating future changes: "I think the deprecated boundary resources and those types of things become far more important as design principles than a road map" (#4).

### 6.2.5 Discussion and Outlook

In our multi-paper design science research project, we developed seven theoretically grounded and empirically validated design principles for boundary resources in e-commerce ecosystems—including social, platform, and development boundary resources (Dal Bianco et al., 2014)—and contextualized them for the e-commerce sector (Wulfert, Woroch, Strobel, Seufert, et al., 2022). We contribute to the literature on e-commerce ecosystems by providing prescriptive design knowledge on the boundaries of innovation and transaction platforms (Gawer, 2021b; Hein, Schreieck, Riasanow, et al., 2020). We also provide platform owners with more concrete recommendations for developing concrete design instances from which they can instantiate boundary resources according to their peculiarities (Kuechler and Vaishnavi, 2011; Meth et al., 2015). Following the framework for minimum reusability evaluation and our interview guide (Iivari, Rotvit Perlt Hansen, et al., 2021), our 14 interviewees evaluated the importance and novelty of our overall set of design principles and deemed them important for the orchestration of ecosystem participants in general and third-party developers in particular: "I think that's why it's 100% important to address these issues" (#10). "It's good to go through them and compare our system toward them so we can identify which areas need further improvement and where the gaps are" (#5). However, some interviewees emphasized prioritization regarding importance within the set of principles (#8). Regarding novelty, our interviewees stated that the aggregation of these design principles is novel (#3, #5, #14): "The individual design principles are not new, but summarized in this set, it is a good overview that can be given to ecosystem participants" (#14). Although the actual boundary resource instances are well known for innovation and transaction platforms, they are not always implemented in practice (#7).

Furthermore, we considered our design principles' relevance for platform-internal roles based on interviewees' job descriptions and explanations (Zutshi and Grilo, 2019). The BPO is the interface for ecosystem participants and relies on proper conversation and communication of possible system errors and boundary resource updates (DP1, DP6, DP7). UIDs are concerned mainly with intra- and inter-ecosystem standards and guidelines regarding interfaces (DP3, DP4). Guidelines must be communicated accordingly (DP1). The IPD is responsible for the evolution and maintenance of the platform's technical core (DP2–7). Moreover, our interview analysis revealed that the digital transformation in retail in general and e-commerce in particular remains in its infancy at the company level. Our interviewees stated that incumbent retailers and small and medium-size enterprises often use existing and standardized software (i.e., innovation platforms) provided by software vendors (i.e., Shopify, Magento) (#2, #8). They rarely implement their own technical cores and even more rarely open them up to external developers as innovation platforms, possibly due to strong network externalities in existing innovation platforms and winner-take-all tendencies (Schütte and Wulfert, 2022a). New platforms would need to distinguish themselves from existing ones and identify possible niches (Cennamo, 2021). A potential downside of applying a standardized technical core is that it likely will yield uniform UIs, but interviewees reported that the user interface as a digital storefront is key in distinguishing oneself from competitors (#4, #6). Otherwise, competitors could be distinguished only by their product prices (Dominici, 2009). Our interview analysis also indicated that innovation platforms in e-commerce must parcel their services out to enable single developers and smaller groups of developers to participate in e-commerce ecosystems (#5, #8, #10, #13). Single developers cannot cope with complex workflows in e-commerce alone. This complexity requires collaboration and value cocreation among third-party developers to provide the focal platforms with useful extensions (Huber, Kude, et al., 2017). In line with existing literature on platform envelopment (Eisenmann, Parker, et al., 2011), we also identified a tendency among innovation platforms to integrate successful extensions (#7, #9).

This research project has limitations that offer interesting avenues for future research. The developed design principles for boundary resources in e-commerce ecosystems focus mainly on attracting third-party developers, but we contacted focal platforms for possible interviewees (Guest et al., 2006). Thus, our interviews capture prescriptive knowledge on implemented boundary resources (Chandra et al., 2015), but disregard potential third-party developers' requirements. Future researchers also might try to interview developers in e-commerce ecosystems about their requirements for successful collaboration and value cocreation. Regarding the number of interviews conducted, we followed the literature on qualitative interviews, suggesting sufficiency and saturation as potential indicators (Guest et al., 2006; Hagaman and Wutich, 2017; Seidman, 2019). We achieved a high level of sufficiency by approaching a variety of innovation and transaction platforms in e-commerce ecosystems as measured by size (Table 6.4) and representative interview partners as indicated by their internal roles (Zutshi and Grilo, 2019). We also reached data saturation emphasized by the fact, that our interviewees picked up information that we already documented in previous interviews independently of the platform type (Gawer, 2021b; Guest et al., 2006). However, future studies could analyze single participants further and roles' involvement in the design and implementation of boundary resources in innovation ecosystems. Another important future research direction would be to formulate design principles for other ecosystem participants explicitly, e.g., retailers and content providers (Böttcher, Rickling, et al., 2021). These additional design principles can be accumulated to develop a holistic design theory (Gregor and Jones, 2007) concerning attraction and collaboration between a variety of ecosystem participants (Schirrmacher et al., 2017). Future studies also could add design principles with more concrete design features that can showcase potential instances of prescribed boundary resources and simplify their implementation for platform owners (Kuechler and Vaishnavi, 2011; Meth et al., 2015). Moreover, as our interviewees were working for platforms mainly active in Europe and the U.S., future research may consider e-commerce platforms conducting business in Africa, Asia, and Latin America (Evans and Gawer, 2016). Such studies also could include cultural implications on boundary resources (Evans and Schmalensee, 2016). Although they might not impact technical boundary resources, social design principles (e.g., ecosystem conversion, extension marketplace) require culture-specific adaptations. Future studies also may analyze the need to implement a boundary resource in connection with platform size, as measured by revenue, number of employees, or number of complementors. Although we derived our design principles from an e-commerce context and interviewees working in the e-commerce sector, several interviewees (#3, #10, #11) suggested that the design principles also might apply to (software) ecosystems in other industries (Jansen, 2013). They even mentioned that they would be quite useful for large enterprises. Vom Brocke, Winter, et al. (2020) referred to the application of design knowledge in additional problem spaces as projectability. Although we developed and evaluated the design principles for the e-commerce context, they might be applicable in other contexts. Thus, future studies could evaluate the design principles' applicability in other contexts to improve their fitness in other domains.

### 6.2.6 Conclusion

In this paper, we presented seven design principles for implementation of boundary resources in innovation, transaction, and hybrid platforms in e-commerce environments derived from an interview study with 14 domain experts. Benefitting from the interview analysis, we could improve the design principles' fitness for innovation and transaction platforms in our multi-paper design's research approach. Implementing boundary resources using our design principles would generate collaboration among ecosystem participants in general and value cocreation of third-party developers in particular. While DP1 mainly focuses on transaction platforms involving a diverse set of ecosystem participants, DP2, DP3, and DP5 are concerned with innovation platforms, easing developers' onboarding and widening the dissemination of their extensions. DP4, DP6, and DP7 are relevant for both types of platforms by enabling improved transaction orchestration and extension quality simultaneously. Hybrid platforms need to implement all boundary resource types to attract ecosystem participants in general and third-party developers in particular. We also introduced three internal platform roles responsible for the implementation of boundary resources, namely business process owner, user interface designer, and internal platform developer. Finally, this paper provides utilizable design knowledge in the form of seven design principles for the design of boundary resources in e-commerce ecosystems.

## 6.3 Paper 8: Boundary Resource Management in Innovation Ecosystems: The Case of E-Commerce

**Abstract**

*Major challenges in contemporary information systems development demand new development approaches, such as innovation ecosystems, consisting of a focal innovation platform surrounded by independent developers.*[7] *The attraction of third-party developers and the amplification of generativity in extension development call for sophisticated boundary resources. As demarcation points between an innovation platform and external developers, boundary resources entail a strategic value and need to be managed carefully. However, existing research does not provide an approach for the holistic management of boundary resources from design to retirement. We apply a design science research approach consisting of two design cycles involving a literature review identifying 26 application lifecycles and the analysis of eight innovation platforms in e-commerce. We evaluated the model for boundary resource management in interviews with nine domain experts. We develop an*

[7] This version of the article has been accepted for publication, after peer review but is not the Version of Record and does not reflect post-acceptance improvements, or any corrections. The Version of Record is referenced as follows: T. Wulfert (2023). "Boundary ResourceManagement in Innovation Ecosystems: The Case of E-Commerce." In: *Electronic Markets* 33.27, pp. 1–27

*integrated boundary resource management lifecycle as a specialization of application lifecycle management for the holistic management of boundary resources in innovation ecosystems addressing third-party developers. The boundary resource management lifecycle consists of four layers: boundary resource governance, continuous communication, technical boundary resources, and supplementary boundary resources. It decouples technical and supplementary boundary resources while emphasizing their intertwined nature for ecosystem participants. Owners of innovation platforms can instantiate the reference procedure model for attracting developers.*

| Fact sheet of publication 8 | |
|---|---|
| Title | Boundary Resource Management in Innovation Ecosystems: The Case of E-Commerce |
| Authors | Tobias Wulfert |
| Publication Type | Journal Article |
| Publication Outlet | Electronic Markets |
| VHB-Jourqual3-Ranking | B |
| Status | Published |
| Research Question | How can boundary resources of focal platforms in innovation ecosystems be managed? |
| Research Approach | Literature Review (Vom Brocke, Simons, et al., 2015)<br>Multi-Case Analysis (Yin, 2018)<br>Interview Analysis (Mayring, 2014; Misoch, 2015) |
| Full Citation | T. Wulfert (2023). "Boundary ResourceManagement in Innovation Ecosystems: The Case of E-Commerce." In: *Electronic Markets* 33.27, pp. 1–27 |

## 6.3.1 Introduction

Major challenges in contemporary IS development such as rapid technological advancements, continuously changing requirements (Norbjerg, 2018), increasing customer demands, shortened application release cycles (Babb et al., 2017), more complex application systems, and limited resources demand new development approaches. The transformation of complex and monolithic application systems into innovation platforms that provide basic application services and attract external developers for an innovation ecosystem increases the overall generativity with additional apps, themes, and software integrations (Aulkemeier, Paramartha, et al., 2016; Hein, Schreieck, Riasanow, et al., 2020). This transformation allows for continuous application engineering and delivery (Dittrich et al., 2018). However, it requires the implementation of boundary resources that enable collaboration and value co-creation among different types of actors by translating diverging viewpoints and interfacing an innovation platform's application services (Ghazawneh and Henfridsson, 2013; Star, 2010). Boundary resources as a means for opening an innovation platform propel the ecosystem's generativity, influence the relationship between a platform owner and third-party developers, and thus increase the platform adoption by third-party developers (Eaton et al., 2015; Engert et al., 2022; Miremadi et al., 2023). Examples of boundary resources include APIs, Integrated Development Environments (IDEs), and SDKs with related documentation (Dal Bianco et al., 2014; Eaton et al., 2015), that provide external developers with access to the core functions of the platform (Karhu, Gustafsson, et al., 2018). Platform boundary resources need to be altered and updated continuously to cope with new market developments, create new business opportunities, and protect the platform against exploitation by third parties (Ghazawneh and Henfridsson, 2010; Jansen and van Capelleveen, 2013; Karhu, Gustafsson, et al., 2018). As demarcation points between an innovation platform and ecosystem participants (Santos and Eisenhardt, 2005), boundary resources bring about a strategic value to the platform owner and need to be managed carefully (Bondel et al., 2021; Vester, 2018; Yoo, Henfridsson, et al., 2010). Obtaining control over the boundary resources of an open platform is a key element of a platform owner's ecosystem governance (Baldwin and Woodard, 2009; Boudreau, 2010; Pon et al., 2014; Schilling, 2000).

Following the seminal works by Star and Griesemer (1989) as well as Ghazawneh and Henfridsson (2010), the research interest on boundary resources in the context of digital platforms and their surrounding innovation ecosystem is on the increase. Dal Bianco et al. (2014) analyzed the use of boundary resources in several software ecosystems. Karhu, Gustafsson, et al. (2018) compared the use of boundary resources with the licensing of a platform's intellectual property. While Petrik and Herzwurm (2020) investigated boundary resources for industrial Internet of Things (IoT) scenarios, Wulfert, Woroch, Strobel, Seufert, et al. (2022) focused on the implementation of boundary resources in e-commerce ecosystems. Although boundary resources are generally applicable in innovation ecosystems in different domains (Ghazawneh, 2012), scientific literature on the management of boundary resources is scarce.

Owing to their strategic relevance for innovation platforms (Boudreau, 2007; Yoo, Henfridsson, et al., 2010), a comprehensive management approach that covers the whole lifecycle from the design to the retirement of the boundary resources is necessary to exploit their full potential. A lifecycle model for boundary resources can be used "as a common reference for communication and understanding" (ISO,

2017, p. 25) among ecosystem participants. Since boundary resources involve and are concerned with software artifacts, we apply an application lifecycle perspective for developing a boundary resource management approach. The concept of Application Lifecycle Management (ALM) is made popular in the Information Technology Infrastructure Library (ITIL) with its application management functions (Fischbach et al., 2013). Many application lifecycles in the literature (Hallerstede et al., 2011) are inspired by the application management cycle included in the ITIL framework (AXELOS, 2012). TheALM considers the entire lifecycle of a software artifact by describing how the software development process should take place from the initial development phase up to the delivery phase and subsequent maintenance (Lacheiner and Ramler, 2011). The global market for ALM is expected to grow to $ 5 billion by 2027, with $ 2.8 billion specifically advocated for ALM tools (StrategyR, 2022). The most popular tools for ALM are, among others, Jama Software, Atlassian Jira, Microsoft Azure DevOps, and Visure (Aston, 2022; Jwo et al., 2013).

Although ALM bears a high potential for innovation ecosystems and although ALM seems to be a promising approach for managing boundary resources, the current body of literature does not consider boundary resources. Research investigated several specializations of ALM for, among others, security management (Demchenko et al., 2017; Zeng, Cheng, et al., 2021), data management (Alhassan et al., 2016; Chen and Zhao, 2012; Khatri and Brown, 2010), cloud application monitoring (Aversa et al., 2015), cloud application provisioning (Benfenatki et al., 2017), and sensor networks (Kabac et al., 2016). Moreover, existing research only focuses on single lifecycle phases, such as retirement (Yu, Chang, et al., 2017), or investigates APIs as specific artifacts. Several practitioner-based lifecycle approaches exist for API management (e.g., Ebert, 2012; Medjaoui et al., 2021; Rossberg, 2014; Vester, 2018; Weir, 2019; Zulian and Bouza, 2021), but without identifying an agreed upon approach (Bondel et al., 2021).

We therefore go beyond an isolated view on APIs and take a broader perspective on boundary resources for innovation platforms that include technical (e.g., APIs, SDKs) and non-technical (e.g., documentation) artifacts. With this approach, we cover the whole lifecycle of a boundary resource from design to retirement. We provide literature on ALM, with a dedicated lifecycle model on boundary resources required for inter-company and ecosystem-wide collaboration and value co-creation (Engert et al., 2022; Wulfert, Woroch, Strobel, Seufert, et al., 2022). While platform owners gain a customizable reference procedure model for the management of boundary resources and the attraction of third-party developers, these developers can anticipate boundary resource evolutions with regard to the reference model. We make use of initial but roughly designed processes for boundary resource management (e.g., Ghazawneh and Henfridsson, 2010; Hein, Weking, et al., 2019; Star,

2010) and provide a comprehensive reference procedure model. Furthermore, we also consider organizational aspects to ensure an adequate degree of platform openness and ecosystem generativity (Blaschke and Brosius, 2018; Boudreau, 2010). Against this backdrop, we raise the following research question:

> **RQ:** How can boundary resources of focal platforms in innovation ecosystems be managed?

To address this research question, we apply a design science research approach (Takeda et al., 1990; Vaishnavi, Kuechler, et al., 2019). Our design science research approach involves a literature synthesis on ALM in general and boundary resource management in particular (Vom Brocke, Simons, et al., 2015; Webster and Watson, 2002), analyzing the boundary management of selected innovation platforms in e-commerce from the theoretical lens of ALM (Niederman and March, 2019; Yin, 2018). In addition, our design science research approach also involves an evaluation of these findings, with the analysis of nine expert interviews to check our developed lifecycle model and to gain further insights on internal boundary resource development (Misoch, 2015). We chose the e-commerce domain for our boundary resource analysis, as several types of participants are orchestrated using digital technologies in e-commerce ecosystems (Wulfert, Woroch, Strobel, Seufert, et al., 2022). Furthermore, collaborative and distributed software development supports diverse electronic business models (Alt, 2020; Aulkemeier, Paramartha, et al., 2016; Timmers, 1998). We propose the boundary resource management lifecycle as a reference procedure model for owners of innovation platforms to implement sophisticated boundary resources and to attract additional third-party developers with their extensions. We provide a detailed reference procedure model for the management of boundary resources to the literature on ALM. Moreover, we emphasize the role of supplementary social boundary resources to enable the use of technical boundary resources by third-party developers.

The remainder of this research paper proceeds with an elicitation on related literature on: innovation platforms as the technical foundation of e-commerce ecosystems, existing approaches for boundary resource management, and ALM introduced as the theoretical lens of this research. Second, we elaborate on our design science research approach, incorporating a systematic literature review, a multi-case study, and an evaluation based on an interview analysis. Third, we present our boundary resource management lifecycle by describing the derived layers and single lifecycle stages for technical boundary resources and by mapping them with related supplementary boundary resources. Finally, we discuss our results, provide a research agenda for potential avenues for future research, and briefly present our conclusions.

## 6.3.2 Related Literature

### 6.3.2.1 Innovation Platforms in E-Commerce Ecosystems

Facilitated by the ongoing digitalization of the retail sector, (parts of) retail and wholesale functions (e.g., bridging, information, storing) are performed digitally to a certain degree in e-commerce (Laudon and Traver, 2019). In e-commerce, transactions involving information, products, and services are executed using information and communication technology (Choi et al., 1997; Nath et al., 1998; Zwass, 1996). E-commerce ecosystems are conceptualized as manifestations of digital business ecosystems in the context of e-commerce, centered around a focal transaction platform (Wulfert, Woroch, Strobel, Seufert, et al., 2022; Yi and Ming, 2011). The transaction platform orchestrates several types of independent ecosystem participants, such as manufacturers, customers, and service providers (Böttcher, Rickling, et al., 2021), who share a common interest, depend on each other, and co-create value (Blaschke, Haki, et al., 2018; Iansiti and Levien, 2004a; McIntyre and Srinivasan, 2017; Wareham et al., 2014). Benefitting from direct and indirect network effects (Armstrong, 2006; Shapiro and Varian, 1998), the value of an e-commerce ecosystem for each participant increases with additional participants joining (Hinz et al., 2020). The purpose of an e-commerce ecosystem is the exchange of physical and digital products for the benefit of a customer (Adner, 2017; Wulfert, Woroch, Strobel, Seufert, et al., 2022).

The technology-supported transactions in e-commerce ecosystems are enabled by innovation platforms that provide the necessary application services for the exchange of goods between and value co-creation among ecosystem participants (Aulkemeier, Schramm, et al., 2016; Blaschke, Riss, et al., 2019; Wulfert, Woroch, Strobel, Seufert, et al., 2022). Hence, innovation platforms form the technological infrastructure for e-commerce ecosystems (Wulfert and Karger, 2022b). While Haki, et al. (2019, p. 1) termed innovation platforms as the "technical core" of a surrounding ecosystem, Baldwin andWoodard (2009, p. 19) defined an innovation platform as "a set of stable components that supports variety and evolvability in a system by constraining the linkages among the other components." By benefitting from these linkages, third-party developers can extend the features of an innovation platform with a variety of extensions, which they achieve by taking advantage of the innovation platform's architectural modularity and the technical and organizational components' re-configurability (Jansen and van Capelleveen, 2013; Tiwana, Konsynski, and Bush, 2010). These extensions make use of the application services provided by the innovation platform via a set of boundary resources, which are implemented by the platform owner (Ghazawneh and Henfridsson, 2010). By implementing a set of boundary resources as part of its governance function (Boudreau, 2010), the platform owner defines the openness of and the access to the innovation platform (Boudreau, 2010; Karhu, Gustafsson, et al., 2018; Tiwana, 2014). A certain degree of platform openness invokes positive network effects and increases generativity for the ecosystem (Blaschke and Brosius, 2018; Karhu, Gustafsson, et al., 2018).

Tiwana, Konsynski, and Bush (2010, p. 675) emphasized that third-party developers can use these stable components as an "extensible codebase [...] that provides core functionality shared by the modules that interoperate with it and the interfaces through which they interoperate." While the stable components form the core of the innovation platform, the third-party modules resemble the periphery (Staykova and Damsgaard, 2015). The platform core and periphery are evolvable, with only the boundary resource specifications that should remain stable over a longer period of time to minimize necessary extension adjustments for the developers (Baldwin and Woodard, 2009). According to Zutshi and Grilo (2019), a digital platform is composed of six different layers, namely business, user interaction, development, integration, data, and the information technology layer. Blaschke, Haki, et al. (2019) mentioned four layers of digital platforms. They described the business layer as service dimension and called the user interaction the ecosystem dimension that allows interaction with the platform core. The platform core includes application services and boundary resources (i.e., extensible codebase) necessary for the development of platform extensions. The platform core is executed by a digital infrastructure.

#### 6.3.2.2 Boundary Resource Management

The concept of boundary resources stems from the social sciences, in which "boundary objects" (Star and Griesemer, 1989, p. 393) function as "demarcation points" (Santos and Eisenhardt, 2005, p. 491) between different social worlds (Santos and Eisenhardt, 2005; Star and Griesemer, 1989). Boundary resources enable the collaboration and co-creation of value among different types of actors by translating between diverging viewpoints (Star, 2010), and are defined as "software tools and regulations that serve as the interface for the arm's-length relationship between the platform owner and the application developer" (Ghazawneh and Henfridsson, 2013, p. 175). Since boundary resources serve as interfaces between different ecosystem participants, they move within companies' organizational boundaries (Santos and Eisenhardt, 2005). Hein et al. conceptualize a set of boundary resources as "affordances" (Hein, Schreieck, Riasanow, et al., 2020, p. 91) of the digital platform. These affordances form the boundary and the demarcation between the digital platform core and third-party developers (i.e., complementors) providing platform extension and amplifying generativity. Hence, boundary resources enable collaboration and value co-creation among different ecosystem participants (Autio, 2022; Hein, Weking, et al., 2019; Wulfert, Woroch, Strobel, Seufert, et al., 2022).

Innovation platforms in e-commerce ecosystems provide boundary resources that support the design, implementation, and provision of third-party developers' extensions (Ghazawneh and Henfridsson, 2013). Boundary resources lower entrance barriers to the ecosystem for new developers (Alves et al., 2017; Hein, Schreieck, Riasanow, et al., 2020) and facilitate access to the focal innovation platform's core resources (Baldwin and Woodard, 2009; Eaton et al., 2015). The set of boundary resources provided influences the innovation platform's openness and the associated ecosystem's generativity (Ghazawneh and Henfridsson, 2010; Karhu, Gustafsson, et al., 2018; Parker, Van Alstyne, and Choundary, 2016).

The concept of boundary resources includes a variety of manifestations (Eaton et al., 2015). Star and Griesemer (1989) introduced repositories, ideal type, coincident boundaries, and standardized forms as boundary resource types. Dal Bianco et al. (2014) described application boundary resources as interfaces for the third-party extensions' interaction with the focal platform, development boundary resources as facilitating the third-party extensions' design and implementation, and social boundary resources as enabling knowledge transfer and the coordination of extension development and boundary resource usage (Carlile, 2002). Application and development boundary resources provide access to the integration and development layer introduced by Zutshi and Grilo (2019). Karhu, Gustafsson, et al. (2018) developed a set of 16 boundary resources for open digital platforms. Building upon the publications of Ghazawneh (2012) and Dal Bianco et al. (2014)t, Wulfert, Woroch, Strobel, Seufert, et al. (2022) synthesized 14 distinct boundary resources. These boundary resources can be grouped using the boundary resource layers of Dal Bianco et al. (2014). These boundary resources can be grouped using the boundary resource layers of Dal Bianco et al. (2014). Social boundary resources involve documentation, roadmaps, registration forms, forums, support, and hackathons. While application boundary resources include APIs and workflows, development boundary resources include SDKs, software libraries, software repositories, debugging tools, prototypes, and mockups. We utilize Dal Bianco et al.'s (2014) conceptualization and differentiate technical boundary resources (i.e., application and development boundary resources) and supplementary social boundary resources. As suggested by Hein, Schreieck, Riasanow, et al. (2020), we integrate both perspectives in our boundary resource management lifecycle.

Existing approaches for developing and managing boundary resources in the context of innovation platforms highlight an iterative approach. Star (2010) introduced a model for describing the evolution of boundary resources consisting of standardization and residualization phases. While standardization describes a process in which certain aspects of the boundary object are standardized for a wider application, residualization refers to a process of manipulation of the boundary resource for the specific purpose and use of a group of users. Hein, Weking, et al., (2019) adapted this initial model, with an additional state for the boundary resource as a result of the standardization. If a boundary resource is successfully standardized and not transferred into a residual, it can become part of standardized Information Technology (IT) infrastructure. Ghazawneh and Henfridsson (2010) emphasized a platform governance perspective in their iterative boundary resource development, balancing platform control and ecosystem generativity.

#### 6.3.2.3 Application Lifecycle Management as Theoretical Lens

ALM was introduced as a complement to Product Lifecycle Management (PLM) for the development, maintenance, and optimization of application systems in organizations (Kääriäinen, 2011; Stark, 2011). While PLM is concerned with hardware lifecycles, the focus of ALM is on software (Kääriäinen, 2011). This requires an integration of both concepts for smart products (Beverungen, Müller, et al., 2019; Deuter and Rizzo, 2016). In analogy to lifecycles of species in biology (Schuh and Rudolf, 2015), ALM covers the whole software lifecycle from requirements engineering to the deprecation of the application (Stark, 2015). (Stark, 2015). In contrast to this broad understanding of ALM, several research papers focus on software development and their investigations end with the deployment of the software artifact. Software configuration management is seen as a building block of ALM (Kääriäinen, 2011); it provides a repository for software components, supports development activities, and controls the overall development (Estublier, 2000). ALM goes beyond these software development management activities, managing artifacts and documentation that is created along the entire lifecycle of an application, and ALM tools integrate them in overarching tools (Kääriäinen, 2011). In this research paper, we apply the broad perspective of ALM as a theoretical lens (Niederman and March, 2019). This perspective complements application development and operation involving, among other phases, the design, implementation, maintenance, and optimization in a continuous cycle (Table A.18).

Although the term application lifecycle management is widely used in software engineering and management, an agreed-upon definition still does not exist (Lacheiner and Ramler, 2011; Pirklbauer et al., 2009). ALM involves "the coordination of activities and the management of artifacts (e.g., requirements, source code, test cases) during the software product's lifecycle" (Kääriäinen and Välimäki, 2009, p. 150). Rossberg (2014) emphasized that an application's lifecycle starts with first thoughts on application development and ends with the application's actual retirement, with possible changes along the application's runtime. Besides the interplay between application development and operation, ALM involves governance as a third major activity (Tüzün et al., 2019). Pirklbauer et al. (2009) suggested an integrated perspective (i.e., concepts, management, engineering) on ALM and emphasized the role of ALM in business-IT alignment. Chanda and Foggon (2013) focused on coordination and management activities for people, investment, and technical artifacts over the entire lifecycle from design to retirement. We theoretically ground the design science research process in this broad understanding of ALM.

Popular application lifecycles are organized along the following phases, which might carry different names: design, build, deploy, operate, and optimize (AXELOS, 2012). ALM therefore emphasizes the interaction between application development and operation, and describes the interaction as a continuous cycle in which the completed development of a version is followed by deployment and optimization. This subsequently leads to new requirements for the next development iteration and the possible decommissioning of the existing application. We provide an overview of ALM phases covered in selected research papers in Table A.8 in the Appendix in the Electronic Supplementary Material. This phased approach guided the development of the initial boundary resource lifecycle (Table 6.7).

### 6.3.3 Design Science Research Approach

This paper aims to provide a theoretically grounded and empirically validated reference procedure model for boundary resource management in platform-centered innovation ecosystems. We identified a need for a holistic management of boundary resources as an important building block of innovation ecosystems in e-commerce (Boudreau, 2010; Santos and Eisenhardt, 2005). Developing and applying a designed artifact to better understand a problem and generate knowledge is the paradigm underlying design science research (Hevner, 2007; Hevner, March, et al., 2004). The boundary resource management lifecycle as a developed artifact should support owners of innovation platforms in developing and implementing proper boundary resources to attract and orchestrate third-party developers (problem space) (Drechsler and Hevner, 2018; Vom Brocke, Winter, et al., 2020). The attraction and governance of third-party developers should increase the focal platform's overall generativity (Blaschke and Brosius, 2018). The boundary resource management lifecycle as a reference model is a meta-artifact (Gregor and Jones, 2007) that can be instantiated for the peculiarities of specific innovation platforms and that can guide boundary resource implementation (solution space). For the development of our meta-artifact, we follow strategy one as described by Iivari (2015).

We applied the design science research approach by Vaishnavi, Kuechler, et al. (2019), focusing on the design phase, and developed a lifecycle model for the holistic management of boundary resources in innovation ecosystems. In total, we performed two design cycles in which the latter design iteration is informed by the previous one for an incremental development (Baskerville et al., 2019; Diederich et al., 2020; Norbjerg, 2018; Rosemann and Vessey, 2008). The theoretical foundation of our lifecycle model lies in the ALM literature. Hence, we conducted a systematic literature review to identify relevant lifecycle phases that involve boundary resources in the first design cycle (a) (Vom Brocke, Simons, et al., 2015). We informed the theory-based lifecycle with a multi-case analysis on selected innovation platforms in e-commerce ecosystems and their communication of boundary resource developments in the second design cycle (b) (Yin, 2018). We evaluated the enhanced lifecycle with nine interviews with domain experts employed at innovation platforms in e-commerce. The evaluation was used to investigate the designed artifact's applicability and to further inform the design process with data on the boundary resources' internal design and implementation (Alsaawi, 2014; Myers and Newman, 2007).

To ensure a holistic management of boundary resources from design to retirement, we apply a lifecycle management perspective throughout the two design cycles (ISO, 2017; Kääriäinen and Välimäki, 2009). Since our perspective on boundary resources is focused on a broad understanding of the lifecycle concept, the litera-

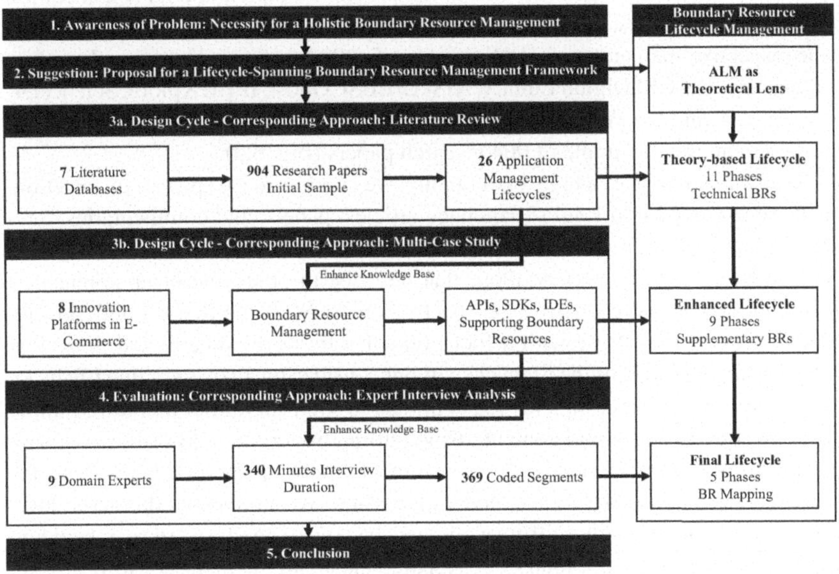


**Fig. 6.8** Overview of the Design Science Research Approach

ture review (i.e., coding of the papers), the case data, and the interview analysis are investigated through a theoretical lens developed during the elaboration on ALM (Niederman and March, 2019). This especially involves the elicitation on the lifecycle phases and their impact on boundary resources. In the following sections, we describe in more detail how the three methods intertwine conceptually to develop the lifecycle model for boundary resource management, and we provide details on the intermediary lifecycle models. The overall design science research approach is visualized in Fig. 6.8.

#### 6.3.3.1 Systematic Literature Review

The first design cycle aims to analyze the existing literature on ALM in general and their relation to boundary resources in particular to derive a theory-based lifecycle model for the management of boundary resources. A systematic literature search offers an appropriate framework to ensure the results' traceability, systematicity, and reproducibility, with the literature search involving the application of process and quality criteria (Cram et al., 2020). In line with these principles, we apply Vom Brocke et al.'s (2015) approach for literature analysis in combination with that of Webster and Watson (2002). The foundation of the literature search was the conceptualization of the object under consideration, i.e., the application lifecycle management as established in the previous chapter, and its consideration of boundary resources. We conducted a broad search for application lifecycles and their respective management. On September 5, 2022, we searched seven literature databases (i.e., ACM Digital Library, AISeL, EBSCOhost, IEEE Xplore, ScienceDirect, Scopus, and the Web of Science) for title, abstract, and keywords. Our search resulted in an initial sample of 904 research papers (Fig. 6.9).

To ensure an appropriate level of quality, we conducted two phases of selection. First, we excluded non-English language articles, panels, and commentaries. Second, we applied additional quality criteria to the selection process. We excluded the following research papers: those that describe the lifecycle of an application at runtime, those with a narrow perspective on ALM, those that do not consider boundary resources, those with a focus on software development, and those that focus on single phases of the lifecycle without considering an overarching lifecycle. We followed Bandara et al.'s (2015) approach to filter the initial literature sample based on title, abstract, and content, using our quality criteria. After the exclusion of duplicates, our analysis led to 18 papers in which the application lifecycle forms a broad perspective in the subject under discussion. We augmented this set of literature with literature resulting from a one-way backward and forward search adding eight research papers for a total of 26 research papers in the final set (Vom Brocke, Simons, et al., 2015). The team of researchers analyzed the articles within the sample independently using full-text screening, and they coded the relevant text passages to extract the information on lifecycle phases. We applied the generic application lifecycle phases (AXELOS, 2012) as initial coding scheme for the extraction of knowledge on the phases, and iteratively derived the theory-based lifecycle model. We summarized the research articles identified and analyzed, in a concept matrix presented in Appendix A.6 (Webster and Watson, 2002) in the Electronic Supplementary Material. The identified research articles covered a period of 14 years, with the first article published in 2008. The number of lifecycle phases varies between three and ten, with an average of six lifecycle phases (Table A.18).

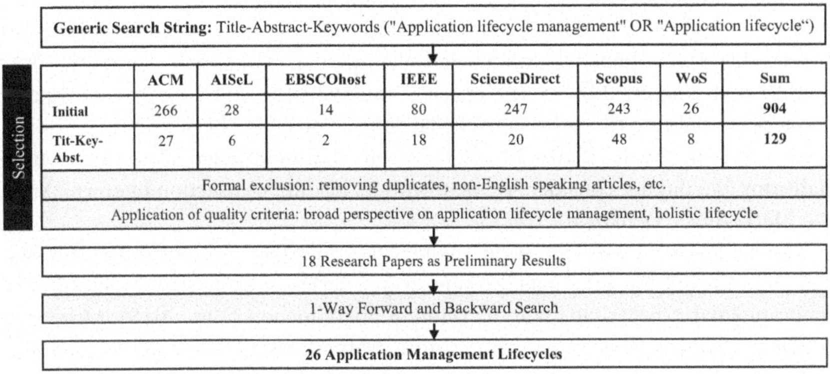


**Fig. 6.9** Overview of the Literature Selection Process

The theory-based boundary resource lifecycle consisted of a total of 11 lifecycle phases derived from the literature review approach (Table A.18). These phases follow each other successively, with the last phase (i.e., governance) leading to the first phase of another cycle run (i.e., requirements). This structure follows the structure of existing application lifecycles (e.g., AXELOS, 2012; Duda et al., 2022; Ebert, 2012; Khabou et al., 2019). Since ALM is concerned with the management of a software artifact, the theory-based lifecycle phases focused on technical boundary resources, such as APIs and IDEs (AXELOS, 2012; Bondel et al., 2021; Vester, 2018). Supplementary boundary resources are rarely discussed as results of single phases (e.g., test and release phase) or main activity of the communication phase (Kääriäinen, 2011; Matthes et al., 2009). However, the management of boundary resources does not only involve technical boundary resources (Dal Bianco et al., 2014; Hein, Schreieck, Riasanow, et al., 2020; Star, 2010). Therefore, we enhanced the theory-based lifecycle with knowledge derived from innovation platforms successfully attracting third-party developers and a high generativity indicated by the number of extensions available.

### 6.3.3.2 Multi-Case Study

To enhance the theory-based lifecycle with prescriptive knowledge derived from real-world cases, we conducted a multi-case analysis on eight innovation platforms (i.e., BigCommerce, Magento Commerce, OXID eShop, Salesforce Commerce Cloud, SAP Commerce Cloud, Shopify, Shopware, and WooCommerce) in e-commerce ecosystems (Eisenhardt, 1989b; Yin, 2018). These innovation platforms form the center of larger innovation ecosystems in the context of e-commerce (Table 6.5).[8] They provide the necessary application services for business models in e-commerce (Aulkemeier, Paramartha, et al., 2016; Timmers, 1998; Wulfert and Karger, 2022b) and provide an extensible codebase for third-party developers enabling the provision of platform extensions (Jansen, 2013; Tiwana, Konsynski, and Bush, 2010). As we focus our multi-case analysis on innovation platforms, we did not consider Amazon with its leading transaction platform Amazon Marketplace (Amazon, 2022).[9]

The multi-case study approach enabled us to derive, more broadly, information on boundary resource management from a series of cases that might differ in environmental aspects but share a common phenomenon (Yin, 2018). Moreover, selecting multiple cases is paramount in identifying cross-case patterns, and represents the basis for comparative analysis (Eisenhardt, 1989b; Van Aken, 2004). It is an essential approach to transcend the specificity of each individual case and abstract generally applicable knowledge that benefits the class of the artifact (Gregor, Müller, et al., 2013; Lee, Pries-Heje, et al., 2011). Hence, we could analyze the "same phenomenon in [...] in a variety of situations" (Yin, 1981, p. 101). The shared phenomenon of our eight cases is their success in e-commerce measured by the number of supported e-commerce websites (Builtwith, 2022; Datanyze, 2021), their design as an extensible codebase enabling third-party developers to design third-party extensions (Tiwana, Konsynski, and Bush, 2010), and as a provider of

[8] The asterisks in Table 6.5 have the following meaning: $*$ = Year of publication of the SAP Cloud Platform and $**$ = Values refer to the OXID eSales AG.

[9] Although Amazon is sometimes described as a hybrid platform in research because of its subsidiary Amazon Web Services providing cloud services to professional as well as private customers (Amazon, 2022; Schütte and Wulfert, 2022a), Amazon does not provide its e-commerce system, powering the Amazon Marketplace, for use at standalone sites or the involvement of third-party developers. For the Amazon Marketplace, only APIs for sellers are available (Github, 2023).

application services for e-commerce businesses (Wulfert and Karger, 2022b). We have selected these platforms to derive knowledge from successful boundary management and go "beyond a single success story" (Chandra Kruse and Seidel, 2017, p. 186).

For the platform analyses, we leveraged a plethora of primary and secondary data sources to collect architectural information on each case that is publicly available (Hänninen et al., 2018). First, we searched for information on the official websites, including textual information, graphics, and videos of the eight innovation platforms. In this part of the analysis, we tried to limit ourselves to objective information. We critically analyzed advertising texts or content that aimed at making a profit and did not include them in our analysis if we had any doubts. We also analyzed information and texts from other media that dealt with the eight platforms, such as newspapers, industry journals, and practitioner magazines. The case data is analyzed through the theoretical lens of lifecycle management for boundary resources (Niederman and March, 2019), with the objective of synthesizing a holistic management model.

For the development of the enhanced boundary resource management lifecycle, we aggregated the information on boundary resource management derived from our multi-case analysis in a cross-case approach (Yin, 2018) and updated the theory-based lifecycle accordingly. In particular, the case analysis revealed a number of potential supplementary social boundary resources, such as quality criteria, status cockpit, and deprecation schedule, that augment the design, development, publication, and maintenance of technical boundary resources (Adobe, 2022c; SAP, 2022c; Shopify, 2022e). Consequently, we were able to decouple the lifecycle of technical and supporting boundary resources while retaining a lifecycle phase-based relation between them (Fig. 6.10). Moreover, the multi-case analysis emphasized the significance of boundary resource governance and communication when it comes to the management of boundary resources in innovation ecosystems (BigCommerce, 2022b; OXID, 2021). We therefore introduced the layered structure of the boundary resource management lifecycle, with governance and communication as core layers. We also introduced the description continuous to the communication phase in order to linguistically emphasize the significance of the communication between the owner of a focal platform in innovation ecosystems and third-party developers providing platform extensions (Jansen, 2013; Staykova and Damsgaard, 2015). The number of lifecycle phases for the management of boundary resources was reduced to nine, with boundary resource governance and continuous communication as core activities.

**Table 6.5** Meta-Information on the Selected Innovation Platforms

| | Establishment | Number of employees | Total turnover (million euro) | Customers |
|---|---|---|---|---|
| Big Commerce | 2009 | 1300 | 219.9 | Fujitsu, Ben & Jerry's, Gillette, General Electric |
| Adobe Commerce | 2008 | 375 | 107.1 | HP, Liebherr, Shoebacca, Rossignol |
| OXID eShop | 2003** | 51–200** | – | AIDA, Bitburger, Hellweg, Mercedes Benz |
| Salesforce Commerce Cloud | 1999 | 73000 | 26900 | Adidas, Amazon Web Services, Barclays, Coca Cola |
| SAP Commerce Cloud | 2012* | 105000 | 27804 | Tesco, Casey's, Moen, The Entertainer |
| Shopify | 2006 | 10000 | 46848 | Freelatics, Babbel, Heinrich, Five Skincare |
| Shopware | 2004 | 300 | 73 | Aston Martin, Borussia Dortmund, Melitta, Thyssenkrupp |
| Woo Commerce | 2008 | – | – | Chaka Khan, All Blacks, Weber, Singer |

#### 6.3.3.3 Expert Interviews

We evaluated the enhanced boundary resource management lifecycle with a series of interviews with domain experts (Schreier, 2014). Within the interviews, we addressed the implementation and management of boundary resources on selected innovation platforms in the context of e-commerce. We chose to conduct semi-structured interviews to directly address the knowledge holders within the domain. In total, we conducted nine expert interviews with an average duration of 38 minutes (Table 6.6). To create a broad and diverse knowledge base, we generalized information from a diverse set of interviewees (Iivari, Rotvit Perlt Hansen, et al., 2021). Within the data collection process, we contacted a total of 46 innovation platforms based in European Union (EU), Latin America, and the United States (US) through

**Table 6.6** Expert Interview Overview

| # | Platform Internal Role | Staff | Revenue (in $) | Boundary Lifecycle Phases | Duration |
|---|---|---|---|---|---|
| 1 | User interface developer | 4,700 | 800 M | 4.1, 4.2, 4.4, 4.6, 4.7 | 46 min |
| 2 | Internal platform developer | 2,500 | 5 M | 4.1, 4.2, 4.3, 4.2, 4.7 | 20 min |
| 3 | Internal platform developer | 49,000 | 17.1 B | 4.3, 4.4, 4.6, 4.7 | 20 min |
| 4 | Business process owner | 270 | 28 M | 4.1, 4.2, 4.4, 4.6, 4.7 | 58 min |
| 5 | Business process owner | 107,000 | 28 B | 4.1, 4.3, .4.5 | 42 min |
| 6 | Business process owner | 107,000 | 28 B | 4.1, 4.5 | 42 min |
| 7 | Internal platform developer | 40 | n. a. | 4.1, 4.4, 4.7 | 33 min |
| 8 | Internal platform developer | 107,000 | 28 B | 4.1, 4.4, 4.5 | 34 min |
| 9 | User interface developer | 107,000 | 28 B | 4.1, 4.2, 4.3 | 45 min |

publicly available information of the companies and profiles of platform representatives on professional social networks such as LinkedIn. Although we mainly contacted platform representatives at management level, we were forwarded to interviewees directly involved in the design and development of (technical) boundary resources. We classified the interviewees by their company internal role according to innovation platform layers identified by Zutshi and Grilo (2019), and by related roles, such as business process owner, user interface designer, and internal platform developer (Wulfert, Woroch, Strobel, Seufert, et al., 2022). We could acquire an almost equal number of roles for the interviews (Table 6.6). As requested by the interviewees, we did not mention any company names.

Our semi-structured interviews were based on a previously known guide to ensure that all conversations covered a similar range of questions regarding the management of boundary resources while allowing the greatest possible flexibility in exploring explicit knowledge (Merton and Kendall, 1946; Misoch, 2015). The interview guide was created based on our research question, using ALM as the theoretical lens, and the enhanced boundary resource management lifecycle. We were able to discuss all lifecycle phases of the enhanced lifecycle and go into further detail for the supplementary boundary resources. The expert interviews were conducted via electronic communication media, between December 2021 and June 2022. Depending on the interviewees' preferences, the interviews were conducted either in English or in German. Interviewees 2 and 7 provided additional textual answers for our interview guide. We introduced each interviewee to the topic and the purpose of the study at the beginning of the interview (Alsaawi, 2014).

For the knowledge extraction, the interviews were anonymized and transcribed to perform a qualitative content analysis (Nunkoosing, 2005), which aims at systematically analyzing an object of communication regardless of its form (Schreier, 2014). Specifically, we used the content structuring approach as a form of deductive category assignment to extract information according to predefined categories (Mayring, 2014). We used our enhanced boundary resource lifecycle as the initial categorization system and iteratively adapted and expanded this scheme with our nine expert interviews. We reached a stable set of codes after the first five interviews. We chose a single sentence as the coding unit, thus ensuring the highest possible level of detail, while a paragraph of an interview can be considered as a context unit (Mayring, 2014). Multiple codes can also be applied to a single sentence if the sentence has several meanings. We used MAXQDA for the coding (Krippendorff, 2013) and the subsequent construction of a quantitative graphical representation (Kuckartz and Rädiker, 2022). In total, we coded 369 segments (Fig. 6.8). We include leading quotes from the interviews in the results section (Yin, 2018).

The coding of the interview transcripts resulted not only in an aggregation of the lifecycle phases for boundary resource management, but also in a more detailed allocation of supplementary boundary resources to these phases as depicted in Fig. 6.10. The small-stepped and detailed phase descriptions seemed not to be applicable in practice where, for instance, plan, design, and development activities happen even iteratively yet in an agile manner (#2). In total, the final boundary resource management lifecycle consists of five phases for technical boundary resources (Fig. 6.10). Two interviewees made an even broader segmentation of the boundary resource management lifecycle, describing an internal perspective concerned with the design and development of boundary resources, an extern perspective when boundary resources can be used by third-party developers, and a release phase as transition between the internal and external phase (#2, #4). Moreover, the evaluation led to several additions to the supplementary boundary resources and changes in the mapping to boundary resource management phases. For instance, our interviewees frequently reported the necessity for developer-requested changes such that we introduced change requests as supplementary boundary resource (#1, #3, #7). Another example would be the mapping of the supplementary boundary resource "terms of use" that was initially mapped to the release phase during the multi-case analysis. However, the interview coding revealed strong evidence for an even earlier consideration of terms of use already in the design phase and even in the governance in which external responsibilities are determined (#2, #7).

The iterative development of our boundary resource management lifecycle resulted in two intermediary lifecycles and a final boundary resource management lifecycle (Fig. 6.10). We summarize the development process in Table 6.7.

**Table 6.7** Overview of the Development Process of our Boundary Resource Management Lifecycle

| | Boundary Resource Management Lifecycle | | |
|---|---|---|---|
| | Theory-Based | Enhanced | Final |
| Input data | 26 application management lifecycles | 8 innovation platforms in e-commerce | 9 interviews with domain experts |
| Method for data collection and analysis | Systematic literature analysis (Vom Brocke, Simons, et al., 2015) and concept matrix (Webster and Watson, 2002) | Multi-case analysis (Yin, 2018) | Semi-structured interviews (Misoch, 2015) and qualitative content analysis (Mayring, 2014) |
| Lifecycle stages | 11 (requirements, plan, design, development, test, release, maintenance, optimization, deprecation, communication, governance) | 9 (requirements, plan, design, development, test, release, maintenance, optimization, deprecation), 2 cross-stage activities (communication, governance) | 5 (plan and design, development and test, release, maintenance and optimization, deprecation), 2 cross-cross stage activities (Boundary resource governance, continuous communication) |
| Development notes | – Focus on technical boundary resources,— Boundary resource development and operation | – Supplementary social boundary resources,— Mapping of boundary resources,— Layered structure | – Aggregation of lifecycle stages,— Additional supplementary social boundary resources,— Updated mapping of boundary resources |

### 6.3.4 Boundary Resource Management Lifecycle

Based on our design science research approach, which is theoretically grounded in the current body of literature on ALM and involves empirical data from eight case analyses triangulated by nine interviews with domain experts, we developed a reference procedure model for the management of boundary resources in innovation ecosystems for the domain of e-commerce. We take a lifecycle perspective to ensure an adequate management of boundary resources from design to retirement (AXELOS, 2012; ISO, 2017). The boundary resource management lifecycle resembles a reference procedure model for the particular context of e-commerce ecosystems (Becker and Delfmann, 2007; Schütte, 1998). The core of the boundary resource management lifecycle is formed by governance activities, illustrated with a light grey shading, spanning the whole lifecycle, and includes portfolio management, responsibility management, business alignment, and the determination of the lifespan for each boundary resource (Chappell, 2008; Vester, 2018). The core of the boundary resource management lifecycle is surrounded by activities for enabling a bidirectional communication with ecosystem participants, such as third-party developers and service providers, across the entire lifecycle depicted in turquoise. As suggested

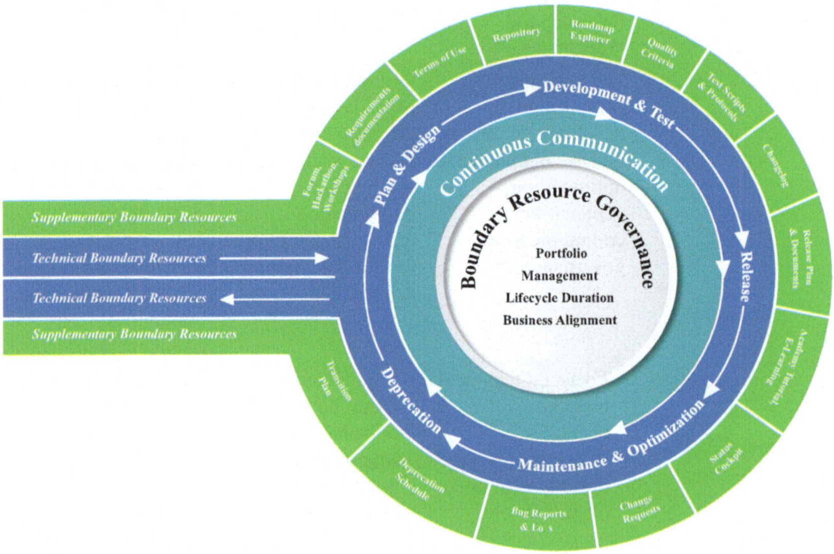


**Fig. 6.10** Boundary Resource Management Lifecycle

by the International Organization for Standardization (ISO) (2017), our proposed boundary resource management lifecycle is "organized into stages." We shaded the lifecycle for managing technical boundary resources in green. The lifecycle for technical boundary resources is divided into five phases, namely plan and design, development and test, release, maintenance and optimization, and deprecation. We also mapped major supplementary boundary resources to each lifecycle stage of the technical boundary resources in the outer circle shaded in blue (Fig. 6.10). The differentiation of technical and supplementary boundary resources is in line with findings from our case and interview analysis (e.g., Shopify, 2022b, #7, #10). Their intertwined nature is reflected by visually integrating both circles (Fig. 6.10).

#### 6.3.4.1 Boundary Resource Governance

The owner of the focal platform in a platform-centered ecosystem "sets, and often enforces, the governance rules" (Adner, 2017, p. 48). These rules are direct and indirect measures to control innovation ecosystems and their participants (Tiwana, 2014). As part of the ecosystem governance that involves "platform regulation including contractual, technological and information design" (Boudreau and Hagiu, 2009, p. 187), boundary resource governance defines the portfolio of necessary boundary resources in coordination with business requirements, the necessary boundary resources' lifetime, overall procedures, and responsibilities (Tüzün et al., 2019; Vester, 2018). The portfolio represents the entire set of past, present, and future boundary resources. The coordination of business requirements with the boundary resource portfolio defines the openness of the platform (Boudreau, 2010; Hein, Schreieck, Wiesche, et al., 2016). *"Before designing new boundaries, internal and external responsibilities need to be clarified and communicated. Who is responsible for what? Which services do we need to provide as a platform provider? Which application services are open for developers*[10]*?"* (#9)[11]. Boundary resource governance can also promote compatibility of an innovation platform's boundary resources with other platforms, thus increasing the reach for third-party developers (Spaeth and Niederhöfer, 2022).

The governance of boundary resources is an overarching task for the management of single boundary resources, and orchestrates development and maintenance tasks (Chappell, 2008; Tüzün et al., 2019). Our analyses revealed three types of platform release cycles that also involve the boundary resources. While the majority

[10] This and the following quotes are illustrative extracts taken from the expert interviews conducted.

[11] The numbers in brackets refer to the expert interview overview in Table 2 and indicate single interviewees.

of analyzed platform cases favor scheduled quarterly releases with fixed boundary resource lifecycles (Adobe, 2022d; OXID, 2021; Shopify, 2022e), unscheduled release cycles with prior announcements are also implemented in practice (#1, #2). Another approach introduced by interviewees #4 and #7 prescribes a continuous evolution of boundary resources without the necessity for deprecation; however, interviewee #4 also highlighted backward compatibility as a major prerequisite. *"We generally operate on a pattern of backward compatibility. Eventually, there's something that can't be, but we can really try not to deprecate things"* (#4). *"To avoid versioning, we ensure that our API is backward compatible. This is part of our API contract. This also results in a number of API fields that are unused."* (#7)

In this regard, breaking and non-breaking changes were introduced by several innovation platforms (BigCommerce, 2022b). While breaking changes usually misplace communication with external applications and result in a new API version, non-breaking changes can be implemented into production systems without any interference for developers. However, third-party developers are encouraged to "write code against our APIs that will not break if an endpoint begins returning additional fields" (BigCommerce, 2022b). Adobe also highlights backward-incompatible changes in a dedicated section of their developer blog (Adobe, 2022a). OXID (2022d) uses a three-tier application landscape consisting of a development, maintenance, and legacy environment. The legacy environment is supported with updates until the next maintenance version is released. The development environment is one version ahead of the maintenance version that is productively used by customers and extension developers. SAP Commerce Cloud makes use of development and quality assurance environments before a boundary resource is made available in productive systems (#5, #9). Four interviewees also reported on a pipeline of staging systems with intermediary quality gates (#1, #5, #6, #8).

Versioning is a common approach that owners of innovation platforms use in order to introduce new and updated APIs while continuing old ones for a predefined transition period (BigCommerce, 2022a; SAP, 2022d; Shopify, 2022e; WooCommerce, 2022c). Versioning can also be used to map possible dependencies between technical boundary resources and external tools (Adobe, 2022e; Shopware, 2022d). Semantic versioning can also give meaning to version numbers by emphasizing major, minor, and patch versions in a single string (Preston-Werner, 2022).

#### 6.3.4.2 Continuous Communication

Continuous communication deals with the continuous information of developers and additional ecosystem participants, with general information about boundary resources (Priyadarsini et al., 2020). Developers depend on timely information about the status of technical boundary resources, as the extensions they develop make use of application and development services provided via boundary resources that are provided by the innovation platform. Communication in this context means publishing supplementary boundary resources to ecosystem participants, and includes information such as news, advantages and disadvantages, projects, roadmaps, new services, and other information (Vester, 2018). This activity is neither tied to a lifecycle phase nor to a single boundary resource release (AXELOS, 2012). Developers and additional ecosystem participants are frequently informed about the status of the boundary resources independently of single lifecycle phases and boundary resource releases. *"One of the biggest challenges is always communicating."* (#1)

Supplementary social boundary resources are used for the continuous communication layer of our boundary resource management lifecycle (Dal Bianco et al., 2014). Communication should be ahead with regard to technical boundary resources' time of actual release cycles to allow third-party developers to cope with upcoming changes (Schwarz and Legner, 2020). Depending on the type of technical boundary resource and lifecycle stage, it is possible to create different supplementary boundary resources and publish them for communication (Dal Bianco et al., 2014; Hein, Schreieck, Riasanow, et al., 2020; Schwarz and Legner, 2020). Besides extensive textual documentation (OXID, 2022a; SAP, 2022d; WooCommerce, 2022f), supplementary boundary resources can also involve developer forums (WooCommerce, 2022e), animated guides, videos (SAP, 2022a), communities and forums (Shopify, 2022d), and hackathons (BigCommerce, 2022g). Boundary resource parameters and documentation are often available via internal or external repositories such as GitHub (BigCommerce, 2022c; Github, 2022a,b,c). Product release notes often contain information on boundary resources for developers (Salesforce, 2022e). *"If you are active in this ecosystem, then you get information on all channels and, additionally, if you are active as a developer on that platform, then there is a special mailing [list]"* (#4). Platform providers can also offer certification programs for partners and developers (WooCommerce, 2022a) and even academies for training developers in the use of boundary resources, including online training courses and blended learning (OXID, 2022b). Smaller platforms also favor direct contact for the communication of changes on boundary resources. *"We also go to our developers in some cases. If we have only a few developers who are now using an old resource, just as we want to deprecate it, then before I tell them it was in the release notes, I'd rather go and talk to them"* (#4).

Technical boundary resources include development and application boundary resources (Dal Bianco et al., 2014) that are necessary for the development of platform extensions by third-party developers. The management of these boundary resources involves all stages, from requirements management and design to their deprecation (Deuter and Rizzo, 2016; Patni, 2017). In the following sections, we integrate the supplementary boundary resources used in the boundary resource management lifecycle stages.

#### 6.3.4.3 Plan and Design

The plan and design stage begins with the determination of requirements for boundary resources (Hallerstede et al., 2011; Vester, 2018). During our analysis, we identified business requirements and developer demands as major sources of requirements. While the former reflects the innovation platform's strategic focus as part of the governance activity's business alignment, the latter requires a close communication with the developers of platform extensions (Bondel et al., 2021; Hallerstede et al., 2011). Developers can make demands for new boundary resources via workshops, hackathons, feature forums, and direct contact with business process owners (BigCommerce, 2022f; Bondel et al., 2021). The governance decision needs to be reflected in order to balance openness and control in boundary resource design (#5). Demand management deals with the collection, analysis, and processing of requirements demanded by developers. These requirements can be collected via the previous iteration of the lifecycle for an existing boundary resource and proactively collected for a new boundary resource (Mohagheghzadeh and Lindman, 2022). *"During the project work, we start with a discovery research phase that allows us to map out the business requirements, the user needs, and from that, we create platform features and boundary resources"* (#2). The requirements definition involves the formulation of functional and non-functional requirements. While the former describes the behavior of the boundary resource as demarcation objects between the focal platform and third-party extensions, the latter defines the intended performance with regard to service-level agreements (Akana, 2020).

After the requirements have been identified and documented (Bondel et al., 2021), they can be translated into feature specifications, business cases can be derived for them, and dedicated boundary resources fulfilling these requirements can be planned (Deuter and Rizzo, 2016; Hallerstede et al., 2011). The boundary resource specification is created and harmonized with the endpoint structure and behavior (Bondel et al., 2021). For technical boundary resources traversing another iteration of the lifecycle, it is necessary to determine which parts of the boundary resource must be updated (#3). Technical boundary resources can only be made available for a specific group of developers, for instance, for gold and silver developers. It is important to consider developers' flexibility and offer technical boundary resources for a number of programming languages to allow developers to flexibly use their favorite language when implementing extensions (Salesforce, 2022b). Since boundary resources and the associated terms of use are inseparable, this aspect is a relevant supplementary boundary resource (#2, #9). The focus of supplementary boundary resources, such as guidelines and regulations, is on accumulating and organizing necessary information.

#### 6.3.4.4 Development and Test

The previously designed technical boundary resources are implemented and tested (Bondel et al., 2021), while supplementary boundary resources are finalized in this stage (Chappell, 2008). The process of developing and testing can take an iterative form (#3). Since contemporary IS are complex, software artifact development including boundary resources is conducted in teams requiring shared repositories and code orchestration (#4). *"For development, each boundary resource has its own repository"* (#2). The development of extensions usually takes place in preconfigured platform environments and integrates necessary boundary resources (Salesforce, 2022a). Before their release to production environments, boundary resources can be flagged as beta. *"We often release boundary resources as a beta first. And then it's kind of a beta component for three months, and during beta we're allowed to deprecate and change things"* (#4). The development of technical boundary resources can be steered by a roadmap overview for products (Shopware, 2022c; WooCommerce, 2022b) and by a detailed roadmap explorer for new boundary resources (Salesforce, 2022d; SAP, 2022e) that can be communicated to ecosystem participants. *"We provide a roadmap on our documentation site, where we share information about features we are implementing or planning to implement"* (#2).

Testing of technical boundary resources involves the investigation of functionality, performance, and behavior (Bondel et al., 2021). Automated and standardized tests of boundary resources ensure sophisticated testing for each boundary resource and each version (#1). This testing ensures a higher quality of extensions and stability of technical boundary resources (Priyadarsini et al., 2020, #3, #9). *"Our marketplace template for e-commerce includes a suit for automated tests. Both units and end-to-end tests involving our boundary resources"* (#2). Innovation platforms usually formulate quality criteria for boundary resource implementation and testing (Adobe, 2022c) and provide a rich set of methods and tools for testing extensions that are also used for testing boundary resources, such as environments for testing API calls and responses (Adobe, 2022f; SAP, 2022c). Testing involves unit tests, integration tests, and dedicated tests for the usage of technical boundary resources (#7). Staging in sandbox environments to test extensions and the correct use of API calls is a common approach for innovation platforms (BigCommerce, 2022e). *"We have a dedicated, free-of-charge staging environment, where third-parties can develop and test their platform extensions manually and automatically"* (#2). Supplementary boundary resources are checked for their harmonization with their technical counterparts and overall communication strategy (#8).

#### 6.3.4.5 Release

In the release stage, previously developed and tested technical boundary resources are deployed into production environments and can be accessed by developers (Chappell, 2008). This release is accompanied by the publishing of supplementary boundary resources, such as release plans, documents, and changelogs, enabling and supporting the use of technical components. In addition to these written documents, multi-media tutorials and e-learning material can be provided (OXID, 2022a; SAP, 2022d; WooCommerce, 2022f). The platform owner needs to ensure that external developers accept and use the technical boundary resources (Hallerstede et al., 2011). To coordinate release activities and harmonize boundary resource types, our interviewees consider a deployment strategy mandatory. *"You definitively need to implement a deployment for internal and external coordination"* (#5). The release of technical boundary resources is often part of continuous deployment pipelines (Bondel et al., 2021; Priyadarsini et al., 2020). For the release of boundary resources and features, innovation platforms can implement a transportation route from development and quality assurance environments to production systems, that needs to be followed, with dedicated quality gates between these environments (#6, #8). The platform provider can also decide to release more complex boundary resources in several iterations in subsequent releases (Tang et al., 2011). In addition to the release of a new stable version, innovation platform owners may also grant access to pre-release versions to provide developers with the ability to test the developed extensions' compatibility with forthcoming versions of the innovation platform (WooCommerce, 2022d).

#### 6.3.4.6 Maintenance and Optimization

In the maintenance and optimization stage, the boundary resources are operated in production environments (Bondel et al., 2021; Tang et al., 2011). This stage includes tasks, such as monitoring the status of technical boundary resources and collecting, categorizing, and evaluating developers' feedback (Chung and Kim, 2010). The task is therefore the structured analysis of all feedback (Deuter and Rizzo, 2016). *"Metrics collected automatically during platform use serve as evidence of the design and implementation quality of existing and future boundary resources"* (#7). Our multi-case analysis revealed that, for maintenance tasks, status pages for monitoring the APIs' and developer services' operability (BigCommerce, 2022d; Shopify, 2022f) and sophisticated cockpits for monitoring the status of single APIs that are even customizable for the APIs that developers use (OXID, 2022c; Salesforce, 2022c) help minimize the effort for bug fixing and optimizations (#3, #7). *"Providing a status page for the platform and the boundary resources used is an important trust-building and support-reducing measure"* (#7). Platform providers also provide tools and functions for fixing bugs in using boundary resources (Shopware, 2022a). *"APIs can report their status, utilization, and utilizing ecosystem participants"* (#4). Boundary resource downtime and bugs can be communicated in dedicated forums (WordPress, 2022). The behavior of boundary resources can be reproduced by implementing logging mechanisms (SAP, 2022b). *"We do monitor, for example, the status of our systems and APIs as well. If we see some issues with a technical boundary resource, we have a process in place that we go investigate and make sure we fix it before it becomes a system critical issue impacting operation"* (#1).

While maintenance activities aim at correcting issues of boundary resources in use, optimizations try to avoid problems and improve the performance of boundary resources. Depending on the requested changes, feedback can be used to incrementally optimize the boundary resource, or the feedback can be implemented in a future lifecycle in a new boundary resource version (Hallerstede et al., 2011). Incremental optimizations are deployed into the current version of the boundary resource (Chung and Kim, 2010). Optimizations can affect the innovation platform's openness and control characteristics positively, but also negatively. Therefore, it is part of change management to analyze an optimization's potential benefits and risk for the overall ecosystem. Platform providers introduced a priority concept for boundary resource optimizations, indicating the necessity of updating and the vulnerability to external fraud (Adobe, 2022b). Since optimizations may involve numerous application services, a team of internal platform developers needs to handle optimizations (#1).

#### 6.3.4.7 Deprecation

In the deprecation stage, the unnecessary, unsafe, and outdated versions of boundary resources are indicated for deprecation and retired (Bondel et al., 2021; Shopify, 2022e). Deprecation practices vary depending on the type of boundary resource. Deprecating technical boundary resources require a migration plan to upgrade to a new or updated resource (Shopware, 2022b; Tang et al., 2011). Deprecations are announced in developer forums (Shopify, 2022c) and dedicated boundary resource release notes (Shopify, 2022a). *"We mark deprecated features in the documentation along with links to documentation topics that describe how to achieve the functionality after the deprecation"* (#2). The deprecation stage includes the planning of deprecation schedules, the implementation of transition plans, and the management of adequate communication of the deprecated boundary resource to developers (Shopify, 2022e). *"For new [boundary resource] releases, there are always release notes, and deprecations are announced via various channels"* (#3). Boundary resources of older innovation platform versions are likely to be deprecated after a transition period (Shopify, 2022c). *"Whenever new boundary resources are implemented, what we usually do is start a phase-over process where we start deprecating the old module and move it into the new module in a phased approach, shifting the utilization to the new resource"* (#1). If a boundary resource is indicated as deprecated, it can be unpublished and removed in a following release (Chung and Kim, 2010; Patni, 2017; Shopify, 2022e; Shopware, 2022b). Deprecating technical boundary resources is a more complicated endeavor for smaller platforms, who tend to keep old but still used boundary resources. *"Unfortunately, we also have to keep many old boundary resources because they are still in use. We can't migrate them"* (#1). Keeping old building blocks of a boundary resource can also be used to ensure backward compatibility (#1). If backward compatibility is propagated, boundary resources can only be deprecated if they are no longer used at all (#4).

### 6.3.5 Discussion

We developed a theoretically grounded and empirically validated lifecycle model for the management of boundary resources in innovation ecosystems. Innovation ecosystems with sophisticated boundary resources and a critical mass of developers can cope with the contemporary challenges of traditional IS development. Boundary resources can be differentiated between technical boundary resources and supplementary boundary resources (Dal Bianco et al., 2014; Ghazawneh and Henfridsson, 2013). We developed the reference procedure model (Becker and Delfmann, 2007) in a design science research approach consisting of two design cycles. The first design cycle utilized a review of the existing ALM literature as input for deriving suitable lifecycle stages of boundary resource management (Vom Brocke, Simons, et al., 2015). The second design cycle investigated eight case studies of innovation platforms in e-commerce to derive boundary resource management activities and supplementary boundary resources for each lifecycle stage (Yin, 2018). We evaluated these results with the analysis of interviews with nine domain experts employed at innovation platforms in e-commerce (Misoch, 2015). We developed two intermediary lifecycle models (i.e., theory-based lifecycle, enhanced lifecycle) to come up with the final boundary resource management lifecycle consisting of five lifecycle phases (Fig. 6.10). With our reference procedure model, we not only depict the lifecycle of boundary resources in innovation ecosystems but also prescribe (i.e., theory for design and action) how owners of innovation platforms should manage these boundary resources (Gregor, 2006). The "choice of lifecycle models for the design process influences what it is that is designed" (Clemmensen et al., 2008, p. 137). Mapping supplementary boundary resources to technical boundary resources' lifecycle stages gives platform owners guidance on which documents to create and publish (Dal Bianco et al., 2014; Ghazawneh and Henfridsson, 2010). In this regard, third-party developers can be addressed directly.

#### 6.3.5.1 Critical Appraisal

We chose e-commerce as the domain of interest as innovation ecosystems in e-commerce are very dynamic in nature, with participants (e.g., customers, developers) joining and departing (McIntyre and Srinivasan, 2017; Wulfert, Woroch, Strobel, Seufert, et al., 2022). This requires a diverse set of application services to support the variety of business models, such as electronic shops, digital marketplaces and electronic auctions (Alt, 2020; Timmers, 1998; Yoo and Jang, 2019), and has a continuous stream of new extensions and updates (Ghazawneh and Henfridsson, 2013; Medjaoui et al., 2021). Since we abstracted from the peculiarities of eight cases and nine additional interviews that include "a range of possible real situations" (Kosiol, 1964, p. 758), it is possible to consider our boundary resource management lifecycle as a reference procedure model (Delfmann, 2006; Fettke and Loos, 2002; Schütte, 1998; Vom Brocke, 2015). The boundary resource management lifecycle also contains variants for a number of activities and decisions that can be selected when instantiating and implementing the reference model for a particular company (Schütte, 1998; Vom Brocke, 2015). For instance, we reported on a versioning and a backward compatible approach for establishing an updated version of an already existing technical boundary resource (Adobe, 2022a, #4, #7). Moreover, the reference model is designed in a flexible manner to be changed and extended for the customization implementation at a particular company.

Besides the mapping of technical boundary resources and supplementary social boundary resources emphasized in our boundary resource management cycle, we integrated boundary resource governance and continuous communication as fundamental activities guiding the development and maintenance of boundary resources (Eaton et al., 2015; Ghazawneh and Henfridsson, 2013; Vester, 2018; Wulfert, Woroch, Strobel, Seufert, et al., 2022). To highlight the necessity of governance and communication for the management of boundary resources, we applied a layered approach for the lifecycle model. Boundary resource governance forms the inner layer in which fundamental decisions regarding the boundary resources of an innovation platform, such as the boundary lifecycle duration, versioning standards, responsibilities, and the portfolio of boundary resources, are made (Boudreau, 2010; Ghazawneh and Henfridsson, 2013). The continuous communication layer involves activities relevant for the communication with ecosystems participants (e.g., third-party developers) (Vester, 2018). The continuous communication is responsible for the dissemination of the boundary resources across the innovation ecosystem. Moreover, the interviewees emphasized the importance of adequate communication between platform owners and third-party developers regarding the management of boundary resources (#1, #2, #4, #9). Therefore, we integrated a dedicated layer for continuous communication along the lifecycle of a boundary resource (Bondel et al., 2021; Kääriäinen and Välimäki, 2008; Priyadarsini et al., 2020). We amplified the communication aspect with the mapping of dedicated supplementary boundary resources for the support of the communication of technical boundary resources (Dal Bianco et al., 2014). The outer layers are formed by technical and supplementary boundary resources that resemble the demarcation points for the information and involvement of ecosystem participants (Dal Bianco et al., 2014; Wulfert, Woroch, Strobel, Seufert, et al., 2022).

Our multi-case and interview analysis revealed a diverging behavior of platform owners about updating technical boundary resources. While the first group of platforms uses strict versioning, the second group propagates a high level of backward compatibility, avoiding new versions. The former approach makes changes and new developments transparent but potentially requires effort in updating extensions to new versions and adapting the use of tools for developers. The latter approach allows for a parallel usage of boundary resources for different developments and reduces possible adaption efforts for developers but may increase potential boundary resource overheads (#2, #4, #7). A possible explanation for the two different approaches can be found in the varying amounts of power the platform owner can wield on ecosystem participants (Gawer, 2021b; Kretschmer, et al., 2022). Introducing new versions of boundary resources while simultaneously deprecating the old version force developers to use the new version after a defined transition period (Star, 2010). Platforms with lower amounts of power in their innovation ecosystems would need to provide backward compatibility. The amount of power and control over the ecosystem also prescribes whether the platform owner can define boundary resources on its own or tune distribution by ecosystem participants (Eaton et al., 2015). We account for the ecosystem participants' integration in the boundary resource management lifecycle with the continuous communication layer, with possibilities to post requirements, and by enabling a dialog of business process owners or internal platform developers and developers.

For the development of the boundary resource management lifecycle, we differentiated between technical and supplementary boundary resources (Dal Bianco et al., 2014). These two types include the possible boundary resources as described by Karhu, Gustafsson, et al. (2018), Star (2010), and Wulfert, Woroch, Strobel, Seufert, et al. (2022). We argue that supplementary boundary resources are often bound to a lifecycle stage and a specific version of a technical boundary resource. We, therefore, provided a mapping of major supplementary boundary resources per lifecycle stage of the technical boundary resource. If the technical boundary resource is created, updated, and deprecated, the explaining supplementary boundary resource also needs to be adapted.

Although our boundary resource management lifecycle seems to have an iterative form, single stages can also be organized in an agile manner (Nerur and Balijepally, 2007; Shore et al., 2022). An example of an agile realization can be the development and test stage, with incremental development of the software artifact and increment tests (#2). These increments can be arranged for the completed boundary resource (Priyadarsini et al., 2020).

#### 6.3.5.2 Scientific and Managerial Contribution

We contribute to the current research on ALM by providing a reference procedure model for managing boundary resources. The integrated lifecycle management approach for a holistic management of boundary resources accounts for the strategic role of boundary resources in innovation ecosystems (Bondel et al., 2021; Vester, 2018; Yoo, Henfridsson, et al., 2010), addresses third-party developers, and fosters an ecosystem's generativity. We provide a novel tool for analyzing and prescribing collaboration in innovation ecosystems. We provide further insights on the role of boundary resource governance and continuous communication as important activities in managing boundary resources. Moreover, our boundary resource management lifecycle emphasizes the focal innovation platform's role for implementing boundary resources while controlling and governing value co-creation. The reference procedure model's development process also highlights the role of innovation platforms in e-commerce. Boundary resources are especially important in electronic markets, enabling digital collaboration and value co-creation. We decoupled technical and supplementary social boundary resources in line with existing research (Dal Bianco et al., 2014; Engert et al., 2022; Ghazawneh and Henfridsson, 2013; Hein, Schreieck, Riasanow, et al., 2020; Wulfert, Woroch, Strobel, Seufert, et al., 2022), but emphasized their intertwined nature for ecosystem participants over the whole boundary resource management lifecycle. Technical boundary resources require a supplementary counterpart for efficient ecosystem-wide application. Within our lifecycle model, supplementary social boundary resources are mapped to technical boundary resources per lifecycle stage.

While platform providers can customize this reference model for their individual peculiarities, developers can reproduce the lifecycle of technical boundary resources, anticipate the development of future boundary resources, and benefit from associated supplementary boundary resources. Although the potential impact of continuous software engineering for the software industry is recognized in research and practice, especially for cloud environments (Dittrich et al., 2018; Wilson, 2017), the role of technical boundary resources remains unchanged as stable demarcation points between innovation platforms and third-party developers (Baldwin and Woodard, 2009). However, the accelerated evolution of innovation platforms as a core of innovation ecosystems often causes adjustments for previously stable boundary resources. Hence, the boundary resource lifecycle can be used as a toolkit for boundary resource management when it comes to implementing and communicating changes in interfaces in proper documentation. The reference procedure model for managing boundary resources can therefore be used to cope with evolving innovation and cloud environments in which extensions are "released in small increments and constantly modified" (Norbjerg, 2018, p. 72). The reference procedure model can be applied as an important tool for the development of ecosystems surrounding innovation platforms in e-commerce.

#### 6.3.5.3 Limitations

This study entails some limitations, which provide opportunities for future research. Since the literature on boundary resource management is sparse, we started our design science approach with an investigation of the ALM literature to identify relevant lifecycle stages. Although technical boundary resources are software artifacts (Dal Bianco et al., 2014), it might be disputable whether all stages are relevant for the management of boundary resources. We therefore conducted an additional design cycle to refine the lifecycle for boundary resource management with insights from practice derived in a multi-case analysis. However, innovation platforms did not provide detailed information on pre-release lifecycle stages that were considered during our analysis as internal. For information on these internal phases, we especially benefitted from the evaluation of the enhanced lifecycle during our interview analysis. For our interview analysis, we were able to obtain interviews with nine domain experts employed at innovation platforms that provide application services for transaction platforms in e-commerce and online retailers (Aulkemeier, Schramm, et al., 2016; Wulfert and Karger, 2022b). Unfortunately, our interviewees demanded that the results of the interview analysis be published anonymously without any reference to the innovation platform. Therefore, we could not draw any direct relations or conclusions from an evaluation of the comparison of case studies and interviews. Instead, we used the interview analysis to evaluate the results of our multi-case analysis.

Our design science research approach involves qualitative research methods for data collection and data analysis. As the results of qualitative research are often criticized as subjective and biased by the investigator (e.g., Beuving and De Vries, 2015; Guba and Lincoln, 1982; Miles et al., 2014; Straub and Gefen, 2004), research rigor with regards to reliability and validity needs to be ensured for qualitative research endeavors (Golafshani, 2003; Miles et al., 2014; Recker, 2013).[12] To ensure the construct validity, internal validity, external validity, and reliability of our research project, we applied the tactics put forth by Guba (1981) and Ying et al. (2018) for an increased research rigor. We ensured construct validity by establishing a chain of evidence triangulating multiple data sources in the multi-case analysis (Yin, 2018) and our interviewees verified interview transcript to avoid misunderstandings and misinterpretations (Shenton, 2004). To increase the internal validity of our research project, we provided transparency on our research procedure, allowed for data triangulation across the applied methods, and depicted linkages among the methods applied. Moreover, we tried to match patterns (i.e., lifecycle phases) derived from existing literature with our case observations (Yin, 2018). Following Hänninen et al. (2018), we used a variety of secondary data sources for our multi-case analysis. We acquired the interviewees based upon a random sampling approach among the population of existing innovation platforms and approached a total of 46 platforms with only nine platform offering volunteers for interviews. We elaborated on the purpose of the interviews, provided additional information a priori to the interviewees, and offered the possibility for refusing the interview (Guba, 1981; Shenton, 2004). With regards to the number of interviewees we applied sufficiency and data saturation as indicators (Guest et al., 2006; Hagaman and Wutich, 2017; Seidman, 2019). While sufficiency was achieved by contacting a variety of innovation platforms in terms of size, revenue, and provided services, and by reaching interviewees with different positions (Table 6.6), data saturation was already achieved after the seventh interview with two additional interviews in the pipeline. Data saturation was measured by the degree of new information added and changes to the coding scheme caused by the consecutive interviews (Guest et al., 2006). The degree of additional information was already low when analyzing the seventh interview and high saturation of information on boundary resources in general and boundary resource management stages in particular was achieved, despite the platforms' diverging peculiarities. External validity was ensured by abstracting from the peculiarities of analyzed case companies and interviewee information. Hence, our boundary resource management lifecycle approach might be implemented for a sophisticated boundary management for technical boundary resources and supplementary social boundary resources in other domains of interest. Although we conducted our multi-case analysis with a replication logic (Yin, 2018), the external validity of our study could be further increased

[12] Several scholars have proposed guidelines to address the subjective and interpretive nature of qualitative research and increase research rigor (e.g., Beuving and De Vries, 2015; Lincoln and Guba, 1985; Merrick, 1999; Myers and Newman, 2007; Shanks, 2002; Walsham, 1995; Walsham 2006). These researchers refer to confirmability, dependability, credibility, and transferability as quality criteria (e.g., Miles et al., 2014; Shenton, 2004).

by analyzing innovation platforms and consulting interviewees in other domains. Reliability was enriched by providing supplementary information on our research endeavor (e.g., concept matrix, intermediary lifecycles). By providing transparency on the intermediary lifecycle models, we maintained a consistent chain of evidence throughout our design science research project (Yin, 2018). Among the team of authors, we periodically reflected on the research design, execution, and results and prepared research diaries for the design science research project (Guba, 1981; Yin, 2018).

#### 6.3.5.4 Research Agenda

Drawing on existing calls for research on boundary resources and platform governance (e.g., de Reuver et al., 2017; Hein, Schreieck, Riasanow, et al., 2020; Schüler and Petrik, 2021; Yoo, Henfridsson, et al., 2010) and our research on boundary resource management, we distilled three areas for future research (i.e., innovation ecosystems, boundary resource management, types of boundary resources) and suggested questions guiding future research, namely (Table 6.8).

The first research area is concerned with future research on boundary resources on (innovation) ecosystem level. Despite first attempts to conceptualize the concept of generativity in platform contexts (e.g., Blaschke and Brosius, 2018; Fürstenau, Baiyere, et al., 2023; Jarvenpaa and Standaert, 2018) and to analyze its effect on third-party developer adoption (Miremadi et al., 2023), the impact of boundary resource provision on ecosystem generativity remains underrepresented (Hein, Schreieck, Riasanow, et al., 2020). In this regard, the role of boundary resources in the dichotomy between platform openness and platform control is worthwhile investigating (e.g., Benlian et al., 2015; Spaeth and Niederhöfer, 2022; Yoo, Henfridsson, et al., 2010). Another possible avenue for future research would be an analysis of the influence of boundary resources on different types of ecosystem participants (Eisenmann, Parker, et al., 2009). The influence of proper boundary resource implementation on third-party developers and their dyadic relationship with platform owners would also be worthwhile investigating. In particular, providing transparency about boundary resource development and provisioning might mitigate distributed tuning of boundary resources by other ecosystem participants (Eaton et al., 2015).

The second research area deals with boundary resource management-analyzed in this research paper-in particular. Adding to the boundary resource management lifecycle, future research might derive design guidelines for implementing boundary resources for innovation platform owners and abstract from domain peculiarities of existing research (e.g., Engert et al., 2022; Wulfert, Woroch, Strobel, Seufert, et al., 2022). Although we chose e-commerce as the application domain for our research on the management of boundary resources to achieve a high fitness for one domain, the lifecycle might be applicable in other domains (e.g., agriculture). Hence, a possible avenue for future research would be to test the boundary resource management lifecycle in other domains and increase its projectability while retaining a high fitness (Vom Brocke, Winter, et al., 2020). Hence, future research could discuss the management of boundary resources with additional experts from other domains and third-party developers. Future research could also evaluate the boundary resource management lifecycle's benefits and performance by applying it in selected companies and customizing it to the peculiarities of these cases (Hevner, March, et al., 2004; Yin, 2018). The performance could be evaluated using appropriate indicators, such as the number of third-party developers, the number of extensions, developer satisfaction, and interface downtime. Another possible avenue for future research would be a more detailed analysis of the core phases boundary resource governance and continuous communication. Analogous to ALM, which is typically supported in practice using dedicated tools depending on the lifecycle stage or integrated solutions (Aston, 2022; Jwo et al., 2013), boundary resource management needs to be supported with IS. Future research might investigate existing ALM solutions for the coverage of boundary resources and implement potential extensions for these tools. An important avenue for future research would be the development of a tool supporting the different boundary resource types' entire lifecycle in an integrated manner.

The third research area identified is concerned with different types of boundary resources (e.g., Dal Bianco et al., 2014; Ghazawneh and Henfridsson, 2013). It could be worth investigating for which innovation platforms and technical boundary resources backward compatibility is required and for which new versions discarding old ties are sufficient. A possible explanation for this might by the platform owner's relative power within the surrounding ecosystem (Hein, Schreieck, Wiesche, et al., 2016; Hurni, Huber, and Dibbern, 2022; Kretschmer, et al., 2022). Moreover, future research could investigate the relationship between technical and supplementary boundary resources in more detail. Existing investigations distinguished these boundary resource types, but did not discuss their relationship (e.g., Dal Bianco et al., 2014; Ghazawneh and Henfridsson, 2013). Moreover, future research could strive for deriving archetypes of boundary resources to enable a more focused management and implementation. We summarize the research areas with potential research questions for future research endeavors in Table 6.8.

**Table 6.8** Future Research Agenda

| Field of action | Exemplary research questions |
|---|---|
| Innovation ecosystems | What is the impact of boundary resources on platform openness? How can boundary resources be managed to ensure platform control How do boundary resources influence ecosystem generativity? Which ecosystem participants can engage in an innovation platform's boundary resource management? How can third-party developers tune boundary resources? |
| Boundary resource management | What are specific design guidelines for implementing boundary resources in innovation platforms? What are the benefits of implementing boundary resource management lifecycle? What principles do exist for boundary resource governance and continuous communication? How is boundary resource management implemented across domains? Which ALM tools support boundary resource management? How can boundary resource management be supported with dedicated software tools? |
| Types of boundary resources | In which scenarios is backward compatibility for technical boundary resources required? How can a proper mapping between technical boundary resources and supplementary boundary resources be achieved? Which boundary resource archetypes do exist for innovation platforms? |

### 6.3.6 Conclusion

We developed a reference procedure model for the management of the whole lifecycle of boundary resources in innovation ecosystems from the perspective of focal platform owners. The boundary resource management lifecycle is structured in four layers: boundary resource governance, continuous communication, technical boundary resources, and associated supplementary boundary resources. With this research, we add value to the literature on ALM with a special focus on technical boundary resources and supplementary social boundary resources. The ALM perspective was especially applied for technical boundary resources and extended for this particular type of software artifacts. Supplementary boundary resources are mapped according to these lifecycle stages to support boundary resources' ecosystem-wide application. The lifecycle for the management of technical boundary resources consists of the following stages: plan and design, development and test, release, maintenance and optimization, and deprecation. Major supplementary boundary resources include written documentation, repositories, changelogs, tutorials, status reports, and transition plans. The boundary resource management lifecycle supports owners of innovation platforms in organizing boundary resources and orchestrating third-party developers. Third-party developers can anticipate potential changes in boundaries resources made by the platform owner applying our model to boundary resources used. We contribute to the current body of knowledge on ALM by providing a sophisticated reference procedure model for the management of technical and supplementary boundary resources. In the same vein, we enhance existing research with a mapping of supplementary boundary resources to technical boundary resources' lifecycle stages while highlighting their intertwined nature.

# 7 Platforms as Orchestrators in E-Commerce Ecosystems

## 7.1 Paper 9: How Giants Rule the World: A Qualitative Comparative Analysis of Dynamic Capabilities for E-Commerce Ecosystem Orchestration

**Abstract**
*E-commerce is the digital backbone of the global goods economy and the most important economic sector with a turnover of $ 7.4 trillion in 2024.*[1] *Predominant in the sector are platform-centered ecosystems that generate value by orchestrating participants' value co-creation and benefiting from resulting network effects. So far, research has lacked any conceptualization of capabilities for orchestration in ecosystems. We combined a systematic literature review with a multi-case study of five e-commerce giants in an inductive approach for the derivation of five dynamic orchestration capabilities and as foundation for a Qualitative Comparative Analysis (QCA) fueled by a case survey of 100 e-commerce ecosystems to develop six successful orchestration types for full assortment and sub-assortment-oriented e-commerce-ecosystems. Based on these novel types, this paper addresses the interplay between centrality and density as determinants for the power position of focal platform owners, thus expanding the knowledge of orchestration for e-commerce ecosystems.*

[1] This version of the article is currently under revision at the Journal of Strategic Information Systems.

**Supplementary Information** The online version contains supplementary material available at https://doi.org/10.1007/978-3-658-45198-1_7.


T. Wulfert, *Selected Perspectives on Platforms in E-Commerce Ecosystems*, https://doi.org/10.1007/978-3-658-45198-1_7

| Fact sheet of publication 9 | |
|---|---|
| Title | How Giants Rule the World: A Qualitative Comparative Analysis of Dynamic Capabilities for E-Commerce Ecosystem Orchestration |
| Authors | Tobias Wulfert |
| | Robert Woroch |
| | Gero Strobel |
| Publication Type | Journal Article |
| Publication Outlet | The Journal of Strategic Information Systems |
| VHB-Jourqual3-Ranking | A |
| Status | Under Review—2nd Round |
| Research Question | What are the necessary capabilities for ecosystem orchestration? |
| Research Approach | Qualitative Comparative Analysis (Park, Fiss, et al., 2020), Literature Review (Vom Brocke, Simons, et al., 2015), Case Study (Yin, 2018), and Case Survey (Larsson, 1993) |
| Full Citation | T. Wulfert, R. Woroch, and G. Strobel (2024). "How Giants Rule the World: A Qualitative Comparative Analysis of Dynamic Capabilities for E-Commerce Ecosystem Orchestration." In: *Journal of Strategic Information Systems* Forthcoming |

### 7.1.1 Introduction

Digital business ecosystems are of increasing interest in research (Adner, 2017; Baumann, 2022a; Feller et al., 2012; Jacobides et al., 2018; Tsujimoto et al., 2018) and have already conquered numerous economic sectors such as Electronic Commerce (e-commerce) (Cennamo, 2021; Gawer, 2021b; White et al., 2007). Viewed from a structural perspective (Adner, 2017), e-commerce ecosystems multilaterally co-create product and service offerings for the (final) customers (Hein, Weking et al., 2019; Tsujimoto et al., 2018; Wulfert, Woroch, Strobel, Seufert, et al., 2022). E-commerce ecosystems typically evolve around a focal transaction platform that acts as virtual locus for value creation and exchange and is owned by a dominant platform owner (Gawer, 2021b; Wulfert, Woroch, Strobel, Seufert, et al., 2022). The most successful companies in e-commerce (e.g., Amazon, Walmart, Alibaba) operate transaction platforms[2] (i.e., digital marketplace) and orchestrate surrounding ecosystem participants benefitting from the economic effects of two-sided markets (e.g., matching and subsidizing selected participants) (Merton, 2021; Parker, Van

[2] We used terms digital marketplace and transaction platform synonymously in this article. Transaction platforms are also described as digital marketplaces in the literature on e-commerce (Hagiu and Wright, 2015a,b; Koch, 2010; Täuscher and Laudien, 2018; Wulfert, Seufert, et al., 2021; Yoo and Jang, 2019).

Alstyne, and Choundary, 2016; Wulfert, Seufert, et al., 2021). These network effects led to winner-takes-all-tendencies in e-commerce (Hagiu, 2007; Karhu and Ritala, 2021; Ruutu et al., 2017) and amplified the evolution of "gigantic" platform owners (Parker, Van Alstyne, and Choundary, 2016; Spaeth and Niederhöfer, 2022).

Due to the transformative and disruptive impact of transaction platforms and surrounding digital business ecosystems on realizing the value proposition (e.g., Lanamäki, et al., 2020; Rahrovani, 2020), value is no longer created in isolation but co-created by multiple ecosystem participants (de Reuver et al., 2017; Iansiti and Levien, 2004a). Therefore, competition has shifted from the level of single companies to the ecosystem level in recent years (Amit and Zott, 2021; Cennamo, 2021; Rietveld and Schilling, 2021; Zhao, von Delft et al., 2020). As an effective orchestration of the value co-creation within ecosystems results in increased outputs compared to disintermediated scenarios (Autio, 2022; Hiller et al., 2022), ecosystem orchestration arises as foundation of a sustainable competitive advantage. Platform owners require specific dynamic capabilities in dynamic environments, such as e-commerce, for a proper ecosystem orchestration (Kazan, Tan, et al., 2018; Teece et al., 1997).

Besides Jacobides et al.'s (2018) proposition for research on valuable resources and capabilities for platform owners in dynamic ecosystem contexts (e.g., e-commerce), Paquin's and Howard-Grenville's (2013) suggestion for an investigation of ideal levels of ecosystem orchestration, and Rietveld's and Schilling's (2021) demand for future research on orchestration configurations, research on dynamic capabilities for successful ecosystem orchestration in dynamic environments remains scarce. So far, research on dynamic capabilities focused on innovation ecosystems (e.g., Foss et al., 2023; Linde et al., 2021) and only a few studies exist that examined capabilities necessary in ecosystems (e.g., Ganesh et al., 2004; Isckia et al., 2020; Koch, 2010; Sirmon et al., 2011). Based on our observation of winner-takes-all tendencies in e-commerce and absent research on dynamic capabilities necessary for ecosystem orchestration, we pose the following research question:

**RQ:** What are the necessary capabilities for ecosystem orchestration?

We developed six orchestration types with five dynamic capabilities as part of a three-stage mixed-method approach to answer this research question (Park, Fiss, et al., 2020). Our starting point was a systematic literature analysis (Webster and Watson, 2002) in combination with a cross-case-based multi-case study (Yin, 2018) of the five largest e-commerce ecosystems. The dynamic capabilities (Teece, 2014) derived from this served as the coding foundation (Corbin and Strauss, 1990) for a QCA (Ragin, 2008) conducted on the grounds of a case survey of 100 e-commerce marketplaces divided by their type of assortment (Larsson, 1993). The outcome of the crisp-set Qualitative Comparative Analysis (csQCA) were 12 capability configurations from which we were able to inductively derive six orchestration types by drawing on the raw case data and prevailing e-commerce ecosystem literature based on the ecology of configurations analysis (Park, Fiss, et al., 2020) and guided by network theory (Rowley, 1997).

Our contribution to Information Systems (IS) research is twofold. First, we derived a set of dynamic capabilities for ecosystem orchestration that owners of focal platforms should possess for successfully establish and develop e-commerce ecosystems. The five dynamic capabilities are supported by 12 second-order themes and 44 first-order constructs that could guide platform owners' ecosystem orchestration activities. Second, we suggested six types of dynamic capability configurations that can be taken on by platform owners in ecosystems that are characterized by different centrality and density situations. Furthermore, we show based on our qualitative comparative analysis of 100 e-commerce platforms that a moderate degree of orchestration in e-commerce ecosystems is beneficial for the owner of the focal platform in most cases. In situations with a low centrality and density, a low degree of orchestration might be in favor of the platform owner.

## 7.1.2 Theoretical Background

### 7.1.2.1 Orchestrating E-Commerce Ecosystems

E-commerce ecosystems can be conceptualized as digital business ecosystems in the context of e-commerce whose overall purpose is the exchange of products and services (Choi et al., 1997; Wulfert, Woroch, Strobel, Seufert, et al., 2022; Zwass, 1996). Ecosystem participants are bound by common goals (Adner, 2017; Wareham et al., 2014) and involved in "multilaterally co-created product and service offerings" (Burkhalter, 2020, p. 10). As the overarching value proposition of an e-commerce ecosystem is co-created, the value creation needs to be carefully orchestrated to "materialize" (Adner, 2017; Foss et al., 2023; Isckia et al., 2020) and requires a set of boundary resources in electronic environments (Engert et al., 2022; Wulfert, Woroch, Strobel, Seufert, et al., 2022; Zapadka et al., 2022). E-commerce ecosystems usually evolve around a focal transaction platform that acts as a virtual locus through which participants conduct retail transactions (Turban, Whiteside, et al., 2017; Wulfert and Karger, 2022b).

These transaction platforms connect organizations and individuals from different markets and social groups to form a dynamic ecosystem (Corallo, 2007; Gawer, 2021b). Owners of these focal platforms orchestrate the overall ecosystem

(Eisenmann, Parker, et al., 2009) as a "keystone firm" (Iansiti and Levien, 2004a), including the value co-creation through, among other things, setting and enforcing governances rules (e.g., access rights, fees, transaction structure) (Adner, 2017; Hein, Schreieck, Wiesche, et al., 2016). The orchestration of ecosystems can be understood as "the set of deliberate, purposeful actions undertaken by the hub firm as it seeks to create value […] and extract value […] from the network" (Dhanasai and Parkhe, 2006, p. 659). These actions involve the collection, processing, and exchange of information via digital means "to guarantee the most efficient and effective use and allocation of the [ecosystem's] resources" (Burkhalter, 2020, p. 10) which "requires a range of capabilities from an orchestrator to enable the creation and extraction of value from the network" (Addo, 2022, p. 362). To analyze these capabilities required for orchestrating e-commerce ecosystems, we take on a dynamic capability perspective (Sambarmurthy et al., 2003; Teece et al., 1997).

#### 7.1.2.2 Dynamic Capabilities for Ecosystem Orchestration

The concept of dynamic capabilities augments the static orientation of the resource-based view by highlighting the notion of evolving and altered resources and capabilities as a response to dynamic market environments to achieve a sustainable competitive advantage (Barreto, 2010; Peteraf et al., 2013; Steininger et al., 2022; Teece et al., 1997). Dynamic capabilities enable and guide the evolution of ordinary capabilities (Helfat, Finkelstein, et al., 2007; Teece, 2014) and can emerge over time from ordinary capabilities (Sambarmurthy et al., 2003). They are the source of competitive advantage to "integrate, build, and reconfigure internal and external competencies to address rapidly changing environments" (Teece et al., 1997, p. 516).[3] Research on digital business ecosystems has so far investigated dynamic capabilities for service innovation (Lütjen et al., 2019), for start-ups to become ecosystem leaders (Feng et al., 2019), and for an increased value capture by focal platforms (Helfat and Raubitschek, 2018). There is also initial research related to dynamic capabilities in the field of e-commerce with the research of dynamic capabilities for building and developing electronic businesses and the role of dynamic capabilities during the lifecycle of the central platform of an ecosystem (Daniel and Wilson, 2003; Teece, 2017). We build on this research and address a holistic ecosystem perspective with reference to Teece's (2007) tripartite dynamic capabilities scheme (i.e., sensing, seizing, reconfiguring/transforming) for the development

[3] Dynamic capabilities are defined by other sources as "firm's processes that use resources—specifically the processes to integrate, reconfigure, gain and release resources—to match and even create market change" (Eisenhardt and Martin, 2000, p. 1107).

of dynamic capabilities as mandatory stepstones (Koch, 2010; Sambarmurthy et al., 2003) for the successful orchestration of e-commerce ecosystems.

### 7.1.3 Methodology

The objective of this paper is to provide a theoretically grounded and empirically validated set of dynamic capabilities which are necessary for platform owners to orchestrate their ecosystems and to grow into economically successful e-commerce giants. To gain insights into the dynamic capabilities of e-commerce ecosystems, we follow an inductive three-step approach (Park, Fiss, et al., 2020) and combine a systematic literature review (Vom Brocke, Simons, et al., 2009) as well as a multi-case study (Yin, 2018) as theoretical grounding with a case survey (Larsson, 1993) of 100 marketplaces as a foundation for a QCA (Fig. 7.1) (Fiss, 2007; Ragin, 2008).

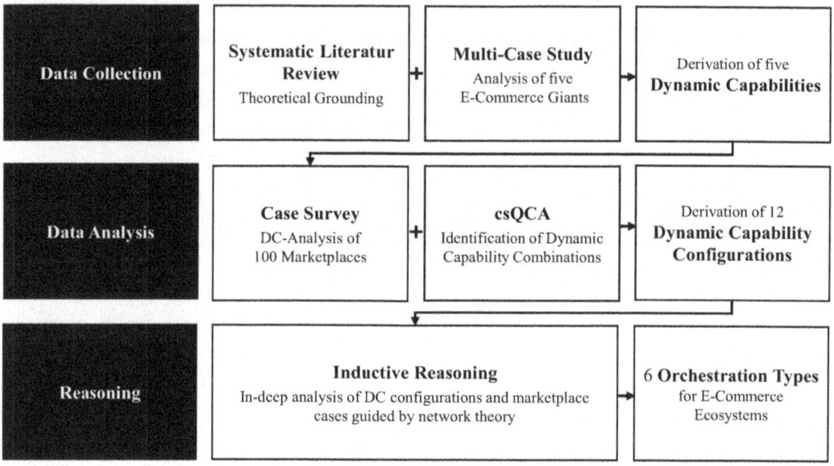


**Fig. 7.1** Overview of the Research Process

#### 7.1.3.1 Data Collection

In the first phase of the data collection, we aimed to identify the key participants within e-commerce ecosystems based on a systematic analysis of existing literature. We followed the methodological approach of Webster and Watson (2002) in combination with Vom Brocke, Simons, et al. (2009), to ensure adequate process quality (Cram et al., 2020). We performed the search, in six common databases with a total of 9,186 initial hits based the object of consideration (i.e., ecosystem) and the domain in which it is considered (e-commerce). To guarantee an appropriate level of quality, we apply further exclusion criteria (Bandara et al., 2015) and excluded all research that was not original (panels, commentaries, etc.), non-English-language, as well as purely technological (e.g., addressing technological aspects for development and realization of e-commerce offerings). Based on these criteria and the reviewing of the articles, we selected 86 relevant publications. These set was coded independently by three researchers deductively based on Eisenmann et al.'s (2009) four archetypal ecosystem participants in three waves to expand and adapt the original code set to ten roles in e-commerce ecosystems.

In the second step of the data collection, we extend the theoretical grounding with a multi-case study originating from the roles and identify the platform owner's core capabilities within e-commerce ecosystems. Multi-case studies are especially well suited for a holistic and in-depth investigation of contemporary and complex phenomenon's (Dubé and Paré, 2003; Yin, 2018). Since we wanted to produce prescriptive design knowledge (Chandra Kruse, Seidel, and vom Brocke, 2019) that enables the successful orchestration of e-commerce ecosystems (Autio, 2022), we collected knowledge from five[4] internationally operating, leading e-commerce ecosystems (i.e., Amazon, eBay, MercadoLibre, Tmall, Walmart) with a broad spatial and cultural differentiation (Appendix A.7 in the Electronic Supplementary Material). This sample size provides a convenient empirical grounding for theory building (Eisenhardt, 1989b), even if the cases differ in environmental aspects, they are sufficiently representative due to the similarity of the value proposition and their ecosystem structure.

Guided by general principles of grounded theory (Corbin and Strauss, 1990), the coding of the case study data[5] was performed iteratively, initiated by an open coding cycle in which all coders independently extracted capabilities of the e-commerce giants (e.g., "At Store No8, we incubate startups to help leapfrog where the customer and market will go, well into the future."—Incubate Innovative Ideas) as first-order concepts (Blair, 2015). Building on this, axial coding was conducted to add up the individual capabilities and convert them into second-order codes (e.g., Incubate Innovative Ideas). In a final step, these codes were then unified by selective coding to

[4] More in-depth details about the cases (e.g., company description, marketplace offerings, etc.) were attached in Appendix A.7 in the Electronic Supplementary Material.

[5] Case study data was collected from various publicly available data sources (corporate communication, websites, customer and developer forums, news, blogs, etc.).

form the dynamic capabilities[6] (e.g., Institutional Innovation) of the giants (Fig. 7.2). The central aspect hereby was the unique nature of the capabilities and the triad of sensing, seizing, and transformation (Teece, 2007).

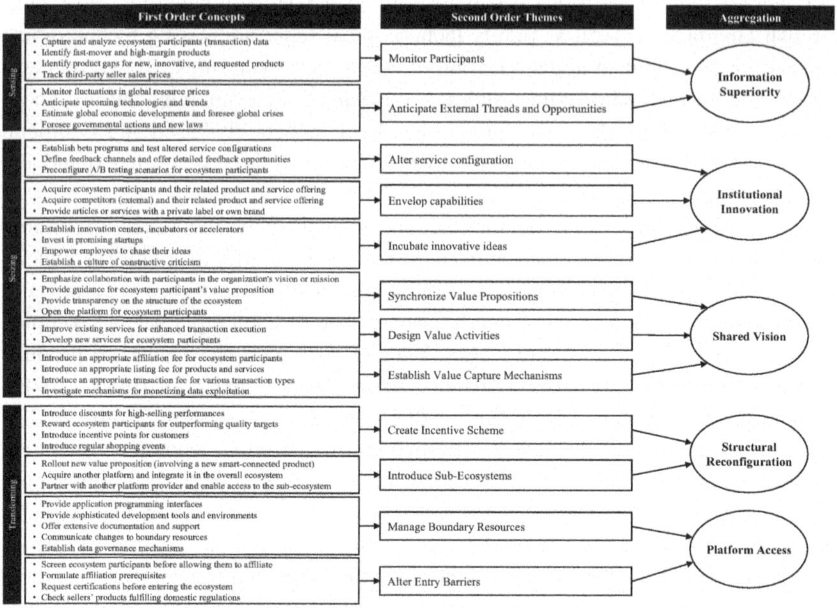


**Fig. 7.2** Dynamic Capabilities of E-Commerce-Ecosystem Giants

### 7.1.3.2 Data Analysis

We base our two-stage analysis on the derived dynamic capabilities and the underlying capability structure in the form of the Gioia diagram (Gioia et al., 2013). We combined a case survey of 100 e-commerce marketplaces with a csQCA to identify capability combinations for orchestrating e-commerce ecosystems. QCA, in particular, is suitable for this kind of analysis of digital phenomena (Bharadwaj et al., 2013; El Sawy, 2003) as it allows to observe multipath relationships between objects of interest by inferring multiple equivalent configurations based on the congruent

[6] A detailed description of the individual Dynamic Capabilities can be found in Appendix A.7 in the Electronic Supplementary Material.

combination of causal conditions (Fiss, 2007; Gresov and Drazin, 1997; Ragin, 2008). QCA is also a well-established research approach in IS research (Böttcher, Weking, et al., 2022; Dawson, Denford, et al., 2016; Koo, et al., 2019; Mattke et al., 2022; Park, Fiss, et al., 2020).

During the case survey, we coded the set membership for all dynamic capabilities and the outcome as part of the calibration. Based on the nature of the object under consideration and the associated binary coding, orchestrators either had the capability (full membership "1") or it was not present (full non-membership "0"). As anchor points for determining the membership score, we use the associated second-order concepts (Fiss, 2011; Park, El Sawy, et al., 2017; Ragin, 2008) of the dynamic capabilities and for the outcome the generated profit of the owner of the focal platform averaged over the last five years.

Before we started constructing the truth table, we checked each condition for its necessity following Ragin's (2008) guidelines. This type of condition is always given within the set if the desired result is achieved. With consistency threshold of 0.90 and a coverage threshold of 0.60 (Mattke et al., 2022), none of the dynamic capabilities could be identified as a necessary condition. After constructing the truth tables (Appendix A.7 in the Electronic Supplementary Material), we were able to perform logical minimization in the form of intermediate and also parsimonious solution (Pappas and Woodside, 2021; Ragin, 2008) without any inconsistencies with a consistency threshold set to 0.75 and a frequency threshold of 1 (Mattke et al., 2022).

#### 7.1.3.3 Interpretation

We started our interpretation based on the outcome of the logical minimization (Table 7.1), where central conditions had to be present in both solutions, whereas peripheral conditions, are present only in the intermediate solution (Fiss, 2011). This is the case for both present and absent constraints. To gain a deeper understanding of the results, we again included all case information in the dissection and, in combination with network theory (Dhanasai and Parkhe, 2006; Meyer and Rowan, 1977; Rowley, 1997; Wassermann and Galaskiewicz, 1994) as the leading metatheory (Park, Fiss, et al., 2020), analyzed the dynamic capabilities for orchestration along the ecosystem dimensions[7] of density and centrality (Fig. 7.3).

[7] While the degree of density describes the relative number of relations within the ecosystem (Rowley, 1997), centrality denotes a participants prominence and power within an ecosystem (Brass and Burckhardt, 1993; Wassermann and Galaskiewicz, 1994).

### 7.1.4 Results

We condensed a total of 12 configurations of dynamic capabilities necessary for successful ecosystem orchestration (Table 7.1) in the csQCA. Since distinct market positioning has a significant impact on a platform's performance (Cennamo and Santalo, 2013), we distinguished the platforms' scope between market generalists providing a diverse assortment (Cennamo, 2021) and specialists providing sub assortments. The solution consistency substantially surpasses the recommended threshold for both outcomes with 86% for the full assortment and 93% for the sub assortment (Pappas and Woodside, 2021). Our configurations also cover most of the cases for full assortment (84%) and for sub assortment (64%). Both quality criteria imply a reasonable degree of robustness in our findings, making the quality of the results comparable to other works in IS research (Lee, Park, et al., 2019; Ma et al., 2023; Tan, Benbasat, et al., 2016).

#### 7.1.4.1 Configurations for Full Assortment Platforms

We identified seven capability configurations for focal platforms accumulating a diverse product and service portfolio (Teller and Elms, 2010; Timmers, 1998). Configuration F1, relies on a combination of structural ecosystem reconfiguration based on institutional innovation as a peripheral condition. The focal platforms thereby elect to strategically restructure the surrounding ecosystem through innovation-driven acquisitions (e.g., start-ups) based on their financial position. The objective is to increase control and ensure a smooth integration of new features into the existing ecosystem. Rakuten pursues this approach by acquiring dozens of platforms (e.g., Buy.com, Kobo, Viber Media), expanding its reach to over 70 services globally, spanning startups and incumbents in fintech, digital content, and communication companies (Rakuten, 2021, 2022). Innovation-driven ecosystem reconfiguration also involves new incentive mechanisms (e.g., meetings with celebrities) to attract additional participants. Rakuten incentivizes both sellers and buyers through measures, such as lower commission fees for expert accounts and ecosystem-wide loyalty points (Young, 2022).

In F2, a shared vision combined with a structural redesign regardless of platform access are the key success factors. Platforms in this configuration propel value co-creation by consciously puzzling ecosystem participants' activities and allowing reconfiguration through strategic partnerships. As part of the shared vision, collecting, analyzing, and sharing data is important to enhance customer experience as well as further to improve the ecosystem's services (Günther et al., 2017). Partnerships with other platforms facilitate structural changes across the ecosystem and increase the participants for both platforms if integrated properly. Fnac cooperates

**Table 7.1** Configurations of Dynamic Capabilities for E-Commerce Ecosystem Orchestration

| Configuration | Full assortment solution | | | | | | | Sub assortment solution | | | | |
|---|---|---|---|---|---|---|---|---|---|---|---|---|
| | F1 | F2 | F3 | F4 | F5 | F6 | F7 | S1 | S2 | S3 | S4 | S5 |
| Information Superiority Institutional Innovation Shared Vision Structural Reconfiguration Platform Access | | | • | ◦ | | | ○ | | | • | ○ | • |
| | · | | | ○ | | ○ | | • | | • | ○ | |
| | | • | ○ | | ○ | · | ○ | | | | | ○ |
| | • | • | | | • | ◦ | ○ | | • | ○ | ○ | ○ |
| | | ○ | • | · | • | • | ○ | ○ | ○ | | • | • |
| Consistency Unique coverage Raw coverage | 0.90 | 1.00 | 0.86 | 0.83 | 1.00 | 1.00 | 0.81 | 1.00 | 1.00 | 1.00 | 1.00 | 0.83 |
| | 0.16 | 0.04 | 0.02 | 0.07 | 0.04 | 0.02 | 0.29 | 0.09 | 0.05 | 0.09 | 0.14 | 0.14 |
| | 0.20 | 0.07 | 0.13 | 0.22 | 0.11 | 0.07 | 0.29 | 0.14 | 0.09 | 0.18 | 0.14 | 0.23 |
| Cases | 3, 4, 6, 9, 10, 13, 16, 37, 47, 74 | 42, 57, 74 | 15, 28, 49, 61, 65, 71, 73 | 2, 11, 15, 18, 20, 23, 28, 49, 61, 65, 66, 73 | 15, 37, 40, 58, 65 | 19, 23, 66 | 34, 43, 51, 53, 64, 69, 70,72, 75, 76, 77, 80, 81, 93, 97, 98 | 48, 55, 88 | 46, 48 | 60, 86 | 41, 44, 99 | 32, 53, 59, 84, 86, 94 |
| **Solution consistency** | 0.86 | | | | | | | 0.93 | | | | |
| **Solution coverage** | 0.84 | | | | | | | 0.64 | | | | |

Black circles "•" indicate the presence of a condition while empty circles "○" indicate its absence. Large circles indicate core conditions, small circles indicate boundary conditions. Empty spaces indicate irrelevance.

with various integration partners ensuring the harmonization of value co-creation (e.g., feed integrators, currency managers, aftersales services) and the simplification of the platform affiliation (Fnac Darty, 2023). They allied with Google to customize search results enhancing their omnichannel presents and achieving better search results (Fnac Darty, 2022; Google, 2023). Further reconfiguration measures in this formation are regular shopping events attracting potential buyers and a reward system for premium members (e.g., Barcia, 2021).

Information superiority and platform access are leveraged for orchestration in configuration F3 without building up capabilities for creating a shared vision. Platform owners in this configuration utilize big data analytics to forecast technological trends and external threats to build up sensing capabilities. This information is a prerequisite to proactively transforming the ecosystem by altering entry barriers. As the primary control mechanism, closed marketplaces set entry barriers relatively high for ecosystem participants. La Redoute, for example, applies artificial intelligence-based trend analysis to increase the accuracy of their assortment planning (Heuritech, 2020). On the other hand, the platform implements a rigorous selection process for potential third-party sellers (La Redoute, 2021) and limits its seller services primarily to transaction processing (Brandsom, 2020).

F4 aims at establishing platform access and augments them with peripheral information capabilities. The focus of information gathering is based on the monitoring of the participants, the analysis of transaction data, and the identification of product gaps. This internal analysis, preferred over the anticipation of external developments, enables the bundling of efforts for the support of existing participants. Capability configuration F4 results in a rather open ecosystem to enable possible external organizations and individuals to join the ecosystem and engage in value co-creation. It is applied by platforms with rather small ecosystems and relatively low gross merchandise volumes to quickly increase the number of participants to reach the maximum benefit from network effects. The platform openness is supported by a high number of boundary resources with detailed information for potential new participants (Shopee, 2023; Takealot, 2023) and low entry barriers. Driven by the low entry barriers, participants are overseen during transactions and excluded from the ecosystem in case of any infringements. Furthermore, exchanged products are controlled for the disregard of intellectual property rights and their conformity with domestic regulations (Coupang, 2023; Tokopedia, 2023).

Transaction platforms that combine structural reconfiguration and platform access capabilities without emphasizing an ecosystem-wide shared vision are involved in F5. Hence, the emphasis of this configuration lies on the ecosystem's transformation with regards to its structure and openness to regulate the participants. Transforming the ecosystem without an overarching shared vision appears as a trial-and-error approach for evaluating additional value propositions, altered entry barriers, and changed boundary resources (Feller et al., 2012). This might lead to uncertainty among ecosystem participants and a relatively unstable ecosystem structure. To mitigate the uncertainty, these platforms consequently communicate changes to their boundary resources (Cdiscount, 2023) and establish data governance mechanisms (Ghosh, 2022). Platforms applying configuration F5 need to partner with IT providers to foster innovation within the ecosystem to increase attractivity for sellers and customers(Nvidia, 2023; Olano, 2021). For instance, Kroger partners with Ocado to growth its digital marketplace in areas where it is absent in stationary retail (Wells, 2021).

F6 focuses on establishing platform access capabilities in conjunction with a peripheral emphasis on maintaining a shared ecosystem vision. Hence, the case companies synchronize their boundary resources and entry barriers with the ecosystem's overall value proposition for the customer (Adner, 2017). This ensures an alignment between customers' product and service demand and supply-side participants' offerings. These platforms strongly rely on service providers for offering additional services to sellers and open their platform accordingly to guarantee service integration (Hepsiburada, 2023a). With an absence of innovation and structural changes in the ecosystem, the platform owner can only steer the value co-creation by allowing the right organizations to participate in the ecosystem and enveloping services from ecosystem participants. Access management is especially implemented by formulating affiliation prerequisites (e.g., membership in a chamber of commerce) and sophisticated screening processes that guarantee sellers' and service providers' reliability (Fnac Darty, 2023; Hepsiburada, 2023b). This configuration also establishes data governance mechanisms allowing value capture from analyzing ecosystem data, providing it to ecosystem participants, and selling it externally (Xebia, 2023).

Platforms in F7 adopt a rather passive role in the ecosystem orchestration resulting in a high degree of freedom for the participants in terms of provided products and services. The platform owners offer the technological foundation with services, such as product search and transaction processing, to conduct retail transactions. This configuration is pursued by platform owners whose business model builds upon classified ads and auctions (e.g., Marktplaats, Aukro). Former online retailers attracting additional third-party sellers and evolving to a digital marketplace to diversify their assortment also favor configuration F7. The platforms only enforce domestic laws to exclude certain product categories or to provide general specifications on product quality but leave trust building to customer ratings. Moreover, there are no Application Programming Interfaces (APIs) leading to a mostly manual interaction with the platform. These platforms waive affiliation fees and capture value by promoting ads (Marktplaats, 2022) and by charging commission fees (Aukro, 2023). It is also worth mentioning that this configuration is used by established platform owners operating in specific geographic locations (e.g., Extra, Catch).

#### 7.1.4.2 Configurations for Sub-Assortment Platforms

Sub-assortment platforms are specifically aligned to a selective category of products (e.g., sports goods, music equipment, books). Due to their limited product range, these companies either serve specific niche markets or specialize in certain product categories that appeal to a more extensive consumer base. We identified five capability configurations for sub-assortment-oriented e-commerce ecosystems.

Transaction platforms applying configuration S1 rely on innovation to enhance their offerings and establish themselves as innovation leaders within the e-commerce ecosystem. The continuous deployment of innovations increases the ecosystem's attractiveness for existing and potentially new participants. Innovative platforms also implement strategic partnerships with pivotal participants altering the ecosystem's service configuration and spreading their culture across the ecosystem. New services go through a structured testing process (e.g., A/B testing) with end users (e.g., D'amato, 2021). Depop's "Think Thrift" campaign exemplifies the company's commitment to raise a culture of curiosity and growth that encourages the emergence of new ideas and services (Depop, 2022).

In S2, platforms drive structural redesign by introducing incentives or new sub-ecosystems, thereby positioning themselves as the transformer. This practice is particularly prevalent in the fashion industry (e.g., Farfetch, Dafiti), where cooperation with other platforms is crucial to offer a diverse range of brands and styles which meet the varying customer preferences. Combined with loyalty programs increasing the ecosystem participants' retention, this approach has proven to be more effective than driving orchestration through platform access practices. For instance, Farfetch cooperates with over 500 online and offline boutiques and supplements this with a graduated membership program as well as personal shopping events (Branson, 2020; Farfetch, 2023).

S3 concentrates on information leadership and exploits this information to drive innovation. Platform owners use data and strategic matchmaking methods for demand planning which is especially crucial in the fashion and technology segment where products have limited high-profit selling periods (Tsay, 2001; Weber and Schütte, 2019a). Zalando, for instance, offers in-depth product-specific sales analyses and budget adjustments to third-party sellers (Zalando, 2021). Analogous to the process of sensing and seizing (Teece, 2007), leveraging information advantage is the initial step in identifying strategically relevant trends and subsequently turning them into marketable innovations. This configuration establishes innovation hubs and invests in startups, prioritizing platform innovation over ecosystem-wide change. The S3 configuration can be found in a variety of cases, such as Zalando and Decathlon, that operate several technology hubs and acquired startups to amplify the ecosystems competitive positioning (Zalando, 2023).

In S4, ecosystem orchestration is exclusively supported by capabilities for managing platform access. This requires the offering of mature boundary resources on application level and documentation level increasing the connectability to the focal platform as well as reducing technical barriers for potential participants (Dal Bianco et al., 2014; Wulfert, Woroch, Strobel, Seufert, et al., 2022). Combining boundary resources with relatively high barriers for entering the ecosystems and offering products should increase the trust in transaction partners as well as ensure the products' authenticity. Hence, platform owners implement rigorous screening processes for both sellers and products. The result of this configuration is a technically open transaction environment restricted by access rules and processes fostering authenticity (Benlian et al., 2015; Broekhuizen et al., 2021). For instance, Netshoes validates sellers through a registration process that requires them to submit multiple business documents (e.g., signed affiliation contract, membership to a domestic chamber of commerce) and complements this by providing comprehensive resources, such as APIs, Software Development Kits (SDKs), developer tools, and webinars (Netshoes, 2023; Olist, 2023).

Data analytics is targeted in S5 for increasing the marketplace's attractiveness to sellers as supply-side participants. Reverb assists sellers by offering estimated pricing based on past competitors transaction and provides demand estimates for popular product categories (Reverb, 2023b). The information focus is combined with comprehensive measures for platform access. In this regard, sensing capabilities are exploited to leverage information for trend detection to purposefully alter platform access mechanisms transforming the overall ecosystem. The informed tuning of platform access mechanisms is intended to increase the number of ecosystem participants and profit from an increased number of transactions. These services are complemented by extensive integration options, reaching from websites over sandbox environments up to whole merchandise management systems (Reverb, 2023a).

### 7.1.5 Combination of Centrality and Density for Ecosystem Orchestration

Based on the derived dynamic capability configurations for successful platform owners in e-commerce ecosystems we propose six orchestration types leading to sustainable competitive advantage depending on their respective ecosystem structure determined by centrality and density (Rowley, 1997). For the construction of the types, we analyzed unique as well as raw coverage and depicted the configurations introduced in an ecology of configurations embodying the structure of e-commerce ecosystems (Fig. 7.3) (Park, Fiss, et al., 2020).[8]

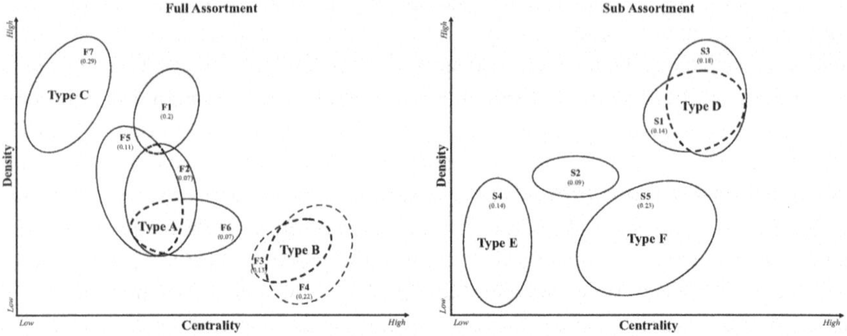


**Fig. 7.3** Ecology of Configurations Embodying Ecosystem Structure

Orchestration type A combines moderate centrality and medium to high density through its focus on altering the ecosystem structure, providing access to the ecosystem, and establishing a shared vision. The provision of boundary resources and reduction of entry barriers increases the number of participants to amplify network effects as well as increase the overall value of the ecosystem for existing and new participants (Boudreau, 2010; Briscoe, Odlyzko, et al., 2006). Through this continuous reconstruction process the ecosystem's density is reduced which in return has a positive impact on the formation of relationships with focal platforms and increases the impact of the orchestrator (Lavie, 2002). A by-product of the strengthened integration is the increased control influence based on the impact on value proposition and the centralization of the ecosystem.

Low density and moderate centrality with a focus on platform access and information superiority in orchestration type B result in an increased number of participants due to lowered entry barriers and more sophisticated boundary resources. Benefitting from the internal and external information advantage in comparison to the other participants, the orchestrator alters the overall ecosystem's value proposition

[8] Configuration S2 is not included in any orchestration type as it differs from the sub-assortment types in terms of dynamic capabilities considered and its relatively low raw coverage with reference to our case sample.

to maintain or strengthen its position of power even in the face of environmental change (Adner, 2017). Ecosystem superiority also enables the focal platform owner to shape the ecosystem according to his vision by granting access to promising external organizations or excluding others (Boudreau and Hagiu, 2009; Ofe and Sandberg, 2023). Synchronizing access with information keeps the density low and fosters the centralization through the set of actions provided by the orchestrator.

In contrast, orchestration type C results from an intentionally low number of orchestration activities and not developing dynamic capabilities for their alteration. Hence, focal platform owners in high density and low centrality e-commerce ecosystems take on a relatively passive role in ecosystem orchestration (Huber, Kude, et al., 2017; Wareham et al., 2014). Passivity fosters the interconnection between other participants, that use the focal transaction platform as a virtual locus for transaction execution and the platform owner focuses on transaction-based value capturing. The source of sustainable competitive advantage is the propelled ecosystem participants' independence and flexibility (Blaschke and Brosius, 2018; Boudreau, 2010). This advantage might be further increased in providing innovations (e.g., advanced recommendation algorithms, same-day delivery) or easing and improving the execution of retail transactions. For a more active role, the orchestrator, can envelop capabilities previously provided by other ecosystem participants (e.g., own brands, additional retail-related services).

Orchestration type D draws on the focal platform's innovation capabilities in combination with the platform's information superiority. This unique capability combination allows for tailoring ecosystem-wide innovation to the participants' requirements as well as economic and technological developments. This combination also compensates a relatively consistent ecosystem structure in terms of incentive scheme and sub-ecosystems which leads to relatively high centrality and density characteristic. The success relies on the focal platform's innovativeness, its control over the ecosystem and the associated threat to envelop ecosystem participants' services, as well as change of the ecosystem's service configuration. Enveloping ecosystem participants' services is likely if the participant provides a lucrative service offering pivotal to the overall ecosystem's value proposition or if the participants does not conform with the overall ecosystem's agenda. Altering the ecosystem's service configuration might also shift participants' share of the value co-creation and consequently the value capture. The focal platform's innovativeness could also result in a change of the assortment type either including additional product categories or focusing on parts of the assortment. The focal platform's focus on innovation for ecosystem orchestration requires multiple relations among the ecosystem participants for harmonizing their value co-creation.

The focus of orchestration type E lies on the management of platform access to influence the degree of openness, which is determined as the number of participants, as well as the likelihood of external organizations and individuals to join the ecosystem. The objective of the orchestrator is the attraction of valuable participants and the amplification of the value co-creation to foster network effects, while excluding participants that do not conform with regulations and standards. Focusing the orchestration on access management provides the participants with opportunities to provide their own ideas, products, and services to the ecosystem and profiting from non-existing value capturing of the focal platform as well as the relatively low entry barriers. This type is based on a low density and relatively low centrality reducing relations among ecosystem participants while inheriting a passive orchestration role. Orchestrators further increase their competitive advantage by creating and harmonizing a shared vision as well as aligning all participants to a multilateral value co-creation.

The combination of information superiority and platform access capabilities to achieve sustainable competitive advantage (comparable to orchestration type B) characterizes orchestration type F. This capability set is successful in e-commerce ecosystems with a low density and moderate to high centrality. The orchestrator focusses on platform access and streamlining the relations to the platform which decreases the necessity for relations among the ecosystem participants. With increased access to the focal platform via low entry barriers and sophisticated boundary resources, communication as well as value exchange is conducted via the focal platform as virtual locus. The manifestation of information capabilities enables the alignment of boundary resources and entry barriers to ecosystem participants' requirements and environmental developments (Feller et al., 2012). While the ecosystem participants are independent in synchronizing their value co-creation to a huge degree, the transaction execution and product or service exchange needs to be centralized on the focal platform, resulting in a relatively high degree of ecosystem centrality.

#### 7.1.5.1 Theoretical Contribution

In this work, we shed light on orchestration as a necessary process for the success of digital business ecosystems in general and e-commerce ecosystems in particular. A focal platform owner orchestrates a diverse set of e-commerce participants to capture a larger share of value (Foss et al., 2023; Iansiti and Levien, 2004a) and provide sustainable competitive advantage for the overall ecosystem (Rietveld and Schilling, 2021). With regards to our focus on dynamic capabilities for ecosystem orchestration in e-commerce, the theoretical contribution of this research is threefold:

Frist, we drew on existing research on generic participants types (Eisenmann, Parker, et al., 2009) and extend research on participants in retail (e.g., Böttcher, Rickling, et al., 2021; Wulfert, Woroch, Strobel, Schoormann, et al., 2023) to the specific peculiarities of e-commerce (Nath et al., 1998; Zwass, 1996) by providing ten generic participant types for e-commerce ecosystems (Appendix A.7 in the Electronic Supplementary Material). Second, we explain which dynamic capabilities are required for focal platform owners for the effective orchestration of e-commerce ecosystems. Furthermore, we provided insights on calls for research on necessary capabilities for ecosystem orchestration in dynamic environments (Jacobides et al., 2018) and extended the knowledge base on capabilities in the context of dynamic ecosystems with a focus on (retail) transactions (Gawer, 2021b). We intentionally chose e-commerce ecosystems as object of investigation as they are subject to continuous change with participants joining and departing as well as dynamically adjusted product and service assortments. Although we focused our research on e-commerce ecosystems, we argue that the dynamic capabilities identified could also propel ecosystem orchestration and, thus, be a source of sustainable competitive advantage in other types of digital business ecosystems. Following Gioia's (2013) approach, we also introduced second-order themes and first-order constructs as milestones for the establishment of the orchestration capabilities. Third, we go beyond the mere identification of dynamic capabilities for ecosystem orchestration and propose six orchestration types for how to successfully orchestrate e-commerce ecosystems. These orchestration types resemble ideal capability configuration for specific e-commerce environments (Paquin and Howard-Grenville, 2013; Rietveld and Schilling, 2021). We differentiated e-commerce ecosystems based on their structure along the dimensions of ecosystem centrality and ecosystem density (Rowley, 1997). We suggested successful orchestration types involving dynamic capability combinations of shared vision, structural reconfiguration, and platform access (type A) as well as information superiority and platform access (type B) for e-commerce ecosystem providing full product and service assortments. For sub assortments, dynamic capability combinations of information superiority and institutional innovation (type D) as well as information superiority and platform access (type F) are potentially successful. These combinations usually cover (parts of) Teece's (2007) tripartite scheme of dynamic capabilities (i.e., sensing, seizing, transforming). The raw coverages of these orchestration types indicate a high success rate for these combinations (Table 7.1). Besides these active orchestration types, our results also support existing studies that investigated digital marketplaces taking on a passive role in orchestration and intra-marketplace competition (Feller et al., 2012; Koch, 2010; Täuscher and Laudien, 2018). In total, 27 cases showed a positive outcome for ecosystem domination even if several dynamic capabilities are absent (type C, type E). These digital marketplaces benefit from the ecosystem participant's engagement in multilateral value co-creation and distributed orchestration resulting from the freedom granted by the focal platform owner (Kretschmer, et al., 2022; Täuscher and Laudien, 2018). Our research on dynamic capabilities in e-commerce ecosystems also extends the literature on platform-based and ecosystem-level competition (Dou and Wu, 2021; Kretschmer, et al., 2022; Rietveld and Schilling, 2021; Zhao, von Delft et al., 2020) by proposing dynamic capabilities as a potential source of sustainable competitive advantage (Teece et al., 1997).

#### 7.1.5.2 Practical Implications

The effective orchestration by the owner of the focal platform within the e-commerce ecosystem amplifies the multilateral value co-creation among the participants (Adner, 2017; Burkhalter, 2020). This orchestration requires a set of capabilities (Addo, 2022; Autio, 2022) which derive a competitive advantage in dynamic e-commerce environments (Turban, Whiteside, et al., 2017; Zwass, 1996). With our dynamic capabilities and underlying structure of constructs (Gioia et al., 2013), we provide a threefold contribution to all participants in the ecosystem: (1) Focal platform owners as orchestrator can apply the capabilities to assess their competitive situation in comparison with other ecosystems, to identify potential shortcomings in their capability configuration, and develop additional capabilities for ecosystem orchestration. This will increase their competitiveness on company level and the competitiveness of the overall ecosystem. (2) Based on an assessment of the ecosystem's structure (Rowley, 1997), focal platform owners can also check whether they apply a potentially successful orchestration type and adjust their capabilities following the guidance provided by the twelve configuration types. (3) Ecosystem participants benefit from the participant overview by classifying themselves and other partners to harmonize their value creation and identify possible opportunities for value creation and capture. Furthermore, they can also make use of the capability types to anticipate potential strategic actions by the owner of the focal platform and prepare accordingly.

#### 7.1.5.3 Limitations and Future Research

Aside from the scientific contributions and practical implications, this research is not free of limitations. First, the synthesis of the participant types in e-commerce ecosystems might be biased by the authors' individual judgement and experiences in the field. To reduce potential author bias in the coding of participant types in e-commerce ecosystems, we coded the relevant research papers in three waves and discussed issues among the team of authors (Okoli, 2015). Second, as with all qualitative studies, the empirical data collected was subject to interpretation during the analysis and might be biased in multi-case analysis and case survey. For our case analysis and case survey, we relied on publicly available information. To account for potential bias by the authors, we triangulate our analysis with secondary information whenever possible (Yin, 2018). As the reliability of case surveys depends on the number of cases identified for the research endeavor (Larsson, 1993), we chose a large number of cases (n = 100) to achieve relevant insights and a high statistical significance (Greckhamer, Misangyi, et al., 2013). These shortcomings can be further omitted when combining a case survey approach with a csQCA (Henfridsson and Bygstad, 2013; Rivard and Lapointe, 2012), capable of analyzing larger sample sizes allowing for empirically sound theorizing (Greckhamer, Furnari, et al., 2018; Soto Setzke et al., 2020). Third, we relied on profit and number of participants as indicators for the success of e-commerce ecosystems. It might be also worthwhile considering the lifecycle stage of each focal platform and its surrounding e-commerce ecosystem as it might also influence the manifestation of the orchestration capabilities for future research (Muzellec et al., 2015; Teece, 2017).

While we focused our analysis on dynamic capabilities necessary for ecosystem orchestration on the context of e-commerce, the dynamic capabilities might also be relevant in other types of digital business ecosystems, such as innovation ecosystems and urban data ecosystems. Future research might evaluate the application of the dynamic capabilities identified with second-order themes and first-order constructs in other ecosystem types. As such, the orchestration types could be reevaluated for other context in future studies. Another fruitful avenue for future research would be an in-depth analysis of the white spots identified in the depiction of the ecology of configurations embodying the ecosystem structure (Park, Fiss, et al., 2020). We could not identify configurations for a high ecosystem centrality in both full-assortment and sub-assortment ecosystems suggesting a more nuanced orchestration by the owner of the focal platform might be more successful than extensive interference. Moreover, a further analysis on capability configurations for orchestration for low-centrality and low-density ecosystems (i.e., full assortment) as well as low-centrality and high-density ecosystems (i.e., sub assortment) could provide additional insights on orchestration and ecosystem structure. Moreover, future research might focus on more concrete processes for implementing the dynamic capabilities. The derived second-order and first-order constructs can already provide first insights on the actual development of dynamic capabilities.

### 7.1.6 Conclusion

Orchestration is the key challenge for focal platforms within an ecosystem in leading the ecosystem more efficiently, reducing transaction costs and making it more successful. This paper paves the way for the orchestration of e-commerce ecosystems by focal platforms as orchestrators with five dynamic capabilities and six orchestration types successful for different ecosystem characteristics along the ecosystem dimensions of centrality and density, which enables different directions for the future of ecosystem orchestration.

## 7.2 Paper 10: Four Shades of Customer: How Value Flows in Fintech Ecosystems

**Abstract**
*The financial sector is undergoing a massive transformation, with new technology-driven players challenging established mechanisms and transforming the sector into a fast-moving market.*[9] *With the gradual transition from a scale economy to a platform-driven network economy, enterprise networks are gaining strategic importance. Despite the growing interest in Financial Technology (fintech)'s, research has so far lacked a conceptualization of value creation in fintech ecosystems. Therefore, this research paper aims to analyze key players, value creation activities, and value streams based on the analysis of the business models of payment services, personal financial management, robo-advisory, peer-to-peer lending, trading, and cryptocurrency. We present a holistic value network for the fintech ecosystem based on a structured literature review and analysis of 171 fintech companies. We were able to show that fintech platform orchestrates multiple market sides and that customers take four distinct roles at the center of the ecosystem when using fintech services.*

[9] This version of the article has been accepted for publication, after peer review but is not the Version of Record and does not reflect post-acceptance improvements, or any corrections. The Version of Record is referenced as follows: R. Woroch et al. (2022) "Four Shades of Customer: How Value Flows in Fintech Ecosystems." In: *Proceedings of the 43rd International Conference on Information Systems (ICIS2022)*. Copenhagen, Denmark, pp. 1–17

| Fact sheet of publication 10 | |
|---|---|
| Title | Four Shades of Customer: How Value Flows in Fintech Ecosystems |
| Authors | Robert Woroch |
| | Gero Strobel |
| | Tobias Wulfert |
| Publication Type | Conference Proceeding |
| Publication Outlet | International Conference on Information Systems |
| VHB-Jourqual3-Ranking | A |
| Status | Published |
| Research Question | How is value created and exchanged within the fintech ecosystem? |
| Research Approach | Literature Review (Vom Brocke, Simons, et al., 2015) and Case Study (Yin, 2018) |
| Full Citation | R.Woroch et al. (2022). "Four Shades of Customer: How Value Flows in Fintech Ecosystems." In: *Proceedings of the 43rd International Conference on Information Systems (ICIS2022)*. Copenhagen, Denmark, pp. 1–17 |

### 7.2.1 Introduction

Fintech is an acronym resembling an amalgamation of finance and technology, comprising the combination of disruptive technologies (Artificial Intelligence, Big Data, Internet of Things, Distributed Ledger Technology, etc.) with established business activities (e.g., payments, investments, financing), resulting in new products, services, and business processes (Imerman and Fabozzi, 2020; Puschmann, 2017). Additionally, the term fintech also refers to innovators and disruptors in the financial sector who propose more flexibility than established financial services companies through innovative business models by leveraging ubiquitous internet communication and automated information processing (Gomber, Koch, et al., 2017). Fintech business models recently experienced significant market growth, as reflected in global cumulative investments of $ 210.1 billion in 2021, with more than half of the investment coming from venture capital ($ 114.9 billion), more than doubling from 2018 ($ 53.2 billion) (Ruddenklau, 2022).

Consequently, the financial services industry experiences a massive transformation in which emerging technological actors challenge established financial activities and transform the sector into a fast-moving market by changing business functions. New market players, often start-ups, adopt digital business strategies and join forces in corporate networks to create sustainable value (Huebner et al., 2019; Schmidt et al., 2018a). Innovative business models include offerings in payment services, digital wealth, lending, or capital markets, increasingly expanded by further innovations in the insurance and real estate sectors (i.e., insurtech and proptech) (Huebner et al., 2019; Schmidt et al., 2018b). In particular, new payment providers (e.g., Alipay, Klarna) have gained traction due to rapid developments in mobile technologies, both online and in brick-and-mortar stores, by turning the payment process into a convenient user experience through contactless authentication, alternative "buy-now-pay-later" payment modalities, and novel technologies (especially blockchain) (Deufel and Kemper, 2018; Pousttchi, 2008; Ruddenklau, 2022). This leads to full automation of processes, allowing customers access at any time, from any location, and at any speed (Verbovetska, 2019). Thus, fintech companies, leverage their opportunities by designing business models from a customer-centric perspective and offering convenient, accessible, and efficient services to a technology-savvy clientele (Eickhoff et al., 2017; Gomber, Koch, et al., 2017; Verbovetska, 2019). Furthermore, stricter regulations in the financial services market, low interest rates and high cost pressure are noticeably impeding established companies, thus providing opportunities for new, lean and agile companies (Torno et al., 2021).

Using ubiquitous digital technologies within financial services requires the development of new digital business models while considering changing customer behavior and additional ecosystem participants (Eickhoff et al., 2017). Both startups and incumbents seek new ways to realize successful offerings by enhancing or completely transforming the customer experience (Gomber, Kauffman, et al., 2018). The accompanying bundling of competencies of different market players leads to significant changes in the business, legal and technological capabilities of individual actors and ultimately to a shift from an economy of scale to a platform-driven network economy (Schmidt et al., 2018a). These emerging fintech ecosystems have strategic relevance and contain numerous interconnections affected by their actors (suppliers, distributors, service providers, etc.) (Iyer and Basole, 2016; Schmidt et al., 2018b). Hence, a sound understanding of those ecosystems with independent actors and value flow between them is required with respect to the design and refinement of fintech business models (Gordijn, Akkermans, and van Vliet, 1999). In the field of fintech research, a few studies address value networks. These papers describe blockchain ecosystems (Riasanow, Burckhardt, et al., 2018) or analyze bitcoin mining profitability using value networks (Derks et al., 2018). Riasanow, Flötgen, et al. (2018) compare the traditional with fintech ecosystems emphasizing on innovation patterns. Additionally, multi-sided platforms in the Insurtech sector are examined (Pousttchi and Gleiss, 2019) or the ecosystem of emerging financial regulatory reporting systems is analyzed (Floetgen, Gomm, et al., 2020). Riasanow, Jäntgen, et al. (2021) adopt another viewpoint and examine similarities in digital transformation of different platform ecosystems. Nevertheless, knowledge about value creation in fintech ecosystems is limited and requires further research (Puschmann and Shiba, 2021). So far, there is no research that addresses a view from inside fintech platforms to the surrounding ecosystem, while focusing on key value-creation activities. Against this backdrop, we pose the following research question:

**RQ:** How is value created and exchanged within the fintech ecosystem?

To address this research question, we analyze the value network of multiple fintech business models based on a multi-case study with 171 companies operating in the business models of payments, personal finance management, robo-advice, Peer-to-Peer (P2P) lending, trading, and cryptocurrency by deriving a generic ecosystem map around these platforms. The paper is structured as follows. First, the theoretical background describes the digital transformation of the financial sector, the role of platforms as enablers to new ecosystems, as well as an approach for ecosystem modeling. Second, the research method for conducting the multi-case study and deriving the value network is presented. Afterwards, the value network of the fintech ecosystem is outlined by means of particular platforms, providing insights into the configuration of the roles, value streams and core value activities involved, until finally the entire value network around these platforms is presented. Based on the empirical investigation, key aspects of the value network are then analyzed as well as limitations of the investigation are discussed. Finally, we briefly summarize our results and provide an outlook for further research.

## 7.2.2 Theoretical Background

### 7.2.2.1 Digital Transformation of the Financial Sector

The financial industry has already adapted emerging technologies for core banking systems and opening them up for end customers via self-services (ATM, online banking, etc.) in the early days of the digital age, leading to the industry's status as an IT-savvy sector that rapidly embraces technological innovations (Beinke et al., 2018; Eickhoff et al., 2017; Huebner et al., 2019). Despite above-average IT investments within the banking sector (Gopalan et al., 2012), the digital transformation of business models and processes in the financial sector has been poorly driven, thus paving the way for the new fintech movement (Alt, Beck, et al., 2018; Gopalan et al., 2012). In this regard, the fintech movement was driven by four influencing factors: (1) an accelerated adoption of innovative hardware- and software-based IT solutions (smartphones, social media, platform-based software ecosystems, etc.),

(2) the emergence of new non-bank companies that brought a creative, dynamic as well as digital mindset to the financial world with a start-up mentality, (3) new expectations of increasingly informed "digital natives" and accompanying changes in bank customers' behavior toward multi-bank relationships, and (4) increasing pressure on traditional financial services companies due to regulatory restrictions as a result of the 2007 financial crisis (Alt, Beck, et al., 2018; Alt and Puschmann, 2012). In addition, fintech innovation has been driven further by substantial amounts of venture capital, services with value propositions that are distinctly different from traditional banks due to their innovative nature, and business processes accessible without spatial and temporal constraints (e.g., bank business hours) (Gomber, Kauffman, et al., 2018). Fintech refers to a revolution of the financial sector based on the congruence of different technologies (artificial intelligence, big data, platforms, distributed ledger, etc.), a customer-centric perspective, new business models, and profound process changes as well as service transformations (Alt, Beck, et al., 2018; Puschmann and Shiba, 2021). Although the term "fintech" aims at different concepts in extant literature, it usually includes the three dimensions innovation object (business model, product/service, organization, process, or system), degree of innovation (disruptive or incremental), and innovation scope (inter- or intraorganizational) (Puschmann, 2017). Drawing on Eickhoff et al. (2017, p. 2) we conceptualize fintechs as organizations "that operate at the intersection of (i) financial products and services and (ii) information technology [and] are usually (iii) relatively new companies (often startups) with (iv) their own innovative product or service offerings." Fintechs operate digital platform business models to facilitate interactions between actors from two or more previously independent markets (Eisenmann, Parker, et al., 2006; Langley and Leyshon, 2021; Staub et al., 2021). These actors form digital business ecosystems surrounding fintech platforms with the purpose of managing (end) customers' financial assets (Adner, 2017; Floetgen, Mitterer, et al., 2021). Although some fintechs have a full banking license (e.g., N26, Klarna), most of these companies offer their services without a banking license and instead partnership with traditional banks or banking-as-a-service provider (e.g., Solarisbank, Railsbank) for providing services that require a full banking license or to access regulatory and risk management expertise (Floetgen, Mitterer, et al., 2021; Gimpel and Röglinger, 2018).

#### 7.2.2.2 Platforms as Enabler of Fintech Ecosystems

Fintech platforms act as intermediaries between multiple participants in fintech ecosystems (Floetgen, Mitterer, et al., 2021) (Floetgen et al. 2021; Xie et al. 2021). Hence, they connect participants from previously independent markets and allow financial transactions between them exploiting the two-sided market paradigm

(Armstrong, 2006; Rochet and Tirole, 2003; Rochet and Tirole, 2006). The reintermediation effects of platform business models in fintech result in consolidation tendencies and a high capitalization of these businesses (Langley and Leyshon, 2021). This also involves strong market externalities and winner-takes-all tendencies (Hinz et al., 2020; Parker and Van Alstyne, 2005; Rietveld and Schilling, 2021). Successful fintech platforms should possess the following characteristics (Dhar and Stein, 2018): open access, standardized domain-specific business processes, and technological foundation. The degree of openness determines which participants can use the platform (Benlian et al., 2015). The standardized execution of financial transactions on technological enablers allows for more convenient self-services processes for customers compared to traditional banking (Pousttchi and Gleiss, 2019). 2018 term platforms with these characteristics "complete" and give electronic exchanges as well as robo-advisors as examples. Also, business models that realize payments or peer-to-peer lending are notably in line with the platform business model, as they mediate multi-sided relationships between their users while coordinating network effects of connectivity (Langley and Leyshon, 2021). Because of their flexible and highly customized technological infrastructures, platform business models can quickly scale up with rising demands (Choudary, 2015). With the amalgamation of two-sided market business models (i.e., transaction) and technological infrastructure for existing and future markets (i.e., innovation), fintech platforms can be classified as hybrid platforms (Gawer, 2021b; Schütte and Wulfert, 2022a).

#### 7.2.2.3 Value Modelling of Ecosystems

The business model aims to bridge the gap between strategy and business processes by describing the dimensions value proposition, value architecture, value network, and value finance while abstracting from real-world companies through conceptualizing business model instances or business model types (Al-Debei and Avison, 2010; Osterwalder, Pigneur, and Tucci, 2005). The value network is seen here as "a way in which an organization enables transactions through coordination and collaboration among parties and multiple companies" (Al-Debei and Avison, 2010, p. 366). Focusing on value activities-a core concept of business models-through which value or benefits are generated within the ecosystem facilitates the subsequent formulation of requirements for business processes and system architectures, which require a more elaborate view (Eickhoff et al., 2017; Gordijn, Akkermans, and van Vliet, 2000; Pousttchi, 2008). In this context, numerous modeling languages (e.g., $e^3$-value, Causal Loop Diagram, Strategic Business Model Ontology) exist for modeling value networks with different levels of understanding (John et al., 2017). The $e^3$-value ontology, originally from computer science, is used across domains and hence suited to represent organizational ecosystems in which actors create complementary value to achieve a common goal around a digital platform (de Reuver et al., 2017; John et al., 2017; Pousttchi and Gleiss, 2019). Although the focus of e$^3$-value is on actors, value objects and value activities as well as associated exchange relationships, in addition to purely network-related aspects-similarly to the geometric-oriented Business Model Canvas-supply-, customer- and finance-related aspects are partly considered as well (Gordijn, Osterwalder, et al., 2005; Osterwalder and Pigneur, 2010). As a result, the e$^3$-value ontology provides a basis for aligning business processes in line with the business model (Solaimani and Bouwman, 2012) (Table 7.2).

**Table 7.2** Notation of the Role-Based e$^3$-Value Modeling Language

| Element | Description | Symbols |
|---|---|---|
| Actor | Specific economic entity functioning as a market participant (e.g. actual person, company) | 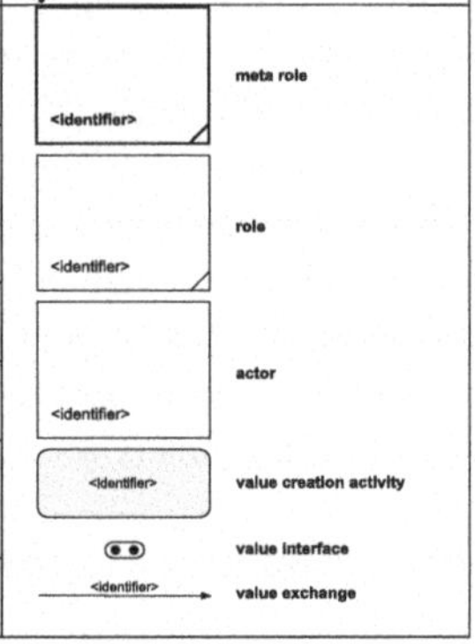 |
| Role | Symbolic market participant with value-adding activities, where one or more roles can be assigned to an actual market participant. | |
| Meta-role | Summary of similar roles (for model simplification) | |
| Value creation activity | A profit- or benefit-increasing activity of an actor or role | |
| Value interface | Value exchange grouping to an atomic value exchange consisting of at least one input and output value port | |
| Value exchange | Value link among actors or (meta-)roles | |



The e$^3$-value ontology thus represents a suitable tool for analyzing the value network around fintech platforms (Pousttchi and Gleiss, 2019; Riasanow, Burckhardt, et al., 2018). To illustrate the value exchange relationships within fintech ecosystems, the notation following Pousttchi (2008) is used, which enables generalization through use of roles. Table 1 indicates elements of the ee$^3$-value notation used in this study (Gleiss et al., 2021; Pousttchi, 2008). In this context, roles represent a central concept of reference models, as they remain stable across different business model instances, while actors represent independent economic entities acting as market participants (Pousttchi, 2008). For abstraction purposes, meta-roles are further used to aggregate similar roles (Gleiss et al., 2021). As mentioned above, value creation activities generate a benefit or added value and can be assigned to an actor or a (meta-)role. Value interfaces, which consist of one or more value ports, act as either an input or an output for an exchange of value, receiving or sending value between entities via the value ports within the value interfaces (Pousttchi, 2008). The resulting value exchange provides a link between the or actors or (meta-)roles and delineates the exchanged value (Gordijn, Akkermans, and van Vliet, 2000).

### 7.2.3 Research Method

Based on the dynamic and emerging nature of the subject under consideration, we use a mixed-method approach in two stages within the scope of our analysis. We combine a systematic literature search with a multi-case study to derive value networks in the fintech sector. First, we use the systematic literature search, to identify the knowledge anchored in the literature corpus and build a theoretical lens from it for the case study (Niederman and March, 2019). Second, we complement these findings with a multiple-case study that uses comparative analysis to identify relevant components of value networks (e.g., roles, value streams) (Yin, 2018). This dual approach we propose is explicitly suited for theory building, as the systematic literature review ensures the incorporation of existing knowledge for theory building (Webster and Watson, 2002), and the multiple-case study complements it through its high "likelihood of generating new theory" (Eisenhardt, 1989b, p. 546).

**Systematic literature search:** The search is based on the methodological approach of Webster and Watson (2002) in combination with Vom Brocke, Simons, et al. (2015) with the objective of developing a theoretical lens for the multi-case study. With reference to the object under consideration and the objective, the search string "("Fintech" OR "Financial technology") AND ("taxonomy" OR "classification")" was used in three well-known databases (i.e., AiSeL, Scopus, IEEE) with initial 418 hits. To ensure an appropriate level of quality, additional quality criteria were added to the search. Non-English language articles, panels and commentaries, purely technical articles and articles with a pure business model focus were excluded. Based on title and abstract as well as the quality criteria, 29 matching publications were identified. The articles were independently analyzed by full text screening and coded. Based on the analysis, 12 final articles were identified, and a forward and backward search was performed to include a broad range of business model types, identifying two more articles, bringing the total number to 14. Following Webster and Watson, the codes were aggregated into six fintech business models in a concept matrix (Table 7.3).

**Table 7.3** Fintech Business Model Types

| Source | Payment | PFM | Robo-Advice | P2P lending | Trading | Cryptocurrency | Study purpose |
|---|---|---|---|---|---|---|---|
| Eickhoff et al. (2017) | • | | • | • | • | • | Fintech business models archetypes |
| Gomber, Koch, et al. (2017) | • | | • | • | • | • | Novel and innovative business functions in digital finance |
| Beinke et al. (2018) | • | | | • | • | • | Fintech business models archetypes for start-ups using blockchain |
| Gimpel et al. (2018) | • | • | • | • | • | • | Non-functional characteristics of consumer-oriented fintech start-ups |
| Gomber, Kauffman, et al. (2018) | • | • | • | • | • | • | Forces of innovation, disruption and transformation in fintech |
| Schmidt et al. (2018a) | • | • | | • | | • | Collaborative business models in ecosystem of fintechs and banks |
| Schmidt et al. (2018b) | • | • | | • | | • | Data-driven business models in fintech |
| Das (2019) | • | • | • | • | • | • | Primary areas in fintech |
| Huebner et al. (2019) | • | • | • | • | • | • | Fintechs intermediating functions archetypes |
| Verbovetska (2019) | | | • | • | | | Financial services through the lens of process virtualization theory |
| Caragea et al. (2020) | • | | • | • | | | Fintech innovations |
| Imerman and Fabozzi (2020) | • | • | • | • | • | • | Fintech innovations |
| Puschmann and Shiba (2021) | • | | • | • | • | | Fintech-driven solutions for sustainability |
| Torno et al. (2021) | | • | | | • | • | Mobile personal finance applications archetypes |

**Multi-Case Study:** We explicitly chose a multiple case study because it allows knowledge to be gathered on a broader basis from a set of cases that differ in environmental aspects but share a common phenomenon (Yin, 2018). As a result, selecting multiple cases is critical for identifying cross-case patterns and the basis for comparative analysis (Eisenhardt, 1989b). Furthermore, multiple cases are necessary to go beyond the specificity of each case and to gain generally applicable insights that benefit the artifact class (Gregor, Müller, et al., 2013): Value networks in fintech sector. Based on the theoretical lens developed through the literature review, a search was conducted on Crunchbase and 171 relevant companies were identified. Referring to the scientific idea and the accumulation of knowledge, we provide the case list (*https://bit.ly/3vDrK3u*). The analysis of the companies is based on publicly available data sources (e.g., description of the portfolio of offerings, FAQ sections, developer portals, press releases) and was conducted in five phases following Peppard and Rylander (2006) to shape the value network. The starting point of the analysis was the classification of the companies and their business models into corresponding value networks (e.g., Payment, Robot-Advisory) based on the theoretical lens. For the respective business model instances (Osterwalder, Pigneur, and Tucci, 2005), the actors with influence on value generation were then identified and abstracted into roles (Table A.26). By using the role concept, a reference model is created that remains stable across the individual use cases (Pousttchi, 2008). Subsequently, the value creation activities and the associated vale exchanges were derived for the roles (Peppard and Rylander, 2006; Pousttchi, 2008) and, based on this, the corresponding value network was modeled for a subdomain. This was repeated for all six subsectors and finally consolidated into a fintech-wide value map (Fig. 7.4).

### 7.2.4 Result: Value Creation in Fintech

Understanding the mechanisms of value creation through digital platforms and the underlying exchange relationships within the fintech industry first requires an overview of the relevant participants. The starting point for the multi-case study was the identified market segments and business models from the literature, thus a generic value network was derived for each market segment. According to Peppard and Rylander (2006), the entities relevant for value creation as well as value flows between them were identified (Gleiss et al., 2021). For a more detailed understanding of the market mechanisms, the service portfolio was further analyzed and related value-creating activities were derived (Gordijn, Akkermans, and van Vliet, 1999; Pousttchi, 2008). These generic partial value networks for each fintech sector serve as a foundation for constructing an overall value network spanning the fintech market.

#### 7.2.4.1 Payment

The first approach to value creation in the fintech market is the analysis of platforms with a service portfolio including payment applications and services in both the online and brick-and-mortar sectors (Eickhoff et al., 2017; Riasanow, Flötgen, et al., 2018).[10] A payment service includes the initiation, authorization, clearing, and confirmation of money transfers, which enables the delivery of goods and services (Zhong, 2009). As a multi-faceted platform, the payment service provider typically operates without regulatory restrictions in order to perform its value-added activities relying on existing regulated infrastructures as well as financial products offered by third-party market participants holding a banking license (de Reuver et al., 2017; Huebner et al., 2019). In line with the purpose of value network modeling, the study aims to capture a comprehensive view of exchange linkages between participants (Gordijn, Akkermans, and van Vliet, 1999), thus the analysis encompasses a variety of scenarios in the fintech market. Initially, the prerequisite for value creation is the desire of the payer of purchasing goods or using services, making the relationship between payer and payee a central one. The purchase of products or services on the customer side and the offering on the seller side thus represent central value creation activities that lead to an exchange of goods or services in exchange for payment requiring a payment process (Lowry et al., 2006). The payment service provider offers mobile applications to perform the transaction, thus unbounding the payment from native environment allowing payment scenarios such as online, in-person, and peer-to-peer payments (Cai et al., 2019; Kazan and Damsgaard, 2016). Hence, both online and stationary merchants with physical points of sale rely on the same payment service, therefore, due to overlapping value streams and for the purpose of generalization, online and stationary merchants are aggregated into the meta-role of payee. Due to multiple driving factors, including fast processing, increased security, and low cost—especially for merchants—in-person payments, which were previously fixed to cash or card payments, are increasingly gaining acceptance (Deufel and Kemper, 2018; Li, 2018; Moghavvemi et al., 2021). To illustrate the structure of the value network, Table 3 presents example value linkages for fintech payment services (Table 7.4).

---

[10] The payment processing is considered as a relevant sub-ecosystem of e-commerce ecosystems. It is responsible for the processing of payments of e-commerce transactions. The payment ecosystem is depicted in the upper-left area in Fig. 7.4.

**Table 7.4** Excerpt from the Fintech Market Value Network

| From entity | Value exchange | To entity |
|---|---|---|
| Payment service provider | Convenient payment; Expense transparency; Convenient transfer; anonymous payment | Payer |
| Payer | Debit interest and fees; Transaction; Money; Data; New users through word of mouth | Payment service provider |
| Payment service provider | Customer contract | Lender |

When executing the payment, the payer usually has a number of options, which affect the value flow. Direct payment debits the payer's bank account immediately after the purchase, thus the transaction is an indirect value-add for the payment service provider as he is authorized to process the payment, resulting in a direct added value in form of transaction fee paid by the merchant, receiving in return the outsourced payment process (e.g., Klarna or Revolut). The payment service provider platform thus pursues a multi-sided revenue model in which the payer triggers the transaction and provides the necessary payment data, while the transaction fee is charged to the payee (Schüritz et al., 2017). In case the payer chooses an additional pay later option (payment on account, installment payment, etc.), further value is generated for the payment service provider in terms of interest charges and other fees (account maintenance, late fees, etc.) (e.g., Opy or Zip). The required installment loans are usually realized by a banking partner as a lender, since the payment service provider itself does not have any banking license (e.g., Affirm or Opy). Regardless of payment option choice, the payer experience convenient payment processing as a result of the transactions' speed, simplicity, and additional services (Deufel and Kemper, 2018; Liu, Abhishek, et al., 2016). To complement the core service, payment service provider issue their customers with a credit card, increasing the platform's customer base—for both in-person and online payment transactions—while leveraging established credit card payment infrastructure as a two-sided market strategy (e.g., Klarna or Revolut) (Gawer, 2021b). Typically, four actors (payer, payee, card issuer and acquirer) are involved in the processing of a credit card payment (Plateaux et al., 2018), therefore the card issuer as well as the acquirer are considered in the value network. To ensure positive network effects, this involves subsidizing the payer's side and financing the payment by the payee through a transaction fee charged by the acquirer, who in exchange authorizes payment (Cash and Tsai, 2018; Kazan and Damsgaard, 2016). An issuer responsible for physical or digital card issuance holds the necessary banking license, has a direct customer relationship with the payer and profits from transaction fees (e.g., Curve or Zilch). Since the card issuance process is fully automated by payment service provider, this also employs virtual one-time-use credit cards, whose validity expires after one transaction (e.g., Zip). Both lender and issuer are combined into a meta-role as a bank license partner to allow generalization and specialization. Payment service providers offer further services, such as P2P money transfer between the digital wallets of the platform participants (e.g., PayPal or Revolut), which increases the number of users on the payer side. As a result, platform attractiveness on the payee side grows, leading ultimately to indirect network effects (de Reuver et al., 2017; Staykova and Damsgaard, 2021). Further value activities are prepaid offers that enable anonymous online payment by replenishing the payer's digital wallet at the retailer's location and providing the payment service provider with its infrastructure (e.g., paysafecard). Moreover, payers receive services to control their expenses (reports, budgets, etc.), which gives them more transparency and control over their expenses, while the payment service provider gains additional customers' data (interests, payment willingness, etc.) (e.g., KakaoPay or Klarna). Lastly, customer loyalty programs (reward offers, vouchers, cashback promotions, etc.) are offered to strengthen customers' engagement with the payment platform ecosystem as well as matching payers and payees (e.g., Klarna, PhonePe or Rakuten).

#### 7.2.4.2 Personal Finance Management

Personal Finance Management (PFM) services empower customers to plan, monitor and organize their financial resources while increasingly being integrated into online banking portals of financial institutions or offered as dedicated services by platform providers (Laks, et al., 2013; Nüesch et al., 2015; Sachse et al., 2012). Due to changing customer loyalty towards utilization of multiple banking providers for different financial purposes, a demand for PFM offerings that aggregate data from different sources using multi-vendor integration is increasing (Sachse et al., 2012). Value propositions typically include transaction categorization (rent, groceries, entertainment, etc.), budgeting, advanced financial analytics, future forecasting, and advice on lower-priced products (insurance, loans, mobile plans, etc.) based on personal transactions (Alfaro et al., 2019; Torno et al., 2021). Ideally, PFM service provider support the optimization of consumers' financial situation by presenting, managing, and optimizing their personal financial and insurance situation (Busch et al., 2022; Fischer, 2021). For realizing the value proposition, the PFM service provider supplies the customer, acting as finance manager in this context, with reports on his financial situation (categorized income and expenses, budget utilization, savings made, etc.) and in some cases combines these with a credit score, which he inquiries from a national credit agency (e.g., Credit Karma). To determine the credit score, credit agencies primarily use past loan repayments from various lenders as base data (Liu, Land, et al., 2006; Tan and Phan, 2016). The PFM service provider receives the necessary account and transaction data from credit card issuers or deposit account providers, aggregated in the Bank License Partner meta-role. For this, banks either provide standardized interfaces (e.g., in Europe based on PSD2) or collaborate with the PFM service provider to offer a credit card or current account as part of the service to the finance manager (e.g. Deutsche Bundesbank) (Fischer, 2021). The PFM service provider's service is financed by subscription fees and or by commissions received for intermediating financial products and services offered by partner service providers. In some cases, PFM service providers offer loans that are designed to improve credit scores through regular repayments and a delayed disbursement of the loan amount once the loan is paid off in full (e.g., Brigit or Self). The offering is complemented by reward programs which are implemented by the PFM service provider itself (e.g., Verabank) as well as an affiliate advertising platform (e.g., Cardlytics for Kora) by providing incentives (cashback promotions, rebates, etc.) for the finance controller to take advantage of beneficial offerings (e.g., Kora or VeraBank).

#### 7.2.4.3 Robo-Advice

Prior to the fintech revolution, investing was a manual process guided by a human advisor assessing a customer's investment goals, investing the customer's money accordingly, and managing his or her portfolio. Robo-advisory is a new, automated way for (private) customers to invest their money (Jung, Dorner, Glaser, et al., 2018). Robo-advisor service provider offer digital financial advisory services that do not require human-to-human interaction, but shift the focus to human-computer interaction by guiding the client through an automated investment advisory process (Jung, Dorner, Glaser, et al., 2018). It manages a customer's portfolio automatically without any human intervention exploiting a set of rules and artificial intelligence (Fein, 2015). A questionnaire is used to determine the investors' objectives and the responses analyzed through an algorithm, determining and implementing a suitable investment strategy (Beketov et al., 2018; Fein, 2015). In addition, the robo-advisor is capable of continuously monitoring and managing the portfolio and either recommending appropriate investments or executing them automatically (Fein, 2015; Jung, Dorner, Weinhardt, et al., 2018). In literature, two groups of robo-advisors are described: static and dynamic ones (Jung, Dorner, Weinhardt, et al., 2018). While the former only rebalances the portfolio in the case of deviation of the identified goal, the latter provides the possibility to interfere at any time (Jung, Dorner, Weinhardt, et al., 2018). The client can always adjust his investment strategy which leads to an adjustment of the portfolio (Jung, Dorner, Glaser, et al., 2018). An investor participating in a robo-adivsory platform wants to invest his or her money and seeks for a qualitative management of his financial assets. Hence, the customer of a robo-advisor provides customer information and the money for the investments (e.g., M1 Finance). The robo-advisory involves three major value creation activities (i.e., customer assessment, automatic and human portfolio management) (e.g. Nucoro). By exploiting the customer information, the robo-advisor assesses the customer (e.g., by using a questionnaire) to identify a suitable investment category with appropriate assets. This category is used for the automated portfolio management (e.g., Nutmeg). The robo-advisor adds assets provided by an independent asset provider to the portfolio of the customer, manages the money transfer and the investments. This management of investments is the main cause for fees of a robo-advisor and can vary in the case of Nutmeg between 0.62% and 1.02%. The asset provider manages different asset classes such as real estate, exchange traded funds, art, and cryptocurrencies. Further financial services are offered to the customer by the financial institute such as banking or pension and further assets provision by an asset manager. Therefore, a financial institute collaborates with the robo-advisor so the advisor can offer more financial services in exchange of a transaction fees and the financial institute can extend its own customer base. A financial supervisory authority supervises the robo-advisor and the financial institute. Further, it provides them with licences, and they pay a fee to the authority (e.g., Liquid is supervised by the BaFin) (Dai, 2021). In addition to those identified actors it was noticeable that many Business-to-Customer (B2C) companies, such as Nutmeg or M1 Finance, do not offer a robo-advisor service with automatic investment services but instead rely on humans for portfolio management activities. However, Business-to-Business (B2B) provider companies, such as Nucoro or Bambu, provide robo-advisor services, automatically managing customers' portfolios and investing money.

#### 7.2.4.4 Peer-to-Peer Lending

In a traditional banking context, a borrower who needs money seeks out to a bank or a credit company to ask for a loan. P2P lending service provider avoid banks or credit companies and function as intermediaries to match lenders money with borrowers need for money (Berger and Gleisner, 2009; Lee, Lee, et al., 2011). This enables the borrower to receive a loan with a lower interest rate (Herzenstein et al., 2008; Klafft, 2008). On the other hand, the lender is able to invest his or her money with a predefined interest rate (Klafft, 2008). This also results in lower transactional cost because of the lack of any financial company due to peers exchanging money directly (Klafft, 2008; Lee, Lee, et al., 2011). The lending without the involvement of a financial institute results in a high degree of information asymmetry between lender and borrower that is partially outweight by the P2P platform (Lee, Lee, et al., 2011; Suryono et al., 2019). To prevent this asymmetry many P2P lending platforms (e.g., Auxmoney) recommend the borrower to provide several information to possible lenders. Such information could be the loans purpose, debts or delinquencies of the borrower (Klafft, 2008). When a customer needs to borrow money, he or she registers to a P2P lending Service Provider where a credit request has to be filled out by the applicant and then he or she receives a loan offer, which can be accepted or declined by the applicant (e.g., Auxmoney). The platform also checks the received information and can accept or decline the credit. When accepting the loan, the platform offers the loan as investing opportunity to the lenders (e.g., Prosper). Those then can decide if they want to invest or lend their money to the borrower. If lenders invest their money into a borrower, the borrower receives the money on his banking account and normally has to pay it back monthly starting from an agreed due date (e.g., Prosper). This money then is received by the lender as return. All financial transactions are handled with the help of a deposit bank which provides those services. But they are normally not involved with the loan offering.

#### 7.2.4.5 Trading

The concept of the neo-broker is comparable to that of neo-broking and describes trading in stocks and shares without a physical presence (Imerman and Fabozzi, 2020). Instead, the focus is primarily on digital communication that is handled via smartphone apps or websites (Imerman and Fabozzi, 2020). In this new form of stock trading, the customer invests his money through the Trading service provider in shares, exchange traded funds or other products at the appropriate trading times of the connected marketplace (stock exchange, market maker) in a self-determined manner without an advisor (Frölich and Lembach, 2021). Particular incentive of this new trading opportunity is, in addition to the independence, the price of the service. In contrast to traditional banks, investors are often charged very little or no fees at all (Frölich and Lembach, 2021). This is made possible by the principle of payment by order flow, in which a broker specifically settles all customer orders via a market maker or trading venue and receives a corresponding commission for this, which again arises from the spread as the bid/ask spread of the trading venue. Another difference and aspect of the price difference is the banking license, in contrast to classic brokers neo-brokers do not have this, but cooperate with a custodian bank, which provides the settlement accounts and corresponding banking services.

#### 7.2.4.6 Cryptocurrency

Distributed ledger technology enables decentralized, digital trust verification by consensus without the involvement of a central party (Gomber, Kauffman, et al., 2018; Kannengießer st al., 2019). This is based on the operation of highly available distributed database (ledger) stored on peer-to-peer distributed nodes (i.e., storage and computing devices) and protected against tampering by cryptographic concepts (Sunyaev et al., 2021; Wenke et al., 2022). The most prominent conceptual instance of distributed ledger technology is blockchain, which comprises a tamper-proof transaction history forming a foundation for transaction transparency (Verbovetska, 2019; Wenke et al., 2022). Cryptocurrencies (e.g., Bitcoin, Ethereum) are predominantly an implementation of the blockchain concept and act as a digital representation of a medium of exchange and store of value. Thereby, cryptocurrencies guarantee the transaction flow and the creation of additional currency units based on cryptography (Eigelshoven, Ullrich, and Parry, 2021; Kannengießer st al., 2019). In contrast to traditional fiat currencies, managed by an institution recognized as trustworthy (e.g., central bank), cryptocurrencies are issued by private developers (Das, 2019; Eigelshoven, Ullrich, and Parry, 2021). Cryptocurrency Exchange service provider (e.g., Coinbase, MoonPay) offer trading services for exchange of traditional fiat currencies and cryptocurrencies. From the customer's perspective, the exchange is an investment in a digital asset, with the cryptocurrency portfolio managed in a wallet (Eigelshoven, Ullrich, and Parry, 2021; Jorgensen and Beck, 2022). A miner, which is part of the distributed ledger technology system, verifies the transaction on the blockchain network and receives a reward (block reward or transaction fee) in return (Eigelshoven, Ullrich, and Bender, 2020). Here, the miner is responsible for validating and storing the distributed data as well as its integrity (Rückeshäuser, 2017). Furthermore, specialized hardware (e.g., ASCI miners) and mining pools are often utilized for efficient mining, which requires the provision of appropriate hardware and software (Eigelshoven, Ullrich, and Bender, 2020; Sedlmeir et al., 2020). Similarly, different hardware and software-based mechanisms (cold and hot wallet) are provided for storing the cryptocurrency (Jorgensen and Beck, 2022). Since mining cryptocurrencies and verifying transactions is energy-intensive due to complex calculations (Eigelshoven, Ullrich, and Bender, 2020), miners benefit from discounted electricity rates or partnerships with power suppliers to reduce mining costs.

#### 7.2.4.7 Overview of the Value Network

Building on the above-described sub-ecosystems of fintech platforms, the respective actors were analyzed, aggregated and corresponding value creation activities as well as resulting value flows were summarized into an e$^3$-value value network, which represents the generic corporate ecosystem (Fig. 7.4).

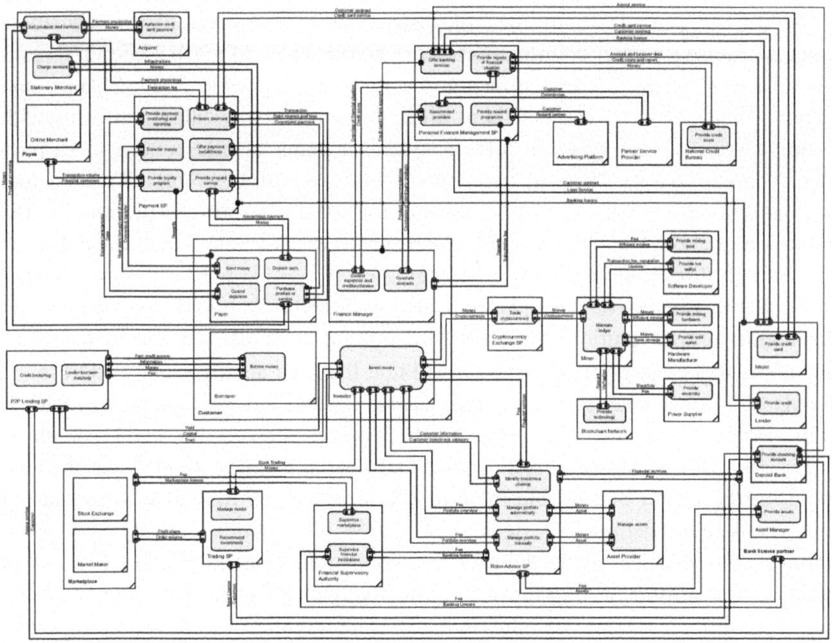

**Fig. 7.4** Generic Fintech Value Network

The customer was aggregated into a meta-role fulfilling four different roles (i.e., payer, financial manager, borrower, and investor) within the fintech ecosystem depending on the service used. It also becomes evident that the fintech platforms rely significantly on several bank-licensed partners in order to realize their offerings within the financial market. The aggregated roles and their attribution to the platforms is presented from Table A.26 in the appendix (Appendix A.8 in the Electronic Supplementary Material).

### 7.2.5 Discussion

Fintech platforms orchestrate different types of participants from multiple market sides. As multi-sided markets, they are able to benefit from direct and indirect network effects to propel their business (Armstrong 2006). However, platform business models require a proper coordination of the market sides to reach a certain threshold of actors for the network effects to occur (Janiesch et al., 2020; Shapiro and Varian, 1998). Our value analysis in fintech ecosystems revealed platforms with three to five different types of roles involved in financial transactions (e.g., payment, robo-advisory). These financial transactions requiring several different actors pose additional coordination requirements for focal platforms. Additionally, the attraction of enough actors on each market side to create enough traction and reach a minimum threshold is already challenging on two-sided markets (Schirrmacher et al., 2017). It gets even more complicated with additional market sides involved demanding proper ignition strategies. Although the customer is at the center of our fintech ecosystems, prior research proved that these strategies usually address the supply side (e.g., asset provider, bank license partners) (e.g., Eisenmann, Parker, et al., 2006; Parker and Van Alstyne, 2016; Parker, Van Alstyne, and Choundary, 2016). In addition, the ecosystem shows that fintech business models put the customer at the center and enable them to take on multiple roles due to increasing multi-bank relationships (Alt, Beck, et al., 2018), which offers entirely new opportunities. For instance, using digital technologies, any customer can easily adopt both the role of borrower and investor, effectively stepping into the role of a bank with little effort. Despite insights into the value creation of fintech ecosystems, the research has some limitations. The case study is driven by established offerings in the fintech market, bringing platform-based services widely adopted in this sector to the forefront of our analysis. As the example of cryptocurrency demonstrated, emerging technologies such as blockchain or artificial intelligence can disrupt proven business models in the financial sector, resulting in an impact on the value creation network. By automating through smart contracts, blockchain technology also holds potential to reduce necessary trust between business partners, leading to disintermediation where intermediaries are gradually removed from the ecosystem (Beinke et al., 2018; Huebner et al., 2019). Thus, future research projects include an investigation exploring the effects of disruptive technologies on fintech business models (e.g. venture capital, asset securitization) and associated changes in value flows (Chen, Zhang, et al., 2022). Moreover, our multi-case study especially considered multi-sided platforms entering the financial sector. Nevertheless, incumbents increasingly provide digitally supported financial services competing with new entrants. For instance, selected banks offer personal finance management tools as additional services (Gomber, Koch, et al., 2017). Thus, they are possible interesting objects of investigation. Modeling the value flows we concentrated on external connections between actors in fintech ecosystems and neglected intra-actor value flows. The design of our multi-case study did not reveal any information on intra-value creation and the structure of value creation within case companies selected. In addition, some support services (e.g., fraud detection, debt collection) were excluded from the analysis, since our study aimed to analyze the services from an end customer's perspective. This provides an opportunity for further research and the associated incorporation of a B2B perspective into the ecosystem. Besides, potential connections between Fintech service provider as well as participants from other domains were not considered and therefore topics could be further explored within an interdependency analysis (Adner, 2017). In this context, the emerging areas insurtech and proptech, which exhibit overlaps with the fintech industry (Das, 2019; Imerman and Fabozzi, 2020), may also be analyzed from a value network perspective and integrated into the value network. Furthermore, the derivation and modeling of roles, value creation activities, and value flows is prone to potential modeler bias. Although we tried to balance the level of abstraction, level of value creation, and degree of value flows between actors, the actual model was created subjectively by single persons. The synchronization of abstraction was established in regular meetings among the author team and a final workshop aggregating the overall ecosystem value model.

### 7.2.6 Conclusion and Outlook

The emergence of fintech technologies companies through ubiquitous digital technologies is transforming the entire financial industry and marks a new era of business models in the financial services sector. Through competence bundling of different market participants, the financial sector is shifting from an economy of scale to a platform-driven network economy, which is why corporate networks are becoming increasingly strategically important. Based on a literature analysis, we identified six major areas in fintech (i.e., alternative payment services, personal finance management, robo-advisory, peer-to-peer lending, trading, cryptocurrency). These areas consist of a focal platform orchestrating independent participants from multiple markets. We identified 28 roles (e.g., customers, service and bank license provider) in a multiple case study approach involving 171 single cases. We also derived major value creation activities (e.g., process payment, recommend providers, manage portfolio automatically) of these actors and the value flows (e.g., money, information, assets) between them. We depicted the overall fintech ecosystem with actors, value creation activities, and value flows using the $e^3$-value modeling notation. The customer meta-role forms the center of the fintech ecosystem, where the customer can take on four roles in a fintech ecosystem (i.e., payer, finance manager, borrower, investor). Value for the focal customer is created and exchanged through fintech platforms forming multi-sided markets. They provide advanced financial service to the customer by intermediating between the customer and additional fintech actors. Although we already cover six major areas in our fintech ecosystem map, future research may investigate roles, value creating activities, and value flows in additional fintech areas such as insurance or property as well as further business cases, empowered by distributed ledger technology. A further evaluation of the customer role in fintech ecosystems could result in principle for customer-centered designs of platforms in this domain with an increased customer experience. Future research might also further detail single actors within fintech ecosystems and their internal value creation activities and value flows. In the same vein, an analysis of incumbent financial companies (e.g., banks, financial advisors) might be worthwhile investigating for providing a more holistic overview on the fintech ecosystem. Considering the multi-sidedness platforms in fintech ecosystems and the number of startups in this sector, future research could also provide dedicated ignition strategies for successfully establishing platforms that reach a minimum threshold of participants. As we focused on the identification of roles and value in fintech ecosystems, an important avenue for future research would be an analysis of the impact of different technologies applied within the six areas or fintech in general on value creation. In conclusion, the paper provides an overall view of the key areas, roles relationships and value flows in the fintech sector and aggregates these into a consolidated value network.

# Discussion 8

In the following section, we briefly summarize the results of the research papers included in this thesis, discuss the theoretical and managerial contributions of this cumulative dissertation Ågerfalk and Karlsson, 2020; Whetten, 1989, evaluate our research against the guidelines for design science research (Hevner, March, et al., 2004, p. 82 ff.), reflect on the potential limitations of this thesis, and explain potential avenues for future research on platforms in general and on e-commerce ecosystems in particular.

## 8.1 Summary of Key Findings

In this cumulative dissertation, we investigated platforms in the context of e-commerce. We differentiated three platform types, namely innovation platforms, transaction platforms, and hybrid platforms. We analyzed platforms as virtual environments for product and service exchanges among different participant types in e-commerce ecosystems. Platforms in e-commerce require a set of boundary resources to facilitate the multilateral value creation for the (final) customer. The value creation can be amplified if the platform owner possesses a set of dynamic capabilities for proper ecosystem orchestration. Our investigation first concentrated on the conceptualization of platforms in e-commerce and architectural considerations. We then focused on a platform's boundary resources in e-commerce and the necessary capabilities for orchestrating the different ecosystem participants.

---

**Supplementary Information** The online version contains supplementary material available at https://doi.org/10.1007/978-3-658-45198-1_8.



T. Wulfert, *Selected Perspectives on Platforms in E-Commerce Ecosystems*, https://doi.org/10.1007/978-3-658-45198-1_8

In the first research paper (P1), we reviewed the application of the concept of platforms in e-commerce research (Wulfert and Karger, 2022a). Starting from our conceptualization of innovation platforms, transaction platforms, and hybrid platforms, we investigated a total of 7,463 platform-related research papers published in major conferences, journals, and books between 1993 and 2021. Besides the descriptive publication information reported (e.g., number of publications, performance analysis), we identified three distinct periods of platform research in e-commerce (Appendix A.4 in the Electronic Supplementary Material), four topical clusters, and numerous avenues for future research. The first phase (Platform 1.0) spans from 1993 to 2004 and is concerned with the emergence of the Internet facilitating initial electronic business models. In this phase, platforms are seen as the technological foundation for conducting e-commerce. The second phase (Platform 2.0) spans from 2005 to 2014 and is concerned with new business models in e-commerce from a transaction platform perspective. This phase focuses on advanced architectures and services for conducting e-commerce as well as the first social commerce considerations. The third phase (Platform 3.0) spans from 2015 to 2021 and focuses on social commerce, the exploitation of new technologies in e-commerce, and their integration in existing and new platforms. The first[1] topical cluster deals with platform business models and their components. The second topical cluster is concerned with social commerce and related technologies, while the third cluster describes innovation platforms as the technological infrastructure of e-commerce. The focus of the fourth topical cluster is on the architecture of innovation platforms as well as transaction platforms and their role in larger ecosystems. We also derived a number of research questions worthwhile pursuing for future research on platforms in e-commerce. The avenues for future research are summarized in Table 4.9.

In the second research paper (P2), we adopted an architecture lens (Niederman and March, 2019) and investigated innovation platforms and their role in e-commerce (ecosystems) in more detail (Wulfert and Karger, 2022b). We developed an architecture framework as a systematic representation of platforms in e-commerce and partitioned e-commerce ecosystems from an architectural perspective (Fig. 4.7). We described innovation platforms as providing the necessary application services for a variety of business models in e-commerce. In line with the literature describing innovation platforms as "extensible codebase" (Tiwana and Konsynski, 2010, p. 675), we introduced a number of boundary resources that can be used by third-party developers to provide platform extensions. These extensions

[1] The numbering of the topical clusters does not imply a hierarchy. We applied numerical denominations for reasons of reference to Fig. 4.5. The topical clusters were numbered in clockwise order.

can be provided to other ecosystem participants via extension markets (Jansen, 2013). Transactions platforms (i.e., platform business models) are among the business models that can be supported with application services. Moreover, we modeled a dedicated infrastructure layer that provides computing power and storage for the innovation platform. We also found out that the infrastructure services are often provided by the software provider taking on multiple roles and that successful actors in e-commerce are likely to implement their own application systems and make use of standardized cloud services for infrastructure services. Based on the first iteration of an architecture framework for platforms in e-commerce, we suggested a second design iterations deriving further insights from additional perspectives such as actor-network theory and structuration theory as well as considering platform ignition strategies and different governance modes employed by the platform owners (Fig. 4.8).

In the third research paper (P3), we developed a taxonomy (Nickerson, Varshney, et al., 2013) for describing transaction platforms in e-commerce (Wulfert, Seufert, et al., 2021). Our taxonomy for classifying transaction platforms in e-commerce consists of 21 dimensions with 99 characteristics in total (Fig. 4.10). We grouped the dimensions following the two-sided market theory (Armstrong, 2006; Rochet and Tirole, 2003), the participants they relate to (supply- and demand-side), and general marketplace dimensions (Eisenmann, 2008, p. 35 ff.). We suggested a taxonomy that makes use of the economic literature on multi-sided markets and network effects focusing on the matchmaking between and orchestration of the distinct market sides (Caillaud and Jullien, 2003; Rochet and Tirole, 2003) aided by aspects of governance as major building blocks of digital marketplaces (Boudreau, 2010; Hein, Schreieck, Wiesche, et al., 2016). The transaction platform group consists of the dimensions of article type, business orientation, centralization, owner behavior, price discovery, geographic focus, and subsidy side. The the group describing the demand side involves the dimensions of market access, fee structure, affiliation type, review system, network effect amplifiers, and provided services, while the group delineating supply-side participants adds the boundary resource dimension. We also applied the taxonomy to a total of 44 transaction platforms in the European Union (EU) and the United States (US) and derived potential avenues for future developments of these platforms.

In the fourth research paper (P4), we focused on the focal transaction platform owner's dual role in selling his own articles within the e-commerce ecosystems (Wulfert and Schütte, 2022). We derived architectural requirements resulting from this dual role for Information Systems (IS) (i.e., innovation platforms) providing application services for these focal transaction platforms. In total, we derived nine architectural requirements from a retailer's dual role on transaction platforms for the respective innovation platforms (Table 5.1). These requirements stem from the retailer's dual role, additional participant types, the affiliation of participants, the matchmaking between the different participant types, the variety of services provided, the innovation services provided, the smart services supported, the aggregated assortment of third-party sellers, and boundary resources. Moreover, we proposed four conceptual architectural patterns modeled in ArchiMate (Open Group, 2016) as a solution for the requirements (i.e., matchmaking for participants, innovation platform services, boundary resources, and aggregated assortment).

In the fifth research paper (P5), we used these architectural requirements and architectural patterns to analyze existing reference architectures for IS for the fulfillment of their requirements. We derived the reference architecture from the body of literature. In total, we investigated 13 reference architectures for the fulfillment of the seven aggregated architectural requirements (i.e., additional types of participants, affiliation to the transaction platform, matching engine, diversity of services, innovation platform services, aggregated assortment, and sophisticated boundary resources). Our reference architecture analysis (Angelov, Grefen, et al., 2012) revealed that none of the reference architectures under consideration fulfilled all seven requirements. Hence, we concluded a potential research gap existed in reference architectures for innovation platforms supporting transaction platforms in e-commerce.

In the sixth (P6) and the seventh (P7) research papers, we applied a multi-paper design science research approach (Vom Brocke, Winter, et al., 2020) to develop a set of design principles to prescribe the implementation of a focal platform's boundary resources in e-commerce (Wulfert, Strobel, et al., 2024; Wulfert, Woroch, Strobel, Seufert, et al., 2022). In P6, we derived six meta-requirements (Walls et al., 1992) and 19 design principles for boundary resources in e-commerce ecosystems (Table 6.2, Fig. 6.5). Applying these design principles should standardize boundary resources among focal platforms in e-commerce ecosystems and, thus, unify the development of extensions for multiple platforms and allow for developers' multi-homing. In P7, we made use of the 19 design principles and evaluated them against the experts' opinions on their reusability (Iivari, Rotvit Perlt Hansen, et al., 2021). These interviews with domain experts also allowed us to further specialize the design principles for innovation platforms and transaction platforms and determine which roles are involved in their design and implementation (Table 6.4). While innovation platforms require all seven condensed design principles (i.e., ecosystem conversation, extension marketplace, ecosystem extension, standards and guidelines, quality assurance, analysis and monitoring, and boundary resource evolution) identified for the attraction of third-party developers and extension development, ecosystem conversation, standards and guidelines, analysis and monitoring, and boundary resource evolution are important design principles for transaction platforms' relation to third-party sellers and service providers. Moreover, we identified Business Process Owners (BPOs), Internal Platform Developers (IPDs), and User Interface Designers (UIDs) as relevant roles involved in the design and implementation of boundary resources on the focal platform (Fig. 6.7).

In the eighth research paper (P8), we developed a reference procedure model for the management of boundary resources throughout their entire lifecycle on an innovation platform in the context of e-commerce ecosystems (Wulfert, 2023). We emphasized the strategic role of boundary resources for innovation platforms attracting third-party developers and potentially increasing the ecosystem's overall generativity. The boundary resource management lifecycle consists of the four layers of boundary resource governance, continuous communication, technical boundary resources, and supplementary boundary resources (Fig. 6.10). While boundary resource governance is concerned with the management of the portfolio of boundary resources, the determination of the boundary resources' lifespan, and the business alignment of the boundary resources, continuous communication activities ensure proper communication and collaboration with third-party developers. These core activities span the entire lifecycle of a boundary resource and can be seen as overarching. The lifecycle of technical boundary resources involves the phases of plan and design, development and test, release, maintenance, and optimization, as well as deprecation. It is augmented by a set of supplementary social boundary resources. We highlighted the intertwined nature of technical and supplementary boundary resources for ecosystem participants.

In the ninth research paper (P9), we investigated ecosystem orchestration in e-commerce ecosystems as a key success factor for increased value creation for the owner of the focal platform and the overall ecosystem (Wulfert, Woroch, and Strobel, 2024). We acknowledged that effective ecosystem orchestration requires dedicated capabilities from the focal platform owner (Addo, 2022; Autio, 2022). In particular, we provided a set of five dynamic capabilities necessary with the related structure (Gioia et al., 2013) for ecosystem orchestration in the context of e-commerce. This structure consists of 12 second-order themes and 44 first-order concepts (Fig. 7.2). Furthermore, we derived six successful orchestration types depending on the product and service assortment provided by the ecosystem and the ecosystem's structure. The dynamic capabilities in combination with their mapping to ecosystem types could be a source of sustainable competitive advantage (Teece et al., 1997) for the owners of the focal platform within e-commerce ecosystems and increase the overall competitiveness of the ecosystem (Fig. 7.3).

In the tenth research paper (P10), we analyzed Financial Technology (fintech) ecosystems as sub-ecosystems of e-commerce ecosystems responsible for payment processing (Woroch et al., 2022). We identified payment services, Personal Finance Management (PFM), robo-advisory, Peer-to-Peer (P2P) lending, trading, and cryptocurrency as major business model types in fintech. We derived a generic fintech value network applying $e^3$-value notation (Peppard and Rylander, 2006; Pousttchi, 2008). In particular, we focused on key ecosystem participants, value-creation activities, and value streams among key participants (Fig. 7.4). Based on the generic value network, we identified that platform business models are common in the fintech industry involving the orchestration of multiple market sides. Moreover, we discovered that customers can take four distinct roles at the center of the ecosystem when using fintech services.

## 8.2 Theoretical Contributions

In this cumulative dissertation, we shed light on platforms and their surrounding business ecosystems in e-commerce. We contributed to the general body of literature on platforms and ecosystems as well as to domain-specific research in the context of e-commerce. An e-commerce ecosystem is a network of independent participants centered around a focal transaction platform multilaterally exchanging products and services for the (final) customer (Wulfert, 2023; Wulfert, Woroch, Strobel, Seufert, et al., 2022; Yi and Ming, 2011). Besides the conceptualization of platforms in e-commerce, we further investigated two major pillars for amplifying the value creation of e-commerce ecosystems. As a first pillar, we prescribed the design of boundary resources for focal platforms in e-commerce ecosystems to allow e-commerce transactions and attract third-party developers with potential platform extensions. As a second pillar, we investigated ecosystem orchestration as a lever for more efficient resource usage in e-commerce ecosystems. Hence, the theoretical contribution of this cumulative dissertation is threefold.

First, we contribute an e-commerce-specific conceptualization to the literature on platforms and ecosystems. We introduced innovation platforms and transaction platforms to e-commerce and differentiated their roles in the value creation of e-commerce ecosystems.[2] Innovation platforms provide the necessary application services for conducting e-commerce transactions and provide the extensible

[2] We also acknowledge platforms can take on the innovation platform and transaction platform role simultaneously. We describe these platforms as hybrid platforms in line with recent publications (e.g., Cusumano, Gawer, and Yoffie, 2019; Gawer, 2021a,b).

codebase for extension development (Aulkemeier, Paramartha, et al., 2016; Baldwin and Woodard, 2009; Tiwana, Konsynski, and Bush, 2010; Wulfert and Karger, 2022b). Transaction platforms resemble virtual locations of commerce facilitating e-commerce transactions among a variety of actors (Täuscher and Laudien, 2018; Timmers, 1998; Turban, Whiteside, et al., 2017; Wulfert and Schütte, 2021). They can be considered a successful business model in e-commerce (Abdelkafi et al., 2019; Cusumano, Gawer, and Yoffie, 2019; Gawer, 2021b). We developed a platform architecture framework to differentiate these two platform types, highlighting their intertwined nature, and positioning them in e-commerce ecosystems (Wulfert and Karger, 2022b, p. 8). The framework also includes different actors of an e-commerce ecosystem and the technical infrastructure providing computing power for innovation platforms. Moreover, we emphasized the role of innovation platforms in e-commerce and the importance of third-party developers for increasing the ecosystem's generativity. We also analyzed reference architectures in the context of e-commerce for their fit with the peculiarities of innovation platforms and called for further research on the development of reference architectures supporting platform business models in e-commerce (Wulfert and Busch, 2022; Wulfert and Schütte, 2022).[3] For the description and development of transaction platforms, we developed a taxonomy with 21 dimensions and 99 characteristics that considered the economic circumstances of two-sided markets (Armstrong, 2006; Rochet and Tirole, 2003; West, 2003) and involved amplifiers for network effects.

Second, we extend the current body of literature on platform boundary resources with prescriptive design knowledge for the design (Wulfert, Strobel, et al., 2024; Wulfert, Woroch, Strobel, Seufert, et al., 2022) and management (Wulfert, 2023) of boundary resources in e-commerce ecosystems. We conceptualized boundary resources as boundary objects that behave as interfaces between a focal platform and ecosystem participants on the social, application, and development levels (Ghazawneh, 2012; Star and Griesemer, 1989). We proposed design knowledge for the boundary resources customized for innovation platforms and transaction platforms as the nexus of e-commerce ecosystems (Dal Bianco et al., 2014; Ghazawneh and Henfridsson, 2013; Wulfert, Strobel, et al., 2024; Wulfert, Woroch, Strobel, Seufert, et al., 2022). In particular, we provided a set of scientifically grounded and empirically validated design principles for boundary resource sets involving different roles for their design and implementation (Wulfert, Strobel, et al., 2024; Wulfert, Woroch, Strobel, Seufert, et al., 2022). For the holistic management of the entire lifecycle of

[3] The criteria for the reference architecture analysis (Angelov, Grefen, et al., 2012) resulted from a retailer's dual role operating a transaction platform while simultaneously selling their own articles (Wulfert and Schütte, 2022).

boundary resources, we developed a reference procedure model as a novel artifact in a design science research approach (Wulfert, 2023, p. 10 ff.). We further explicated the differentiation of boundary resource types (e.g., Dal Bianco et al., 2014; Engert et al., 2022; Foerderer, Kude, Schuetz, et al., 2019; Karhu, Gustafsson, et al., 2018) and suggested technical boundary resources with supplementary social boundary resources for innovation platforms.

Third, we contribute to the literature on ecosystem orchestration and value creation in business ecosystems (Woroch et al., 2022; Wulfert, Woroch, and Strobel, 2024). As we identified ecosystem orchestration as an important lever for the facilitation of value creation in e-commerce ecosystems (Autio, 2022; Hiller et al., 2022; Mann et al., 2022), we proposed five dynamic capabilities necessary for proper ecosystem orchestration (Wulfert, Woroch, and Strobel, 2024).[4] These dynamic capabilities are supported by 12 second-order themes and 44 first-order concepts following the approach by Gioia et al. (2013). Furthermore, we proposed six successful ecosystem orchestration types depending on the assortment type and the structure of the ecosystem. While we differentiated between full assortment and sub-assortment-oriented e-commerce-ecosystems, the ecosystem structure is denoted by the two dimensions of ecosystem centrality and ecosystem density (Brass and Burckhardt, 1993; Rowley, 1997; Wassermann and Galaskiewicz, 1994). We also contribute to the literature on value creation in business ecosystems by disentangling the value creation activities and value flows in payment ecosystems (Woroch et al., 2022).[5] We identified payees, payers, acquirers, payment service providers, and bank license partners as major participant types in payment ecosystems (Woroch et al., 2022, p. 10). Payment processing in e-commerce can be embedded in larger fintech ecosystems (Eickhoff et al., 2017; Floetgen, Mitterer, et al., 2021). Moreover, we derived insights into the role of transaction platforms in fintech ecosystems. Transaction platforms in the fintech industry orchestrate three to five different participant types but focus on the customers' value increase. They enable customers to take on four different roles due to increasing multi-bank relationships (Alt, Beck, et al., 2018). We summarize the scientific contributions of this cumulative dissertation in Fig. 8.1.

[4] For the organization of the dynamic capabilities, we applied Teece's (2007) tripartite scheme of dynamic capabilities for sensing, seizing, and transforming in dynamic environments.

[5] We consider payment ecosystems as a particular sub-ecosystem of e-commerce ecosystems.

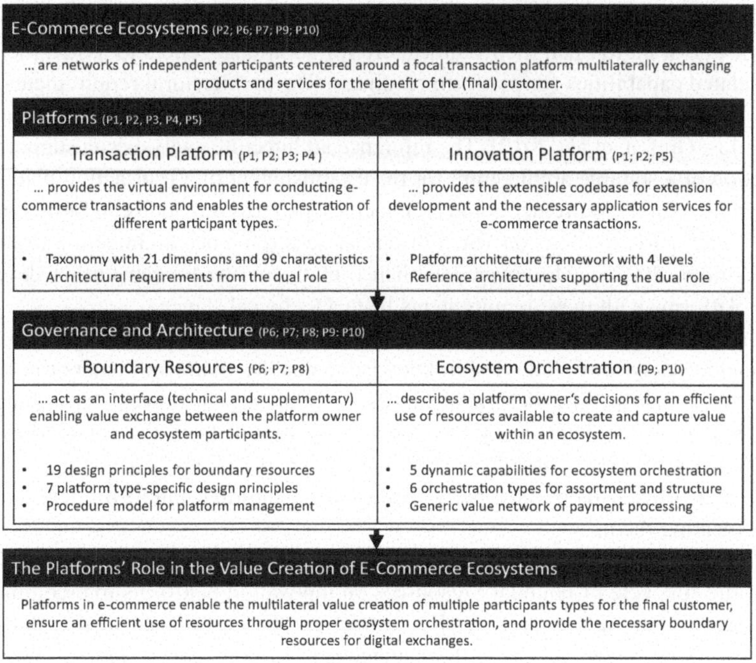


**Fig. 8.1** Contributions on Platform Research in E-Commerce

## 8.3 Managerial Contributions

Besides the contributions to the literature mentioned above, this cumulative dissertation also provides guidance and recommendations for focal platform owners and other participants in e-commerce ecosystems.

First, the platform architecture framework provides guidance for the selection of a suitable innovation platform for transaction platforms and third-party sellers. Moreover, third-party developers may use our architecture framework to position themselves adequately in such ecosystems and discover new business opportunities.

Second, the developed taxonomy for describing transaction platforms in e-commerce ecosystems can be used by platform owners to identify possible directions for the future development of the platform. In particular, we suggested five examples for developing a transaction platform involving measures to attract additional participants and value capture mechanisms depending on its lifecycle stage.

Third, the architectural requirements for innovation platforms can be applied by sellers who want to open their online store for additional sellers to assess their IS and related capabilities for platform readiness. The architectural requirements can be used to identify potential gaps in the current IS and decrease time-to-market (Angelov, Grefen, et al., 2012). The reference architecture analysis can support the selection of a suitable architecture for the establishment of a transaction platform. The architectural requirements and architectural patterns provided can further highlight which parts to focus on when instantiating a reference architecture (Schütte and Wulfert, 2022b). Moreover, established platform owners can further develop their platforms with these requirements from a technical stance.

Fourth, we provided guidance for the design of boundary resources in e-commerce ecosystems. The design knowledge enables platform owners to design and implement specific design instances of boundary resources depending on their requirements. Third-party developers may benefit from the standardization of boundary resources based on common design principles. They might multi-home and provide their third-party extension on multiple platforms without re-implementing them.

Fifth, we suggested a reference procedure model for the holistic management of the entire lifecycle of boundary resources on innovation platforms in e-commerce ecosystems. This reference procedure model can be customized to the platform owner's peculiarities and guide the management of boundary resources. We also offered guidance for platform owners on which supplementary boundary resources should support which lifecycle stage and technical boundary resource.

Sixth, we provided transaction platform owners with dynamic capabilities for proper ecosystem orchestration in e-commerce. We further operationalized these capabilities with additional second-order themes and first-order constructs. These capabilities can be implemented by the platform owners in e-commerce ecosystems to improve ecosystem orchestration. When implemented adequately, the ecosystem orchestration can ensure more efficient resource use in the ecosystem and, thus, increase the overall value creation. Moreover, we offered seven orchestration types for e-commerce ecosystems providing a full assortment and five for sub-assortment ecosystems. We further highlighted six successful orchestration types depending on the ecosystem's structure (i.e., ecosystem centrality and ecosystem density). These orchestration types can be used as blueprints for a platform owner's proper ecosystem orchestration. On the one hand, the orchestration type can be useful for transaction platforms in the ignition stage for the determination of a suitable type. On the other hand, they can be used by incumbents to evaluate their ecosystem orchestration and guide a potential re-configuration.

Seventh, we provided an overview of payment ecosystems necessary for payment processing in e-commerce. We suggested considering the customers' further roles in addition to the payer. Furthermore, the generic value network of the fintech industry can assist in determining the market position for existing and future ecosystem participants.

## 8.4 Evaluation of the Design Science Research Endeavors

Hevner, March, et al. (2004, p. 82 ff.) introduced seven guidelines for design science in IS research. Although these guidelines are not without critique (e.g., Venable et al., 2016; Zelewski, 2007), they are often applied to support and evaluate design science research endeavors (Niehaves and Becker, 2006, p. 3584). These guidelines suggest pathways for researchers and increase the rigor of a research project (Venable et al., 2016; Venable, 2006; Zelewski, 2007).

Hevner et al.'s (2004, p. 82 ff.) first guideline (*Design as an Artifact*) denoted that the output of a design science research endeavor should be "a purposeful IT artifact created to address an important organizational problem." The authors further explained that "artifacts constructed in design science research are rarely full-grown IS [...]" but "[...] innovations that define the ideas, practices, technical capabilities, and products [...]" (Hevner, March, et al., 2004, p. 83). Following the output types of design science research presented by Vaishnavi, Kuechler, et al. (2019, p. 16 ff.), we classified the main artifacts developed within the research papers of this cumulative dissertation.[6] The first research paper included a bibliometric study of platform research in e-commerce and provided an overview of past, present, and future research on platforms (Wulfert and Karger, 2022a). In this research paper, we introduced innovation platforms and transaction platforms as constructs in e-commerce. In the second research paper, we developed an architecture framework shaping innovation platforms and transaction platforms in e-commerce (Wulfert and Karger, 2022b). The platform architecture framework can be classified as a model positioning platform types in e-commerce and depicting the relations among them (Vaishnavi, Kuechler, et al., 2019, p. 17). In the third research paper, we derived a taxonomy for describing transaction platforms in e-commerce with a focus on the two-sidedness and network effects (Wulfert, Seufert, et al., 2021). Taxonomies are also considered as "models" (Nickerson, Muntermann, et al., 2009, p. 2) in design science research. In the fourth research paper, we developed architectural patterns as part of an IS reference architecture for platforms in e-commerce. The architectural

[6] For an overview on artifact types in design science research, see Table 2.3.

patterns are considered as (parts of) an architecture depicting "[h]igh level structures of systems" (Vaishnavi, Kuechler, et al., 2019, p. 16). We analyzed existing reference architectures in the literature for the fulfillment of the developed patterns in the fifth research paper.[7] In the sixth and seventh research papers, we developed a set of design principles for boundary resources in e-commerce ecosystems (Wulfert, Woroch, Strobel, Seufert, et al., 2022). Vaishnavi, Kuechler, et al. (2019, p. 19) mentioned design principles as a dedicated artifact type that guides the design of design instances. In the eighth research paper, we presented a boundary resource management lifecycle. As the management lifecycle described activities for the management of technical boundary resources and related supplementary boundary resources, it can be characterized as a method (Vaishnavi, Kuechler, et al., 2019, p. 16). In the ninth research paper, we developed capability configurations for the orchestration of successful e-commerce ecosystems. Following Vaishnavi, Kuechler, et al. (2019, p. 19) we suggested a framework consisting of five dynamic capabilities and related second-order constructs and first-order themes.[8] In the tenth research paper, we derived a value network using the $e^3$-value notation (Gordijn, Osterwalder, et al., 2005; Pousttchi and Gleiss, 2019) to analyze value creation in payment ecosystems as a sub-ecosystem of e-commerce ecosystems. The value network can be characterized as a model containing "[s]ets of propositions or statements expressing relationships between constructs" (Vaishnavi, Kuechler, et al., 2019, p. 16). We depict an overview of the main artifacts' types per research paper in Table 8.1.

**Table 8.1** Overview of Artifact Types

| # | Reference | Artifact | Artifact Type |
|---|---|---|---|
| P1 | Wulfert and Karger (2022a) | Platform Types | Constructs |
| P2 | Wulfert and Karger (2022b) | Platform Architecture Framework | Model |
| P3 | Wulfert, Seufert, et al. (2021) | Taxonomy for Transaction Platforms | Model |
| P4 | Wulfert and Schütte (2022) | Architectural Patterns | (Reference) Architecture |

(Continued)

[7] The fifth research paper does not include an artifact in the sense of design science research (Hevner, March, et al., 2004; March and Smith, 1995; Vaishnavi, Kuechler, et al., 2019).

[8] However, we need to stress that the framework lacks appropriate evaluation.

**Table 8.1** (Continued)

| # | Reference | Artifact | Artifact Type |
|---|---|---|---|
| P5 | Wulfert and Busch (2022) | n.a. | n.a. |
| P6 | Wulfert, Woroch, Strobel, Seufert, et al. (2022) | Design Principles for Boundary Resources | Design Principles |
| P7 | Wulfert, Strobel, et al. (2024) | Design Principles for Boundary Resources | Design Principles |
| P8 | Wulfert (2023) | Boundary Resource Management Lifecycle | Method |
| P9 | Wulfert, Woroch, and Strobel (2024) | Capabilities for Ecosystem Orchestration and Orchestration Types | Framework |
| P10 | Woroch et al. (2022) | Value Creation Framework for Payment Ecosystems | Model |

For the second guideline (*Problem Relevance*), Hevner,March, et al. (2004, p. 84) suggested that the objective of IS research is to "enable the development and implementation of technology-based solutions to heretofore unsolved and important business problems." Design science research supports "this goal through the construction of innovative artifacts aimed at changing the phenomena that occur" (Hevner, March, et al., 2004, p. 84). We already highlighted the relevance of platforms in the economy and the relevance of research on platforms and ecosystems in e-commerce in general, and on boundary resources in particular, in Chap. 1. Furthermore, the introductions of the research papers of this cumulative dissertation also contain statements on the particular problem's relevance for practice. Transactions platforms have changed the structures (Miremadi et al., 2023, p. 1) and have been established as successful intermediaries in many industries (Gawer, 2021b, p. 7 ff.).[9] E-commerce is one such sector, with an expected overall revenue of $ 7.4 trillion in 2025 (EMarketer, 2022). Platforms orchestrating larger ecosystems are considered important levers for increased value creation and are able to capture the lion's share of this value (Zhao, von Delft et al., 2020, p. 1).[10] Increasing revenues, sales, and customers in e-commerce require adequate information services to support these elec-

[9] We also refer to Stallkamp and Schotter (2021) for the impact of platforms on entire industries.

[10] Multiple transactions platforms, such as Amazon, Facebook, Whats App, and YouTube, have billions of participants on their demand sides (Manyika et al., 2016, p. 6).

tronic transactions (EMarketer, 2022; Grewal, Roggeveen, et al., 2017; Kollmann, 2022). Innovation platforms can provide these application services for retailers and platform providers. The market volume of innovation platforms in e-commerce is expected to reach $ 7.4 trillion in 2025 (AppsRunTheWorld, 2021; Datanyze, 2021). Most of these innovation platforms provide dedicated boundary resources for third-party developers, expanding the feature set of these platforms through the provision of third-party extensions (Baldwin and Woodard, 2009; Jansen, 2013; Staykova and Damsgaard, 2015).

Moreover, recent research called for domain-specific research on digital platforms (de Reuver et al., 2017; McIntyre and Srinivasan, 2017) and more intensive research on the design and control of boundary resources to provide platform owners with new capabilities for ecosystem governance (de Reuver et al., 2017; Hein, Schreieck, Riasanow, et al., 2020; Yoo, Henfridsson, et al., 2010). Hevner,March, et al. (2004, p. 85) also suggested that the relevance of every design science research endeavor needs to be evaluated "with respect to the constituent community." In IS research, this community is formed by practitioners developing, operating, and working with IS. In our interview series with domain experts (e.g., Wulfert, 2023; Wulfert and Karger, 2022b; Wulfert, Strobel, et al., 2024), we also discussed the relevance of boundary resources for innovation platforms and transaction platforms in e-commerce. We present exemplary quotes from these interviews:[11]

> "Yeah, overall, I think [the design principles for boundary resources in e-commerce ecosystems] absolutely address real and important problems" (#4).
> "I would say all of [the design principles] are relevant in our professional life." (#5)
> "Using the design principles and understanding platforms streamlines developer engagement, therefore I value [the design principles for platforms in e-commerce] as highly important" (#7).

In the third guideline (*Design Evaluation*), Hevner,March, et al. (2004, p. 85 ff.) suggested that the "utility, quality, and efficacy of a design artifact must be rigorously demonstrated." We evaluated our artifacts designed using demonstration, informed argument, and interviews. We demonstrated our architecture framework for platforms in e-commerce in an application for selected platforms in e-commerce (Wulfert and Karger, 2022b). The taxonomy for describing transaction platforms in e-commerce was applied to characterize 44 transaction platforms (Wulfert, Seufert, et al., 2021). We evaluated the architectural patterns for transaction platforms using

[11] As requested by the interviewees, we did not mention any company names for the quotes. Instead, we will reference the interviewee number as presented in Wulfert, Strobel, et al. (2024).

informed arguments from the literature on two-sided markets and IS reference architecture (Wulfert and Schütte, 2022). While the design principles for boundary resources in the e-commerce ecosystem were evaluated using informed arguments in Wulfert, Woroch, Strobel, Seufert, et al. (2022), we further evaluated these design principles in a series of interviews with domain experts to increase their fitness and projectability (Vom Brocke, Winter, et al., 2020). The boundary resource management lifecycle was also evaluated in a series of interviews with domain experts. We also evaluated our value models payment ecosystems using informed arguments (Woroch et al., 2022). We provide an overview of the evaluation methods applied for evaluating the developed artifacts within this cumulative dissertation in Table 8.2.

**Table 8.2** Overview of Evaluation Types

| # | Reference | Evaluation Type |
|---|---|---|
| P1 | Wulfert and Karger (2022a) | n.a. |
| P2 | Wulfert and Karger (2022b) | Demonstration |
| P3 | Wulfert, Seufert, et al. (2021) | Demonstration |
| P4 | Wulfert and Schütte (2022) | Informed Argument |
| P5 | Wulfert and Busch (2022) | n.a. |
| P6 | Wulfert, Woroch, Strobel, Seufert, et al. (2022) | Informed Argument |
| P7 | Wulfert, Strobel, et al. (2024) | Interview |
| P8 | Wulfert (2023) | Interview |
| P9 | Wulfert, Woroch, and Strobel (2024) | n.a. |
| P10 | Woroch et al. (2022) | Informed Argument |

In the fourth guideline (*Research Contributions*), Hevner, March, et al. (2004, p. 87) requested "clear contributions in the areas of the design artifact, design construction knowledge [...], and/or design evaluation knowledge [...]." This serves to advance existing scientific knowledge (Zelewski, 2007). The authors evaluated a research contribution in design science research along the three characteristics of novelty, generality, and significance of the designed artifact. We advanced the knowledge of platforms and ecosystems in the specific domain of e-commerce.[12] The specialization of existing research (e.g., Wulfert and Karger, 2022b; Wulfert

[12] In doing so, we addressed calls for domain-specific research on platforms and boundary resources (de Reuver et al., 2017; Hein, Schreieck, Riasanow, et al., 2020; McIntyre and Srinivasan, 2017).

and Schütte, 2022; Wulfert, Woroch, and Strobel, 2024) and development of new artifacts for the context of e-commerce (e.g., Wulfert, 2023; Wulfert, Seufert, et al., 2021; Wulfert, Woroch, Strobel, Seufert, et al., 2022) can be considered as novel. We further detailed research on boundary resources in general and on e-commerce ecosystems in particular. With reference to our series of interviews with domain experts (#10, #12), the derived design principles and the boundary resource management lifecycle were deemed to be generally applicable in other software-related domains. The application of these artifacts might entail significance for research and practice.[13] Following the taxonomy by Gregor (2006, p. 619 ff.), we can classify our contribution according to the following types: analysis (1), explanation (2), prediction (3), explanation and prediction (4), and design and action (5). With our bibliometric analysis, we describe the past, present, and future of platform research in e-commerce (Wulfert and Karger, 2022a). Our platform architecture framework describes the relations of platforms in e-commerce ecosystems (Wulfert and Karger, 2022b). The taxonomy developed for describing different types of transaction platforms in e-commerce (Wulfert, Seufert, et al., 2021) serves the description objective of providing transparency about an aspect in IS research (Gregor, 2006, p. 619). The architectural patterns can provide guidance for implementing IS supporting platform business models (Wulfert and Schütte, 2022). In Wulfert and Busch (2022), we analyzed existing reference architectures in e-commerce for the support of platform business models. The design principles for the implementation of boundary resources on focal platforms in e-commerce ecosystems provided guidance for design and action (Wulfert, Strobel, et al., 2024; Wulfert, Woroch, Strobel, Seufert, et al., 2022). The boundary resource management lifecycle describes a typical lifecycle of technical boundary resources and provides guidance for design and action for platform owners (Wulfert, 2023). The dynamic capabilities for ecosystem orchestration describe the dynamic capabilities necessary for an orchestration in e-commerce ecosystems (Wulfert, Woroch, and Strobel, 2024). The orchestration types developed can be considered a contribution to design and action, prescribing the capabilities required for a successful orchestration of participants in e-commerce ecosystems depending on particular ecosystem structures. The value model for payment ecosystems described the value co-creation within fintech ecosystems in an abstract manner and explained the relations among ecosystem participants (Woroch et al., 2022). We summarize the contribution of the research papers within this cumulative dissertation in Table 8.3.

[13] For a more detailed elaboration on the scientific and managerial contribution of this cumulative dissertation, please refer to Section 8.2 and Section 8.3 as well as the single research papers.

**Table 8.3** Overview of Scientific Contributions

| # | Reference | Contribution | Type |
|---|---|---|---|
| P1 | Wulfert and Karger (2022a) | Overview of platform research in e-commerce | 1 |
| P2 | Wulfert and Karger (2022b) | Describe relations of platform types in e-commerce | 1 |
| P3 | Wulfert, Seufert, et al. (2021) | Describing transaction platforms in e-commerce and indicate future actions | 1,5 |
| P4 | Wulfert and Schütte (2022) | Architectural patterns | 5 |
| P5 | Wulfert and Busch (2022) | Analysis of e-commerce reference architectures | 1 |
| P6 | Wulfert, Woroch, Strobel, Seufert, et al. (2022) | Design principles for boundary resources in e-commerce | 5 |
| P7 | Wulfert, Strobel, et al. (2024) | Design principles for boundary resources in e-commerce | 5 |
| P8 | Wulfert (2023) | Boundary resource management lifecycle | 1,5 |
| P9 | Wulfert, Woroch, and Strobel (2024) | Capabilities and configurations for the orchestration of e-commerce ecosystems | 2, 5 |
| P10 | Woroch et al. (2022) | Value creation structure of payment ecosystems | 1,2 |

In the fifth guideline (*Research Rigor*), Hevner, March, et al. (2004, p. 87 f.) demanded "the application of rigorous methods in both the construction and evaluation" of the artifact designed. As Hevner, March, et al. (2004, p. 88) suggested to ground the construction and evaluation of the design artifact in existing research, we often synthesized the body of literature in structured literature reviews (e.g., Woroch et al., 2022; Wulfert, Woroch, Strobel, Seufert, et al., 2022). In general, structured literature reviews are useful for setting the foundation at the beginning of research endeavors (e.g., Okoli and Schabram, 2010; Rowe, 2014; Webster and Watson, 2002). We further derived insights from e-commerce practice using well-established research approaches for the exploration of new concepts and theories. We especially applied multi-case analyses and interview analyses (Eisenhardt, 1989b; Misoch, 2015; Yin, 2018) to abstract and generalize knowledge from e-commerce practice. We abstracted from the peculiarities of concrete platforms in e-commerce to generate knowledge for a class of systems (i.e., platforms in e-commerce)(Gregor, Müller, et al., 2013; Lee, Pries-Heje, et al., 2011). We often framed our single research endeavors in design science research approaches (e.g., Wulfert, 2023; Wulfert and Karger, 2022b; Wulfert and Schütte, 2022; Wulfert, Woroch, Strobel, Seufert, et al., 2022). An overview of the methods used in the individual research papers was provided in Section 2.3 and in detail in the individual research papers.[14] For each research paper, we provided reasons for the application of the selected method and their applicability to the concrete research problem. We provide an overview of the scientific approaches applied within the research papers of this cumulative dissertation in Table 8.4.

**Table 8.4** Overview of Research Approaches

| # | Reference | Research Approach |
|---|---|---|
| P1 | Wulfert and Karger (2022a) | Bibliometric analysis (Broadus, 1987; Donthu, Kumar, Mukherjee, et al., 2021) |
| P2 | Wulfert and Karger (2022b) | DSR approach (Hevner, March, et al., 2004) involving a multi-case analysis (Yin, 2018) and qualitative content analysis of interview data (Mayring, 2014) |
| P3 | Wulfert, Seufert, et al. (2021) | Taxonomy development (Nickerson, Varshney, et al., 2013) |
| P4 | Wulfert and Schütte (2022) | DSR approach (Peffers et al., 2007) |
| P5 | Wulfert and Busch (2022) | Systematic literature review (Vom Brocke, Simons, et al., 2009) |
| P6 | Wulfert, Woroch, Strobel, Seufert, et al. (2022) | DSR approach (Hevner, March, et al., 2004) involving a systematic literature review (Vom Brocke, Simons, et al., 2009) and a multi-case analysis (Yin, 2018) |
| P7 | Wulfert, Strobel, et al. (2024) | Qualitative content analysis (Mayring, 2014) of a series of interviews (Misoch, 2015) |
| P8 | Wulfert (2023) | DSR approach involving (Vaishnavi, Kuechler, et al., 2019) involving a systematic literature review (Vom Brocke, Simons, et al., 2015) and multi-case analysis (Yin, 2018) |
| P9 | Wulfert, Woroch, and Strobel (2024) | Systematic literature review (Vom Brocke, Simons, et al., 2009), multi-case analysis (Yin, 2018), case survey (Larsson, 1993), and csQCA (Park, Fiss, et al., 2020) |
| P10 | Woroch et al. (2022) | Systematic literature review (Vom Brocke, Simons, et al., 2015) and multi-case analysis (Yin, 2018) |

[14] For an overview of the single research endeavors and their integration in the overarching design science research approach (Peffers et al., 2007, p. 54), see Fig. 2.1.

For an overview of the different evaluation types, we refer to Table 8.2.[15] Moreover, research rigor is based on consistent definitions, internal consistency, and formal language representation of the subject matter (Zelewski, 2007, p. 78). We strived for a high internal consistency regarding the terms and concepts used. Hence, we introduced the concepts of innovation platform, transaction platform, hybrid platform, business ecosystem, and e-commerce ecosystem in Chap. 3. Even if we used synonyms for focal terms in single research papers, we annotated them with footnotes accordingly.[16]

In the sixth guideline (*Design as a Search Process*), Hevner,March, et al. (2004, p. 88) described design science as "inherently iterative." Simon (1996, p. 128 ff.) depicted the iterative nature of design science research as a generate and test cycle. The aim of design science research endeavors is not to develop the best and optimal designs but to "produce feasible, good designs that can be implemented in the business environment" (Hevner, March, et al., 2004, p. 88).[17] In this regard, problems can be simplified and decomposed to develop satisfactory solutions for smaller real-world problems (Hevner, March, et al., 2004; Simon, 1996). The process of "satisficing" (Simon, 1996, p. 28 ff.) aims at developing starting points for further designs without specifying the totality of possible designs and solutions. In this cumulative dissertation, we first focused on platforms as the nexus of e-commerce ecosystems (Wulfert and Karger, 2022a,b; Wulfert, Seufert, et al., 2021). We conceptualized innovation platforms and transaction platforms as major platform types in e-commerce and further analyzed potential architectural patterns (Wulfert and Busch, 2022; Wulfert and Schütte, 2022). We then investigated the focal platform's role within e-commerce ecosystems and value creation in these ecosystems in general. Finally, we analyzed the interaction between the owner of the focal platform and other e-commerce ecosystem participants. We provided design principles for boundary resources propelling the interaction between the owner of the focal platform and other e-commerce ecosystem participants in general and third-party developers in particular. As the management of these boundary resources is a crucial activity for the success of the focal platform and the surrounding ecosystem (Boudreau, 2010;

[15] For further information, refer to the explanations for guideline three (Design Evaluation).

[16] For instance, we used the term "multi-sided market" for transaction platforms in e-commerce in Wulfert, Seufert, et al. (2021). As we could not update the published research papers for this cumulative dissertation, we tried to increase internal consistency by annotating terms deviating from the understanding of this cumulative dissertation introduced in Chap. 3.

[17] We support the objective for "good" solutions in design science research with our meta-theoretical assumptions (Goldkuhl, 2012a,b). Knowledge in pragmatism is produced "through making" and needs to be applicable to be "true" (Creswell and Creswell, 2022; Lee and Nickerson, 2010; Rescher, 2000).

Yoo, Henfridsson, et al., 2010), we further proposed a boundary resource management lifecycle consisting of activities necessary for the proper management of technical boundary resources and supplementary boundary resources. We especially tested our designed artifacts in a series of semi-structured interviews with domain experts from innovation platforms and transaction platforms in e-commerce (e.g., Wulfert and Karger, 2022b; Wulfert, Strobel, et al., 2024).

Regarding the seventh guideline (*Communication of Research*), Hevner,March, et al. (2004, p. 90) demanded that design science research needs to be "presented both to technology-oriented as well as management-oriented audiences." We published the research papers of this cumulative dissertation in renowned IS journals and IS conference proceedings for scholars. Among others, we successfully published two research papers in the Electronic Markets journal (Wulfert, 2023; Wulfert, Woroch, Strobel, Seufert, et al., 2022) and presented our research at the International Conference on Information Systems (Woroch et al., 2022). We also presented our research results to practitioners in preparation and debriefing of our interview sessions with representatives from leading platforms in e-commerce. We summarize the publication media of our research papers in Table 8.5.

**Table 8.5** Overview of Publication Media

| # | Reference | Publication Medium |
|---|---|---|
| P1 | Wulfert and Karger (2022a) | Corporate Ownership & Control |
| P2 | Wulfert and Karger (2022b) | Pacific Asia Conference on Information Systems |
| P3 | Wulfert, Seufert, et al. (2021) | Hawaii International Conference on System Sciences |
| P4 | Wulfert and Schütte (2022) | SN Computer Science |
| P5 | Wulfert and Busch (2022) | International Conference on Enterprise Information Systems |
| P6 | Wulfert, Woroch, Strobel, Seufert, et al. (2022) | Electronic Markets Journal |
| P7 | Wulfert, Strobel, et al. (2024) | Electronic Commerce Research (under review) |
| P8 | Wulfert (2023) | Electronic Markets Journal |
| P9 | Wulfert, Woroch, and Strobel (2024) | The Journal of Strategic Information Systems (under review) |
| P10 | Woroch et al. (2022) | International Conference on Information Systems |

## 8.5 Limitations and Avenues for Future Research

Despite the aforementioned theoretical and managerial contributions, this thesis also involves some limitations that open up potential avenues for future research. We also take on selected avenues for future research from the research papers and unfold them in more detail. We focus on four topics for the elaboration of limitations and opportunities for future research. We describe methodological considerations, architectural considerations, e-commerce ecosystem development, and the applicability of our domain-specific research to other domains and platform research in general.

This cumulative dissertation includes research papers in which qualitative research methods (e.g., systematic literature review, multi-case analysis, semi-structured interviews, qualitative content analysis) are applied (Table 8.4). Subjective interpretations and investigator bias are often described as hazards in qualitative research (e.g., Beuving and De Vries, 2015; Guba and Lincoln, 1982; Miles et al., 2014; Straub and Gefen, 2004). Although we applied several methods for improving reliability and validity in our research endeavors (e.g., Guba, 1981; Miles et al., 2014), our interpretations and research results might be prone to investigator bias.[18] We tried to match patterns from the literature with our observations and discussed our interpretations among the teams of authors (Shenton, 2004). However, we cannot guarantee objectivity with regard to our interpretations and research results. Therefore, further research is invited to verify our research results independently and apply them to varying contexts to prove generalizability.

In addition to a potential investigator bias, several research papers within this thesis apply a systematic literature review (Table 8.4) following Vom Brocke, Simons, et al. (2009), Vom Brocke, Simons, et al. (2015), and Webster and Watson (2002). The application of this scientific approach in the context of IS research needs to be discussed.[19] None of the aforementioned approaches defined and restricted what is included in the literature that is screened. The authors seemed to imply a focus on scientific literature but did not explicate this scope. Hence, we focused the body of literature searched on scientific literature–mainly from scientific journals and conference proceedings–and deliberately excluded white papers and practitioner reports. We also included additional quality criteria when filtering our literature

[18] Please refer to Section 2.3.2 for a detailed elaboration on the methods applied for increasing research rigor in our research papers.

[19] In essence, approaches for systematic literature reviews presuppose a cumulative process for creating knowledge, in which not only the past knowledge needs to have the same meta-theoretical assumptions but also knowledge developed in the future needs to follow these assumptions.

sample (Bandara, et al., 2015; Cram et al., 2020), since the mere identification of literature for a selected search query does not necessarily result in high-quality research papers. As the systematic literature review approaches emphasize process quality rather than the content-based quality of the research results, we augmented the literature-based results with findings from real-world cases in multi-case studies (Yin, 2018) and interview data (Myers and Newman, 2007). Exploratory multi-case analyses were also applied, if the body of literature on a particular topic was deemed to be sparse or even non-existent (Yin, 2018).

We introduced different perspectives for analyzing platforms discussed in the literature (Gawer, 2014; Reillier and Reillier, 2017; Sun, Gregor, et al., 2015; Thomas et al., 2014). For our research focus on innovation platforms, transaction platforms, and hybrid platforms in e-commerce, we chose five different theoretical lenses for the research endeavors (Niederman and March, 2019). The perspectives taken on for the analysis of platforms in e-commerce are enterprise architecture (Ahlemann, 2012; Angelov, Grefen, et al., 2012; Cloutier et al., 2010), boundary resource (Dal Bianco et al., 2014; Eaton et al., 2015; Ghazawneh and Henfridsson, 2013), application lifecycle (AXELOS, 2012; Kääriäinen, 2011; Stark, 2011), dynamic capabilities (Eisenhardt and Martin, 2000; Teece et al., 1997), and value creation perspective (Adner, 2017; Vargo and Lusch, 2014; Vargo, Maglio, et al., 2008). In this regard, we needed to limit our results to these five perspectives and did not provide a holistic design theory for platforms in e-commerce (Gregory and Muntermann, 2011; Kuechler and Vaishnavi, 2008; Walls et al., 1992). Therefore, we encourage future research to go beyond our five perspectives and analyze platforms in e-commerce from multiple facets (Wulfert and Karger, 2022a).

Moreover, this cumulative dissertation involves a number of research papers in which we developed prescriptive design knowledge for "a class of artifacts" (Sein et al., 2011, p. 66). These design principles include abstract knowledge (Chandra et al., 2015; Gregor, Chandra Kruse, et al., 2020; Hevner and Chatterjee, 2010) for the design, implementation, and management of boundary resources (Wulfert, 2023; Wulfert, Strobel, et al., 2024; Wulfert, Woroch, Strobel, Seufert, et al., 2022). As the meta-requirements and design principles for boundary resources are developed for arbitrary innovation platforms and transaction platforms in e-commerce, they need to be further instantiated for use on a particular platform. Hence, a platform owner might apply the design principles developed for the design of its boundary resources, but he or she has to further specialize them to fit the particular requirements of that platform. As the delta between a design principle and the particular design instance at a platform might be huge, Meth et al. (2015, p. 807) proposed the derivation of additional intermediary design features to accelerate and simplify the design of concrete design instances. Therefore, we encourage future research to further instantiate the design principles developed and derive design features as an additional intermediary step toward concrete instances. Another potentially interesting avenue for future research would be to evaluate the prescriptive design knowledge derived and implement boundary resources, the boundary resource management lifecycle, and capabilities for ecosystem orchestration accordingly.[20] While the application of the prescriptive design knowledge requires customization to the peculiarities for the selected organization (Hevner, March, et al., 2004; Vaishnavi, Kuechler, et al., 2019; Vom Brocke, 2015), it enables testing of the performance of the prescriptive design knowledge on the level of instances against appropriate indicators that can further inform the design knowledge.[21] Besides our focus on technical and supplementary social boundary resources (Dal Bianco et al., 2014), future research can further investigate the role of "human boundary spanners" (Wei et al., 2020, p. 533) in the e-commerce ecosystem. It might be interesting to investigate how human boundary spanners, such as salespeople and account managers, can influence the attraction and retention of ecosystem participants. Wulfert, Strobel, et al. (2024) identified BPO, IPD, and UID as relevant roles in the design and implementation of boundary resources on platforms in e-commerce but disregard their role in bridging organizational boundaries. Moreover, future research might analyze the interplay of human boundary spanners and technical boundary resources (Wei et al., 2020, p. 533). Human boundary spanners can be part of and support supplementary social boundary resources (Karhu, Gustafsson, et al., 2018; Wulfert, 2023; Wulfert, Woroch, Strobel, Seufert, et al., 2022).

Furthermore, in a number of research papers, we applied the theoretical lens concept to narrow the focus of our research endeavors.[22] Our platform framework is developed from an enterprise architecture perspective (Wulfert and Karger, 2022b, p. 3). For developing design principles that standardize the demarcations among single participants and their IS in e-commerce ecosystems, we applied boundary resources as theoretical lens (Wulfert, Woroch, Strobel, Seufert, et al., 2022, p. 1817). We used the framework for minimum reliability evaluation (Iivari, Rotvit Perlt Hansen, et al., 2021) as a lens for evaluating the design principles for bound-

[20] Due to time restrictions in the revision cycles, we have left the application of our design principles for boundary resources and the implementation of the boundary resource management lifecycle for future research.

[21] Potential performance indicators in the context of boundary resources could include the number of third-party developers, the number of extensions, developer satisfaction, and interface downtime. The number of matches and matching fee generated may be possible performance indicators for implemented orchestration capabilities.

[22] It should be noted here, that the term "theoretical lens" and its use in IS research are highly disputed (Niederman and March, 2019, p. 4f.).

ary resources in the interview study (Wulfert, Strobel, et al., 2024, p. 9). However, the downside of applying theoretical lenses in research is intentionally hiding information from reality (Niederman and March, 2019). Although we tried to specify the theoretical lenses applied in as much detail as possible, we did not provide any explicit explanation of the objective of an applied theoretical lens. We relied on the explanations provided by Niederman and March (2019) instead of clearly explicating the objectives and limitations of applying a theoretical lens. Nevertheless, the implicit use of the theoretical lens concept is very common in IS research (Niederman and March, 2019, p. 17). Future research can apply additional lenses to our research endeavors and examine the research problems from other perspectives. For instance, Wulfert and Karger (2022b, p. 11) suggested advancing the platform architecture framework with investigations focusing on actor-network theory and applying a sociological perspective.

Another suitable theoretical lens would be the use of principal-agent theory (e.g., Buxmann, Diefenbach, et al., 2011; Eisenhardt, 1989a; Pavlou et al., 2007; Tatnall, 2005). As e-commerce ecosystems consist of multiple independent participants affiliated with a focal platform for the exchange of value, there are multiple relations between the focal platform owner and the ecosystem participants. These relations show characteristics of principal-agent constellations (Pavlou et al., 2007, p. 111) involving hidden information possessed and hidden action taken by the agent (Eisenhardt, 1989a, p. 59 f.). In the case of e-commerce ecosystems, the platform owner can be considered as the agent in multiple relationships with different participant types. Future research could investigate the different relations of the focal platform owner with other ecosystem participants through this particular lens. However, principal-agent theory is usually only applied to dyadic relationships in e-commerce (e.g., Pavlou et al., 2007). Hence, an interesting avenue for future research would be to extend dyadic principal-agent relations to multiple relations with special emphasis on the role of the focal platform owner within its ecosystem. As contracts between principals and agents form the unit of analysis in principal-agent relations (Eisenhardt, 1989a, p. 58), future research could analyze contracts between participants in e-commerce ecosystems to identify antecedents and the mitigation of hidden information and hidden action in e-commerce.

Next, we discuss the architectural requirements derived from the literature and the selected architectural patterns developed. The list of architectural requirements stemming from a retailer's dual role as seller and transaction platform owner is neither comprehensive nor complete.[23] However, our literature analysis revealed

[23] The dual role is also likely to occur in the social commerce context with the seller also consuming products from a transaction platform. A recent study by Hawlitschek, Teubner,

that even this incomplete list of requirements is not fully supported by existing reference architectures for e-commerce (Wulfert and Busch, 2022). In the same vein, we only modeled selected architectural patterns that could not be puzzled together without further effort. Therefore, we would engage in future research to develop a reference architecture for innovation platforms in e-commerce supporting the major business processes of a transaction platform.

Instead of focusing on IS reference architectures, other factors, such as motivation elements and value creation, can be modeled to improve collaboration and communication within an e-commerce ecosystem. We unfolded the e-commerce ecosystem's overall value proposition for the (final) customer (Adner, 2017) and indicated the participation of different ecosystem participants in this joint value creation (Wulfert and Dennhardt, 2023, p. 3 ff.). Although we derived the value creation structure of payment processing as a specific sub-ecosystem in e-commerce (Woroch et al., 2022), we did not depict the value creation in e-commerce ecosystems in further detail and remained on a more abstract level. Therefore, future research could investigate the value creation in e-commerce ecosystems, in more detail including key ecosystem participants, their value-creation activities, and value streams among key participants (Peppard and Rylander, 2006). The $e^3$-value modeling notation could be an appropriate vehicle for providing a detailed analysis of the value creation and value flows in e-commerce ecosystems (Pousttchi, 2008; Pousttchi and Gleiss, 2019).

As the multilateral value co-creation in e-commerce ecosystems requires the collaboration and communication of a number of independent actors (Adner, 2017; Böttcher, Rickling, et al., 2021; Wulfert, Woroch, Strobel, Seufert, et al., 2022), harmonizing processes and standardizing communication can make the resource usage more efficient (Koch, 2010; Mann et al., 2022). Tsai, Zdravkovic, et al. (2022) investigated the use of well-established modeling approaches (e.g., Unified Modeling Language (UML), BPMN, $e^3$-value) in research endeavors on digital business ecosystems. The authors (Tsai, Zdravkovic, et al., 2022, p. 14) emphasized that the majority of research papers identified did not represent the elements of digital business ecosystems (i.e., actors, role, capability, relationship, digital component). For this purpose, future research might develop a domain-specific modeling language (Frank, 2013) that incorporates the peculiarities of e-commerce ecosystems and highlights value creation.[24] The goal of such a domain-specific modeling language

and Gimpel (2018) revealed that more than 90 percent of sellers in social commerce take on a dual role.

[24] Existing process modeling approaches, such as the Event-Driven Process Chain and the Business Process Model and Notation, allow to annotate responsibilities for activities but do not emphasize the simultaneous involvement of multiple actors in value creation.

is to create a uniform base for discourse, enabling a better understanding of the involved parties from research and practice. A domain-specific modeling language allows the translation of essential business and technological concepts into linguistic elements (Strobel, 2021, p. 4817). This enables the participants of e-commerce ecosystems to focus on the content of the concepts without having to create their own modeling concepts themselves (Frank, 2013, p. 133 f.).

Another interesting avenue for future research would be the investigation of development paths for e-commerce ecosystems as a whole and single ecosystem participants in particular considering environmental, economic, and social developments. In this regard, potential strategies for the development of digital business ecosystems in general and e-commerce ecosystems, in particular, could be investigated. While literature analyzing ignition strategies for the establishment of ecosystems based on the two-sided market concept exist (e.g, Parker and Van Alstyne, 2016; Reillier and Reillier, 2017; Schirrmacher et al., 2017), research on the development of ecosystems beyond the ignition stage is sparse (e.g., Cennamo, 2021). Therefore, future research may conduct longitudinal case studies on successful digital business ecosystems and derive potential strategies for the owner of the focal platform and ecosystem participants (Eisenmann, Parker, et al., 2009; Street and Ward, 2012). The evolution of a digital business ecosystem might be described in analogy to Darwin's evolution theory (Darwin, 1859). In analogy to biological ecosystems, the following types of strategies may be considered. First, ecosystem extension involves, for instance, the cooperation with and acquisition of additional ecosystem participants.[25] Second, the increase of ecosystem capabilities, such as the provision of the central platform with new features, would be another strategy. [26] Third, ecosystem innovation involves the development of new capabilities and innovations for the ecosystem.[27]

Another interesting avenue for future research would be to investigate the role of the e-commerce ecosystem in general and of the owners of focal platforms in the e-commerce ecosystem in fostering sustainability. Wulfert, Woroch, Strobel, Schoormann, et al. (2023) already identified six sustainability initiative types implemented by owners of focal transaction platforms in e-commerce ecosystems. As the authors only investigated a small sample of companies and only focused on transaction platforms as ecosystem participants, future research might further analyze

[25] This strategy is described as selection in biological ecosystems.

[26] This is described as reproduction in biological ecosystems.

[27] This resembles the variation strategy in biological ecosystems.

sustainability measures in e-commerce ecosystems by consolidating additional focal platforms and considering additional ecosystem participants (e.g., manufacturers, sellers, service providers) for the analysis. It is also worthwhile investigating which types of sustainable development initiatives are combined by the owners of focal platforms in e-commerce ecosystems.

Future developments may also involve rapid changes in the technologies available for conducting commerce transactions. As our understanding of e-commerce included electronically-supported commerce transactions involving physical and digital products and services via the Internet involving different transaction partners, the business models, and commerce experiences can be altered by new technologies. In light of current technological developments, such as mixed reality applications and the metaverse, e-commerce is becoming more immersive (Jeong et al., 2022; Schöbel and Tingelhoff, 2023). The metaverse is expected to support or even replace existing transaction platforms (EHI, 2022). Hence, future research might investigate the economics of transaction platforms (e.g., network effects) in the metaverse. It might also be worthwhile investigating which business models are profitable for sellers in general, and platform owners in particular, in the metaverse. Moreover, future research could consider the nature of boundary resources in these settings (Dwivedi et al., 2022). The virtual representation of the assortment and even whole stores requires altered and additional boundary resources for ecosystem participants.

For investigations on the role of platforms in the e-commerce ecosystem, we always took on the perspective of the focal platform owner. However we implicitly considered other ecosystem participants, such as third-party sellers, customers, and third-party developers, when analyzing e-commerce ecosystems from an affiliation perspective and a structural perspective (Adner, 2017). In line with Thomas et al.'s (2014, p. 4) argument that "the platform [... becomes the focus of strategic maneuvering," we highlighted that the owner of the focal platform has the power to govern and orchestrate the ecosystem stemming from its pivotal position within the ecosystem. However, an interesting avenue for future research could be the analysis of the developments of e-commerce ecosystems or sub-ecosystems from the perspective of other participant types. Such a multi-perspective investigation could provide additional insights into platforms and ecosystems in e-commerce. For the elicitation on transaction platforms as focal participants in e-commerce ecosystems (Wulfert, Woroch, Strobel, Seufert, et al., 2022), we draw on the concept of two-sided markets (Armstrong, 2006; Rochet and Tirole, 2003; Rochet and Tirole, 2006; Wright, 2003). However, we are well aware that e-commerce ecosystems consist of more than two market sides. Only a few studies acknowledge the multi-sidedness of transaction platforms (e.g., Evans and Schmalensee, 2016; Evans and Noel, 2008; Staykova and Damsgaard, 2015). The conceptualization of transaction platforms usually focuses on two market sides to reduce complexity (e.g., Armstrong, 2006; Boudreau, 2021; Dou and Wu, 2021; Rochet and Tirole, 2003; Tucker and Zhang, 2010). Therefore, we would encourage future research to intentionally conceptualize multi-sided markets and test assumptions attributed to two-sided markets in scenarios with multiple market sides.

Digital business ecosystems in general and e-commerce ecosystems in particular emerge and develop in an unforeseeable manner. These developments are influenced by endogenous and external factors. Because of their complex and global networks of independent but intertwined participants and a multitude of interacting processes (Jacobides et al., 2018), the evolution of these ecosystems and the behavior of single participants in them can hardly be predicted (Hannah, Tidhar, et al., 2021; Harrison et al., 2007). Therefore, computationally intensive theory construction approaches, such as simulation (Miranda et al., 2022) can provide a suitable vehicle to address the complexity and dissect the evolution of digital business ecosystems and their participants. Recent studies in IS research have shown that agent-based simulations are especially suitable for the analysis of complex adaptive systems (i.e., two-sided markets, ecosystems) (Davis et al., 2007; Schalowski and Barrot, 2019; Schmid et al., 2021). Therefore, an interesting avenue for future research would be the development of simulation models for digital business ecosystems involving traditional and artificial intelligence approaches, the prediction of the future development of existing ecosystems, and the prediction of potential new focal platforms with their surrounding ecosystems. These simulation models might also provide further insights into multiple roles taken on simultaneously by ecosystem participants, such as platform owners competing with third-party sellers in selected product areas (Wulfert and Schütte, 2022; Zhu and Liu, 2018).

Lastly, we elaborate on the applicability of our research results to other domains and on the generalizability of these results to platform research in general. In this cumulative dissertation and the integrated research papers, we focused on e-commerce as the domain of investigation and developed the artifacts, such as the architecture framework, design principles for boundary resources, and the boundary resource management lifecycle, for application in this particular domain. We also mentioned a number of e-commerce peculiarities in Chap. 1 and 3.5. However, we have shown that platforms and business ecosystems in e-commerce inherit a majority of characteristics from general research on platforms and business ecosystems. Therefore, our contributions also enrich the body of literature on platforms as well as business ecosystems and could potentially be transferred to other application domains.[28] Future research could evaluate the applicability of our developed artifacts in other domains. In this regard, another interesting avenue for future research would be a comparison of platforms in different application domains to derive possible extensions for the general platform concept. de Reuver et al. (2017, p. 129 f.) described the comparability discussion in the context of defining the scope of the platform concept. It would also be worthwhile using longitudinal studies to investigate how platforms in practice have evolved over time and further adapting the platform concept to economic developments.

[28] The applicability of our research to other application domains is also supported by the fact that the anonymous reviewers of some research papers pointed to the potential generalizability of our developed artifacts (e.g., Wulfert, 2023; Wulfert and Karger, 2022b; Wulfert, Woroch, and Strobel, 2024; Wulfert, Woroch, Strobel, Seufert, et al., 2022).

# 9 Conclusion

Following heightened research interest in platforms and globally increasing e-commerce revenues, the objective of this cumulative dissertation was the domain-specific investigation of platforms in e-commerce. We investigated platforms in e-commerce from enterprise architecture, boundary resource, application lifecycle, dynamic capabilities, and value creation perspectives. Based upon an in-depth analysis of general platform research and research on platforms in the context of e-commerce, the investigation of multiple innovation platforms (e.g., Magento Commerce, Shopify, WooCommerce) and transaction platforms (e.g., Amazon Marketplace, Etsy, Walmart Marketplace) in e-commerce, and interviews with multiple domain experts (i.e., business process owners, user interface designers, internal platform developers) on platforms and boundary resources in e-commerce ecosystems, we emphasized how owners of focal platforms in e-commerce ecosystems shape ecosystem structure and participants. For our analyses, we applied a narrow conceptualization of e-commerce and unfolded the e-commerce ecosystem from an affiliation and structural perspective.

We identified innovation platforms, transaction platforms, and their hybrid combinations as key concepts in e-commerce. While innovation platforms provide the necessary application services for e-commerce transactions and an extensible codebase for third-party developers' extensions, transaction platforms provide a virtual environment for e-commerce transactions between multiple actors. We highlighted their intertwined nature in a platform architecture framework for e-commerce ecosystems. In particular, we emphasized the significance of innovation platforms for powering transaction platform business models and the attraction of third-party developers. We also suggested a taxonomy for describing the current state of and future directions for transaction platforms in e-commerce ecosystems.



T. Wulfert, *Selected Perspectives on Platforms in E-Commerce Ecosystems*,
https://doi.org/10.1007/978-3-658-45198-1_9

We further described boundary resources as a key pillar for value creation in e-commerce enabling electronic transactions across the boundaries of single organizations. Boundary resources were understood as boundary objects that act as interfaces between a focal platform in e-commerce ecosystems and additional participants on the social, application, and development levels. We provided prescriptive design knowledge in the form of design principles for the standardization and harmonization of boundary resources in e-commerce ecosystems. We suggested a total of 19 design principles for boundary resources as interfaces between the focal transaction platform and third-party sellers' IS and development resources for third-party developers. Moreover, we prescribed a reference procedure model for the management of the entire lifecycle of boundary resources.

We denoted ecosystem orchestration as a second key pillar for value creation in e-commerce ecosystems. Proper ecosystem orchestration ensures efficient resource usage within the ecosystem and propels value creation for the (final) customer. Ecosystem orchestration requires a set of dynamic capabilities from the owner of the focal platform in e-commerce ecosystems. We proposed information superiority, institutional innovation, shared vision, structural reconfiguration, and platform access as dynamic capabilities for the orchestration of e-commerce ecosystems. We argued that a set of capabilities depending on assortment type and ecosystem structure needs to be implemented for the overall ecosystem to be successful. Moreover, we disentangled the value creation for payment ecosystems as a particular sub-ecosystem of e-commerce ecosystems.

# Bibliography

Abbu, H. R., D. Fleischmann, and P. Gopalakrishna (2021). "The Digital Transformation of the Grocery Business—Driven by Consumers, Powered by Technology, and Accelerated by the COVID-19 Pandemic". In: *World Conference on Information Systems and Technologies (WorldCIST 2021)*. Vol. 1367 AISC. Terceira Island, Azores, Portugal, pp. 1–12.

Abdelkafi, N., C. Raasch, A. Roth, and R. Srinivasan (2019). "Multi-sided platforms". In: *Electronic Markets* 29.4, pp. 553–559.

Abdollahi, G. and U. Leimstoll (2011). "A classification for business model types in e-commerce". In: Detroit, Michigan, USA, pp. 756–769.

Abhishek, V., K. Jerath, and Z. J. Zhang (2016). "Agency selling or reselling? Channel structures in electronic retailing". In: *Management Science* 62.8, pp. 2259–2280.

Ackerman, M. S., L. F. Cranor, and J. Reagle (1999). "Privacy in E-commerce: Examining user scenarios and privacy preferences". In: *Proceedings of the 1st ACM conference on Electronic commerce*. New York, NY, USA: ACM, pp. 1–8.

Ackroyd, A. T. (2023). *Alibaba Reorganizes To Unlock Value*. URL: https://www.alizila.com/alibaba-reorganizes-to-unlock-value/ (visited on 05/10/2023).

Acs, Z. J., E. Stam, D. B. Audretsch, and A. O'Connor (2017). "The lineages of the entrepreneurial ecosystem approach". In: *Small Business Economics* 49.1, pp. 1–10.

Adam, N. R. and Y. Yesha (1996). "Electronic commerce: An overview". In: *Electronic Commerce: Current Research Issues and Applications*. Ed. by N. R. Adam and Y. Yesha. Berlin, Germany: Springer. Chap. 2, pp. 5–12.

Addo, A. (2022). "Orchestrating a digital platform ecosystem to address societal challenges: A robust action perspective". In: *Journal of Information Technology* 37.4, pp. 359–386.

Addo, I. D., S. I. Ahamed, and W. C. Chu (2014). "A reference architecture for high-availability automatic failover between PaaS cloud providers". In: *Proceedings of the 1st International Conference on Trustworthy Systems and Their Applications (TSA2014)*. Taichung, Taiwan, pp. 14–21.

Adner, R. (2006). "Match your innovation strategy to your innovation ecosystem". In: *Harvard Business Review* 84.4, pp. 1–11.

Adner, R. (2012). *The Wide Lens: A New Strategy for Innovation*. New York City, USA: Portfolio/Penguin.

Adner, R. (2017). "Ecosystem as Structure: An Actionable Construct for Strategy". In: *Journal of Management* 43.1, pp. 39–58.



T. Wulfert, *Selected Perspectives on Platforms in E-Commerce Ecosystems*,
https://doi.org/10.1007/978-3-658-45198-1

Adner, R. and R. Kapoor (2010). "Value creation in innovation ecosystems: How the structure of technological interdependence affects firm performance in new technology generations". In: *Strategic Management Journal* 31.3, pp. 306–333.

Adobe (2022a). *Backward-incompatible changes highlights*. URL: https://developer.adobe.com/commerce/php/development/backward-incompatible-changes/highlights/ (visited on 10/18/2022).

Adobe (2022b). *Code Contriubtions*. URL: https://developer.adobe.com/commerce/contributor/guides/code-contributions/ (visited on 10/18/2022).

Adobe (2022c). *Definition of done*. URL: https://developer.adobe.com/commerce/contributor/guides/code-contributions/definition-of-done/#tco-analysis (visited on 10/18/2022).

Adobe (2022d). *Release schedule*. URL: https://experienceleague.adobe.com/docs/commerce-operations/release/schedule.html?mt=false (visited on 10/18/2022).

Adobe (2022e). *System requirements*. URL: https://experienceleague.adobe.com/docs/commerce-operations/installation-guide/system-requirements.html?mt=false (visited on 10/18/2022).

Adobe (2022f). *Web API Functional Testing*. URL: https://developer.adobe.com/commerce/webapi/get-started/web-api-functional-testing/ (visited on 10/18/2022).

Aerts, A., J. Goossenaerts, D. K. Hammer, and J. C. Wortmann (2004). "Architectures in context: On the evolution of business, application software, and ICT platform architectures". In: *Information and Management* 41.6, pp. 781–794.

Ågerfalk, P. J. and F. Karlsson (2020). "Artefactual and empirical contributions in information systems research". In: *European Journal of Information Systems* 29.2, pp. 109–113.

Ahlemann, F. (2012). *Strategic enterprise architecture management: Challenges, best practices, and future developments. Management for professionals*. Berlin [u. a.]: Springer.

Ahlemann, F., C. Legner, and D. Schäfczuk (2012). "Introduction". eng. In: *Strategic enterprise architecture management: challenges, best practices, and future developments*. Ed. by F. Ahlemann, E. Stettiner, M. Messerschmidt, and C. Legner. Heidelberg [u. a.]: Springer-Verlag. Chap. 1, pp. 1–33.

Ahmed, M. (2020). "Web Based Seller Support System: Amazon Overview". In: *Proceedings of the IEEE Conference on e-Learning, e-Management and e-Services (IC3E2020)*. Kota Kinabalu, Malaysia, pp. 104–109.

Aichele, C. and M. Schönberger (2016). *E-Business. Eine Übersicht für erfoglreiches B2B und B2C*. Wiesbaden, Germany: Springer Fachmedien.

Akaka, M. A., S. L. Vargo, and R. F. Lusch (2012). "An exploration of networks in value cocreation: A service-ecosystems view". In: *Review of Marketing Research* 9, pp. 13–50.

Akana (2020). *How to Define API Requirements*. URL: https://www.akana.com/blog/api-requirements-what-consider (visited on 10/26/2022).

Akatkin, Y., O. Karpov, V. Konyavskiy, and E. Yasinovskaya (2017). "Digital economy: Conceptual architecture of a digital economic sector ecosystem". In: *Business Informatics* 2017.4, pp. 17–28.

Alami, D., M. Rodríguez, and S. Jansen (2015). "Relating health to platform success: Exploring three E-commerce Ecosystems". In: *Proceedings of the 2015 European Conference on Software Architecture Workshops (ECSAW2015)*. Dubrovnik, Croatia, pp. 1–6.

Albayati, H., S. K. Kim, and J. J. Rho (2020). "Accepting financial transactions using blockchain technology and cryptocurrency: A customer perspective approach". In: *Technology in Society* 62, pp. 1–20.

Albers, M., C. M. Jonker, M. Karami, and J. Treur (2004). "Agent Models and Different User Ontologies for an Electronic Market Place". In: *Knowledge and Information Systems* 6.1, pp. 1–41.

Albert, H. (1978). *Traktat über rationale Praxis*. Tübingen, Germany: J. C. B. Mohr.

Albert, S. and D. Whetten (1995). "Organizational identity". In: *Research in Organizational Behavior*. Ed. by M. Staw. 7th ed. Greenwich, USA: JAI Press, pp. 263–295.

Aldrich, H. (1971). "Organizational Boundaries and Inter-organizational Conflict". In: *Human Relations* 24.4, pp. 279–293.

Aldrich, H. and D. Herker (1977). " Boundary Spanning Roles and Organization Structure". In: *Academy of Management Review* 2.2, pp. 217–230.

Aldrich, H. E. (1979). *Organizations and Environments*. New York City, USA: Prentice Hall.

Aldrich, H. E. and M. Ruef (2006). *Organizations evolving*. 2nd ed. Los Angeles [u. a.]: SAGE, pp. 1–330.

Alfaro, E. et al. (2019). "BBVA's Data Monetization Journey". In: *MIS Quarterly Executive* 18.2, pp. 117–128.

Alharahsheh, H. H. and A. Pius (2020). "A Review of key paradigms: positivism VS interpretivism". In: *Global Academic Journal of Humanities and Social Sciences* 2.3, pp. 39–43.

Alhassan, I., D. Sammon, and M. Daly (2016). "Data governance activities: an analysis of the literature". In: *Journal of Decision Systems* 25, pp. 64–75.

Alibaba Group (2022). *Alibaba Cloud Provides Secure and Reliable Services-Alibaba Cloud*. URL: https://www.alibabacloud.com/de/go_global/go-global (visited on 07/28/2022).

Alsaawi, A. (2014). "A critical review of qualitative interviews". In: *European Journal of Business and Social Sciences* 3.4, pp. 149–156.

Alshomali, M. A., J. R. Hamilton, J. Holdsworth, and S. Tee (2017). "GitHub: Factors influencing project activity levels". In: *Proceedings of The 17th International Conference on Electronic Business (ICEB2017)*. Dubai, UAE, pp. 116–124.

Alt, R. (2020). "Electronic Markets on business model development". In: *Electronic Markets* 30.3, pp. 405–411.

Alt, R., R. Beck, and M. T. Smits (2018). "FinTech and the transformation of the financial industry". In: *Electronic Markets* 28.3, pp. 235–243.

Alt, R. and T. Puschmann (2012). "The rise of customer-oriented banking—electronic markets are paving the way for change in the financial industry". In: *Electronic Markets* 22.4, pp. 203–215.

Alt, R. and H. D. Zimmermann (2019). "Electronic Markets on platform competition". In: *Electronic Markets* 29.2, pp. 143–149.

Alter, S. (2008). "Defining information systems as work systems: Implications for the IS field". In: *European Journal of Information Systems* 17.5, pp. 448–469.

Alves, C., J. Oliveira, and S. Jansen (2017). "Software Ecosystems Governance-A Systematic Literature Review and Research Agenda". In: *Proceedings of the 19th International Conference on Enterprise Information Systems (ICEIS 2017)*. Vol. 3. Porto, Portugal, pp. 26–29.

Amazon (2019). *Amazon Annual Report 2018*. URL: https://s2.q4cdn.com/299287126/files/doc_financials/annual/2018-Annual-Report.pdf (visited on 05/15/2023).

Amazon (2021). *Amazon Annual Report 2020*. URL: https://s2.q4cdn.com/299287126/files/doc_financials/2021/ar/Amazon-2020-Annual-Report.pdf (visited on 05/15/2023).

Amazon (2022). *Amazon Annual Report 2021*. URL: https://s2.q4cdn.com/299287126/files/doc_financials/2022/ar/Amazon-2021-Annual-Report.pdf (visited on 01/18/2023).

Amazon (2023). *Amazon.com announces fourth quarter 2022 results*. URL: https://s2.q4cdn.com/299287126/files/doc_financials/2022/q4/Q4-2022-Amazon-Earnings-Release.pdf (visited on 06/06/2023).

Amelio, A. and B. Jullien (2012). "Tying and freebies in two-sided markets". In: *International Journal of Industrial Organization* 30.5, pp. 436–446.

Amit, R. and C. Zott (2021). *Business model innovation strategy: Transformational concepts and tools for entrepreneurial leaders*. Hoboken, New Jersey, USA: Wiley.

An, J. et al. (2020). "A lightweight blockchain-based model for data quality assessment in crowdsensing". In: *IEEE Transactions on Computational Social Systems* 7.1, pp. 84–97.

Anderson, P. and M. L. Tushman (1986). "Technological Discontinuities and Organizational Environments". In: *Administrative Science Quarterly* 31.3, pp. 439–465.

Angelov, S., P. Grefen, and D. Greefhorst (2009). "A classification of software reference architectures: Analyzing their success and effectiveness". In: *Proceedings of the Joint Working IEEE/IFIP Conference on Software Architecture & European Conference on Software Architecture*. Cambridge, UK: IEEE, pp. 141–150.

Angelov, S., P. Grefen, and D. Greefhorst (2012). "A framework for analysis and design of software reference architectures". In: *Information and Software Technology* 54.4, pp. 417–431.

Angelov, S., J. J. M. Trienekens, and P. Grefen (2008). "Towards a method for the evaluation of reference architectures: Experiences from a case". In: *Proceedings of the European Conference on Software Architecture (ECSA2008)*. Ed. by R. Morrison, D. Balasubramaniam, and K. Falkner. Paphos, Cyprus: Springer, pp. 225–240.

Anwar, M. J. and A. Q. Gill (2019). "A review of the seven modelling approaches for digital ecosystem architecture". In: *Proceedings of the 21st IEEE Conference on Business Informatics (CBI2019)*. Moscow, Russia: IEEE, pp. 94–103.

AppsRunTheWorld (2021). *Top 10 eCommerce Software Vendors, Market Size and Market Forecast 2020–2025*. URL: https://www.appsruntheworld.com/top-10-ecommerce-software-vendors-and-market-forecast/ (visited on 07/15/2022).

APQC (2019). *Amcerican Productivity & Quality Center Process Classification Framework (PCF)—Retail PCF*. URL: https://www.apqc.org/resource-library/resource-listing/apqc-process-classification-framework-pcf-retail-pcf-pdf-version (visited on 06/15/2021).

Aria, M. and C. Cuccurullo (2017). "bibliometrix : An R-tool for comprehensive science mapping analysis". In: *Journal of Informetrics* 11.4, pp. 959–975.

Armstrong, M. (2006). "Competition in two-sided markets". In: *RAND Journal of Economics* 37.3, pp. 668–691.

Armstrong, M. and J. Wright (2007). "Two-sided Markets, Competitive Bottlenecks and Exclusive Contracts". In: *Economic Theory* 32.2, pp. 353–380.

Arnott, D. (2006). "Cognitive biases and decision support systems development: a design science approach". In: *Information Systems Journal* 16.1, pp. 55–78.

Arora, A., S. Bansal, et al. (2019). "Measuring social media influencer index- insights from facebook, Twitter and Instagram". In: *Journal of Retailing and Consumer Services* 49, pp. 86–101.

Arora, A., R. Longo, et al. (2022). *Becoming indispensable: Moving past e-commerce to NeXT commerce.* URL: https://www.mckinsey.com/capabilities/growth-marketing-and-sales/our-insights/becoming-indispensable-moving-past-e-commerce-to-next-commerce#/ (visited on 08/22/2023).

Arreola González, A., M. Pfaff, and H. Krcmar (2019). "Value Modeling for Ecosystem Analysis". In: *Computers* 8.3, pp. 1–15.

Asadullah, A., I. Faik, and A. Kankanhalli (2018a). "Digital Platforms: A Review and Future Directions". In: *Proceedings of the 22nd Pacific Asia Conference on Information Systems (PACIS 2018).* Yokohama, Japan, pp. 248–262.

Asadullah, A., I. Faik, and A. Kankanhalli (2018b). "Evolution Mechanisms for Digital Platforms: A Review and Analysis across Platform Types". In: *39th International Confernece on Information Systems (ICIS2018).* San Francisco, pp. 1–9.

Ashforth, B. E. and F. Mael (1989). "Social Identity Theory and the Organization". In: *Academy of Management Review* 14.1, pp. 20–39.

Aston, B. (2022). *10 Best ALM Tools For 2022.* URL: https://thedigitalprojectmanager.com/tools/alm-tools/ (visited on 10/25/2022).

Atluri, V., M. Dietz, and N. Henke (2017). *Competing in a world of sectors without borders.* URL: https://www.mckinsey.com/capabilities/quantumblack/our-insights/competing-in-a-world-of-sectors-without-borders (visited on 07/17/2023).

Aukro (2023). *Ceník poplatků a služe | Aukro.cz.* URL: https://aukro.cz/stranka/cenik (visited on 06/28/2023).

Aulkemeier, F., R. Daukuls, et al. (2016). "Sales forecasting as a service: A cloud based pluggable e-commerce data analytics service". In: vol. 2. Rome, Italy, pp. 345–352.

Aulkemeier, F., M. E. Iacob, and J. van Hillegersberg (2017). "An architectural perspective on service adoption: A platform design and the case of pluggable cross-border trade compliance in e-commerce". In: *Journal of Organizational Computing and Electronic Commerce* 27.4, pp. 325–341.

Aulkemeier, F., M. E. Iacob, and J. van Hillegersberg (2019). "Platform-based collaboration in digital ecosystems". In: *Electronic Markets* 29.4, pp. 597–608.

Aulkemeier, F., M. A. Paramartha, M. E. Iacob, and J. van Hillegersberg (2016). "A pluggable service platform architecture for e-commerce". In: *Information Systems and e-Business Management* 14.3, pp. 469–489.

Aulkemeier, F., M. Schramm, M.-E. Iacob, and J. van Hillegersberg (2016). "A Service-Oriented E-Commerce Reference Architecture". In: *Journal of theoretical and applied electronic commerce research* 11, pp. 26–45.

Autio, E. (2022). "Orchestrating ecosystems: a multi-layered framework". In: *Innovation: Organization and Management* 24.1, pp. 96–109.

Autio, E., S. Nambisan, L. D. W. Thomas, and M. Wright (2018). "Digital affordances, spatial affordances, and the genesis of entrepreneurial ecosystems". In: *Strategic Entrepreneurship Journal* 12.1, pp. 72–95.

Avdiji, H., D. Elikan, S. Missonier, and Y. Pigneur (2020). "A design theory for visual inquiry tools". In: *Journal of the Association for Information Systems* 21.3, pp. 695–734.

Aversa, R., N. Panza, and L. Tasquier (2015). "An Agent-Based Platform or Cloud Applications Performance Monitoring". In: *Proceedings of the 9th International Conference on Complex, Intelligent, and Software Intensive Systems.* Santa Catarina, Brazil: IEEE, pp. 535–540.

AXELOS (2012). *ITIL Service Operation*. London, GB: AXELOS Publishing.

Ba, S., D. Ke, J. Stallaert, and Z. Zhang (2012). "Comparing the quality of customer service in 3D virtual worlds to web-based service". In: *Lecture Notes in Business Information Processing* 108 LNBIP, pp. 234–247.

Babb, J., J. Nørbjerg, D. J. Yates, and L. J. Waguespack (2017). "The Empire Strikes Back: The End of Agile as we Kno it?" In: *Proceedings of the 40th Information Systems Research Seminar in Scandinavia*. Halden, Norway, pp. 44–59.

Baghdadi, Y. (2016). "A framework for social commerce design". In: *Information Systems* 60, pp. 95–113.

Bai, T. et al. (2017). "An Experimental Study of Text Representation Methods for Cross-Site Purchase Preference Prediction Using the Social Text Data". In: *Journal of Computer Science and Technology* 32.4, pp. 828–842.

Bailey, K. D. (1994). *Typologies and Taxonomies. An Introduction to Classification Techniques*. Thousand Oaks, California: Sage publications.

Bailey, K. D. (1984). "A three-level measurement model". In: *Quality and Quantity* 18.3, pp. 225–245.

Baker, A. D., H. van Dyke Parunak, and K. Erol (1999). "Agents and the Internet: Infrastructure for mass customization". In: *IEEE Internet Computing* 3.5, pp. 62–69.

Baker, H. K., N. Pandey, S. Kumar, and A. Haldar (2020). "A bibliometric analysis of board diversity: Current status, development, and future research directions". In: *Journal of Business Research* 108, pp. 232–246.

Bakos, J. Y. (1991). "A strategic analysis of electronic marketplaces". In: *MIS Quarterly* 15.4, pp. 295–310.

Bakos, Y. (1998). "The Emerging Role of Electronic Marketplaces on the Internet". In: *Communications of the ACM* 41.8, pp. 35–42.

Bakos, Y. and E. Katsamakas (2008). "Design and ownership of two-sided networks: Implications for Internet platforms". In: *Journal of management information systems* 25.2, pp. 171–202.

Balachandran, B. M. and M. Mohammadian (2015). "Development of a Fuzzy-based Multi-agent System for E-commerce Settings". In: *Procedia Computer Science* 60, pp. 593–602.

Balaji, M. S., K. W. Khong, and A. Chong (2016). "Determinants of negative word-of-mouth communication using social networking sites". In: *Information and Management* 53.4, pp. 528–540.

Baldwin, C. Y. and K. B. Clark (2000). *Design Rules: The Power of Modularity*. Cambridge: MIT Press.

Baldwin, C. Y. and J. C. Woodard (2009). "The architecture of platforms: a unified view". In: *Platforms, Markets and Innovation*. Ed. by A. Gawer. Cheltenham, UK: Edward Elgar, pp. 19–44.

Baldwin, C. Y. (2008). "Where do transactions come from? Modularity, transactions, and the boundaries of firms". In: *Industrial and Corporate Change* 17.1, pp. 155–195.

Bandara, W. et al. (2015). "Achieving rigor in literature reviews: Insights from qualitative data analysis and tool-support". In: *Communications of the Association for Information Systems* 37, pp. 154–204.

Banker, Mitra, Sambamurthy, and Mitra (2011). "The Effects of Digital Trading Platforms on Commodity Prices in Agricultural Supply Chains". In: *MIS Quarterly* 35.3, p. 599.

Bapst, F. and C. Vanoirbeek (1999). "XML documents production for an electronic platform of requests for proposals". In: *Proceedings of the 18th IEEE Symposium on Reliable Distributed Systems*. Lausanne, Switzerland, pp. 1–11.

Barbosa, A. O., A. Santana, S. Hacks, and N. Von Stein (2019). "A taxonomy for enterprise architecture analysis research". In: *Proceedings of the 21st International Conference on Enterprise Information Systems (ICEIS2019)*. Vol. 2. Heraklion, Greece, pp. 493–504.

Barcia, A. (2021). *The Best Live Shopping Campaigns from Online Retailers*. URL: https://blogen.influence4you.com/the-best-live-shopping-campaigns-from-online-retailers/ (visited on 06/28/2023).

Barney, J. (2015). "Firm resources and sustained competitive advantage". In: *International Business Strategy: Theory and Practice* 17.1, pp. 283–301.

Barreto, I. (2010). "Dynamic Capabilities: A Review of Past Research and an Agenda for the Future". In: *Journal of Management* 36.1, pp. 256–280.

Barrett, M., E. Oborn, W. J. Orlikowski, and J. A. Yates (2012). "Reconfiguring boundary relations: Robotic innovations in pharmacy work". In: *Organization Science* 23.5, pp. 1448–1466.

Barth, K., M. Hartmann, and H. Schröder (2015). *Betriebswirtschaftslehre des Handels*. Vol. 7., überar. Wiesbaden, Germany: Springer Gabler.

Barton, T. (2014). "E-Business". In: *E-Business mit Cloud Computing*. Ed. by T. Barton. Wiesbaden, Germany: Springer Fachmedien. Chap. 2, pp. 3–23.

Baskerville, R., M. Kaul, J. Pries-Heje, and V. Storey (2019). "Inducing Creativity in Design Science Research". In: *Proceedings of the 14th International Conference on Design Science Research in Information Systems and Technology (DESRIST2019)*. Worcester, MA, pp. 3–17.

Basole, R. C. (2009). "Structural analysis and visualization of ecosystems: a study of mobile device platforms". In: *Proceedings of the 15th Americas Conference on Information Systems (AMCIS2009)*. San Francisco, California, USA, pp. 1–10.

Basole, R. C. and M. G. Russell (2015). "Understanding Business Ecosystem Dynamics: A Data-Driven". In: *ACM Transactions on Management Information Systems* 6.2, pp. 1–32.

Bass, L., P. Clements, and R. Kazman (2003). *Software architecture in practice*. Boston, USA: Addison-Wesley Professional.

Battleson, D. A. et al. (2016). "Achieving dynamic capabilities with cloud computing: An empirical investigation". In: *European Journal of Information Systems* 25.3, pp. 209–230.

Baumann, S., ed. (2022a). *Handbook on Digital Business Ecosystems: Technolgies, Markets, Business Models, Management, and Societal Challenges*. Cheltenham, UK: Edward Elgar Publishing.

Baumann, S. (2022b). "Introduction to the Handbook on Digital Busienss Ecosystems: Strategies, Platforms, Technologies, Governance and Societal Challenges". In: *Handbook on Digital Busienss Ecosystems: Strategies, Platforms, Technologies, Governance and Societal Challenges*. Ed. by S. Baumann. Cheltenham, UK: Edward Elgar. Chap. 1, pp. 1–9.

Baumann, S. and M. Leerhoff (2022). "Networks, platforms, and digital business ecosystems: mapping the development of a field". In: *Handbook on Digital Busienss Ecosystems: Strategies, Platforms, Technologies, Governance and Societal Challenges*. Ed. by S. Baumann. Cheltenham, UK: Edward Elgar. Chap. 2, pp. 11–24.

Bawack, R. E. et al. (2021). "Influencer is the New Recommender: Insights for Enhancing Social Recommender Systems". In: *Lecture Notes in Computer Science (including subseries*

*Lecture Notes in Artificial Intelligence and Lecture Notes in Bioinformatics)* 12896 LNCS, pp. 681–691.

Bechara, J. P. and A. H. Van de Ven (2007). "Philosophy of science underlying engaged scholarship". In: *Engaged scholarship: a guide for organizational and social research*. Ed. by A. H. Van de Ven. Oxford, UK: Oxford University Press. Chap. 2, pp. 36–70.

Becker, J. and P. Delfmann (2007). *Reference Modeling. Efficient Information Systems Design Through Reuse of Information Models*. Heidelberg: Physica-Verlag.

Becker, J., R. Holten, R. Knackstedt, and B. Niehaves (2003). "Forschungsmethodische Positionierung in der Wirtschaftsinformatik: Epistemologische, ontologische und linguistische Leitfragen". In: *Arbeitsberichte des Instituts für Wirtschaftsinformatik. Universität Münster* 93, pp. 1–39.

Becker, J. and B. Niehaves (2007). "Epistemological perspectives on IS research: a framework for analysing and systematizing epistemological assumptions". In: *Information Systems Journal* 17.2, pp. 197–214.

Becker, J. and R. Schütte (2004). *Handelsinformationssysteme : domänenorientierte Einführung in die Wirtschaftsinformatik*. 2., vollst. Frankfurt am Main: Redline Wirtschaft.

Beinke, J. H., D. Nguyen, and F. Teuteberg (2018). "Towards a Business Model Taxonomy of Startups in the Finance Sector using Blockchain". In: *Proceedings of the 39th International Conference on Information Systems (ICIS2018)*. San Francisco, USA, pp. 1–17.

Beketov, M., K. Lehmann, and M. Wittke (2018). "Robo Advisors: quantitative methods inside the robots". In: *Journal of Asset Management* 19.6, pp. 363–370.

Bekiaris, M. and P. Papanastasiou (2021). "A bibliometric analysis of family business: Insights from interdisciplinary studies". In: *Corporate governance: A search for emerging trends in the pandemic times*. Ed. by S. Hundal, A. Kostyuk, and D. Govorun. Virtus Interpress, pp. 59–63.

Benbasat, I., D. K. Goldstein, and M. Mead (1987). "The Case Research Strategy in Studies of Information Systems". In: *MIS Quarterly* 11.3, pp. 369–386.

Benbasat, I. and R. Weber (1996). "Research Commentary: Rethinking "Diversity" in Information Systems Research". In: *Information Systems Research* 7.4, pp. 389–399.

Benbasat, I. and R. W. Zmud (1999). "Empirical Research in Information Systems: The Practice of Relevance". In: *MIS Quarterly* 23.1, pp. 3–16.

Benedict, M. (2018). "Modelling Ecosystems in Information Systems—A Typology Approach". In: *Multikonferenz Wirtschaftsinformatik (MKWI2018)*. Lüneburg, Germany, pp. 453–464.

Benfenatki, H. et al. (2017). "MADONA: a method for automated provisioning of cloud-based component-oriented business applications". In: *Service Oriented Computing and Applications* 11.1, pp. 87–100.

Benlian, A., D. Hilkert, and T. Hess (2015). "How open is this platform? The meaning and measurement of platform openness from the complementors' perspective". In: *Journal of Information Technology* 30.3, pp. 209–228.

Berendes, C. I. et al. (2020). "Usage of Local Online Platforms in Retail : Insights from retailers' expectations". In: *Proceedings of the Twenty-Eighth European Conference on Information Systems (ECIS2020)*. Marrakesh, Morroco, pp. 1–11.

Berger, S. C. and F. Gleisner (2009). "Emergence of Financial Intermediaries in Electronic Markets: The case of Online P2P Lending". In: *Business Research* 2, pp. 39–65.

Bergman, M., K. Lyytinen, and G. Mark (2007). "Boundary objects in design: An ecological view of design artifacts". In: *Journal of the Association for Information Systems* 8.11, pp. 546–568.

Berne, E. (1966). *The structure and dynamics of organizations and groups*. New York City, USA: Grove Press.

Berners-Lee, T., D. Dimitroyannis, A. J. Mallinckrodt, and S. McKay (1994). "World Wide Web". In: *Computers in Physics* 8.3, p. 298.

Bertolino, A., P. Inverardi, P. Pelliccione, and M. Tivoli (2009). "Automatic synthesis of behavior protocols for composable web-services". In: *Proceedings of the 12th Joint European Software Engineering Conference and 17th ACM SIGSOFT Symposium on the Foundations of Software Engineering (ESEC-FSE2009)*. New York City, NY, USA, pp. 141–150.

Beuving, J. and G. De Vries (2015). *Doing Qualitative Research: The Craft of Naturalistic Inqiry*. Amsterdam, Netherlands: Amsterdam University Press.

Beverungen, D., D. Kundisch, and N. Wünderlich (2020). "Transforming into a platform provider: strategic options for industrial smart service providers". In: *Journal of Service Management* 32.4, pp. 507–532.

Beverungen, D., O. Müller, et al. (2019). "Conceptualizing smart service systems". In: *Electronic Markets* 29.1, pp. 7–18.

Bharadwaj, A., O. A. El Sawy, P. A. Pavlou, and N. Venkatraman (2013). "Digital Business Strategy: Toward a Next Generation of Insights". In: *MIS Quarterly* 37.2, pp. 471–482.

Bhatt, A. M., A. Goldberg, and S. B. Srivastava (2022). "A Language-Based Method for Assessing Symbolic Boundary Maintenance between Social Groups". In: *Sociological Methods and Research* 51.4, pp. 1681–1720.

Bhatti, A. et al. (2020). "E-commerce trends during COVID-19 Pandemic". In: *International Journal of Future Generation Communication and Networking* 13.2, pp. 1449–1452.

Bican, P. M. and A. Brem (2020). "Digital Business Model, Digital Transformation, Digital Entrepreneurship: Is There A Sustainable "Digital"?" In: *Sustainability* 12.13, pp. 1–15.

BigCommerce (2022a). *About Our APIs*. URL: https://developer.bigcommerce.com/docs/3eea0e31b4c6a-about-our-ap-is (visited on 10/18/2022).

BigCommerce (2022b). *API Best Practices*. URL: https://developer.bigcommerce.com/docs/a0ddf08933ce4-api-best-practices (visited on 10/18/2022).

BigCommerce (2022c). *BigCommerce Examples & Starters*. URL: https://developer.bigcommerce.com/dev-library/ (visited on 10/18/2022).

BigCommerce (2022d). *BigCommerce Status*. URL: https://status.bigcommerce.com/ (visited on 10/18/2022).

BigCommerce (2022e). *Create a Sandbox Store*. URL: https://developer.bigcommerce.com/docs/a8d9c927e48b5-create-a-sandbox-store (visited on 10/18/2022).

BigCommerce (2022f). *Feature Request*. URL: https://support.bigcommerce.com/s/topic/0TO130000001RKPGA2/feature-request?language=en_US (visited on 10/18/2022).

BigCommerce (2022g). *Knowledge Base*. URL: https://support.bigcommerce.com/s/knowledge?language=en_US (visited on 10/18/2022).

Billewar, S. R. et al. (2021). "The rise of 3D E-Commerce: the online shopping gets real with virtual reality and augmented reality during COVID-19". In: *World Journal of Engineering* 19.2, pp. 244–253.

Blaikie, N. (1993). *Approaches to Social Enquiry*. Ed. by Blackwell. Oxford, UK.

Blair, E. (2015). "A reflexive exploration of two qualitative data coding techniques". In: *Journal of Methods and Measurement in the Social Sciences* 6.1, pp. 14–29.

Blank, S. and B. Dorf (2012). *The startup owner's manual: The step-by-step guide for building a great company*. Hoboken, New Jersey: John Wiley & Sons.

Blaschke, M. R. and M. Brosius (2018). "Digital Platforms: Balancing Control and Generativity". In: *Proceedings of the 39th International Conference on Information Systems (ICIS2018)*. San Francisco, USA, pp. 1–9.

Blaschke, M. R., K. Haki, S. Aier, and R. Winter (2018). "Capabilities of Digital Platform Survival: Insights from a Business-to-Business Digital Platform". In: *Proceedings of the 39th International Conference on Information Systems (ICIS2018)*. San Francisco, USA, pp. 1–17.

Blaschke, M. R., K. Haki, S. Aier, and R. Winter (2019). "Taxonomy of digital platforms: A platform architecture perspective". In: *Proceedings of the 14th International Conference on Wirtschaftsinformatik (WI2019)*. Siegen, Germany, pp. 1–15.

Blaschke, M. R., U. Riss, K. Haki, and S. Aier (2019). "Design principles for digital value co-creation networks: a service-dominant logic perspective". In: *Electronic Markets* 29.3, pp. 443–472.

Blind, K. and A. Mangelsdorf (2016). "Motives to standardize: Empirical evidence from Germany". In: *Technovation* 48–49, pp. 13–24.

Bogers, M., J. Sims, and J. West (2019). "What Is an Ecosystem? Incorporating 25 Years of Ecosystem Research". In: *Academy of Management Proceedings* 2019.1, pp. 1–25.

Bondel, G., A. Landgraf, and F. Matthes (2021). "API Management Patterns for Public, Partner, and Group Web API Initiatives with a Focus on Collaboration". In: *26th European Conference on Pattern Languages of Programs*. New York, NY, USA: ACM, pp. 1–17.

Borgh, M., M. Cloodt, and A. G. L. Romme (2012). "Value creation by knowledge-based ecosystems: evidence from a field study". In: *R&D Management* 42.2, pp. 150–169.

Bosch, J. (2009). "From Software Product Lines to Software Ecosystems". In: *Proceedings of the 13th international software product line conference*. Carnegie Mellon University, pp. 111–119.

Böttcher, T., L. Rickling, et al. (2021). "Towards the Digital Self-Renewal of Retail: The Generic Ecosystem of the Retail Industry". In: *Proceedings of the 16th Internationale Tagung Wirtschaftsinformatik (WI2021)*. Essen, Germany, pp. 1–9.

Böttcher, T. P., J. Weking, et al. (2022). "Pathways to digital business models: The connection of sensing and seizing in business model innovation". In: *The Journal of Strategic Information Systems* 31.4, pp. 1–22.

Boudreau, K. (2007). "Does Opening a Platform Stimulate Innovation? The Effect on Systemic and Modular Innovations". In: *MIT Sloan Research Paper No. 4611–06*, pp. 1–33.

Boudreau, K. (2010). "Open platform strategies and innovation: Granting access vs. devolving control". In: *Management Science* 56.10, pp. 1849–1872.

Boudreau, K. and A. Hagiu (2009). "Platform Rules: Multi-sided Platforms as Regulators". In: *Platforms, Markets and Innovation*. Ed. by A. Gawer. Cheltenham: Edward Elgar, pp. 163–191.

Boudreau, K. J. and L. B. Jeppesen (2015). "Unpaid crowd complementors: The platform network effect mirage". In: *Strategic Management Journal* 36.12, pp. 1761–1777.

Boudreau, K. J. (2017). "Platform Boundary Choices & Governance: Opening-Up While Still Coordinating and Orchestrating". In: *Entrepreneurship, Innovation, and Platforms*. Ed. by

A. Furman, B. Silverman, and S. Stern. Bingley, West Yorkshire, UK: Emerald Group Publishing Limited, pp. 227–297.

Boudreau, K. J. (2021). "Promoting platform takeoff and self-fulfilling expectations: Field experimental evidence". In: *Management Science* 67.9, pp. 5953–5967.

Brackmann, C., M. Hütsch, and T. Wulfert (2023). "Identifying Application Areas for Machine Learning in the Retail Sector". In: *SN Computer Science* 4.5, p. 426.

Braik, W., F. Morandat, J.-R. Falleri, and X. Blanc (2016). "Real time streaming pattern detection for ecommerce". In: Pisa, Italy, pp. 916–922.

Brandsom (2020). *Selling on la redoute—explanation, tips en tricks*. URL: https://blog.brandsom.nl/-eng/selling-on-la-redoute-explanation-tips-tricks (visited on 06/28/2023).

Brandwijk, N. and M. de Reuver (2023). "Do negative externalities affect platform membership conditions? The impact of moral and regulatory legitimacy on platform governance". In: *Proceedings of the 31st European Conference on Information Systems (ECIS 2023)*. Kristiansand, Norway, pp. 1–18.

Branson, A. (2020). *The Farfetch Personal Shopping Service You never knew you needed*. URL: https://www.americangirlintokyo.com/post/the-farfetch-personal-shopping-service-you-never-knew-you-needed-until-now (visited on 06/27/2023).

Brass, D. J. and M. E. Burckhardt (1993). "Potential power and power use: An investigation of structure and behavior". In: *Academy of Management Journal* 36, pp. 441–470.

Bravetti, M., R. Lucchi, G. Zavattaro, and R. Gorrieri (2004). "Web services for E-commerce: Guaranteeing security access and quality of service". In: *Proceedings of the ACM Symposium on Applied Computing (SAC2004)*. Nicosia, Cyprus, pp. 1–7.

Breidbach, C. F., D. Antons, and T. O. Salge (2016). "Seamless Service? On the Role and Impact of Service Orchestrators in Human-Centered Service Systems". In: *Journal of Service Research* 19.4, pp. 458–476.

Brenner, W. and S. Breuer (2001). "Elektronische Marktplätze—Grundlagen und strategische Herausforderungen". In: *Internet & Co. im Handel*. Ed. by D. Ahlert, J. Becker, P. Kenning, and R. Schütte. 2. Wiesbaden, Germany: Springer. Chap. 8, pp. 141–160.

Bresnahan, T. F. and S. Greenstein (1999). "Technological competition and the structure of the computer industry". In: *The Journal of Industrial Economics* 47.1, pp. 1–40.

Brickell, E., J. Camenisch, and L. Chen (2004). "Direct anonymous attestation". In: Washington DC, USA, pp. 132–145.

Briers, M. and F. C. Wai (2001). "The role of actor-networks and boundary objects in management accounting change: A field study of an implementation of activity-based costing". In: *Accounting, Organizations and Society* 26.3, pp. 237–269.

Briscoe, B., A. Odlyzko, and B. Tilly (2006). "Metcalfe's law is wrong-communications networks increase in value as they add members-but by how much?" In: *IEEE Spectrum* 43.7, pp. 34–39.

Briscoe, G. and P. De Wilde (2006). "Digital ecosystems: evolving service-orientated architectures". In: *Proceedings of the 1st international conference on Bio inspried models of network, information and computing systems (BIONETICS2006)*. Cavalese, Italy: ACM, pp. 1–7.

Broadus, R. N. (1987). "Toward a definition of "bibliometrics"." In: *Scientometrics* 12.5-6, pp. 373–379.

Broekhuizen, T. et al. (2021). "Digital platform openness: Drivers, dimensions and outcomes". In: *Journal of Business Research* 122, pp. 902–914.

Brooks, F. P. (1996). "The computer scientist as toolsmith II". In: *Communications of the ACM* 39.3, pp. 61–68.

Brown, R. and C. Mason (2017). "Looking inside the spiky bits: a critical review and conceptualisation of entrepreneurial ecosystems". In: *Small Business Economics* 49.1, pp. 11–30.

Buddeberg, H. (1959). *Betriebslehre des Binnenhandels*. Wiesbaden, Germany: Gabler Verlag.

Budler, M., I. Župič, and P. Trkman (2021). "The development of business model research: A bibliometric review". In: *Journal of Business Research* 135, pp. 480–495.

Bughin, J. R. and N. van Zeebroeck (2017). *New evidence for the power of digital platforms*. URL: https://www.mckinsey.com/capabilities/mckinsey-digital/our-insights/new-evidence-for-the-power-of-digital-platforms# (visited on 07/17/2023).

Builtwith (2022). *eCommerce Usage Distribution in the Top 1 Million Sites*. URL: https://trends.builtwith.com/shop (visited on 03/15/2022).

Bundeskartellamt (2015). *Digitale Ö konomie—Internetplattformen zwischen Wettbewerbsrecht, Privatsphäre und Verbraucherschutz*. URL: https://www.bundeskartellamt.de/SharedDocs/Publikation/DE/Diskussions_Hintergrundpapier/AK_Kartellrecht_2015_Digitale_Oekonomie.pdf?__blob=publicationFile%5C&v=2 (visited on 12/04/2019).

Burkhalter, M., C. Betz, S. Auge-Dickhut, and R. Jung (2021). "Orchestrating Value Co-Creation in Business Ecosystems". In: *Theories of Change: Change Leadership Tools, Models and Applications for Investing in Sustainable Development*. Ed. by K. Wendt. Cham, Switzerland: Springer Nature, pp. 257–291.

Burkhalter, M. (2020). *Allocentric Business Models – An Allocentric Business Model Ontology for the Orchestration of Value Co-Creation using the Example of Financial Service Ecosystems*. Untersiemau, Switzerland: Difo-Druck GmbH.

Burrel, G. and G. Morgan (2019). *Sociological Paradigms and Organisational Analysis: Elements of Sociology of Corporate Life*. 2nd ed. New York City, USA: Routledge.

Busch, J., T. Sengewald, and S. Haag (2022). "Identifying Customer Values of B2C-Fintech Services in the Area of Personal Financial Management". In: *Proceedings of the 17th International Conference on Wirtschaftsinformatik (WI2022)*. Nuernberg, Germany, pp. 1–15.

Butler, P. and J. Peppard (1998). "Consumer purchasing on the Internet:: Processes and prospects". In: *European Management Journal* 16.5, pp. 600–610.

Büttner, J., M. Hesenius, and V. Gruhn (2019). "Facilitating the support of cloud-based service marketplaces". In: *Proceedings of the IEEE International Conference on Services Computing (SCC2019)*. Milan, Italy, pp. 231–233.

Buxmann, P., H. Diefenbach, and T. Hess (2011). "Ökonomische Prinzipien der Softwareindustrie". In: *Die Softwareindustrie: Ö konomische Prinzipien, Strategien, Perspektiven*. Ed. by P. Buxmann, H. Diefenbach, and T. Hess. Berlin, Germany: Springer, pp. 23–64.

Buxmann, P. and T. Hess (2015). *Die Softwareindustrie: Ö konomische Prinzipien, Strategien, Perspektiven*. Wiesbaden, Germany: Springer-Verlag.

Buyya, R., D. Abramson, J. Giddy, and H. Stockinger (2002). "Economic models for resource management and scheduling in grid computing". In: *Concurrency and Computation: Practice and Experience* 14.13-15, pp. 1507–1542.

Cai, X., C. Qi, and C. Li (2019). "From traditional payment to mobile payment—Examining the antecedents and consequences of Hongkonger's mobile payment habit". In: *Proceedings of the 23rd Pacific Asia Conference on Information Systems (PACIS2019)*. Xian, China, pp. 1–8.

Cai, Y. (2012). "Comparative Analysis of the Workflow Modeling". In: *Proceedings of the International Conference on Management of e-Commerce and e-Government (ICMECG2012)*. Beijing, China, pp. 226–229.

Caillaud, B. and B. Jullien (2003). "Chicken & egg: Competition among intermediation service providers". In: *RAND Journal of Economics* 34.2, pp. 309–328.

Can, W. and L. Yong-Quan (2014). "Research on a novel multi-agent negotiation model for E-commerce platform". In: *7th International Conference on Intelligent Computation Technology and Automation*. Changsha, China, pp. 492–495.

Cao, L., K. Mohan, P. Xu, and B. Ramesh (2009). "A framework for adapting agile development methodologies". In: *European Journal of Information Systems* 18.4, pp. 332–343.

Caragea, D. et al. (2020). "Identifying FinTech Innovations Using BERT". In: *Proceedings of the IEEE International Conference on Big Data*. Ed. by Wu X. et al. Atlanta, USA, pp. 1117–1126.

Carlile, P. R. (2002). "A Pragmatic View of Knowledge and Boundaries: Boundary Objects in New Product Development". In: *Organization Science* 13.4, pp. 442–455.

Carlile, P. R. (2004). "Transferring, Translating, and Transforming: An Integrative Framework for Managing Knowledge Across Boundaries". In: *Organization Science* 15.5, pp. 555–568.

Carlsson, S. A. (2010). "Design Science Research in Information Systems: A Critical Realist Approach". In: *Design Research in Information Systems*. Ed. by A. R. Hevner and S. Chatterjee. Boston: Springer. Chap. 15, pp. 209–233.

Carr, N. G. (2003). "IT doesn't matter". In: *Harvard Business Review* 81.5, pp. 41–49.

Cash, A. and H.-J. Tsai (2018). "Readability of the credit card agreements and financial charges". In: *Finance Research Letters* 24, pp. 145–150.

Castro-Schez, J. J., R. Miguel, D. Vallejo, and V. Herrera (2010). "A multi-agent architecture to support B2C e-Marketplaces: The e-ZOCO case study". In: *Internet Research* 20.3, pp. 255–275.

Cater-Steel, A. and S. Grist (2008). "The Definition Dilemma of E-Commerce". In: *Electronic Commerce: Concepts, Methodologies, Tools, and Applications*. Ed. by S. A. Becker. Hershey, New York: Information Science Reference. Chap. 7.16, pp. 2093–2102.

Cavalcante, E. et al. (2011). "Cloud Integrator: Building value-added services on the cloud". In: *Proceedings of the 1st International Symposium on Network Cloud Computing and Applications (NCCA2011)*. Toulouse, France, pp. 135–142.

Caviggioli, F. and E. Ughetto (2019). "A bibliometric analysis of the research dealing with the impact of additive manufacturing on industry, business and society". In: *International Journal of Production Economics* 208, pp. 254–268.

Cdiscount (2023). *Latest API newsletter—Cdiscount—Marketplace powered by OCTOPIA*. URL: https://dev.cdiscount.com/marketplace/?page_id=3199 (visited on 06/28/2023).

Ceccagnoli, M., C. Forman, P. Huang, and D. J. Wu (2012). "Cocreation of Value in a Platform Ecosystem: The Case of Enterprise Software". In: *MIS Quarterly* 36.1, pp. 263–290.

Cennamo, C. (2018). "Building the Value of Next-Generation Platforms: The Paradox of Diminishing Returns". In: *Journal of Management* 44.8, pp. 3038–3069.

Cennamo, C. (2021). "Competing in digital markets: A platform-based perspective". In: *Academy of Management Perspectives* 35.2, pp. 265–291.

Cennamo, C. and J. Santalo (2013). "Platform competition: Strategic trade-offs in platform markets". In: *Strategic Management Journal* 34.11, pp. 1331–1350.

Chaffey, D. (2015). *E-Business and E-Commerce Management. Strategy, Implementation and Practice*. 6th ed. Harlow: Prentice-Hall.

Chanda, S. and D. Foggon (2013). "Application Lifecycle Management". In: *Beginning ASP.NET 4.5 Databases*. Berkeley, CA: Apress, pp. 235–249.

Chandler, A. D. (1962). *Strategy and Structure: Chapters in the History of the Industrial Enterprise*. Cambridge, Massachusetts: MIT Press.

Chandra, L., S. Seidel, and S. Gregor (2015). "Prescriptive Knowledge in IS Research: Conceptualizing Design Principles in Terms of Materiality, Action, and Boundary Conditions". In: *Proceedings of the 48th Hawaii International Conference on System Sciences (HICSS2015)*. Maui, Hawaii, USA: IEEE, pp. 4039–4048.

Chandra Kruse, L. and S. Seidel (2017). "Tensions in design principle formulation and reuse". In: *Proceedings of the 12th International Conference on Design Science Research in Information Systems and Technology (DESRIST 2017)*. Ed. by A. Maedche, J. Vom Brocke, and A. R. Hevner. Karlsruhe, Germany, pp. 180–188.

Chandra Kruse, L., S. Seidel, and S. Purao (2016). "Making use of design principles". In: *Proceedings of the 11th international conference on design science research in information systems (DESRIST2016)*. St. John's, NL, Canada: Springer, pp. 37–51.

Chandra Kruse, L., S. Seidel, and J. vom Brocke (2019). "Design Archaeology: Generating Design Knowledge from Real-World Artifact Design". In: *Proceedings of the 14th International Conference on Design Science Research in Information Systems and Technology (DESRIST2019)*. Worcester, MA, pp. 32–45.

Chang, E., M. West, and M. Hadzic (2006). "A Digital Ecosystem for Extend Logistics Enterprises". In: *Proceedings of the 11th International Workshop on Telework*. Ankara, Turkey, pp. 32–40.

Chapin, S. F. (2011). "The Ecosystem Concept". In: *Principles of terrestrial ecosystem ecology*. Ed. by S. F. Chapin, P. A. Matson, P. Morrison Vitousek, and M. C. Chapin. New York: Springer, pp. 3–22.

Chappell, D. (2008). *What is Application Lifecycle Management?* URL: https://download.microsoft.com/download/9/5/F/95FB4775-F769-477E-B187-6737DC09BC9D/What%20is%20ALM--Chappell.pdf (visited on 10/25/2022).

Chatterjee, D., R. Grewal, and V. Sambamurthy (2002). "Shaping up for E-commerce: Institutional enablers of the organizational assimilation of Web technologies". In: *MIS Quarterly* 26.2, pp. 65–90.

Chen, D. and H. Zhao (2012). "Data Security and Privacy Protection Issues in Cloud Computing". In: *Proceedings of the International Conference on Computer Science and Electronics Engineering (ICCSEE2012)*. London, UK, pp. 647–651.

Chen, L., M. S. Rashidin, et al. (2021). "Determinants of Consumer's Purchase Intention on Fresh E-Commerce Platform: Perspective of UTAUT Model". In: *SAGE Open* 11.2, pp. 1–17.

Chen, M. Y., E. Kiciman, et al. (2002). "Pinpoint: Problem determination in large, dynamic internet services". In: *Proceedings of the International Conference on Dependable Systems and Networks*. Washington DC, USA, pp. 1–10.

Chen, P. P.-S. (1976). "The entity-relationship model: Toward a unified view of data". In: *ACM Transactions on Database Systems* 1.1, pp. 9–36.

Chen, Q., M. Hsu, U. Dayal, and M. Griss (1999). "Multi-agent cooperation, dynamic workflow and XML for E-Commerce automation". In: *HP Laboratories Technical Report* 136, pp. 1–11.

Chen, Q. L. and Y. H. Feng (2016). "The design and application of the online bookstore system.Net based on three tier architecture". In: *Proceedings of the 10th International Conference on Information, Communications and Signal Processing (ICICS2015)*. Singapore, pp. 1–3.

Chen, W. and R. Hirschheim (2004). "A paradigmatic and methodological examination of information systems research from 1991 to 2001". In: *Information Systems Journal* 14.3, pp. 197–235.

Chen, X., Q. Zhang, et al. (2022). "Standardization of Financial Blockchain: Technologies, Challenges, and Future". In: *Proceedings of the 9th IEEE International Conference on Cyber Security and Cloud Computing (CSCloud)/Proceedings of the 8th IEEE International Conference on Edge Computing and Scalable Cloud (EdgeCom)*. Xi'an, China, pp. 30–35.

Chen, Y.-H., S.-H. Chen, and I.-C. Lin (2018). "Blockchain based smart contract for bidding system". In: *Proceedings of 4th IEEE International Conference on Applied System Innovation (ICASI 2018)*. Taiwan, China, pp. 208–211.

Chen, Y. K., F. R. Chiu, and C. J. Yang (2014). "An optimization model for product placement on product listing pages". In: *Advances in Operations Research* 2014, pp. 1–9.

Chesbrough, H., C. Lettl, and T. Ritter (2018). "Value Creation and Value Capture in Open Innovation". In: *Journal of Product Innovation Management* 35.6, pp. 930–938.

Cheung, C., M. Lee, and N. Rabjohn (2008). "The impact of electronic word-of-mouth: The adoption of online opinions in online customer communities". In: *Internet Research* 18.3, pp. 229–247.

Cheung, C. M. and D. R. Thadani (2012). "The impact of electronic word-of-mouth communication: A literature analysis and integrative model". In: *Decision Support Systems* 54.1, pp. 461–470.

Chevalier, J. A., Y. Dover, and D. Mayzlin (2018). "Channels of impact: User reviews when quality is dynamic and managers respond". In: *Marketing Science* 37.5, pp. 688–709.

Chi, J. et al. (2010). "Generic service composition platform for pervasive E-Commerce". In: *Wireless Communications and Mobile Computing* 10, pp. 1366–1377.

Chiu, Y.-L., Y.-T. Hsu, X. Mao, and J.-N. Wang (2021). "An Empirical Study of Holiday Season Discounts: A Comparison Between Third-Party Marketplace Sellers and Fulfilled by Walmart Sellers". In: *SAGE Open* 11.2, pp. 1–11.

Choi, J. P. (2010). "Tying in two-sided markets with multi-homing". In: *The Journal of Industrial Economics* 58.3, pp. 607–626.

Choi, S.-y., A. B. Whinston, and D. Stahl (1997). *The Economics of Electronic Commerce*. Indianapolis, USA: Macmillan Computer Publishing.

Choudary, S. (2015). *Platform Scale. How a New Breed of Startups Is Building Large Empires with Minimum Investment*. New York City, USA: Platform Thining Labs.

Christiaanse, E., T. Van Diepen, and J. Damsgaard (2004). "Proprietary versus internet technologies and the adoption and impact of electronic marketplaces". In: *Journal of Information Systems* 13.2, pp. 151–165.

Chung, K.-S. and S. Kim (2010). "A study on the application lifecycle management over SOA based application hosting platform". In: Phoenix Park, South Korea, pp. 930–933.

Ciborra, C. U. (1996). "The platform organization: Recombining strategies, structures, and surprises". In: *Organization Science* 7.2, pp. 103–118.

Cichocki, C. (2018). "Comparison of Wordpress Woocommerce with Magento Community Edition". In: *Journal of Computer Sciences Institute* 7, pp. 132–137.

Clarysse, B., M. Wright, J. Bruneel, and A. Mahajan (2014). "Creating value in ecosystems: Crossing the chasm between knowledge and business ecosystems". In: *Research Policy* 43.7, pp. 1164–1176.

Clement, M., K. Peteres, and F. Preiß (1999). "Electronic Commerce". In: *Marketing mit interaktiven Medien. Strategien zum Markterfolg*. Ed. by S. Albers. 2. Frankfurt am Main, Germany: FAZ-Inst. für Management-, Markt- und Medieninformationen, pp. 49–64.

Clement, R., D. Schreiber, P. Bossauer, and C. Pakusch (2019). *Internet- Ö konomie. Grundlagen und Fallbeispiele der digitalen und vernetzten Wirtschaft*. 4th ed. Wiesbaden, Germany: Springer.

Clemmensen, T., J. Norbjerg, and A. Joshi (2008). "Information Systems—Why it is a Design Science". In: *Proceedings of the International Conference on Computers and Philosophy*. Laval, France, pp. 126–139.

Cloutier, R. et al. (2010). "The concept of reference architectures". In: *Systems Engineering* 13.1, pp. 14–27.

Coase, R. H. (1937). "The Nature of the Firm". In: *Economica* 4.16, pp. 386–405.

Cobo, M. J., A. G. López-Herrera, E. Herrera-Viedma, and F. Herrera (2011). "An approach for detecting, quantifying, and visualizing the evolution of a research field: A practical application to the Fuzzy Sets Theory field". In: *Journal of Informetrics* 5.1, pp. 146–166.

Cole, R., S. Purao, M. Rossi, and M. Sein (2005). "Being Proactive: Where Action Research meets Design Research". In: *Proceedings of the 26th International Conference on Information Systems (ICIS2005)*. Las Vegas, USA, pp. 325–336.

Collins, P., A. Pluess, and C. Huber (2004). "Developing new organisational practice: The virtual factory". In: *International Journal of Manufacturing Technology and Management* 6.1-2, pp. 155–170.

Combe, C. (2006). *Introduction to E-business. Management and strategy*. Oxford, UK: Butterworth-Heinemann.

Commercetools (2020). *Lösungen für den digitalen Handel von morgen*. URL: https://commercetools.com/de (visited on 08/18/2023).

Companiesmarketcap.com (2023). *Largest Companies by Market Cap*. URL: https://companiesmarketcap.com/ (visited on 05/04/2023).

Computing Research and Education Association of Australasia (2023). *Core Conference Portal—Core 2018*. URL: http://portal.core.edu.au/conf-ranks/?search=ICEIS&by=all&source=CORE2018&sort=atitle&page=1 (visited on 08/18/2023).

Constantinides, P., O. Henfridsson, and G. G. Parker (2018). "Introduction: platforms and infrastructures in the digital age". In: *Information Systems Research* 29.2, pp. 381–400.

Cooper, D. R. and P. S. Schindler (2014). *Business Research Methods*. 12th ed. New York City, USA: McGraw-Hill Irwin.

Cooper, H. M. (1988). "Organizing Knowledge Synthesis: A Taxonomy of Literature Reviews". In: *Knowledge in Society* 1.1, pp. 104–126.

Corallo, A. (2007). "The business ecosystem as a multiple dynamic network". In: *The Digital Business Ecosystem*. Ed. by A. Corallo, G. Passiante, and A. Prencipe. Cheltenham, UK: Edward Elgar Publishing, pp. 11–32.

Corallo, A., G. Passiante, and A. Prencipe (2007). *The Digital Business Ecosystem*. Cheltenham, UK: Edward Elgar Publishing.

Corbin, J. M. and A. Strauss (1990). "Grounded theory research: Procedures, canons, and evaluative criteria". In: *Qualitative Sociology* 13.1, pp. 3–21.

Coupang (2023). *Intellectual Property Rights (IPR) & Coupang Marketplace—Coupang Marketplace*. URL: https://globalsellers.coupang.com/seller-university/intellectual-property-rights/ (visited on 06/28/2023).

Crabb, J. (2014). "The bestbuy.com cloud architecture". In: *IEEE Software* 31.2, pp. 91–96.

Cram, W. A., M. Templier, and G. Paré (2020). "(Re)considering the concept of literature review reproducibility". In: *Journal of the Association for Information Systems* 21.5, pp. 1103–1114.

Creswell, J. W. and J. D. Creswell (2022). *Research design: Qualitative, quantitative, and mixed methods approaches*. 6th ed. London, UK: SAGE Publications Ltd.

Creswell, J. W. and V. L. Plano Clark (2018). *Designing and conducting mixed methods research*. 3rd ed. London, UK: SAGE Publications Ltd.

Croitor, E. and M. Adam (2020). "Perceived Input Control on Digital Platforms: an Empirical Investigation". In: *Proceedings of the 28th European Conference on Information Systems (ECIS2020)*. 5. Marrakesh, Morroco, pp. 1–17.

Crotty, M. (1998). *The foundations of social research: Meaning and perspective in the research process*. London, UK: SAGE Publications Ltd.

Cui, Y., J. Mou, J. Cohen, and Y. Liu (2019). "Understanding information system success model and valence framework in sellers' acceptance of cross-border e-commerce: a sequential multi-method approach". In: *Electronic Commerce Research* 19.4, pp. 885–914.

Culnan, M. J., P. J. McHugh, and J. I. Zubillaga (2010). "How large U.S. companies can use twitter and other social media to gain business value". In: *MIS Quarterly Executive* 9.4, pp. 243–259.

Cusumano, M. A. and A. Gawer (2002). "The Elements of Platform Leadership". In: *MIT Sloan Management Review* 43.3, pp. 51–58.

Cusumano, M. A., A. Gawer, and D. B. Yoffie (2019). *Business of Platforms. Strategy in the Age of Digital Competition, Innovation and Power*. New York, NY: Harper Business.

D'amato, I. (2021). *A/B Test Bucketing using Hashing*. URL: https://engineering.depop.com/a-b-test-bucketing-using-hashing-475c4ce5d07 (visited on 06/28/2023).

Da Silveira, G., D. Borenstein, and F. S. Fogliatto (2001). "Mass customization: Literature review and research directions". In: *International Journal of Production Economics* 72.1, pp. 1–13.

Dai, Q. (2023). "Understanding how platform modularity enhances network effects". In: *Electronic Markets* 40.40, pp. 1–17.

Dai, Q. and R. J. Kauffman (2001). "Business models for Internet-based e-procurement systems and B2B electronic markets: An exploratory assessment". In: *Proceedings of the 34th Annual Hawaii International Conference on System Sciences (HICSS2001)*. Maui, Hawaii, USA, pp. 1–10.

Dai, W. (2021). "Development and Supervision of Robo-Advisors under Digital Financial Inclusion in Complex Systems". In: *Complexity* 2021, pp. 1–12.

Dal Bianco, V., V. Myllärniemi, M. Komssi, and M. Raatikainen (2014). "The Role of Platform Boundary Resources in Software Ecosystems: A Case Study". In: *Proceedings of the*

*11th Working IEEE/IFIP Conference on Software Architecture (WICSA22014)*. Sydney, Australia, pp. 11–20.

Dalvi, H. D., A. Joshi, and N. Shekokar (2016). "Trustworthiness evaluation system in E-Commerce context". In: *Proceedings of the International Conference on Computing Communication Control and Automation (ICCUBEA2016)*. Pune, India, pp. 1–6.

Dan, C. (2014). "Electronic Commerce: State-of-the-Art". In: *American Journal of Intelligent Systems* 4.4, pp. 135–141.

Daniel, D. A. J. and M. J. Meena (2021). "A novel sentiment analysis for amazon data with TSA based feature selection". In: *Scalable Computing* 22.1, pp. 53–66.

Daniel, E. M. and H. N. Wilson (2003). "The role of dynamic capabilities in e-business transformation". In: *European Journal of Information Systems* 12.4, pp. 282–296.

Darwin, C. R. (1859). *On the origin of species by menas of natural selection, or the preservation of favoured races in the struggle for life*. 6th ed. London, United Kingdom: John Murray.

Das, S. and A. K. Kolya (2017). "Sense GST: Text mining & sentiment analysis of GST tweets by Naive Bayes algorithm". In: *Proceedings of the 3rd IEEE International Conference on Research in Computational Intelligence and Communication Networks (ICRCICN2017)*. Kolkata, India, pp. 239–244.

Das, S. R. (2019). "The future of fintech". In: *Financial Management* 48.4, pp. 981–1007.

Datanyze (2021). *E-Commerce Platforms Market Share*. URL: https://www.datanyze.com/market-share/e-commerce-platforms--6 (visited on 01/20/2021).

Davis, J. P., K. M. Eisenhardt, and C. B. Bingham (2007). "Developing Theory Through Simulation Methods". In: *Academy of Management Review* 32.2, pp. 480–499.

Dawson, A., M. Hirt, and J. Scanlan (2016). *The economic essentials of digital strategy. A supply and demand guide to digital disruption*. URL: http://www.mckinsey.com/business-functions/strategy-and-corporate-finance/our-insights/the-economic-essentials-of-digital-strategy (visited on 08/30/2023).

Dawson, G. S., J. S. Denford, and K. C. Desouza (2016). "Governing innovation in U.S. state government: An ecosystem perspective". In: *The Journal of Strategic Information Systems* 25.4, pp. 299–318.

Al-Debei, M. M. and D. Avison (2010). "Developing a unified framework of the business model concept". In: *European Journal of Information Systems* 19.3, pp. 359–376.

De Bont, S. (2018). "Building a closed loop supply chain for a customer-centric and circular economy". In: *Journal of Supply Chain Management, Logistics and Procurement* 1.3, pp. 234–246.

DeCandia, G. et al. (2007). "Dynamo: Amazon's highly available key-value store". In: *Proceedings of twenty-first ACM SIGOPS symposium on Operating systems principles (SOSP2007)*. Stevenson, USA, pp. 205–220.

Deges, F. (2020). *Grundlagen des E-Commerce. Strategien, Modelle, Instrumente*. Wiesbaden, Germany: Springer.

Del Aguila Obra, A. R., S. B. Cámara, and A. P. Meléndez (2003). "Electronic B2B markets as an e-business model. Empirical study in the Spanish construction sector". In: *IFIP Advances in Information and Communication Technology* 105, pp. 667–684.

Delfmann, P. (2006). *Adaptive Referenzmodellierung. Methodische Konzepte zur Konstruktion und Anwendung wiederverwendungsorientierter Informationsmodelle*. Berlin, Germany: Logos Verlag.

Dellarocas, C. (2003). “The Digitization of Word of Mouth: Promise and Challenges of Online Feedback Mechanisms”. In: *Management Science* 49.10, pp. 1407–1424.

Dellermann, D., C. Jud, and K. M. Popp (2016). “Why don’t they join? Analyzing the Nature and Consequences of Complementors Costs in Platform Ecosystems”. In: *Proceedings of the 37th International Converence on Information Systems (ICIS2016)*. Dublin, Ireland, pp. 1–11.

Delteil, B., A. Le, and M. Miller (2020). *Six golden rules for ecosystem players to win in Vietnam*. URL: https://www.mckinsey.com/vn/our-insights/six-golden-rules-for-ecosystem-players-to-win-in-vietnam (visited on 07/17/2023).

Demchenko, Y., F. Turkmen, M. Slawik, and C. de Laat (2017). “Defining Intercloud Security Framework and Architecture Components for Multi-cloud Data Intensive Applications”. In: *Proceedings of the 17th IEEE/ACM International Symposium on Cluster, Cloud and Grid Computing (CCGRID2017)*. Madrid, Spain: IEEE, pp. 945–952.

Deng, Q. and S. Ji (2018). “A Review of Design Science Research in Information Systems: Concept, Process, Outcome, and Evaluation”. In: *Pacific Asia Journal of the Association for Information Systems* 10.1, pp. 1–36.

Depop (2022). *Depop Mission, Benefits, and Work Culture—Indeed.com*. URL: https://uk.indeed.com/cmp/Depop/about (visited on 06/26/2023).

Derave, T. (2019). “A Reference Architecture for Customizable Marketplaces”. In: *Proceedings of the ER 2019 Workshops FAIR, MREBA, EmpER, MoBiD, OntoCom, and ER Doctoral Symposium Papers*. Ed. by G. Guizzardi, F. Gailly, and R. S. Pintangueira Maciel. Salvador, Brazil, pp. 222–230.

Derave, T., T. P. Sales, F. Gailly, and G. Poels (2020). *Towards a Reference Ontology for Digital Platforms*. Vol. 12400 LNCS. New York City, USA: Springer International Publishing, pp. 289–302.

De Reuver, M., C. Sørensen, and R. C. Basole (2017). “The digital platform: a research agenda”. In: *Journal of Information Technology* 33.2, pp. 124–135.

Derks, J., J. Gordijn, and A. Siegmann (2018). “From chaining blocks to breaking even: A study on the profitability of bitcoin mining from 2012 to 2016”. In: *Electronic Markets* 28.3, pp. 321–338.

Derviş, H. (2019). “Bibliometric analysis using Bibliometrix an R Package”. In: *Journal of Scientometric Research* 8.3, pp. 156–160.

Deufel, P. and J. Kemper (2018). “Online Payment Method Selection: The Habitual Choice of Deferring Payment”. In: *Proceedings of the 39th International Conference on Information Systems (ICIS2018)*. San Francisco, USA, pp. 1–17.

Deuter, A. and S. Rizzo (2016). “A Critical View on PLM/ALM Convergence in Practice and Research”. In: *Procedia Technology* 26, pp. 405–412.

Deutsch, M. (1998). *Electronic Commerce. Zwischenbetriebliche Geschäftsprozesse und neue Marktzugänge realisieren*. Wiesbaden, Germany: Springer Fachmedien.

Dewey, J. (1931). *Philosophy and civilization*. New York: Minton, Balch & Co.

Dhanasai, C. and A. Parkhe (2006). “Orchestrating Innovation Networks”. In: *The Academy of Management Review* 31.3, pp. 659–669.

Dhar, V. and R. M. Stein (2018). “Complete and incomplete FinTech platforms”. In: *Journal Of Investment Management* 16.2, pp. 4–16.

Diamantini, C., L. Genga, D. Potena, and E. Storti (2015). "Towards a customizable user-centered model for data analytics". In: *Proceedings of the International Conference on Collaboration Technologies and Systems (CTS2015)*. Atlanta, Georgia, USA, pp. 249–256.

Diao, L. and Y. Ma (2008). "A versatile SOA-based e-Business platform". In: *Proceedings of the International Symposium on Electronic Commerce and Security (ISECS 2008)*. Guangzhou, China, pp. 638–641.

Diederich, S., A. B. Brendel, and L. M. Kolbe (2020). "Designing Anthropomorphic Enterprise Conversational Agents". In: *Business and Information Systems Engineering* 62.3, pp. 193–209.

Dietz, M., H. Khan, and I. Rab (2020). *How do companies create value from digital ecosystems?* URL: https://www.mckinsey.com/business-functions/mckinsey-digital/our-insights/how-do-companies-create-value-from-digital-ecosystems (visited on 07/17/2023).

DigitalCommerce360 (2023). *Global Online Marketplaces Database*. URL: https://www.digitalcommerce360.com/product/online-marketplaces-database/ (visited on 07/14/2023).

Dittrich, Y., J. Nørbjerg, P. Tell, and L. Bendix (2018). "Researching cooperation and communication in continuous software engineering". In: *Proceedings of the 11th International Workshop on Cooperative and Human Aspects of Software Engineering*. New York, NY, USA: ACM, pp. 87–90.

Doherty, N., F. Ellis-Chadwick, and C. Hart (1999). "Cyber retailing in the UK: the potential of the Internet as a retail channel". In: *International Journal of Retail & Distribution Management* 27.1, pp. 22–36.

Doherty, N. F. and F. E. Ellis-Chadwick (2006). "New perspectives in internet retailing: A review and strategic critique of the field". In: *International Journal of Retail & Distribution Management* 34.4-5, pp. 411–428.

Dominici, G. (2009). "From Marketing Mix to E-Marketing Mix: A Literature Overview and Classification". In: *International Journal of Business and Management* 4.9, pp. 17–24.

Dong, H., F. K. Hussain, and E. Chang (2007). "An Integrative view of the concept of Digital Ecosystem". In: *Proceedings of the 3rd International Conference on Networking and Services (ICNS2007)*. Athens, Greece, pp. 42–46.

Donszem (2020). *A Brief Hisotry of Online Shopping*. URL: https://www.odmworld.com/blog/brief-history-of-online-shopping/ (visited on 01/03/2023).

Donthu, N., S. Kumar, D. Mukherjee, et al. (2021). "How to conduct a bibliometric analysis: An overview and guidelines". In: *Journal of Business Research* 133, pp. 285–296.

Donthu, N., S. Kumar, and D. Pattnaik (2020). "Forty-five years of Journal of Business Research: A bibliometric analysis". In: *Journal of Business Research* 109, pp. 1–14.

Doolin, B. and L. McLeod (2012). "Sociomateriality and boundary objects in information systems development". In: *European Journal of Information Systems* 21.5, pp. 570–586.

Doty, D. H. and W. H. Glick (1994). "Typologies As a Unique Form Of Theory Building: Toward Improved Understanding and Modeling". In: *Academy of Management Review* 19.2, pp. 230–251.

Dou, Y. and D. J. Wu (2021). "Platform competition under network effects: Piggybacking and optimal subsidization". In: *Information Systems Research* 32.3, pp. 820–835.

Dowie, J., S. Henningsson, T. Kude, and K. M. Popp (2017). "Merging platform ecosystems in technology acquisitions: A governance perspective". In: *Proceedings of the 25th European Conference on Information Systems (ECIS2017)*. Münster, Germany, pp. 2301–2316.

Drechsler, A. and A. R. Hevner (2018). "Utilizing, Producing, and Contributing Design Knowledge in DSR Projects". In: *Proceedings of the 13th International Conference on Design Science Research in Information Systems and Technology (DESRIST2018)*. Madras, India, pp. 82–97.

Dubé and Paré (2003). "Rigor in Information Systems Positivist Case Research: Current Practices, Trends, and Recommendations". In: *MIS Quarterly* 27.4, pp. 597–636.

Duda, J., S. Oleszek, and K. Santarek (2022). "Product Lifecycle Management (PLM) in the Context of Industry 4.0". In: *Advances in Manufacturing III*. Ed. by J. Trojanowska, A. Kujawińska, J. Machado, and I. Pavlenko. Lecture Notes in Mechanical Engineering. Cham, Switzerland: Springer International Publishing, pp. 171–185.

Dukes, A. and L. Liu (2016). "Online shopping intermediaries: The strategic design of search environments". In: *Management Science* 62.4, pp. 1064–1077.

Durckheim, É., C. Cosman, and M. S. Cladis (2008). *The Elementary Forms of Religious Life (Oxford World's Classics)*. Oxford, UK: Oxford University Press.

Dutton, J. E. and J. M. Dukerich (1991). "Keeping an eye on the mirror: image and identity in organizational adaption". In: *Academy of Management Journal* 34.3, pp. 517–554.

DWDS (2023). *Electronic Commerce*. URL: https://www.dwds.de/wb/Electronic%20Commerce (visited on 01/31/2023).

Dwivedi, Y. K. et al. (2022). "Metaverse beyond the hype: Multidisciplinary perspectives on emerging challenges, opportunities, and agenda for research, practice and policy". In: *International Journal of Information Management* 66, pp. 1–55.

Easley, D. and J. Kleinberg (2010). *Networks, crowds, and markets*. Vol. 8. Cambridge, Massachusetts: Cambridge University Press.

Eaton, B., S. Elaluf-Calderwood, C. Sorensen, and Y. Yoo (2015). "Distributed tuning of boundary resources: the case of Apple's iOS service system". In: *MIS Quarterly* 39.1, pp. 217–243.

eBay (2021a). *API Deprecation Status*. URL: https://developer.ebay.com/docs/api-deprecation (visited on 07/28/2021).

eBay (2021b). *Features Guide—Testing in the Sandbox*. URL: https://developer.ebay.com/devzone/guides/features-guide/default.html#basics/call-sandboxtesting.html (visited on 07/29/2021).

Ebert, C. (2012). *Global Software and IT. A Guide to Distributed Development, Projects, and Outsourcing*. Hoboken, New Jersey: John Wiley & Sons.

Ecomod (2006). *Referenzgeschäftsprozesse und Strategien im E-Commerce*. URL: https://www.wi-inf.uni-duisburg-essen.de/FGFrank/ecomod/index.php (visited on 08/30/2023).

Economides, N. and E. Katsamakas (2006). "Two-sided competition of proprietary vs. open source technology platforms and the implications for the software industry". In: *Management Science* 52.7, pp. 1057–1071.

Eden, S., A.-L. Hoyer, L. Prochaska, and D. Niemer (2021). *Top 100 online marketplaces in 2021: A global in-depth report*. URL: https://ecommercedb.com/reports/top-100-online-marketplaces-in-2021-a-global-in-depth-report/408 (visited on 05/11/2023).

Eggert, I. (2020). "Handel in Zahlen". In: *brand eins. Handelt. Schwerpunkt: Die neue Konsumgesellschaft* 22.11, pp. 106–107.

EHI (2022). *Einschätzung der Absatzmöglichkeiten von Metaverse (2022)*. URL: https://www.handelsdaten.de/einschatzung-der-absatzmoglichkeiten-von-metaverse-2022 (visited on 11/02/2022).

Eickhoff, M., J. Muntermann, and T. Weinrich (2017). "What do FinTechs actually do? A Taxonomy of FinTech Business Models". In: *Proceedings of the 38th International Conference on Information Systems (ICIS2017)*. Seoul, South Korea, pp. 1–19.

Eigelshoven, F., A. Ullrich, and B. Bender (2020). "Ecosystem Types in Information Systems". In: *Proceedings of the Twenty-Eighth European Conference on Information Systems (ECIS2020)*. Marrakesh, Morroco, pp. 1–19.

Eigelshoven, F., A. Ullrich, and D. Parry (2021). "Cryptocurrency Market Manipulation—A Systematic Literature Review". In: *Proceedings of the 42nd International Conference on Information Systems (ICIS2021)*. Austin, Texas, USA, pp. 1–17.

Eisenhardt, K. M. (1989a). "Agency Theory: An Assessment and Review". In: *Academy of Management Review* 14.1, pp. 57–74.

Eisenhardt, K. M. (1989b). "Building Theories from Case Study Research". In: *The Academy of Management Review* 14.4, pp. 532–550.

Eisenhardt, K. M. and J. A. Martin (2000). "Dynamic capabilities: what are they?" In: *Strategic Management Journal* 21.10-11, pp. 1105–1121.

Eisenmann, T., G. Parker, and M. Van Alstyne (2009). "Opening Platforms: How, When and Why?" In: *Platforms, Markets and Innovation*. Ed. by A. Gawer. Cheltenham, UK: Edward Elgar, pp. 131–162.

Eisenmann, T., G. Parker, and M. Van Alstyne (2011). "Platform envelopment". In: *Strategic Management Journal* 32.12, pp. 1270–1285.

Eisenmann, T., G. Parker, and M. W. Van Alstyne (2006). "Strategies for Two-Sided Markets". In: *Havard Business Review* 84.10, pp. 92–101.

Eisenmann, T. R. (2008). "Managing proprietary and shared platforms". In: *California Management Review* 50.4, pp. 31–53.

El Sawy, O. A. (2003). "The IS Core IX: The 3 Faces of IS Identity: Connection, Immersion, and Fusion". In: *Communications of the Association for Information Systems* 12.1, pp. 588–598.

Elango, B. and P. Rajendran (2012). "Authorship trends and collaboration pattern in the marine sciences literature: a scientometric study". In: *International Journal of Information Dissemination and Technology* 2.3, pp. 166–169.

Elmorshidy, A. (2011). "Benefits Analysis of Live Customer Support Chat in E-commerce Websites: Dimensions of a New Success Model for Live Customer Support Chat". In: *Proceedings of the 10th International Conference on Machine Learning and Applications and Workshops (ICMLA2011)*. Honolulu, Hawaii, USA, pp. 325–329.

EMarketer (2022). *Retail Ecommerce sales Worldwide, 2020–2025*. URL: https://www.insiderintelligence.com/%20content/global-ecommerce-forecast-2022#page-report (visited on 07/14/2023).

eMarketer (2020). *Retail Ecommerce sales Worldwide, 2019–2024*. URL: https://www.emarketer.com/chart/242908/retail-ecommerce-sales-worldwide-2019-2024-trillions-change-of-total-retail-sales (visited on 07/14/2023).

Engert, M., J. Evers, A. Hein, and H. Krcmar (2022). "The Engagement of Complementors and the Role of Platform Boundary Resources in e-Commerce Platform Ecosystems". In: *Information Systems Frontiers* 34, pp. 2007–2025.

Engert, M., J. Evers, A. Hein, and H. Krcmar (2023). "Sustaining complementor engagement in digital platform ecosystems: Antecedents, behaviours and engagement trajectories". In: *Information Systems Journal* X.X, pp. 1–35.

Estublier, J. (2000). “Software Configuration Management: A Roadmap”. In: *22nd International Conference on Software Engineering (ICSE2000)*. Ed. by C. Ghezzi, M. Jazayeri, and A. L. Wolf. Limerick, Ireland, pp. 279–289.

Etsy (2021a). *Etsy Open API v3—Etsy Open API v3*. URL: https://developers.etsy.com/documentation/ (visited on 07/27/2021).

Etsy (2021b). *Listing—Etsy-Entwickler*. URL: https://www.etsy.com/developers/documentation/reference/listing (visited on 07/28/2021).

European Commission (2018). *Digital Single Market: Online Platforms*. URL: https://ec.europa.eu/digital-single-market/en/online-platforms-digital-single-market (visited on 03/20/2022).

European Commission (2020). *Kartellrecht: Kommission richtet Mitteilung der Beschwerdepunkte an Amazon wegen Nutzung nicht öffentlicher Daten unabhängiger Verkäufer und leitet zweite Untersuchung der E-Commerce-Geschäftspraxis des Unternehmens ein*. URL: https://ec.europa.eu/commission/presscorner/detail/de/ip_20_2077 (visited on 08/19/2023).

Evans, D. S. and R. Schmalensee (2008). “Markets with Two-sided Platforms”. In: *Issues in Competition Law and Policy (ABA Section of Antitrust Law)* 1.28, pp. 667–693.

Evans, D. S. and R. Schmalensee (2016). *Matchmakers. The New Economics of Multisided Platforms*. Boston, USA: Harvard Business Review Press.

Evans, D. S. (2003). “Some empirical aspects of multi-sided platform industries”. In: *Review of Network Economics* 2.3, pp. 191–209.

Evans, D. S. (2011). *Platform economics: Essays on multi-sided businesses*. Boston, Mass.: Competition Policy International.

Evans, D. S., A. Hagiu, and R. Schmalensee (2008). *Invisible engines. how software platforms drive innovation and transform industries*. Cambridge, Massachusetts: The MIT press.

Evans, D. S. and M. D. Noel (2008). “The analysis of mergers that involve multisided platform businesses”. In: *Journal of Competition Law and Economics* 4.3, pp. 663–695.

Evans, D. S. and R. Schmalensee (2007). “The industrial organization of markets with two-sided platforms”. In: *Competition Policy International* 3.1, pp. 1–37.

Evans, D. S. and R. Schmalensee (2010). “Failure to launch: Critical mass in platform businesses”. In: *Review of Network Economics* 9.4, pp. 1–27.

Evans, P. C. and A. Gawer (2016). *The rise of the platform enterprise: a global survey*. URL: https://www.thecge.net/app/uploads/2016/01/PDF-WEB-Platform-Survey_01_12.pdf (visited on 07/17/2023).

Evans, P. C. and R. C. Basole (2016). “Revealing the API ecosystem and enterprise strategy via visual analytics”. In: *Communications of the ACM* 59.2, pp. 26–28.

Faber, A., M. Riemhofer, S. V. Rehm, and G. Bondel (2019). “A systematic mapping study on business ecosystem types”. In: *Proceedings of the 25th Americas Conference on Information Systems (AMCIS2019)*. Cancun, Mexico, pp. 1–10.

Fabrizio, K. R. and O. Hawn (2013). “Enabling diffusion: How complementary inputs moderate the response to environmental policy”. In: *Research Policy* 42.5, pp. 1099–1111.

Fahimnia, B., J. Sarkis, and H. Davarzani (2015). “Green supply chain management: A review and bibliometric analysis”. In: *International Journal of Production Economics* 162, pp. 101–114.

Fan, S. et al. (2019). “Metapath-guided heterogeneous graph neural network for intent recommendation”. In: Anchorage, USA, pp. 2478–2486.

Faraj, S., S. L. Jarvenpaa, and A. Majchrzak (2011). "Knowledge Collaboration in Online Communities". In: *Organization Science* 22.5, pp. 1224–1239.

Faraoni, M., R. Rialti, L. Zollo, and A. C. Pellicelli (2019). "Exploring e-Loyalty Antecedents in B2C e-Commerce: Empirical results from Italian grocery retailers". In: *British Food Journal* 121.2, pp. 574–589.

Farfetch (2023). *Loyalty-Programm—Access Rewards*. URL: https://www.farfetch.com/de/farfetch-access-programme/ (visited on 06/27/2023).

Farrell, J. and G. Saloner (1985). "Standardization, compatibility, and innovation". In: *The RAND journal of economics* 16.1, pp. 70–83.

Fauzi, A. H. and H. Taylor (2011). "Community trust stores for peer-to-peer e-commerce applications". In: *Communications in Computer and Information Science* 251.1, pp. 428–442.

Feenberg, A. (1991). *Critical Theory of Technology*. New York City, USA: Oxford University Press.

Fein, M. L. (2015). "Robo-Advisors: A Closer Look". In: *SSRN Electronic Journal*, pp. 1–33.

Feller, J., P. Finnegan, J. Hayes, and P. O'Reilly (2012). "'Orchestrating' sustainable crowdsourcing: A characterisation of solver brokerages". In: *The Journal of Strategic Information Systems* 21.3, pp. 216–232.

Feng, N. et al. (2019). "The key role of dynamic capabilities in the evolutionary process for a startup to develop into an innovation ecosystem leader: An indepth case study". In: *Journal of Engineering and Technology Management* 54, pp. 81–96.

Fennell, M. L. and J. A. Alexander (1987). "Organizational Boundary Spanning in Institutionalized Environments". In: *The Academy of Management Journal* 30.3, pp. 456–476.

Ferstl, O. K. and E. J. Sinz (2013). *Grundlagen der Wirtschaftsinformatik*. Vol. 7., aktual. Munich, Germany: Oldenbourg Verlag.

Fettke, P. and P. Loos (2002). "Der Referenzmodellkatalog als Instrument des Wissensmanagements: Methodik und Anwendung". In: *Wissensmanagement mit Referenzmodellen. Konzepte für die Anwendungssystem- und Organisationsgestaltung*. Ed. by J. Becker and R. Knackstedt. Berlin, Heidelberg: Springer, pp. 3–24.

Fichman, R. G. (2004). "Real options and IT platform adoption: Implications for theory and practice". In: *Information systems research* 15.2, pp. 132–154.

Filipowska, A., M. Kaczmarek, and T. Kaczmarek (2011). "Semantic-based delivery of e-services applying the SaaS approach—Poster paper". In: *Proceedings of the International Conference on e-Business (ICE-B2011)*. Seville, Spain, pp. 1–4.

Filistrucchi, L., D. Geradin, E. Van Damme, and P. Affeldt (2014). "Market definition in two-sided markets: Theory and practice". In: *Journal of Competition Law & Economics* 10.2, pp. 293–339.

Fini, R., J. Jourdan, M. Perkmann, and L. Toschi (2023). "A New Take on the Categorical Imperative: Gatekeeping, Boundary Maintenance, and Evaluation Penalties in Science". In: *Organization Science* 34.3, pp. 1090–1110.

Fischbach, M., T. Puschmann, and R. Alt (2013). "Catchword: Service Lifecycle Management". In: *Business & Information Systems Engineering* 5.1, pp. 45–49.

Fischer, M. (2021). *Fintech business models: Applied canvas method and analysis of venture capital rounds*. Business & economics. Berlin and Boston: De Gruyter.

Fiss, P. C. (2007). "A set-theoretic approach to organizational configurations". In: *Academy of Management Review* 32.4, pp. 1180–1198.

Fiss, P. C. (2011). "Building Better Causal Theories: A Fuzzy Set Approach to Typologies in Organization Research". In: *Academy of Management Journal* 54.2, pp. 393–420.

Fitzgerald, M., N. Kruschwitz, D. Bonnet, and M. Welch (2013). "Embracing Digital Technology: A New Strategic Imperative". In: *MIT Sloan Management Review* 55.2, pp. 1–12.

Floetgen, R., S. Gomm, M. Böhm, and H. Krcmar (2020). "Implications of Emerging Financial Regulatory Reporting Frameworks for Digital Platforms Boundary Resources". In: *Proceedings of the 26th Americas Conference on Information Systems (AMCIS2020)*. Salt Lake City, USA.

Floetgen, R. J., N. Mitterer, F. Urmetzer, and M. Böhm (2021). "Platform Ecosystem Structures: Leveraging Platform-based Technology and the Finance Ecosystem for the New Normal". In: *Proceedings of the 25th Pacific Asia Conference on Information Systems (PACIS2021)*. Dubai, UAE, pp. 1–16.

Fnac Darty (2022). *Fnac Darty Forms a New Partnership with Google on Data and Cloud—Fnac Darty*. URL: https://www.fnacdarty.com/en/fnac-darty-signe-un-nouveau-partenariat-avec-google-autour-du-cloud-et-de-la-data-pour-accelerer-sa-trajectoire-digitale-et-lexecution-de-son-plan-strategique-everyday/ (visited on 06/28/2023).

Fnac Darty (2023). *Fnac Darty Marketplace*. URL: https://fnacdartymarketplace.com/en/our-partners/ (visited on 06/28/2023).

Foerderer, J., T. Kude, S. Mithas, and A. Heinzl (2018). "Does platform owner's entry crowd out innovation? Evidence from Google Photos". In: *Information Systems Research* 29.2, pp. 444–460.

Foerderer, J., T. Kude, S. W. Schuetz, and A. Heinzl (2019). "Knowledge boundaries in enterprise software platform development: Antecedents and consequences for platform governance". In: *Information Systems Journal* 29.1, pp. 119–144.

Foglie, D., N. Gelfi, M. Giacomin, and G. Guida (2010). "A Computational Model for Adapting Presentation to Content in Web Interfaces". In: *International Journal on Artificial Intelligence Tools* 19.06, pp. 783–818.

Forliano, C., P. de Bernardi, and D. Yahiaoui (2021). "Entrepreneurial universities: A bibliometric analysis within the business and management domains". In: *Technological Forecasting and Social Change* 165, p. 120522.

Foss, N. J., J. Schmidt, and D. J. Teece (2023). "Ecosystem leadership as a dynamic capability". In: *Long Range Planning* 56.1, pp. 1–16.

Frank, L. (2004). "Architecture for integration of distributed ERP systems and e-commerce systems". In: *Industrial Management and Data Systems* 104.5, pp. 418–429.

Frank, U. (2001). "A Conceptual Foundation for Versatile E-Commerce Platforms". In: *Journal of Electronic Commerce Research (JECR)* 2.2, pp. 48–57.

Frank, U. (2007). "Ein Vorschlag zur Konfiguration von Forschungsmethoden". In: *Wissenschaftstheoretische Fundierung und wissenschaftliche Orientierung der Wirtschaftsinformatik*. Ed. by F. Lehner and S. Zelewski. Berlin, Germany: GITO, pp. 158–185.

Frank, U. (2013). "Domain-specific modeling languages: Requirements analysis and design guidelines". In: *Domain Engineering: Product Lines, Languages, and Conceptual Models*. Ed. by I. Reinhartz-Berger et al. Wiesbaden, Germany: Springer, pp. 133–157.

Frölich, L. and J. Lembach (2021). *Fachartikel—Was Neo-Broker versprechen—und halten*. URL: https://www.bafin.de/SharedDocs/Veroeffentlichungen/DE/Fachartikel/2021/fa_bj_2106_Neo_Broker.html (visited on 05/03/2022).

Frosch, R. A. and N. E. Gallopoulos (1989). "Strategies for Manufacturing". In: *Scientific American* 261.3, pp. 144–152.

Fürstenau, D., D. Anisimova, et al. (2019). "The Digital Platform Otto. de: A Case Study of Growth, Complexity, and Generativity". In: *Proceedings of the 14th International Conference on Wirtschaftsinformatik (WI2019)*. Siegen, Germany, pp. 907–910.

Fürstenau, D., A. Baiyere, et al. (2023). "Extended Generativity Theory on Digital Platforms". In: *Information System Research* 35.5, pp. 1–25.

Galison, P. (1997). *Image and Logic: A Material Culture of Micro-physics*. Chicago, USA: University of Chicago Press.

Galster, M. (2015). "Software reference architectures: related architectural concepts and challenges". In: *Proceedings of the 1st International Workshop on Exploring Component-based Techniques for Constructing Reference Architectures (CobRA2015)*. Montréal, Canada, pp. 5–8.

Galster, M. and P. Avgeriou (2011). "Empirically-grounded reference architectures: a proposal". In: *Proceedings of the joint ACM SIGSOFT conference-QoSA and ACM SIGSOFT symposium-ISARCS on Quality of software architectures-QoSA and architecting critical systems (ISARCS2011)*. Boulder, USA, pp. 153–158.

Ganesh, J., T. R. Madanmohan, P. D. Jose, and S. Seshadri (2004). "Adaptive Strategies of Firms in High-Velocity Environments". In: *Journal of Global Information Management* 12.1, pp. 41–59.

Gang, C. and Y. Weiping (2008). "The design of the collaborative E-Business platform". In: *Proceedings of the International Conference on MultiMedia and Information Technology (MMIT2008)*. Washington DC, United States, pp. 188–191.

Gao, L., J.-S. Han, and H.-H. Kum (2019). "A Review of the Research on E-commerce Ecosystems in China". In: *Journal of Digital Convergence* 17.8, pp. 141–148.

Gao, R. Z., Z. J. Zhang, et al. (2013). "Modelling the Emergence and Evolution of e-Business Ecosystems from a Network Perspective". In: *Studies in Informatics and Control* 22.4, pp. 339–348.

Gawer, A. (2009a). "Platform dynamics and strategies: from products to services". In: *Platforms, Markets and Innovation*. Ed. by A. Gawer. Cheltenham, UK: Edward Elgar, pp. 45–76.

Gawer, A. (2009b). "Platforms, markets and innovation—an introduction". In: *Platforms, Markets and Innovation*. Ed. by A. Gawer. Cheltenham, UK: Edward Elgar, pp. 1–16.

Gawer, A. (2014). "Bridging differing perspectives on technological platforms: Toward an integrative framework". In: *Research policy* 43, pp. 1239–1249.

Gawer, A. and M. A. Cusumano (2002). *Platform Leadership—How Intel, Microsoft, and Cisco drive industry innovation*. Boston, United States: Harvard Busienss School Press.

Gawer, A. and M. A. Cusumano (2008). "How Companies Become Platform Leaders". In: *MIT Sloan Management Review* 49.2, pp. 28–35.

Gawer, A. and M. A. Cusumano (2014). "Industry Platforms and Ecosystem Innovation". In: *Journal of Production Innovation Management* 31.3, pp. 417–433.

Gawer, A. (2021a). "Digital platforms and ecosystems: remarks on the dominant organizational forms of the digital age". In: *Innovation: Organization & Management* 24.1, pp. 110–124.

Gawer, A. (2021b). "Digital platforms' boundaries: The interplay of firm scope, platform sides, and digital interfaces". In: *Long Range Planning* 54.5, pp. 1–16.

Gawer, A. and R. Henderson (2007). "Platform owner entry and innovation in complementary markets: Evidence from Intel". In: *Journal of Economics & Management Strategy* 16.1, pp. 1–34.

Gebhart, N., M. Kruse, and D. Krause (2016). "Gleichteile-, Modul- und Plattformstrategie". In: *Handbuch Produktentwicklung*. Ed. by U. Lindemann. Munich, Germany: Carl Hanser Verlag.

Ghanam, Y., F. Maurer, and P. Abrahamsson (2012). "Making the leap to a software platform strategy: Issues and challenges". In: *Information and Software Technology* 54.9, pp. 968–984.

Ghazawneh, A. (2012). *Towards a Boundary Resources Theory of Software Platforms*. 85. Jönköping, Sweden: ARK Tryckaren AB, pp. 1–207.

Ghazawneh, A. and O. Henfridsson (2010). "Governing third-party development through platform boundary resources". In: *Proceedings of the 31st International Conference on Information Systems (ICIS2010)*. Saint Louis, Missouri, USA, pp. 1–17.

Ghazawneh, A. and O. Henfridsson (2011). "Micro-strategizing in platform ecosystems: A multiple case study". In: *Proceedings of the 32nd International Conference on Information Systems (ICIS2011)*. Shanghai, China, pp. 4202–4220.

Ghazawneh, A. and O. Henfridsson (2013). "Balancing platform control and external contribution in third-party development: the boundary resources model". In: *Information systems journal* 23.2, pp. 173–192.

Ghazawneh, A. and O. Henfridsson (2015). "A Paradigmatic Analysis of Digital Application Marketplaces". In: *Journal of Information Technology* 30.3, pp. 198–208.

Ghosh, S. G. (2022). *Tokopedia instils a 'Privacy by Design' mindset in all Nakama: Leny Suwardi*. URL: https://ciosea.economictimes.indiatimes.com/news/big-data/tokopedia-instils-a-privacy-by-design-mindset-in-all-nakama-leny-suwardi/94354601?redirect=1 (visited on 06/28/2023).

Giachetti, R. E. (2010). *Design of enterprise systems: theory, architecture, and methods*. Boca Raton: CRC Press.

Giaglis, G. M., S. Klein, and R. M. O'Keefe (2002). "The role of intermediaries in electronic marketplaces: developing a contingency model". In: *Information systems journal* 12.3, pp. 231–246.

Gibreel, O., D. A. AlOtaibi, and J. Altmann (2018). "Social commerce development in emerging markets". In: *Electronic Commerce Research and Applications* 27, pp. 152–162.

Giddens, A. (1991). "Structuration Theory: Past, Present and Future". In: *Giddens' Theory of Structuration: A Critical Appreciation*. Ed. by C. G. A. Bryant and D. Jary. London: Routledge, pp. 201–221.

Gieryn, T. F. (1983). "Boundary-Work and the Demarcation of Science from Non-Science: Strains and Interests in Professional Ideologies of Scientists". In: *American Sociological Review* 48.6, pp. 781–795.

Giessmann, A. and K. Stanoevska (2012). "Platform as a Service—A Conjoint Study on Consumer's Preferences". In: *Proceedings of the 33rd International Conference on Information Systems (ICIS2012)*. Orlando, USA, pp. 1–20.

Gilbert, A. (2004). *E-commerce turns 10*. URL: http://news.cnet.com/E-commerce-turns-10/2100-1023_3-5304683.html (visited on 01/03/2023).

Gilder, G. (1993). *Metcalfe's law and legacy*. URL: https://www.discovery.org/a/41/ (visited on 08/01/2023).

Gillespie, T. (2010). “The politics of platforms”. In: *New media & society* 12.3, pp. 347–364.

Gimpel, H., D. Rau, and M. Röglinger (2018). “Understanding FinTech start-ups—a taxonomy of consumer-oriented service offerings”. In: *Electronic Markets* 28.3, pp. 245–264.

Gioia, D. A., K. G. Corley, and A. L. Hamilton (2013). “Seeking Qualitative Rigor in Inductive Research”. In: *Organizational Research Methods* 16.1, pp. 15–31.

GitHub (2021). *Amazon Selling Parnter API Docs*. URL: https://github.com/amzn/selling-partner-api-docs (visited on 07/29/2021).

Github (2022a). *BigCommerce*. URL: https://github.com/bigcommerce (visited on 10/18/2022).

Github (2022b). *OXID eSales AG*. URL: https://github.com/OXID-eSales (visited on 10/21/2022).

Github (2022c). *shopware/platform*. URL: https://github.com/shopware/platform (visited on 10/21/2022).

Github (2023). *Selling Partner API Documentation*. URL: https://github.com/amzn/selling-partner-api-docs?ld=NSGoogle (visited on 01/18/2023).

Gleiss, A., M. Kohlhagen, and K. Pousttchi (2021). “An apple a day—how the platform economy impacts value creation in the healthcare market”. In: *Electronic Markets* 31.4, pp. 849–876.

Golafshani, N. (2003). “Understanding Reliability and Validity in Qualitative Research”. In: *The Qualitative Report* 8.4, pp. 597–606.

Goldkuhl, G. (2004). “Design Theories in Information Systems: A Need for Multi-Grounding”. In: *Journal of Information Technology Theory and Application* 6.2, pp. 59–72.

Goldkuhl, G. (2008). “What Kind of Pragmatism in Information Systems Research?” In: *Proceedings of the Inaugural Meeting of the AIS SIGPRAG inaugural meeting*. Paris, France, pp. 1–8.

Goldkuhl, G. (2012a). “Design Research in Search for a Paradigm: Pragmatism Is the Answer”. In: *Practical Aspects of Design Science*. Ed. by M. Helfert and B. Donnellan. Berlin, Germany: Springer. Chap. 8, pp. 84–95.

Goldkuhl, G. (2012b). “Pragmatism vs interpretivism in qualitative information systems research”. In: *European Journal of Information Systems* 21.2, pp. 135–146.

Goles, T. and R. Hirschheim (2000). “The paradigm is dead, the paradigm is dead...long live the paradigm: the legacy of Burrell and Morgan”. In: *The International Journal of Management Science* 28, pp. 249–268.

Gomber, P., R. J. Kauffman, C. Parker, and B. W. Weber (2018). “On the Fintech Revolution: Interpreting the Forces of Innovation, Disruption, and Transformation in Financial Services”. In: *Journal of Management Information Systems* 35.1, pp. 220–265.

Gomber, P., J.-A. Koch, and M. Siering (2017). “Digital Finance and FinTech: current research and future research directions”. In: *Journal of Business Economics* 87.5, pp. 537–580.

Google (2023). *Fnac Darty Case Study—Google Cloud*. URL: https://cloud.google.com/customers/fnacdarty (visited on 06/28/2023).

Gopalan, S., J. Gaurav, K. Gaurav, and T. Jessica (2012). *Breakthrough IT banking*. Ed. by McKinsey & Company. URL: https://www.mckinsey.com/capabilities/mckinsey-digital/our-insights/breakthrough-it-banking (visited on 10/06/2022).

Gordijn, J., H. Akkermans, and H. van Vliet (2000). “Value based requirements creation for electronic commerce applications”. In: *Proceedings of the 33rd Annual Hawaii International Conference on System Sciences (HICSS2000)*. Maui, Hawaii, USA, pp. 1–10.

Gordijn, J., H. Akkermans, and H. van Vliet (1999). “Business Modelling Is Not Process Modelling”. In: *Proceedings of the Conceptual Modeling for E-Business and the Web (ER2000)*. Ed. by G. Goos. Salt Lake City, Utah, USA: Springer, pp. 40–51.

Gordijn, J., A. Osterwalder, and Y. Pigneur (2005). “Comparing Two Business Model Ontologies for Designing e-Business Models and Value Constellations”. In: *Proceedings of the 18th Bled eConference*. Bled, Slovenia, pp. 1–17.

Gosain, S., A. Malhotra, and O. A. El Sawy (2004). “Coordinating for flexibility in e-business supply chains”. In: *Journal of Management Information Systems* 21.3, pp. 7–45.

Gossain, S. and G. Kandiah (1998). “Reinventing Value: The New Business Ecosystem”. In: *Strategy & Leadership* 26.11, pp. 28–33.

Goswami, A., W. Han, Z. Wang, and A. Jiang (2015). “Controlled experiments for decision-making in e-Commerce search”. In: *Proceedings of the IEEE International Conference on Big Data (BigData2015)*. Santa Clara, CA, USA: IEEE, pp. 1094–1102.

Govil, N. et al. (2020). “Analyzing the Behaviour of Applying Agile Methodologies & DevOps Culture in e-Commerce Web Application”. In: *Proceedings of the 4th International Conference on Trends in Electronics and Informatics (ICOEI2020)*. Tirunelveli, India, pp. 899–902.

Graham, E. (2019). “Boundary maintenance and the origins of trolling”. In: *New Media and Society* 21.9, pp. 2029–2047.

Greckhamer, T., S. Furnari, P. C. Fiss, and R. V. Aguilera (2018). “Studying configurations with qualitative comparative analysis: Best practices in strategy and organization research”. In: *Strategic Organization* 16.4, pp. 482–495.

Greckhamer, T., V. F. Misangyi, and P. C. Fiss (2013). “The Two QCAs: From a Small-N to a Large-N Set Theoretic Approach”. In: *Configurational theory and methods in organizational research*. Ed. by P. C. Fiss. Vol. 38. Research in the Sociology of Organizations. Bingley: Emerald, pp. 49–75.

Greefhorst, D., H. Koning, and H. Van Vliet (2006). “The many faces of architectural descriptions”. In: *Information Systems Frontiers* 8.2, pp. 103–113.

Greene, J. (2007). *Mixed Methods in Social Inquiry*. San Francisco, CA, USA: John Wiley & Sons, Inc.

Gregg, D. G., U. R. Kulkarni, and A. S. Vinzé (2001). “Understanding the Philosophical Underpinnings of Software Engineering Research in Information Systems”. In: *Information Systems Frontiers* 3.2, pp. 169–183.

Gregor, S. (2006). “The nature of theory in Information Systems”. In: *MIS Quarterly* 30.3, pp. 611–642.

Gregor, S. (2009). “Building Theory in the Sciences of the Artificial”. In: *Proceedings of the 4th International Conference on Design Science Research in Information Systems and Technology (DESRIST2009)*. New York, NY, USA, pp. 1–10.

Gregor, S., L. Chandra Kruse, and S. Seidel (2020). “Research perspectives: The anatomy of a design principle”. In: *Journal of the Association for Information Systems* 21.6, pp. 1622–1652.

Gregor, S. and A. R. Hevner (2013). “Positioning and Presenting Design Science Research for Maximum Impact”. In: *MIS Quarterly* 37.2, pp. 337–355.

Gregor, S. and D. Jones (2007). “The anatomy of a design theory”. In: *Journal of the Association for Information Systems* 8.5, pp. 312–335.

Gregor, S., O. Müller, and S. Seidel (2013). “Reflection, abstraction, and theorizing in design and development research”. In: *Proceedings of the 21st European Conference on Information Systems (ECIS2013)*. Utrecht, The Netherlands, pp. 1–12.

Gregory, R. and J. Muntermann (2020). “Research note-Heuristic theorizing: Proactively generating design theories”. In: *Information Systems Research* 25.3, pp. 639–653.

Gregory, R. W. and J. Muntermann (2011). “Theorizing in design science research: Inductive versus deductive approaches”. In: *Proceedings of the 32nd International Conference on Information Systems (ICIS2011)*. Shanghai, China, pp. 888–904.

Gresov, C. and R. Drazin (1997). “Equifinality: Functional Equivalence in Organization Design”. In: *Academy of Management Review* 22.2, pp. 403–428.

Grewal, D. and M. Levy (2007). “Retailing research: Past, present, and future”. In: *Journal of Retailing* 83.4, pp. 447–464.

Grewal, D., A. L. Roggeveen, and J. Nordfällt (2017). “The Future of Retailing”. In: *Journal of Retailing* 93.1, pp. 1–6.

Grieger, M. (2003). “Electronic marketplaces: A literature review and a call for supply chain management research”. In: *European Journal of Operational Research* 144.2, pp. 280–294.

Grönroos, C. (2012). “Conceptualising value co-creation: A journey to the 1970s and back to the future”. In: *Journal of Marketing Management* 28.13-14, pp. 1520–1534.

Große Holtforth, D. (2018). *Machine Learning im E-commerce*. URL: https://ecommerceinstitut.de/machine-learning-im-e-commerce/ (visited on 10/22/2022).

GS1 (2021). *Standards provided by GS1*. URL: https://www.gs1.org/standards (visited on 07/14/2021).

Guan, Y., Q. Wei, and G. Chen (2019). “Deep learning based personalized recommendation with multi-view information integration”. In: *Decision Support Systems* 118, pp. 58–69.

Guba, E. G. (1981). “Criteria for Assessing the Trustworthiness of Naturalistic Inquiries”. In: *Educational technology research and development* 29.2, pp. 75–91.

Guba, E. G. and Y. S. Lincoln (1982). “Epistemological and methodological bases of naturalistic inquiry”. In: *Educational technology research and development* 30.4, pp. 233–252.

Guba, E. G. and Y. S. Lincoln (1994). “Competing paradigms in qualitative research”. In: *Handbook of qualitative research*. Ed. by N. K. Denzin and Y. S. Lincoln. Thousand Oaks, CA, USA: Sage Publications. Chap. 6, pp. 105–117.

Guest, G., A. Bunce, and L. Johnson (2006). “How Many Interviews Are Enough?: An Experiment with Data Saturation and Variability”. In: *Field Methods* 18.1, pp. 59–82.

Guggenberger, T. M. et al. (2020). “Ecosystem Types in Information Systems”. In: *Proceedings of the 28th European Conference on Information Systems (ECIS2020)*. Marrakesh, Morroco, pp. 1–21.

Gui, X., H. Mei, and Y. Wu (2018). “The Evolution of E-Commerce Service Ecosystem”. In: *Proceedings of the 18th International Conference on Electronic Business (ICEB2018)*. Guilin, China, pp. 614–625.

Guilloux, V., J. Locke, and A. Lowe (2013). “Digital business reporting standards: mapping the battle in France”. In: *European Journal of Information Systems* 22.3, pp. 257–277.

Gunasekaran, A., H. B. Marri, R. E. McGaughey, and M. D. Nebhwani (2002). “E-commerce and its impact on operations management”. In: *International Journal of Production Economics* 75.1-2, pp. 185–197.

Günther, W. A., M. H. Rezazade Mehrizi, M. Huysman, and F. Feldberg (2017). "Debating big data: A literature review on realizing value from big data". In: *The Journal of Strategic Information Systems* 26.3, pp. 191–209.
Gupta, R., C. Pathak, and Dagli C.H. (2014). "A machine learning framework for predicting purchase by online customers based on dynamic pricing". In: *Procedia Computer Science* 36.3, pp. 599–605.
Gutiérrez-Salcedo, M. et al. (2017). "Some bibliometric procedures for analyzing and evaluating research fields". In: *Applied Intelligence* 48, pp. 1275–1287.
Habeeb, R. et al. (2021). "Social Commerce Adoption Based UTAUT Model for Consumer Behavior: Iraq Small and Medium Enterprise". In: *Proceedings of the 2nd International Informatics and Software Engineering Conference (IISEC2021)*. Ankara, Turkey, pp. 1–8.
Hagaman, A. K. and A. Wutich (2017). "How Many Interviews Are Enough to Identify Metathemes in Multisited and Cross-cultural Research? Another Perspective on Guest, Bunce, and Johnson's (2006) Landmark Study". In: *Field Methods* 29.1, pp. 23–41.
Hagberg, J., M. Sundstrom, and N. Egels-Zandén (2016). "The digitalization of retailing: an exploratory framework". In: *International Journal of Retail & Distribution Management* 44.7, pp. 694–712.
Hage, J. (1980). *Theories of Organizations*. London, UK: John Wiley & Sons.
Hagel, J. and A. G. Armstrong (1997). *Net Gain. Expanding Markets Through Virtual Communities*. Boston, Mass.: Havard Business School Press.
Hagel, J., J. S. Brown, and L. Davison (2008). "Shaping in a world of constant disruption". In: *Harvard Business Review* 86.10, pp. 1–11.
Hagiu, A. and S. Rothman (2016). "Network Effects Aren't Enough". In: *Harvard Business Review* 94.4, pp. 17–24.
Hagiu, A. and J. Wright (2011). "Multi-Sided Platforms". In: *Havard Business School Working Paper 12–024*.
Hagiu, A. and J. Wright (2015a). "Multi-sided platforms". In: *International Journal of Industrial Organization* 43, pp. 162–174.
Hagiu, A. (2006). "Pricing and commitment by two-sided platforms". In: *RAND Journal of Economics* 37.3, pp. 720–737.
Hagiu, A. (2007). "Merchant or two-sided platform?" In: *Review of Network Economics* 6.2, pp. 115–133.
Hagiu, A. (2009). "Multi-sided platforms: From microfoundations to design and expansion strategies". In: *Harvard Business School Strategy Unit Working Paper* 09–115.
Hagiu, A. and J. Wright (2015b). "Marketplace or reseller?" In: *Management Science* 61.1, pp. 184–203.
Hajli, M. (2013). "A research framework for social commerce adoption". In: *Information Management & Computer Security* 21.3, pp. 144–154.
Hajli, M. N. (2014). "The role of social support on relationship quality and social commerce". In: *Technological Forecasting and Social Change* 87, pp. 17–27.
Hajli, N. (2015). "Social commerce constructs and consumer's intention to buy". In: *International Journal of Information Management* 35.2, pp. 183–191.
Halckenhaeusser, A., F. Mann, J. Foerderer, and P. Hoffmann (2022). "Comparing Platform Core Features with Third-Party Complements. Machine-Learning Evidence from Apple iOS". In: *Proceedings of the 56th Hawaii International Conference on System Sciences (HICSS2022)*. Maui, Hawaii, USA, pp. 6679–6688.

Hallerstede, S. H., F. Danzinger, A. C. Bullinger, and K. M. Möslein (2011). "Akzeptanzorientiertes Application Lifecycle Management". In: *HMD Praxis der Wirtschaftsinformatik* 48.2, pp. 30–40.

Han, B. R., L. Y. Chu, T. Sun, and L. Wu (2020). "Commercializing the Package Flow: Cross-sampling Physical Products Through E-commerce Warehouses". In: *41st International Conference on Information Systems (ICIS 2020)*. Hyderabad, India, pp. 1–16.

Hanani, U., B. Shapira, and P. Shoval (2001). "Information filtering: Overview of issues, research and systems". In: *User Modeling and User-Adapted Interaction* 11.3, pp. 203–259.

Hanke, J., M. Hauser, A. Durr, and F. Thiesse (2018). "Redefining the Offline Retail Experience: Designing Product Recommendation". In: *Proceedings of the 26th European Conference on Information Systems (ECIS2018)*. Portsmouth, UK, pp. 1–14.

Hannah, D. P. and K. M. Eisenhardt (2018). "How firms navigate cooperation and competition in nascent ecosystems". In: *Strategic Management Journal* 39.12, pp. 3163–3192.

Hannah, D. P., R. Tidhar, and K. M. Eisenhardt (2021). "Analytic models in strategy, organizations, and management research: A guide for consumers". In: *Strategic Management Journal* 42.2, pp. 329–360.

Hänninen, M., A. Smedlund, and L. Mitronen (2018). "Digitalization in retailing: multi-sided platforms as drivers of industry transformation". In: *Baltic Journal of Management* 13.2, pp. 152–168.

Hansen, H. and G. Neumann (2007). *Arbeitsbuch Wirtschaftsinformatik. IT-Lexikon. Aufgaben. Lösungen.* 7th ed. Stuttgart, Germany: Lucius & Lucius.

Hanseth, O. and K. Lyytinen (2010). "Design theory for dynamic complexity in information infrastructures: The case of building internet". In: *Journal of Information Technology* 25.1, pp. 1–19.

Hantschke, I. (2013). *Strategisches Management der IT-Landschaft: Ein praktischer Leitfaden für das Enterprise Architecture Management*. Vol. 3. Munich, Germany: Carl Hanser Verlag.

Harrison, J. R., Z. Lin, G. R. Carroll, and K. M. Carley (2007). "Simulation modeling in organizational and management research". In: *Academy of Management Review* 32.4, pp. 1229–1245.

Hart, O. and J. Moore (1990). "Property Rights and the Nature of the Firm". In: *Journal of Political Economy* 98.6, pp. 1119–1158.

Haselhoff, V. J. and M. Harwardt (2022). "Digitale Plattformen—Grundlagen, Herausforderungen und Lösungsansätze". In: *Praxisbeispiele der Digitalisierung*. Ed. by M. Harwardt, P. F.-J. Niermann, A. M. Schmutte, and A. Steuernagel. Wiesbaden, Germany: Springer Fachmedien, pp. 137–158.

Haucap, J. and U. Heimeshoff (2018). "Ordnungspolitik in der digitalen Welt: Ordnungsdefizite und Lösungsansätze". In: *Wirtschaftspolitik im Wandel: Ordnungsdefizite und LösungsansätzeSchriften zu Ordnungsfragen der Wirtschaft 105*. Ed. by J. Thieme and J. Haucap. Oldenburg: De Gruyter, pp. 79–132.

Haucap, J. and T. Wenzel (2011). "Wettbewerb im Internet: Was ist online anders als offline?" In: *Zeitschrift für Wirtschaftspolitik* 60.2, pp. 200–211.

Hawley, A. H. (1986). *Human ecology: A theoretical essay*. London: University of Chicago Press.

Hawlitschek, F., B. Notheisen, and T. Teubner (2020). "A 2020 perspective on "The limits of trust-free systems: A literature review on blockchain technology and trust in the sharing economy" ". In: *Electronic Commerce Research and Applications* 40, pp. 1–22.

Hawlitschek, F., T. Teubner, M. Adam, et al. (2016). "Trust in the sharing economy: An experimental framework". In: *Proceedings of the 37th International Conference on Information Systems (ICIS2016)*. Dublin, Ireland, pp. 1–16.

Hawlitschek, F., T. Teubner, and H. Gimpel (2018). "Consumer motives for peer-to-peer sharing". In: *Journal of Cleaner Production* 204, pp. 144–157.

Hedman, J. and T. Kalling (2003). "The business model concept: Theoretical underpinnings and empirical illustrations". In: *European Journal of Information Systems* 12.1, pp. 49–59.

Hein, A., M. Schreieck, T. Riasanow, et al. (2020). "Digital platform ecosystems". In: *Electronic Markets* 30.1, pp. 87–98.

Hein, A., M. Schreieck, M. Wiesche, and H. Krcmar (2016). "Multiple-Case Analysis on Governance Mechanisms of Multi-Sided Platforms". In: *Proceedings of the Multikonferenz Wirtschaftsinformatik (MKWI2016)*. Ilmenau, Germany, pp. 1613–1624.

Hein, A., J. Weking, et al. (2019). "Value co-creation practices in business-to-business platform ecosystems". In: *Electronic Markets* 29.3, pp. 503–518.

Heinemann, G. (2018). *Der neue Online-Handel. Geschäftsmodelle, Geschäftssysteme und Bechmarks im E-Commerce*. Vol. 9. Wiesbaden, Germany: Springer Gabler.

Heinemann, G., H. M. Gehrckens, M. Täuber, and Accenture GmbH (2019). *Handel mit Mehrwert*. Wiesbaden, Germany: Springer.

Heinrich, L. and D. Stelzer (2009). *Informationsmanagement: Grundlagen, Aufgaben, Methoden*. Munich, Germany: Oldenbourg Wissenschaftsverlag.

Helfat, C. E. and M. B. Lieberman (2002). "The birth of capabilities : market entry and the importance of pre-history". In: *Industrial and Corporate Change* 11.4, pp. 725–760.

Helfat, C. E. (1997). "Know-how and asset complementarity and dynamic capability accumulation: the case of R&D". In: *Strategic Management Journal* 18.5, pp. 339–360.

Helfat, C. E. and R. S. Raubitschek (2018). "Dynamic and integrative capabilities for profiting from innovation in digital platform-based ecosystems". In: *Research Policy* 47.8, pp. 1391–1399.

Helfat, C., S. Finkelstein, et al. (2007). *Dynamic capabilities: Understanding strategic change in organizations*. Malden, MA: Blackwell Publishing.

Henderson, R. M. and K. B. Clark (1990). "Architectural Innovation: The Reconfiguration of Existing Product Technologies and the Failure of Established Firms". In: *Administrative Science Quarterly* 35.1, pp. 9–30.

Henfridsson, O. and B. Bygstad (2013). "The Generative Mechanisms of Digital Infrastructure Evolution". In: *MIS Quarterly* 37.3, pp. 907–931.

Henfridsson, O., J. Nandhakumar, H. Scarbrough, and N. Panourgias (2018). "Recombination in the Open-Ended Value Landscape of Digital Innovation. Information and Organization". In: *Information and Organization* 28.2, pp. 89–100.

Henkel, J. (2006). "Selective revealing in open innovation processes: The case of embedded Linux". In: *Research Policy* 35.7, pp. 953–969.

Henkel, J., S. Schöberl, and O. Alexy (2014). "The emergence of openness: How and why firms adopt selective revealing in open innovation". In: *Research Policy* 43.5, pp. 879–890.

Hennig-Thurau, T. and G. Walsh (2003). “Electronic word-of-mouth: Motives for and consequences of reading customer articulations on the internet”. In: *International Journal of Electronic Commerce* 8.2, pp. 51–74.

Hennig-Thurau, T., K. P. Gwinner, G. Walsh, and D. D. Gremler (2004). “Electronic Word-of-Mouth via Consumer-Opinion Platforms: What Motivates Consumers to Articulate Themselves on the Internet?” In: *Journal of Interactive Marketing* 18.1, pp. 38–52.

Hepsiburada (2023a). *For Our Business Partners*. URL: https://kurumsal.hepsiburada.com/en/business-partners (visited on 06/28/2023).

Hepsiburada (2023b). *How to become a seller*. URL: https://www.hepsiburada.com/nasil-online-satis-yapilir (visited on 06/28/2023).

Hermanns, A. and M. Sauter (1999). “Electronic Commerce—Grundlagen, Potentiale, Marktteilnehmer und Transaktionen”. In: *Management-Handbuch Electronic Commerce: Grundlagen, Strategien, Praxisbeispiele*. Ed. by A. Hermanns and M. Sauter. Munich, Germany: Vahlen.

Hermes, S., J. Kaufmann-Ludwig, et al. (2020). “A Taxonomy of Platform Envelopment: Revealing Patterns and Particularities”. In: *Proceedings of the 26th Americas Conference on Information Systems (AMCIS 2020)*. Salt Lake City, USA, pp. 1–10.

Hermes, S., S. Pfab, et al. (2020). “Digital Platforms and Market Dominance: Insights from a Systematic Literature Review and Avenues for Future Research”. In: *Proceedings of the 24th Pacific Asia Conference on Information Systems (PACIS 2020)*. Dubai, UAE, pp. 1–14.

Hermes, S. and M. Schreieck (2021). “Moving beyond the Build-or-Join Decision : A Multiple Case Study on Multi-Platform Strategies of Incumbent Firms”. In: *Proceedings of the 54th Hawaii International Conference on System Sciences (HICSS 2021)*, pp. 6143–6152.

Heron, J. and P. Reason (1997). “A Participatory Inquiry Paradigm”. In: *Qualitative Inquiry* 3.3, pp. 274–294.

Herzenstein, M., R. L. Andrews, U. M. Dholakia, and E. Lyandres (2008). “The Democratization of Personal Consumer Loans? Determinants of Success in Online Peer-to-peer Lending Communities”. In: *Bulletin of the University of Delaware* 15.3, pp. 1–45.

Hesse, M., T. Teubner, and M. T. P. Adam (2020). “Bring your own stars—the economics of reputation portability”. In: *Proceedings of the 28th European Conference on Information Systems (ECIS2020)*. Marrakesh, Morroco, pp. 1–18.

Hesse, W., G. Barkow, et al. (1994). “Terminologie der Softwaretechnik. Ein Begriffssystem für die Analyse und Modellierung von Anwendungssystemen. Teil I: Begriffssystematik und Grundbegriffe”. In: *Informatik Spektrum* 17.2, pp. 96–105.

Heuritech (2020). *How La Redoute can continue innovating through trend forecasting*. URL: https://www.heuritech.com/company-analysis/la-redoute-french-brand-innovation-trend-forecasting/ (visited on 06/28/2023).

Hevner, A. and S. Chatterjee (2010). *Design Research in Information Systems*. Vol. 22. Integrated Series in Information Systems. New York [u.a.]: Springer, pp. 63–86.

Hevner, A. R. (2007). “A Three Cycle View of Design Science Research”. In: *Scandinavian Journal of Information Systems* 19.2, pp. 87–92.

Hevner, A. R., S. T. March, J. Park, and S. Ram (2004). “Design science in information systems research”. In: *MIS Quarterly* 28.1, pp. 75–105.

Hienerth, C., C. Lettl, and P. Keinz (2014). “Synergies among Producer Firms, Lead Users, and User Communities: The Case of the LEGO Producer–User Ecosystem”. In: *Journal of Product Innovation Management* 31.4, pp. 848–866.

Hiller, S., D. Morar, and D. Petrik (2022). "Creating Value in Additive Manufacturing: Modeling of Ecosystem Determinants". In: *Proceedings of the Thirtieth European Conference on Information Systems (ECIS 2022)*. Timisoara, Romania, pp. 1–17.

Hinterhuber, A. (2002). "Value Chain Orchestration in Action and the Case of the Global Agrochemical Industry". In: *Long Range Planning* 35.6, pp. 615–635.

Hinz, O., T. Otter, and B. Skiera (2020). "Estimating Network Effects in Two-Sided Markets". In: *Journal of Management Information Systems* 37.1, pp. 12–38.

Hirschheim, R. and H. K. Klein (1989). "Four paradigms of information systems development". In: *Communications of the ACM* 32.10, pp. 1199–1216.

Hirschheim, R., H. K. Klein, and K. Lyytinen (1995). *IS development and data modelling: Conceptual and philosophical foundation*. Cambridge, UK: Cambridge University Press.

Holsapple, C. W. and M. Singh (2000). "Electronic commerce: From a definitional taxonomy toward a knowledge-management view". In: *Journal of Organizational Computing and Electronic Commerce* 10.3, pp. 149–170.

Holzmann, T., K. Sailer, and B. R. Katzy (2014). "Matchmaking as multi-sided market for open innovation". In: *Technology Analysis & Strategic Management* 26.6, pp. 601–615.

Homscheid, D., T. Kilian, and M. Schaarschmidt (2015). "Offen versus geschlossen—Welchen Zusammenhang gibt es zwischen Apple iOS- und Android-App-Entwicklern?" In: *Proceedings of the 12th International Conference on Wirtschaftsinformatik (WI2015)*. Osnabrück, Germany, pp. 1191–1205.

Hossain, T., D. Minor, and J. Morgan (2011). "Competing matchmakers: An experimental analysis". In: *Management Science* 57.11, pp. 1913–1925.

Hovorka, D. (2009). "Design Science Research: A Call for a Pragmatic Perspective". In: *Proceedings of SIGPrag Workshop. Sprouts Working Papers on Information Systems*. Prague, Czeck Republic, pp. 1–11.

Hsu, K. K., Li E.Y., and Li H. (2019). "Discussion on the live broadcast of social media and e-commerce". In: *Proceedings of the 19th International Conference on Electronic Business (ICEB2019)*. Newcastle Upon Tyne, UK, pp. 178–185.

Hu, Y. et al. (2018). "Reinforcement learning to rank in E-commerce search engine: Formalization, analysis, and application". In: *Proceedings of the 24th ACM SIGKDD International Conference on Knowledge Discovery and Data Mining (KDD2018)*. London, UK, pp. 368–377.

Huang, M., Y. Zhao, and L. Zhu (2011). "Research for e-commerce platform security framework based on SOA". In: *Proceedings of the 4th International Conference on Biomedical Engineering and Informatics (BMEI2011)*. Shanghai, China, pp. 2171–2174.

Huang, Y., Y. Chai, Y. Liu, and J. Shen (2019). "Architecture of next-generation e-commerce platform". In: *Tsinghua Science and Technology* 24.1, pp. 18–29.

Huang, Z. and M. Benyoucef (2013). "From e-commerce to social commerce: A close look at design features". In: *Electronic Commerce Research and Applications* 12.4, pp. 246–259.

Huang, Z. and M. Benyoucef (2015). "User preferences of social features on social commerce websites: An empirical study". In: *Technological Forecasting and Social Change* 95, pp. 57–72.

Huber, T., T. Hurni, O. Krancher, and J. Dibbern (2022). "How Access to Resources Affects Complementor Innovation in Platform Ecosystems". In: *Digitalization Across Organizational Levels*. Ed. by J. Dibbern et al. Cham: Springer International Publishing, pp. 127–146.

Huber, T. L., T. Kude, and J. Dibbern (2017). "Governance Practices in Platform Ecosystems: Navigating Tensions Between Cocreated Value and Governance Costs". In: *Information Systems Research* 28.3, pp. 563–584.

Huebner, J., D. Vuckovac, E. Fleisch, and A. Ilic (2019). "FINTECHS AND THE NEW WAVE OF FINANCIAL INTERMEDIARIES". In: *Proceedings of the 23rd Pacific Asia Conference on Information Systems (PACIS2019)*. Xian, China, pp. 1–14.

Hurni, T., T. L. Huber, and J. Dibbern (2022). "Power dynamics in software platform ecosystems". In: *Information Systems Journal* 32.2, pp. 310–343.

Hurni, T., T. L. Huber, J. Dibbern, and O. Krancher (2021). "Complementor dedication in platform ecosystems: rule adequacy and the moderating role of flexible and benevolent practices". In: *European Journal of Information Systems* 30.3, pp. 237–260.

Hütsch, M. and T. Wulfert (2022a). "A Structured Literature Review on the Application of Machine Learning in Retail". In: *Proceedings of the 24th International Conference on Enterprise Information Systems (ICEIS2022)—Volume 1*. Online Event, pp. 332–343.

Hütsch, M. and T. Wulfert (2022b). "Design Principles for Machine Learning Marketplaches in Enterprise Systems". In: *Proceedings of the 55th Hawaii International Conference on System Sciences (HICSS2022)*. Maui, Hawaii, USA, pp. 1–10.

Hyrynsalmi, S. and S. M. Hyrynsalmi (2019). "Ecosystem: A Zombie Category?" In: *Proceedings of the IEEE International Conference on Engineering, Technology and Innovation (ICE/ITMC2019)*. Valbonne, France, pp. 1–8.

Hyrynsalmi, S., M. Seppänen, and A. Suominen (2014). "Sources of value in application ecosystems". In: *Journal of Systems and Software* 96, pp. 61–72.

Iansiti, M. and R. Levien (2004a). *The Keystone Advantage—What the New Dynamics of Business Ecosystems Mean for Strategy, Innovation, and Sustainability*. Boston, USA: Harvard Business School Publishing Corporation.

Iansiti, M. and R. Levien (2004b). "Strategy as ecology". In: *Harvard business review* 82.3, pp. 68–78, 126.

IBM (2003). *e-business on demand: A developer's roadmap accessed via waybackmachine.org*. URL: https://web.archive.org/web/20030424095322/http://www-106.ibm.com:80/developerworks/ibm/library/i-ebodov/ (visited on 03/24/2003).

Idu, A., T. V. D. Zande, and S. Jansen (2011). "Multi-homing in the Apple Ecosystem: Why and How Developers Target Multiple Apple App Stores". In: *Proceedings of the 3rd International Conference on Management of Emergent Digital EcoSystems (MEDES2011)*. San Francisco, USA, pp. 122–128.

Iivari, J., R. Hirschheim, and H. K. Klein (1998). "Information Systems Development Approaches and Methodologies". In: *Information Systems Research* 9.2, pp. 164–193.

Iivari, J. (1983). *Contributions to the theoretical foundations of systemeering research and the PICOCO model*. Oulu, Finland: Institute of Data Processing Science, University of Oulu, Unpublished Dissertation.

Iivari, J. (2007). "A paradigmatic analysis of information systems as a design science". In: *Scandinavian Journal of Information Systems* 19.2, pp. 39–64.

Iivari, J. (2003). "The IS Core—VII: Towards Information Systems as a Science of Meta-Artifacts". In: *Communications of the Association for Information Systems* 12, pp. 568–581.

Iivari, J. (2015). "Distinguishing and contrasting two strategies for design science research". In: *European Journal of Information Systems* 24.1, pp. 107–115.

Iivari, J. (2020). “Editorial: A critical look at theories in design science research”. In: *Journal of the Association for Information Systems* 21.3, pp. 502–519.

Iivari, J., M. Rotvit Perlt Hansen, and A. Haj-Bolouri (2021). “A proposal for minimum reusability evaluation of design principles”. In: *European Journal of Information Systems* 30.3, pp. 1–18.

Illik, J. A. (1998). “Electronic Commerce—eine systematische Bestandsaufnahme”. In: *HMD Praxis der Wirtschaftsinformatik* 199, pp. 10–24.

Imerman, M. B. and F. J. Fabozzi (2020). “Cashing in on innovation: a taxonomy of FinTech”. In: *Journal of Asset Management* 21.3, pp. 167–177.

Ireland, R. and A. Liu (2018). “Application of data analytics for product design: Sentiment analysis of online product reviews”. In: *CIRP Journal of Manufacturing Science and Technology* 23, pp. 128–144.

Isckia, T., M. de Reuver, and D. Lescop (2020). “Orchestrating Platform Ecosystems: The Interplay of Innovation and Business Development Subsystems”. In: *Journal of Innovation Economics & Management* 32.2, pp. 197–223.

ISO (2017). *ISO 12207:2017—Software Lifecycle Processes*.

Ivarsson, F. and F. Svahn (2020a). “Becoming a Digital Ecosystem Orchestrator—The Sydved Case”. In: *Proceedings of the 28th European Conference on Information Systems (ECIS2020)*. Marrakesh, Morroco, pp. 1–17.

Ivarsson, F. and F. Svahn (2020b). “Digital and Conventional Matchmaking : Similarities, Differences and Tensions”. In: *Proceedings of the 53rd Hawaii International Conference on System Sciences (HICSS 2020)*. Maui, Hawaii, USA, pp. 5932–5941.

Iyer, B. R. and R. C. Basole (2016). “Visualization to understand ecosystems”. In: *Communications of the ACM* 59.11, pp. 27–30.

Jacobides, M. G., C. Cennamo, and A. Gawer (2018). “Towards a theory of ecosystems”. In: *Strategic Management Journal* 39.8, pp. 2255–2276.

Jagals, M., E. Karger, and F. Ahlemann (2021). “Already grown-up or still in puberty? A bibliometric review of 16 years of data governance”. In: *Corporate Ownership & Control* 19.1, pp. 105–120.

Janiesch, C., C. Rosenkranz, and U. Scholten (2020). “An information systems design theory for service network effects”. In: *Journal of the Association for Information Systems* 21.6, pp. 1402–1460.

Jansen, S. and M. A. Cusumano (2013). “Defining software ecosystems: a survey of software platforms and business network governance”. In: *Software Ecosystems. Analyzing and Managing Business Networks in the Software Industry*. Ed. by S. Jansen, S. Brinkkemper, and M. A. Cusumano. Cheltenham: Edward Elgar, pp. 13–28.

Jansen, S. (2013). “How quality attributes of software platform architectures influence software ecosystems”. In: *Proceedings of the 2013 international workshop on ecosystem architectures (WEA2013)*. Saint Petersburg, Russia, pp. 6–10.

Jansen, S., A. Finkelstein, and S. Brinkkemper (2009). “A sense of community: A research agenda for software ecosystems”. In: *Proceedings of the 31st International Conference on Software Engineering (ICSE2009)*. Vancouver, Canada: IEEE, pp. 187–190.

Jansen, S. and G. van Capelleveen (2013). “Quality review and approval methods for extensions in software ecosystems”. In: *Software Ecosystems: Analyzing and Managing Business Networks in the Software Industry*. Ed. by S. Jansen, S. Brinkkemper, and M. A. Cusumano. Cheltenham, UK: Elgar, pp. 187–217.

Jantsch, E. (1980). *The Self-Organizing Universe*. Oxford, UK: Pergamon.

Jarvenpaa, S. L. and W. Standaert (2018). "Digital probes as opening possibilities of generativity". In: *Journal of the Association for Information Systems* 19.10, pp. 982–1000.

Jax, K. (2006). "Ecological Units: Definitions and Application". In: *The Quarterly Review of Biology* 81.3, pp. 237–258.

Jenkins, A. M. (1985). "Research methodologies and MIS research". In: *Research Methods in Information Systems*. Ed. by E. Mumford, R. Hirschheim, G. Fitz-Gerald, and T. Wood-Harper. Amsterdam: Elsevier. Chap. 6, pp. 97–107.

Jeong, H., Y. Yi, and D. Kim (2022). "An innovative e-commerce platform incorporating metaverse to live commerce". In: *International Journal of Innovative Computing, Information and Control* 18.1, pp. 221–229.

John, T., D. Kundisch, and D. Szopinski (2017). "Visual Languages for Modeling Business Models: A Critical Review and Future Research Directions". In: *Proceedings of the 38th International Conference on Information Systems (ICIS2017)*. Seoul, South Korea, pp. 1–22.

Jones, M. R. and H. Karsten (2008). "Giddens's structuration theory and information systems research". eng. In: *MIS Quarterly* 32.1, pp. 127–157.

Jonkers, H. et al. (2006). "Enterprise architecture: Management tool and blueprint for the organisation". In: *Information Systems Frontiers* 8.2, pp. 63–66.

Jorgensen, K. P. and R. Beck (2022). "Universal Wallets". In: *Business & Information Systems Engineering* 64.1, pp. 115–125.

Jung, D., B. C. Kim, M. Park, and D. W. Straub (2019). "Innovation and Policy Support for Two-Sided Market Platforms: Can Government Policy Makers and Executives Optimize Both Societal Value and Profits?" In: *Information Systems Research* 30.3, pp. 1037–1050.

Jung, D., V. Dorner, F. Glaser, and S. Morana (2018). "Robo-Advisory". In: *Business & Information Systems Engineering* 60.1, pp. 81–86.

Jung, D., V. Dorner, C. Weinhardt, and H. Pusmaz (2018). "Designing a robo-advisor for risk-averse, low-budget consumers". In: *Electronic Markets* 28.3, pp. 367–380.

Jwo, J.-S., T.-S. Hsu, and Y. C. Cheng (2013). "Jumpstarting Application Lifecycle Management: A New Approach with Tool Support". In: *Journal of Infomraiton Science and Engineering* 29, pp. 475–492.

Kääriäinen, J. (2011). *Towards an Application Lifecycle Management Framework*. Oulu, Finland: VTT Publications.

Kääriäinen, J. and A. Välimäki (2008). "Impact of Application Lifecycle Management—A Case Study". In: *Enterprise Interoperability III*. Ed. by K. Mertins, R. Ruggaber, K. Popplewell, and X. Xu. London: Springer London, pp. 55–67.

Kääriäinen, J. and A. Välimäki (2009). "Applying Application Lifecycle Management for the Development of Complex Systems: Experiences from the Automation Industry". In: *Software Process Improvement*. Ed. by R. V. O'Connor et al. Berlin, Germany: Springer Berlin Heidelberg, pp. 149–160.

Kabac, M., C. Consel, and N. Volanschi (2016). "Leveraging Declarations over the Lifecycle of Large-Scale Sensor Applications". In: *Proceedings of the IEEE Conferences on Ubiquitous Intelligence & Computing, Advanced and Trusted Computing, Scalable Computing and Communications, Cloud and Big Data Computing, Internet of People, and Smart World Congress*. Atlanta, GA, USA, pp. 211–219.

Kalakota, R. and A. B. Whinston (1996). *Frontiers of electronic commerce*. Reprinted with corr. Internet, electronic commerce. Reading, Mass.: Addison-Wesley.

Kambil, A. (1995). "Electronic Commerce: Implications of the Internet for Business Practice and Strategy". In: *Business Economics* 30.4, pp. 27–33.

Kanellis, P. and T. Papadopoulos (2009). "Conducting Research in Information Systems: An Epistemological Journey". In: *Information Systems Research Methods, Epistemology, and Applications*. Ed. by A. Cater-Steel and L. Al-Hakim. Hershey: Information Science Reference. Chap. 1, pp. 1–34.

Kang, C. (2021). *Lawmakers, Taking Aim at Big Tech, Push Sweeping Overhaul of Antitrust*. URL: https://www.nytimes.com/2021/06/11/technology/big-tech-antitrust-bills.html (visited on 02/02/2022).

Kannengießer, N., S. Lins, T. Dehling, and A. Sunyaev (2019). "What Does Not Fit Can be Made to Fit! Trade-Offs in Distributed Ledger Technology Designs". In: *Prooceedings of the 52nd Hawaii International Conference on System Sciences (HICSS2019)*. Maui, Hawaii, USA.

Kano, N., N. Seraku, F. Takahashi, and S. ichi Tsuji (1984). "Attractive Quality and Must-Be Quality". In: *Journal of the Japanese Society for Quality Control* 14.2, pp. 147–156.

Kaoud, E., M. A. Abdel-Aal, T. Sakaguchi, and N. Uchiyama (2020). "Design and Optimization of the Dual-Channel Closed Loop Supply Chain with E-Commerce". In: *Sustainability* 12.23, pp. 1–21.

Kapoor, K. et al. (2021). "A socio-technical view of platform ecosystems: Systematic review and research agenda". In: *Journal of Business Research* 128, pp. 94–108.

Kapoor, R. (2018). "Ecosystems: broadening the locus of value creation". In: *Journal of Organization Design* 7.1, pp. 1–16.

Karagiannis, D. and H. Kühn (2002). "Metamodelling platforms". In: *Lecture Notes in Computer Science (including subseries Lecture Notes in Artificial Intelligence and Lecture Notes in Bioinformatics)* 2455, p. 182.

Karger, E., M. Jagals, and F. Ahlemann (2021). "Blockchain for AI Data—State of the Art and Open Research". In: *Proceedings of the 42nd International Conference on Information Systems (ICIS2021)*. Austin, Texas, pp. 1–17.

Karhu, K., R. Gustafsson, and K. Lyytinen (2018). "Exploiting and defending open digital platforms with boundary resources: Android's five platform forks". In: *Information Systems Research* 29.2, pp. 479–497.

Karhu, K. and P. Ritala (2021). "Slicing the cake without baking it: Opportunistic platform entry strategies in digital markets". In: *Long Range Planning* 54.5, pp. 1–23.

Kauffman, R. J. and H. Mohtadi (2004). "Proprietary and open systems adoption in E-procurement: A risk-augmented transaction cost perspective". In: *Journal of Management Information Systems* 21.1, pp. 137–166.

Kawa, A. and M. Walesiak (2019). "Marketplace as a key actor in e-commerce value networks". In: *Logforum* 15.4, pp. 521–529.

Kazan, E. and J. Damsgaard (2016). "Towards A New Market Entry Framework For Digital Payment Platforms". In: *Communications of the Association for Information Systems* 38.1, pp. 761–783.

Kazan, E., C.-W. Tan, et al. (2018). "Disentangling Digital Platform Competition: The Case of UK Mobile Payment Platforms". In: *Journal of Management Information Systems* 35.1, pp. 180–219.

Keller, W. (2017). *IT-Unternehmensarchitektur. Von der Geschäftsstrategie zur optimalen IT-Unterst ü tzung*. 3rd ed. Heidelberg: dpunkt.verlag.

Kellogg, K. C., W. J. Orlikowski, and J. A. Yates (2006). "Life in the trading zone: Structuring coordination across boundaries in postbureaucratic organizations". In: *Organization Science* 17.1, pp. 22–44.

Kenney, M. and J. Zysman (2016). "The Rise of the Platform Economy". In: *Issues in Science and Technology* 32.3, pp. 61–69.

Kenney, M. and J. Zysman (2015). "Choosing a future in the platform economy: the implications and consequences of digital platforms". In: *Kauffman Foundation New Entrepreneurial Growth Conference*. Amelia Island, Florida, USA, pp. 1–23.

Khabou, N., I. Bouassida Rodriguez, and M. Jmaiel (2019). "An overview of a novel analysis approach for enhancing context awareness in smart environments". In: *Information and Software Technology* 111, pp. 131–143.

Khalid, H., E. Shihab, M. Nagappan, and A. E. Hassan (2014). "What Do Mobile App Users Complain About". In: *IEEE Software* 32.3, pp. 70–77.

Khalifa, M. and N. Shen (2005). "Effects of Electronic Customer Relationship Management on Customer Satisfaction—A Temporal Model". In: *Proceedings of the 38th Hawaii International Conference on System Sciences (HICSS2005)*. Maui, HI, USA, pp. 1–10.

Khalil, M. A. T., P. Dominic, M. F. Bin Hassan, and A. Mushtaq (2011). "An In Depth Analysis of Ecosystems & Blueprint of Digital Business Ecosystem (DBE) Framework for Malaysian SMEs". In: *Computer Science Journal* 1.1, pp. 65–78.

Khanra, S., A. Dhir, P. Kaur, and M. Mäntymäki (2021). "Bibliometric analysis and literature review of ecotourism: Toward sustainable development". In: *Tourism Management Perspectives* 37, pp. 1–15.

Khanra, S., A. Dhir, and M. Mäntymäki (2020). "Big data analytics and enterprises: a bibliometric synthesis of the literature". In: *Enterprise Information Systems* 14.6, pp. 737–768.

Khanra, S., A. Dhir, V. Parida, and M. Kohtamäki (2021). "Servitization research: A review and bibliometric analysis of past achievements and future promises". In: *Journal of Business Research* 131, pp. 151–166.

Khatri, V. and C. V. Brown (2010). "Designing data governance". In: *Communications of the ACM* 53.1, pp. 148–152.

Khavul, S. and G. D. Bruton (2013). "Harnessing Innovation for Change: Sustainability and Poverty in Developing Countries". In: *Journal of Management Studies* 50.2, pp. 285–306.

Khern-am-nuai, W., K. Kannan, and H. Ghasemkhani (2018). "Extrinsic versus intrinsic rewards for contributing reviews in an online platform". In: *Information Systems Research* 29.4, pp. 871–892.

Kittlaus, H. B. and P. N. Clough (2009). *Software product management and pricing: Key success factors for software organizations*. Heidelberg: Springer, pp. 1–231.

Klafft, M. (2008). "Online Peer-to-Peer Lending: A Lenders' Perspective". In: *Proceedings of the 2008 International Conference on E-Learning, E-Business, Enterprise Information Systems, and E-Government*. Las Vegas, USA, pp. 371–375.

Klein, L. R. (1993). "CALS+EDI = electronic commerce". In: *Trends in ecology & evolution* 1, pp. 1–15.

Knackstedt, R. (2006). *Fachkonzeptionelle Referenzmodellierung einer Managementunterstützung mit quantiativen und qualitativen Daten. Methodische Konzepte zur Konstruktion und Anwendung*. Berlin, Germany: Logos Verlag.

Knodel, J. and K. Manikas (2015). "Towards a Typification of Software Ecosystems". In: *Lecture Notes in Business Information Processing*. Vol. 210, pp. 60–65.

Knoll, M. and S. Rinderle-Ma (2015). "Plattformen: Eine Einführung". In: *HMD Praxis der Wirtschaftsinformatik* 52.3, pp. 322–336.

Koch, H. (2010). "Developing dynamic capabilities in electronic marketplaces: A cross-case study". In: *The Journal of Strategic Information Systems* 19.1, pp. 28–38.

Kogut, B. and N. Kulatilaka (1994). "Options Thinking and Platform Investments: Investing in Opportunity". In: *California Management Review* 36.2, pp. 52–71.

Koh, T. K. and M. Fichman (2014). "Multi-homing users' preferences for two-sided exchange networks". In: *MIS Quarterly* 38.4, pp. 977–996.

Kollmann, T. (2019). *E-Business. Grundlagen elektronischer Geschäftsprozesse in der Digitalen Wirtschaft*. ger. 7., überar. Wiesbaden, Germany: Springer Gabler.

Kollmann, T. (2022). *Digital Business. Grundlagen von Geschäftsmodellen und -prozessen in der Digitalen Wirtschaft*. Wiesbaden, Germany: Springer Fachmedien.

Kong, Y., Y. Wang, S. Hajli, and M. Featherman (2020). "In Sharing Economy We Trust: Examining the Effect of Social and Technical Enablers on Millennials' Trust in Sharing Commerce". In: *Computers in Human Behavior* 108, pp. 1–27.

Koo, Y., Y. Park, J. Ham, and J.-N. Lee (2019). "Congruent patterns of outsourcing capabilities: A bilateral perspective". In: *The Journal of Strategic Information Systems* 28.4, pp. 1–25.

Koppenhagen, N., O. Gaß, and B. Müller (2012). "Design Science Research in Action—Anatomy of Success Critical Activities for Rigor and Relevance". In: *Proceedings of the 20th European Conference on Information Systems (ECIS2012)*. Barcelona, Spain, pp. 1–12.

Korpela, K. and O. Taipale (2013). "A Framework for Exploring Digital Business Ecosystems". In: *Proceedings of the 46th Hawaii International Conference on System Sciences (HICSS 2013)*. Maui, Hawaii, USA, pp. 1–10.

Koseoglu, M. A. (2016). "Mapping the institutional collaboration network of strategic management research: 1980–2014". In: *Scientometrics* 109.1, pp. 203–226.

Kosiol, E. (1964). "Betriebswirtschaftslehre und Unternehmensforschung. Eine Untersuchung ihrer Standorte und Beziehungen auf wissenschaftstheoretischer Grundlagen". In: *Zeitschrift für Betriebswirtschaft* 34.12, pp. 743–762.

Kotler, P. (1972). "A generic concept of marketing". In: *Journal of Marketing* 36.2, pp. 46–54.

Kotler, P. and F. Bliemel (2001). *Marketing-Management. Analyse, Planung, Umsetzung und Steuerung*. Stuttgart, Germany: Schäffer-Poeschel.

Kotler, P. and K. L. Keller (2016). *Marketing Management*. 16th ed. Boston [u. a.]: Pearson.

Kotusev, S. and S. Kurnia (2020). "The theoretical basis of enterprise architecture: A critical review and taxonomy of relevant theories". In: *Journal of Information Technology* 35.4, pp. 1–41.

Kreger, H. (2003). "Fulfilling the Web services promise". In: *Communications of the ACM* 46.6, pp. 29–30+32–34.

Kretschmer, T., A. Leiponen, M. Schilling, and G. Vasudeva (2022). "Platform ecosystems as meta-organizations: Implications for platform strategies". In: *Strategic Management Journal* 43.3, pp. 405–424.

Krippendorff, K. (2013). *Content Analysis: An introduction to its methodology*. 3rd ed. Los Angeles, USA: SAGE.

Krishen, A. S., Y. K. Dwivedi, N. Bindu, and K. S. Kumar (2021). "A broad overview of interactive digital marketing: A bibliometric network analysis". In: *Journal of Business Research* 131, pp. 183–195.

Kuckartz, U. (2010). *Einführung in die computergestützte Analyse qualitativer Daten*. 3., aktual. Wiesbaden, Germany: Verlag für Sozialwissenschaften.

Kuckartz, U. and S. Rädiker (2022). *Qualitative Inhaltsanalyse. Methoden, Praxis, Computerunterstützung*. 5., überar. Weinheim: Beltz.

Kude, T., J. Dibbern, and A. Heinzl (2012). "Why do complementors participate an analysis of partnership networks in the enterprise software industry". In: *IEEE Transactions on Engineering Management* 59.2, pp. 250–265.

Kude, T., T. Huber, and J. Dibbern (2019). "Successfully Governing Software Ecosystems: Competence Profiles of Partnership Managers". In: *IEEE Software* 36.3, pp. 39–44.

Kuechler, B. and V. Vaishnavi (2008). "Theory development in design science research: Anatomy of a research project". In: *Proceedings of the 3rd International Conference on Design Science Research in Information Systems and Technology (DESRIST2008)*. Atlanta, GA, USA, pp. 1–15.

Kuechler, B. and V. Vaishnavi (2011). "Promoting Relevance in IS Research: An Informing System for Design Science Research". In: *Informing Science: The International Journal of an Emerging Transdiscipline* 14, pp. 125–138.

Kuhn, T. S. (1962). *The Structure of Scientific Revolutions*. Chicago, USA: University of Chicago Press.

Kuk, G. and M. Janssen (2013). "Assembling infrastructures and business models for service design and innovation". In: *Information Systems Journal* 23.5, pp. 445–469.

Kundisch, D. et al. (2022). "An Update for Taxonomy Designers: Methodological Guidance from Information Systems Research". In: *Business and Information Systems Engineering* 64.4, pp. 421–439.

Kwak, J., Y. Zhang, and J. Yu (2019). "Legitimacy building and e-commerce platform development in China: The experience of Alibaba". In: *Technological Forecasting and Social Change* 139, pp. 115–124.

La Redoute (2021). *Marketplace—La Redoute Corporate*. URL: https://www.laredoute-corporate.com/en/marketplace-en/ (visited on 06/28/2023).

Lacheiner, H. and R. Ramler (2011). "Application Lifecycle Management as Infrastructure for Software Process Improvement and Evolution: Experience and Insights from Industry". In: *Proceedings of the 37th EUROMICRO Conference on Software Engineering and Advanced Applications (SEA2011)*. Oulu, Finland: IEEE, pp. 286–293.

Laks, E., A. Even, and Y. Parmet (2013). "Usage Patterns of Internet-Based Banking and BI Services". In: *Proceedings of the 19th Americas Conference on Information Systems (AMCIS2013)*. Chicago, USA, pp. 1–6.

Lamont, M. and V. Molnár (2002). "The Study of Boundaries in the Social Sciences". In: *Annual Review of Sociology* 28.1, pp. 167–195.

Lan, J., Y. Liu, and Y. Chai (2008). "A solution model for service-oriented architecture". In: Chongqing, China, pp. 4184–4189.

Lanamäki, A., K. Väyrynen, S. Laari-Salmela, and M. Kinnula (2020). "Examining relational digital transformation through the unfolding of local practices of the Finnish taxi industry". In: *The Journal of Strategic Information Systems* 29.3, pp. 1–21.

Langley, A., K. Lindberg, et al. (2019). "Boundary Work among Groups, Occupations, and Organizations: From Cartography to Process". In: *Academy of Management Annals* 13.2, pp. 704–736.

Langley, P. and A. Leyshon (2021). "The Platform Political Economy of FinTech: Reintermediation, Consolidation and Capitalisation". In: *New Political Economy* 26.3, pp. 376–388.

Langlois, R. N. (2002). "Modularity in technology and organization". In: *Journal of Economic Behavior & Organization* 49.1, pp. 19–37.

Larsson, R. (1993). "Case Survey Methodology: Quantitative Analysis of Patterns Across Case Studies". In: *Academy of Management Journal* 36.6, pp. 1515–1546.

Laudon, K. C. and C. G. Traver (2019). *E-Commerce 2018: business. technology. society.* Vol. 14. Boston: Pearson Education.

Laudon, K. C. and C. G. Traver (2020). *E-Commerce 2020–2021: Business. technology. society.* 16th ed. Vol. 16. Boston: Pearson Education.

Laumann, E. O., P. V. Marsden, and D. Prensky (1983). "The Boundary Specification Problem in Network Analysis". In: *Applied Network Analysis*. Ed. by R. Burt and M. Minor. Beverly Hills, CA, USA: SAGE. Chap. 1, pp. 18–34.

Lavassani, K. M., B. M. Movahedi, and V. Kumar (2011). "E-marketplaces: Taxonomy of the new collaborative information platform". In: *E-Novation for Competitive Advantage in Collaborative Globalization: Technologies for Emerging E-Business Strategies*. Ed. by H. M. Pattinson and D. R. Low. Hershey, PA, USA: IGI Global. Chap. 2, pp. 16–33.

Lavie, D. (2002). "The competitive advantage of interconnected firms: An extension of the resource-based view". In: *Academy of Management Proceedings* 31.3, pp. 638–658.

Lebow, S. (2021). *Worldwide ecommerce continues double-digit growth following pandemic push to online*. URL: https://www.emarketer.com/content/worldwide-ecommerce-continues-double-digit-growth-following-pandemic-push-online (visited on 04/22/2022).

Lee, A. S. and J. V. Nickerson (2010). "Theory As a Case of Design: Lessons for Design from the Philosophy of Science". In: *Proceedings of the 43rd Hawaii International Conference on System Sciences (HICSS2010)*. Maui, Hawaii: IEEE, pp. 1–8.

Lee, C. P. (2007). "Boundary Negotiating Artifacts: Unbinding the Routine of Boundary Objects and Embracing Chaos in Collaborative Work". In: *Computer Supported Cooperative Work (CSCW)* 16.3, pp. 307–339.

Lee, E., B. Lee, and M. Chae (2011). "Herding Behavior In Online P2P Lending: An Empirical Investigation". In: *Proceedings of the Pacific Asia Conference on Information Systems (PACIS2011)*. Brisbane, Australia, pp. 1–15.

Lee, H. Y. and N. J. Wang (2019). "Cloud-based enterprise resource planning with elastic model:view:controller architecture for Internet realization". In: *Computer Standards and Interfaces* 64, pp. 11–23.

Lee, J. H., M. G. Hancock, and M.-C. Hu (2014). "Towards an effective framework for building smart cities: Lessons from Seoul and San Francisco". In: *Technological Forecasting and Social Change* 89, pp. 80–99.

Lee, J.-N., Y. Park, D. W. Straub, and Yunmo Koo (2019). "Holistic Archetypes of IT Outsourcing Strategy: A Contingency Fit and Configurational Approach". In: *MIS Quarterly* 43.4, pp. 1201–1225.

Lee, J. S., J. Pries-Heje, and R. Baskerville (2011). "Theorizing in Design Science Research". In: *Proceedings of the International Conference on Design Science Research in Information Systems and Technology (DESRIST2011)*. Milwaukee, Wisconsin, USA, pp. 1–16.

Lefebvre, L. A. and É. Lefebvre (2002). "E-commerce and virtual enterprises: Issues and challenges for transition economies". In: *Technovation* 22.5, pp. 313–323.

Leiblein, M. J. (2016). "Platform Innovation". In: *The Palgrave Encyclopedia of Strategic Management*. Ed. by M. Augier and D. J. Teece. London: Palgrave Macmillan UK, pp. 1–4.

Leten, B. et al. (2013). "IP Models to Orchestrate Innovation Ecosystems: IMEC, a Public Research Institute in Nano-Electronics". In: *California Management Review* 55.4, pp. 51–64.

Leung, X. Y., J. Sun, and B. Bai (2017). "Bibliometrics of social media research: A co-citation and co-word analysis". In: *International Journal of Hospitality Management* 66, pp. 35–45.

Levy, M. and R. Hirschheim (2012). "Removing the positivist straight jacket from information systems design science research". In: *Proceedings of the 20th European Conference on Information Systems (ECIS2012)*. Barcelona, Spain, pp. 1–12.

Levy, M., B. A. Weitz, and D. Grewal (2019). *Retailing management*. Tenth edit. New York, NY: McGraw-Hill Education.

Lexico.com (2019). *Platform*. URL: https://www.lexico.com/en/definition/platform (visited on 11/22/2019).

Lexico.com (2023). *E-commerce*. URL: https://www.lexico.com/browse/e-commerce (visited on 01/31/2023).

Li, F.-L., M. Qiu, et al. (2017). "AliMe Assist: An Intelligent Assistant for Creating an Innovative E-Commerce Experience". In: *Proceedings of the Conference on Information and Knowledge Management (CIKM2017)*. New York, NY, USA, pp. 2495–2498.

Li, J. (2019). "Design of B2B E-commerce Platform Based on SOA Architecture". In: *IOP Conference Series: Materials Science and Engineering* 569.3, pp. 1–5.

Li, L., F. Su, W. Zhang, and J.-Y. Mao (2018). "Digital transformation by SME entrepreneurs: A capability perspective". In: *Information Systems Journal* 28.6, pp. 1129–1157.

Li, Q., J. Zhou, et al. (2010). "Business processes oriented heterogeneous systems integration platform for networked enterprises". In: *Computers in Industry* 61.2, pp. 127–144.

Li, S.-T. and L.-Y. Shue (2001). "Towards XML-enabling E-Commerce infomediary: A case study in cargo tracking". In: *Proceedings of the 34th Hawaii International Conference on System Sciences (HICSS2001)*. Maui, Hawaii, USA, pp. 1–10.

Li, W., Y. Badr, and F. Biennier (2012). "Digital ecosystems: Challenges and prospects". In: *Proceedings of the International Conference on Management of Emergent Digital EcoSystems (MEDES2012)*. New York City, USA, pp. 117–122.

Li, X. and X. Ai (2021). "A choice of selling format in the online marketplace with cross-sales supply chain: Platform selling or traditional reselling?" In: *Electronic Commerce Research* 21.2, pp. 393–422.

Li, Y., X. Lu, et al. (2006). "The realization of service-oriented e-Marketplaces". In: *Information Systems Frontiers* 8.4, pp. 307–319.

Li, Y. (2018). "The Strategic Decision on Mobile Payment: A Study on Merchants' Adoption". In: *Proceedings of the 22nd Americas Conference on Information Systems (AMCIS2018)*. San Diego, USA, pp. 1–10.

Li, Z. and T. Penard (2014). "The role of quantitative and qualitative network effects in B2B platform competition". In: *Managerial and Decision Economics* 35.1, pp. 1–19.

Liang, H. (2013). "Research and implementation of SOA-based e-commerce platform". In: *Advances in Intelligent Systems and Computing* 180 AISC, pp. 923–928.

Liang, T.-P. and E. Turban (2011). "Introduction to the special issue social commerce: A research framework for social commerce". In: *International Journal of Electronic Commerce* 16.2, pp. 5–13.

Liang, T.-P., Y.-T. Ho, Y.-W. Li, and E. Turban (2011). "What Drives Social Commerce: The Role of Social Support and Relationship Quality". In: *International Journal of Electronic Commerce* 16.2, pp. 69–90.

Liao, S.-H., R. Widowati, and Y.-C. Hsieh (2021). "Investigating online social media users' behaviors for social commerce recommendations". In: *Technology in Society* 66, pp. 1–19.

Libert, B., Y. Wind, and M. Fenley (2014). "What Airbnb, Uber, and Alibaba have in common". In: *Harvard business review* 11.1, pp. 1–9.

Lilien, G. L. (2016). "The B2B Knowledge Gap". In: *International Journal of Research in Marketing* 33.3, pp. 543–556.

Lim, C., K.-H. c (2018). "From data to value: A nine-factor framework for data-based value creation in information-intensive services". In: *International Journal of Information Management* 39, pp. 121–135.

Lim, Y. H., H. Hashim, et al. (2019). "Blockchain technologies in e-commerce: Social shopping and loyalty program applications". In: *Lecture Notes in Computer Science (including subseries Lecture Notes in Artificial Intelligence and Lecture Notes in Bioinformatics)* 11579 LNCS, pp. 403–416.

Lin, C.-M., H.-S. Yang, and Y. He (2005). "Collaborative e-Business platform based on grid service". In: *Proceedings of the 1st International Conference on Semantics, Knowledge and Grid (SKG2005)*. Beijing, China, pp. 1–3.

Lin, T.-H., C. Gao, and Y. Li (2019). "Cross-platform recommendation for social e-commerce". In: *Proceedings of the 42nd International ACM SIGIR Conference on Research and Development in Information Retrieval (SIGIR2019)*. Paris, France, pp. 515–524.

Lin, X., Y.-W. Zhou, et al. (2020). "Pricing and Product-bundling Strategies for E-commerce Platforms with Competition". In: *European Journal of Operational Research* 283.3, pp. 1026–1039.

Lincoln, Y. S. and E. G. Guba (1985). *Naturalistic Inquiry*. Beverly Hills, CA, USA: Sage Publications.

Linde, L., D. Sjödin, V. Parida, and J. Wincent (2021). "Dynamic capabilities for ecosystem orchestration A capability-based framework for smart city innovation initiatives". In: *Technological Forecasting and Social Change* 166, pp. 1–12.

Lindeman, R. L. (1942). "The Trophic-Dynamic Aspect of Ecology". In: *Ecology* 23.4, pp. 399–417.

Linnhoff-Popien, C., R. Schneider, and M. Zaddach (2018). *Digital Marketplaces Unleashed*. Ed. by C. Linnhoff-Popien, R. Schneider, and M. Zaddach. Berlin, Heidelberg: Springer Berlin Heidelberg.

Liu, H., X. Li, and S. Wang (2021). "A bibliometric analysis of 30 years of platform research: Developing the research agenda for platforms, the associated technologies and social impacts". In: *Technological Forecasting and Social Change* 169.8, pp. 1–24.

Liu, J., V. Abhishek, and B. Li (2016). "The Impact of Mobile Adoption on Customer Omni-Channel Banking Behavior". In: *Proceedings of the 37th International Conference on Information Systems (ICIS2016)*. Dublin, Ireland, pp. 1–10.

Liu, K., L. Land, and V. Edwards (2006). "Knowledge Sharing in Financial Credit Reporting: An Exploratory Case Study in an Australian Context". In: Kuala Lumpur, Malaysia, pp. 112–121.

Loebbecke, C. (1999). "Electronic trading in on-line delivered content". In: *Proceedings of the 32nd Hawaii International Conference on System Sciences (HICSS1999)*. Maui, Hawaii, USA, pp. 1–10.

Loh, L. and N. Venkatraman (1992). "Determinants of Information Technology Outsourcing: A Cross-Sectional Analysis". In: *Journal of Management Information Systems* 9.1, pp. 7–24.

López-Carril, S. et al. (2020). "The Rise of Social Media in Sport: A Bibliometric Analysis". In: *International Journal of Innovation and Technology Management* 17.06, pp. 1–29.

Love, P. E. D., C. Standing, and S. Standing (2010). "A review of research on e-marketplaces 1997–2008". In: *Decision Support Systems* 49, pp. 41–51.

Lovejoy, A. O. (1908). "The Thirteen Pragmatisms". In: *The Journal of Philosophy, Psychology and Scientific Methods* 5.1, pp. 5–12.

Lowry, P. B. et al. (2006). "Online Payment Gateways Used to Facilitate E-Commerce Transactions and Improve Risk Management". In: *Communications of the Association for Information Systems* 17.1, pp. 123–151.

Lu, J. et al. (2015). "Recommender system application developments: A survey". In: *Decision Support Systems* 74, pp. 12–32.

Lusch, R. F. and S. Nambisan (2015). "Service Innovation: A Service-Dominant Logic Perspective". In: *MIS Quarterly* 39, pp. 155–175.

Lütjen, H., C. Schultz, F. Tietze, and F. Urmetzer (2019). "Managing ecosystems for service innovation: A dynamic capability view". In: *Journal of Business Research* 104, pp. 506–519.

Luxem, R. (2000). "Impact of trading digital products on retail information systems". In: *Proceedings of the 33rd Hawaii International Conference on System Sciences (HICSS2000)*. Maui, Hawaii, USA, pp. 248–257.

Luxem, R. and L. H. Ehlers (2000). "Retailing Digital Products: Reconstructing the Traditional RIS-Model". In: *Proceedings of the 8th European Conference on Information Systems (ECIS2000)*. Vienna, Austria, pp. 1–8.

Ma, T. et al. (2023). "Theorising moderation in the configurational approach: A guide for identifying and interpreting moderating influences in QCA". In: *Information systems journal*, pp. 1–26.

Magento (2021a). *About this document—Adobe Commerce Developer Guide*. URL: https://devdocs.magento.com/guides/v2.4/coding-standards/technical-guidelines.html (visited on 07/27/2021).

Magento (2021b). *Backward compatible development—Adobe Commerce Developer Guide*. URL: https://devdocs.magento.com/contributor-guide/backward-compatible-development/ (visited on 07/28/2021).

Magento (2021c). *Contributors—Adobe Commerce Developer Guide*. URL: https://devdocs.magento.com/contributor-guide/contributors.html#/individual-contributors (visited on 07/29/2021).

Magento (2021d). *Home—Magento Forums*. URL: https://community.magento.com/ (visited on 07/27/2021).

Magento (2021e). *Magento Extensions & Themes for Your Online Store—Marketplace.* URL: https://marketplace.magento.com/ (visited on 05/29/2021).
Magento (2021f). *Optimal Development Environment—Adobe Commerce Developer Guide.* URL: https://devdocs.magento.com/guides/v2.4/extension-dev-guide/build/optimal-dev-environment.html (visited on 07/28/2021).
Magento (2021g). *Upcoming releases|Adobe Commerce Developer Guide.* URL: https://devdocs.magento.com/release/ (visited on 07/27/2021).
Magento (2021h). *Zendesk Connector.* URL: https://marketplace.magento.com/wagento-zendesk.html (visited on 05/29/2021).
Makinen, S. J. and O. Dedehayir (2012). "Business ecosystem evolution and strategic considerations: A literature review". In: *Proceedings of the 18th International ICE Conference on Engineering, Technology and Innovation.* Munich, Germany, pp. 1–10.
Malhotra, D. and O. Rishi (2021). "An intelligent approach to design of E-Commerce metasearch and ranking system using next-generation big data analytics". In: *Journal of King Saud University—Computer and Information Sciences* 33.2, pp. 183–194.
Manikas, K. and K. M. Hansen (2013). "Software ecosystems—A systematic literature review". In: *Journal of Systems and Software* 86.5, pp. 1294–1306.
Mann, G., S. Karanasios, and C. F. Breidbach (2022). "Orchestrating the digital transformation of a business ecosystem". In: *Journal of Strategic Information Systems* 31.3, pp. 1–16.
Mantin, B., H. Krishnan, and T. Dhar (2014). "The Strategic Role of Third-Party Marketplaces in Retailing". In: *Production and Operations Management* 23.11, pp. 1937–1949.
Manyika, J. et al. (2016). *Digital Globalization: The New Era of Global Flows.* URL: https://www.mckinsey.com/$%5Csim$/media/mckinsey/businessfunctions/mckinseydigital/ourinsights/digitalglobalizationtheneweraofglobaflows/mgi-digital-globalization-full-report.pdf (visited on 05/04/2023).
March, J. G. and H. A. Simon (1958). *Organizations.* New York City: John Wiley & Sons, Inc.
March, S. T. and G. F. Smith (1995). "Design and natural science research on information technology". In: *Decision Support Systems* 15.4, pp. 251–266.
March, S. T. and V. C. Storey (2008). "Design Science in the Information Systems Discipline: An Introduction to the Special Issue on Design Science Research". In: *MIS Quarterly* 32.4, pp. 725–730.
Marcondes, R. M., L. G. Vasconcelos, and L. A. Baldochi (2017). "Introducing Adaptation Templates to Support the Implementation of Adaptive E-Commerce Applications". In: *Proceedings of the 14th International Conference on e-Business Engineering (ICEBE2017).* Shanghai, China, pp. 243–248.
Marktplaats (2022). *Hoe kan ik mijn advertentie promoten?* URL: https://help.marktplaats.nl/s/article/hoe-kan-ik-mijn-advertentie-promoten (visited on 06/28/2023).
Markus, M. L. and C. Loebbecke (2013). "Commoditized Digital Processes and Business Community Platforms: New Opportunities and Challenges for Digital Business Strategies". In: *MIS Quarterly* 37.2, pp. 649–653.
Martinsons, M. G. and P. K. C. Chong (1999). "The Influence of Human Factors and Specialist Involvement on Information Systems Success". In: *Human Relations* 52.1, pp. 123–152.
Maruping, L. M. and S. Matook (2020). "The evolution of software development orchestration: current state and an agenda for future research". In: *European Journal of Information Systems* 29.5, pp. 443–457.

Masak, D. (2009). "Digitale Öksoysteme". In: *Digitale Ö kosysteme. Serviceorientierung bei dynaonlineh vernetzten Unternehmen*. Ed. by D. Masak. Berlin [u.a.]: Springer. Chap. 6, pp. 209–243.

Massaro, M., J. Dumay, and J. Guthrie (2016). "On the shoulders of giants: undertaking a structured literature review in accounting". In: *Accounting, Auditing & Accountability Journal* 29.5, pp. 767–801.

Matook, S. and I. Vessey (2008). "Types of Business-to-Business E-Marketplaces: The Role of a Theory-Based, Domain-Specific Model". In: *Journal of Electronic Commerce Research* 9.4, p. 260.

Matthes, F., C. Neubert, and A. Steinhoff (2009). "Federated Application Lifecycle Management Based on an Open Web Architecture". In: *Workshop Design for Future—Langlebige Softwaresysteme, GI-Arbeitskreis Langlebige Softwaresysteme (L2S2)*. Karlsruhe, Germany, pp. 1–12.

Mattke, J. et al. (2022). "Qualitative Comparative Analysis (QCA) In Information Systems Research: Status Quo, Guidelines, and Future Directions". In: *Communications of the Association for Information Systems* 50, pp. 208–240.

Mayring, P. (2014). *Qualitative Content Analysis: Theoretical Foundation, Basic Procedures and Software Solution*. Klagenfurt, Austria: Beltz.

McAdams, D. A. (2003). "Identification and Codification of Principles for Functional Tolerance Design". In: *Journal of Engineering Design* 14.3, pp. 355–375.

McAfee, A. and E. Brynjolfsson (2017). *Machine, Platform, Crowd: Harnessing Our Digital Future*. New York City, USA: W W Norton & Co Inc.

McCombs, A. and R. Banh (2009). *The definitive guide to Magento: a comprehensive look at Magento*. Berkeley, California: Apress.

McIntyre, D. P. and A. Srinivasan (2017). "Networks, platforms, and strategy: Emerging views and next steps". In: *Strategic Management Journal* 38.1, pp. 141–160.

McIntyre, D. P. and M. Subramaniam (2009). "Strategy in network industries: A review and research agenda". In: *Journal of Management* 35.6, pp. 1494–1517.

McKinsey (2020). *The conflicted Continent: Ten charts show how COVID-19 is affecting consumers in Europe*. URL: https://www.mckinsey.com/business-functions/marketing-and-sales/our-insights/the-conflicted-continent-ten-charts-show-how-covid-19-is-affecting-consumers-in-europe (visited on 10/10/2022).

Medjaoui, M., E. Wilde, R. Mitra, and M. Amundsen (2021). *Continuous API Management: Making the Right Decisions in an Evolving Landscape*. 2nd ed. Sebastopol, USA: O'Reilly.

Meents, S. and T. Verhagen (2018). "Reducing consumer risk in electronic marketplaces: The signaling role of product and seller information". In: *Computers in Human Behavior* 86, pp. 205–217.

Meier, A. and H. Stormer (2008). *eBusiness & eCommerce*. Berlin [u. a.]: Springer.

Merigó, J. M., A. Mas-Tur, N. Roig-Tierno, and D. Ribeiro-Soriano (2015). "A bibliometric overview of the Journal of Business Research between 1973 and 2014". In: *Journal of Business Research* 68.12, pp. 2645–2653.

Merrick, E. (1999). "An Exploration of Quality in Qualitative Research: Are "Reliability" and "Validity" Relevant?" In: *Using Qualitative Methods in Psychology*. Ed. by M. Kopala and M. Suzuki. Thousand Oaks, California, US: SAGE Publications, Inc., pp. 25–36.

Mertens, D. M. (1998). *Research methods in education and psychology: Integrating diversity with quantitative and qualitative approaches*. Thousand Oaks, CA, USA: SAGE Publications Ltd.

Mertens, P. et al. (2017). *Grundzüge der Wirtschaftsinformatik*. 12th ed. Vol. 53. 9. Wiesbaden, Germany: Springer Gabler.

Merton, K. (2021). *The World's Top Online Marketplaces 2021*. URL: https://www.webretailer.com/b/online-marketplaces/.

Merton, R. K. and P. L. Kendall (1946). "The Focused Interview". In: *American Journal of Sociology* 51.6, pp. 541–557.

Messerschmitt, D. and C. Szyperski (2003). *Software Ecosystem. Understanding an Indispensable Technology and Industry*. Cambdrige: The MIT Press.

Metcalfe, B. (1995). *Metcalfe's Law: A network becomes more valuable as it reaches more users*. URL: http://www.infoworld.com/cgi-bin/displayNew.pl?/metcalfe/bm050696.htm%7D (visited on 10/03/2020).

Meth, H., B. Mueller, and A. Maedche (2015). "Designing a requirement mining system". In: *Journal of the Association for Information Systems* 16.9, pp. 799–837.

Meyer, J. W. and B. Rowan (1977). "Institutional organizations: Formal structures as myth and ceremony". In: *American Journal of Sociology* 80, pp. 340–363.

Meyer, M. and A. Lehnerd (1997). *The power of product platforms. Building value and cost leadership*. New York: The Free Press.

Meyer, M. and R. Sliger (1998). "Product Platforms in Software Development". In: *Sloan Management Review* 40.1, pp. 61–74.

Miles, M. B., A. M. Huberman, and J. Saldana (2014). *Qualitative Data Analysis: A Methods Sourcebook*. 3rd ed. Los Angeles [u. a.]: Sage.

Milgrom, P. and J. Roberts (1990). "Rationalizability, Learning, and Equilibrium in Games with Strategic Complementarities". In: *Econometrica* 58.6, pp. 1255–1277.

Milojicic, D. S. et al. (2002). *Peer-to-peer computing*. URL: https://www.cs.kau.se/cs/education/courses/dvad02/p2/seminar4/Papers/HPL-2002-57R1.pdf (visited on 07/30/2023).

Mineraud, J., O. Mazhelis, X. Su, and S. Tarkoma (2016). "A gap analysis of Internet-of-Things platforms". In: *Computer Communications* 89–90.9, pp. 5–16.

Miranda, S. et al. (2022). "Editor's Comments: Computational Intensive Theory Construction: A Primer for Authors and Reviewer". In: *MIS Quarterly* 46.2, pp. iii–xviii.

Miremadi, I., M. Khoshbash, and M. M. Saeedian (2023). "Fostering generativity in platform ecosystems: How open innovation and complexity interact to influence platform adoption". In: *Research Policy* 52.6, pp. 1–20.

Mishra, S. and A. R. Tripathi (2020). "Literature review on business prototypes for digital platform". In: *Journal of Innovation and Entrepreneurship* 9.1, pp. 1–19.

Misoch, S. (2015). *Qualitative Interviews*. Berlin, Germany: De Gruyter.

Mittendorf, C. (2018). "Collaborative consumption: the role of familiarity and trust among Millennials". In: *Journal of Consumer Marketing* 35.4, pp. 377–391.

Moazed, A. and L. N. Johnson (2016). *Modern Monopolies. What It Takes to Dominate the 21st-Century Economy*. New York: St. Martin's Press.

Moghavvemi, S., T. X. Mei, S. W. Phoong, and S. Y. Phoong (2021). "Drivers and barriers of mobile payment adoption: Malaysian merchants' perspective". In: *Journal of Retailing and Consumer Services* 59, pp. 1–12.

Mohagheghzadeh, A. and J. Lindman (2022). "Shaping Boundary Resources of a Digital Platform: Alliance Co- development Perspective". In: *Proceedings of the 26th Pacific Asia Conference on Information Systems (PACIS2022)*. Taipei, Taiwan, pp. 1–13.

Mohagheghzadeh, A. and D. Rudmark (2017). "Accelerated Tuning of Platform Boundary Resources". In: *Lecture Notes in Business Information Processing*. Vol. 294, pp. 98–110.

Mohagheghzadeh, A. and F. Svahn (2016). "Shifting design capability to third-party developers: An affordance perspective on platform boundary resources". In: *Proceedings of the 22nd Americas Conference on Information Systems (AMCIS2016)*. San Diego, USA, pp. 1–10.

Möller, F., T. M. Guggenberger, and B. Otto (2020). "Towards a Method for Design Principle Development in Information Systems". In: *Proceedings of the 15th International Conference on Design Science Research in Information Systems and Technology (DESRIST 2020)*. Ed. by S. Hofmann, O. Müller, and M. Rossi. Kristiansand, Norway: Springer, pp. 208–220.

Möller, F., T. Schoormann, and B. Otto (2021). " 'Caution: Principle Under Construction' a Visual Inquiry Tool for Developing Design Principles". In: *The Next Wave of Sociotechnical Design*, pp. 223–235.

Moore, J. F. (1993). "Predators and prey: a new ecology of competition". In: *Harvard business review* 71.3, pp. 75–86.

Moore, J. F. (1996). *The Death of Competition: Leadership & Strategy in the Age of Business Ecosystems*. New York City: HarperCollins.

Moore, J. F., J. Palfrey, and U. Gassner (2003). *ICT and Entrepreneurship: Digital Business Ecosystems and the Law*. URL: https://cyber.harvard.edu/bold/devel03/modules/episodeII.html (visited on 06/06/2023).

Moral-Muñoz, J. A., E. Herrera-Viedma, A. Santisteban-Espejo, and M. J. Cobo (2020). "Software tools for conducting bibliometric analysis in science: An up-to-date review". In: *El Profesional de la Informaci ó n* 29.1, pp. 1–20.

Morgan, D. L. (2007). "Paradigms Lost and Pragmatism Regained". In: *Journal of Mixed Methods Research* 1.1, pp. 48–76.

Müller-Hagedorn, L., W. Toporowski, and S. Zielke (2012). *Der Handel. Grundlagen—Management—Strategien*. 2., vollst. Stuttgart, Germany: Kohlhammer.

Murakami Wood, D. and T. Monahan (2019). "Editorial: Platform Surveillance". In: *Surveillance & Society* 17.1/2, pp. 1–6.

Muschkiet, M. et al. (2023). "Unleashing the digital building bricks: A smart service taxonomy for retail". In: *Electronic Markets* 33.1, pp. 1–29.

Mushthofa, D., M. K. Sabariah, and V. Effendy (2018). "Modelling the user interface design pattern for designing Islamic e-commerce website using user centered design". In: *Proceedings of the Human-Dedicated Sustainable Product and Process Design Conference on Materials, Resources, and Energy*. Surakarta, Indonesia, pp. 1–9.

Muzellec, L., S. Ronteau, and M. Lambkin (2015). "Two-sided Internet platforms: A business model lifecycle perspective". In: *Industrial Marketing Management* 45, pp. 139–150.

Myers, M. D. and M. Newman (2007). "The qualitative interview in IS research: Examining the craft". In: *Information and Organization* 17.1, pp. 2–26.

Nachira, F. (2002). *Towards a network of digital business ecosystems: Digital business ecosystems fostering the local development*. URL: http://europa.eu.int/information_society/topics/ebusiness/godigital/sme_research/doc/dbe_%5C%0Adiscussionpaper.pdf (visited on 05/23/2021).

Nachira, F., P. Dini, and A. Nicolai (2007). "A Network of Digital Business Ecosystems for Europe: Roots, Processes and Perspectives". In: *Digital Business Ecosystems*. Ed. by F. LNachira et al. Luxembourg: Office for Official Publications of the European Communities. Chap. 1, pp. 1–20.

Nakagawa, E. Y., P. Oliveira Antonino, and M. Becker (2011). "Reference architecture and product line architecture: A subtle but critical difference". In: *Proceedings of the 5th European Conference on Software Architecture (ECSA2011)*. Essen, Germany, pp. 207–211.

Nakamoto, S. (2008). *Bitcoin: A Peer-to-Peer Electronic Cash System*. URL: https://bitcoin.org/bitcoin.pdf%20(Accessed:%2015.06.2020)(visited on 10/24/2021).

Nambisan, S. (2013). "Information Technology and Product/Service Innovation: A Brief Assessment and Some Suggestions for Future Research". In: *Journal of the Association for Information Systems* 14.4, pp. 215–226.

Nambisan, S., K. Lyytinen, A. Majchrzak, and M. Song (2017). "Digital Innovation Management: Reinventing Innovation Management Research in a Digital World". In: *MIS Quarterly* 41.1, pp. 223–238.

Narayanan, S., A. S. Marucheck, and R. B. Handfield (2009). "Electronic data interchange: Research review and future directions". In: *Decision Sciences* 40.1, pp. 121–163.

Nath, R. et al. (1998). "Electronic Commerce and the Internet: Issues, Problems, and Perspectives". In: *International Journal of Information Management* 18.2, pp. 91–101.

Nemat, R. (2011). "Taking a look at different types of e-commerce". In: *World Applied Programming* 1.2, pp. 100–104.

Nerur, S. and V. Balijepally (2007). "Theoretical Reflections on AGILE DEVELOPMENT METHODOLOGIES". In: *Communications of the ACM* 50.3, pp. 79–83.

Netshoes (2023). *Entenda a API—Grupo Netshoes*. URL: https://developers.netshoes.com.br/api-portal/content/entenda-api (visited on 06/27/2023).

Neufeld, D. (2018). *Shopify's Infrastructure Collaboration with Google*. URL: https://shopify.engineering/shopify-infrastructure-collaboration-with-google (visited on 07/28/2021).

Nickerson, R., J. Muntermann, U. Varshney, and H. Isaac (2009). "Taxonomy Development in Information Systems: Developing a Taxonomy of Mobile Applications". In: *Proceedings of the 17th European Conference on Information Systems (ECIS2009)*. Verona, Italy, pp. 1–12.

Nickerson, R. C., U. Varshney, and J. Muntermann (2013). "A method for taxonomy development and its application in information systems". In: *European Journal of Information Systems* 22.3, pp. 336–359.

Nicola, M. et al. (2020). "The socio-economic implications of the coronavirus pandemic (COVID-19): A review". In: *International Journal of Surgery* 78.1, pp. 185–193.

Nicotra, M., M. Romano, M. Del Giudice, and C. E. Schillaci (2018). "The causal relation between entrepreneurial ecosystem and productive entrepreneurship: a measurement framework". In: *Journal of Technology Transfer* 43.3, pp. 640–673.

Nie, J., S. Luo, and M. K. Xu B. (2019). "Research on the Influential Factors of Blockchain-based traceable products Purchase Intention". In: *Proceedings of 4th IEEE Advanced Information Technology, Electronic and Automation Control Conference (IAEAC2019)*. Chengdu, China, pp. 2758–2766.

Niederman, F. and S. March (2019). "The "theoretical lens" concept: We all know what it means, but do we all know the same thing?" In: *Communications of the Association for Information Systems* 44.1, pp. 1–33.

Niehaves, B. (2007a). "ON EPISTEMOLOGICAL DIVERSITY IN DESIGN SCIENCE—NEW VISTAS FOR A DESIGN-ORIENTED IS RESEARCH?" In: *Twenty Eighth International Conference on Information Systems (ICIS2007)*. Montréal, Canada, pp. 1–13.

Niehaves, B. (2007b). "On Epistemological Pluralism in Design Science". In: *Scandinavian Journal of Information Systems* 19.2, pp. 93–104.

Niehaves, B. and J. Becker (2006). "Epistemological Perspectives on Design Science in IS Research". In: *Proceedings of the 12th Americas Conference on Information Systems (AMCIS2006)*. Acapulco, Mexico, pp. 3579–3589.

Niemann, K. D. (2006). *From Enterprise Architecture to IT Governance. Elements of Effective IT Management*. Wiesbaden, Germany: Vieweg.

Norbjerg, J. (2018). "Design. Building the Solution or Framing a Problem? Reflections on Design Science". In: *Scandinavian Journal of Information Systems* 30.2, pp. 69–74.

Nüesch, R., R. Alt, and T. Puschmann (2015). "Hybrid Customer Interaction". In: *Business & Information Systems Engineering* 57.1, pp. 73–78.

Numberger, S. and C. Rennhak (2005). "The Future of B2C E-Commerce". In: *Electronic Markets* 15.3, pp. 269–282.

Nunamaker, J. F., M. Chen, and T. D. M. Purdin (1990). "Systems Development in Information Systems Research". In: *Journal of Management Information Systems* 7.3, pp. 89–106.

Nunkoosing, K. (2005). "The Problems With Interviews". In: *Qualitative Health Research* 15.5, pp. 698–706.

Nvidia (2023). *Kroger and NVIDIA to Reinvent the Shopping Experience Through State-of-the-Art, AI-Enabled Applications and Services*. URL: https://nvidianews.nvidia.com/news/kroger-and-nvidia-to-reinvent-the-shopping-experience-through-state-of-the-art-ai-enabled-applications-and-services (visited on 06/28/2023).

O'Mahony, S. and R. Karp (2022). "From proprietary to collective governance: How do platform participation strategies evolve?" In: *Strategic Management Journal* 43.3, pp. 530–562.

O'Reilly, T. (2005). "The open source paradigm shift". In: *Perspectives on Free and Open Source Software*. Ed. by J. Feller, B. Fitzgerald, S. A. Hissam, and K. R. Lakhani. Cambridge, Massachusetts: MIT Press, pp. 461–482.

Odusanya, K. et al. (2019). "Understanding the Adoption and Use of E-tail Websites: An Empirical Analysis Based on the Revised UTAUT2 Model Using Risk and Trust Factors". In: *Proceedings of the International Conference on Transfer and Diffusion of IT (TDIT2019)*. Accra, Ghana, pp. 191–211.

OECD (1999). *The Economic and Social Impact of Electronic Commerce*. Paris, France: OECD, p. 168.

OECD (2002). *Measuring the Information Economy*. URL: https://www.oecd.org/digital/ieconomy/1835738.pdf (visited on 01/19/2023).

OECD (2019). *An Introduction to Online Platforms and Their Role in the Digital Transformation*. Paris, France: OECD Publishing.

Ofe, H. A. and J. Sandberg (2023). "The emergence of digital ecosystem governance: An investigation of responses to disrupted resource control in the Swedish public transport sector". In: *Information systems journal* 33.2, pp. 350–384.

Offermann, P., S. Blom, M. Schönherr, and U. Bub (2010). "Artifact Types in Information Systems Design Science—A Literature Review". In: *Global Perspectives on Design Science Research*. Ed. by R. Winter, J. L. Zhao, and S. Aier. Berlin, Heidelberg: Springer, pp. 77–92.

Ogilvy, J. (1996). "Futures studies and the human sciences: the case for normative scenarios". In: *New Thinking for a New Millennium*. Ed. by R. A. Slaughter. London, UK: Routledge, pp. 26–83.

Ojanperä, T. and T. O. Vuori (2021). *Platform Strategy. Transform Your Bsuiness with AI, Platforms and Human Intelligence*. London, UK: Kogan Page.

Okoli, C. (2015). "A guide to conducting a standalone systematic literature review". In: *Communications of the Association for Information Systems* 37.1, pp. 879–910.

Okoli, C. and K. Schabram (2010). "A guide to conducting a systematic literature review of information systems research". In: *Sprouts: Working Papers on Information Systems* 10.26, pp. 1–49.

Olano, G. (2021). *Tower reveals several partnership deals*. URL: https://www.insurancebusinessmag.com/nz/news/breaking-news/tower-reveals-several-partnership-deals-301583.aspx (visited on 06/28/2023).

Olist (2023). *Como vender nos marketplaces Netshoes e Zattini: Passo a passo*. URL: https://olist.com/blog/pt/como-vender-mais/em-marketplaces/como-vender-nos-marketplaces-netshoes-e-zattini/ (visited on 06/27/2023).

Olleros, X. (2008). "The lean core in digital platforms". In: *Technovation* 28.5, pp. 266–276.

OMG (2019). *ARTS Business Process Model for Retail*. URL: https://www.omg.org/retail-depository/arts-bpm/ (visited on 08/21/2023).

Ondrus, J., A. Gannamaneni, and K. Lyytinen (2015). "The impact of openness on the market potential of multi-sided platforms: A ase study of mobile payment platforms". In: *Journal of Information Technology* 30, pp. 260–275.

Open Group (2016). *TOGAF Worldwide*. URL: http://www.opengroup.org/subjectareas/enterprise/togaf/worldwide (visited on 01/02/2021).

Open Group (2019). *ArchiMate 3.1 Specification*. URL: https://pubs.opengroup.org/architecture/archimate3-doc/ (visited on 03/02/2020).

Oren, S. S. and S. A. Smith (1981). "Critical Mass and Tariff Structure in Electronic Communications Markets". In: *The Bell Journal of Economics* 12.2, p. 467.

Orlikowski, W. J. and J. J. Baroudi (1991). "Studying Information Technology in Organizations: Research Approaches and Assumptions". In: *Information Systems Research* 2.1, pp. 1–28.

Orlikowski, W. J. and C. S. Iacono (2001). "Research Commentary: Desperately Seeking the "IT" in IT Research—A Call to Theorizing the IT Artifact". In: *Information Systems Research* 12.2, pp. 121–134.

Österle, H. et al. (2010). "Memorandum zur gestaltungsorientiertenWirtschaftsinformatik". In: *Gestaltungsorientierte Wirtschaftsinformatik: Ein Plädoyer für Rigor und Relevanz*. Ed. by H. Österle, R. Winter, and W. Brenner. Nuremberg, Germany: Inforwerk ag, pp. 1–6.

Österle, H. et al. (2011). "Memorandum on design-oriented information systems research". In: *European Journal of Information Systems* 20.1, pp. 7–10.

Osterwalder, A. and Y. Pigneur (2002). "An e-business model ontology for modeling e-business". In: Bled, Slovenia, pp. 75–91.

Osterwalder, A. and Y. Pigneur (2010). *Business model generation: A handbook for visionaries, game changers, and challengers*. Hoboken, New Jersey: John Wiley & Sons.

Osterwalder, A., Y. Pigneur, G. Bernarda, and A. Smith (2014). *Value proposition design: How to create products and services customers want*. Hoboken, New Jersey: John Wiley & Sons.

Osterwalder, A., Y. Pigneur, and C. L. Tucci (2005). "Clarifying Business Models: Origins, Present, and Future of the Concept". In: *Communications of the Association for Information Systems* 16.7, pp. 1–25.

Otto SE (2021a). *Hackathon der Otto Group IT & Microsoft Azure*. URL: https://www.otto.de/jobs/technology/techblog/artikel/hackathon-der-otto-group-it-microsoft-azure.php?highlight=hackathon (visited on 07/27/2021).

Otto SE (2021b). *OTTO Market API documentation*. URL: https://api.otto.market/docs/08_Receipts/v2/receipts-interface.html#/ (visited on 07/27/2021).

Otto SE (2021c). *Tech-Blog*. URL: https://www.otto.de/jobs/technology/techblog/ (visited on 07/27/2021).

Oxford English Dictionary (2023). *Artifact*. URL: https://www.oed.com/dictionary/artefact_n?tab=factsheet#38730868 (visited on 08/15/2023).

OXID (2021). *Roadmap*. URL: https://www.oxid-esales.com/blog/roadmap-oxid-eshop-2021/ (visited on 07/11/2021).

OXID (2022a). *Development*. URL: https://docs.oxid-esales.com/developer/en/latest/development/index.html (visited on 10/21/2022).

OXID (2022b). *Know-how from the OXID Academy*. URL: https://www.oxid-esales.com/en/resources/academy/ (visited on 10/21/2022).

OXID (2022c). *OXID eshop bugtrack*. URL: https://bugs.oxid-esales.com/view_all_bug_page.php (visited on 10/21/2022).

OXID (2022d). *Release Plan*. URL: https://docs.oxid-esales.com/en/releases/release-plan.html (visited on 10/21/2022).

Oxid AG (2021a). *Jetbrains PhpStorm—OXID eShop developer documentation 6.1.0 documentation*. URL: https://docs.oxid-esales.com/developer/en/6.1/getting_started/ide/phpstorm.html (visited on 07/28/2021).

Oxid AG (2021b). *OXID Market Connect—einfach und schnell Produkte in Marktplätze ausspielen*. URL: https://www.oxid-esales.com/shopsystem/pimp-my-shop/oxid-market-connect/ (visited on 07/27/2021).

Pagani, M. (2013). "Digital business strategy and value creation: framing the dynamic cycle of control points". In: *MIS Quarterly* 37.2, pp. 617–632.

Panjaitan, M. et al. (2019). "Examining generation X experiences on using e-commerce: Integrating the technology acceptance model and perceived risks". In: *Journal of Physics: Conference Series* 1361.1, pp. 1–6.

Papachristos, G. and G. Van De Kaa (2021). "A System Dynamics Model of Standards Competition". In: *IEEE Transactions on Engineering Management* 68.1, pp. 18–32.

Pappas, I. O. and A. G. Woodside (2021). "Fuzzy-set Qualitative Comparative Analysis (fsQCA): Guidelines for research practice in Information Systems and marketing". In: *International Journal of Information Management* 58, pp. 1–23.

Paquin, R. L. and J. Howard-Grenville (2013). "Blind Dates and Arranged Marriages: Longitudinal Processes of Network Orchestration". In: *Organization Studies* 34.11, pp. 1623–1653.

Paré, G., M. C. Trudel, M. Jaana, and S. Kitsiou (2015). "Synthesizing information systems knowledge: A typology of literature reviews". In: *Information and Management* 52.2, pp. 183–199.

Park, N. K., J. M. Mezias, and J. Song (2004). "A resource-based view of strategic alliances and firm value in the electronic marketplace". In: *Journal of Management* 30.1, pp. 7–27.

Park, S. C., M. Keil, J. U. Kim, and G. W. Bock (2012). "Understanding overbidding behavior in C2C auctions: An escalation theory perspective". In: *European Journal of Information Systems* 21.6, pp. 643–663.

Park, Y., O. A. El Sawy, and P. C. Fiss (2017). "The Role of Business Intelligence and Communication Technologies in Organizational Agility: A Configurational Approach". In: *Journal of the Association for Information Systems* 18.9, pp. 648–686.

Park, Y., P. C. Fiss, and O. A. El Sawy (2020). "Theorizing the Multiplicity of Digital Phenomena: The Ecology of Configurations, Causal Recipes, and Guidelines for Applying QCA". In: *MIS Quarterly* 44.4, pp. 1493–1520.

Parker, G. and M. Van Alstyne (2005). "Two-Sided Network Effects: A Theory of Information Product Design". In: *Management Science* 51.10, pp. 1494–1504.

Parker, G. and M. Van Alstyne (2016). "Platform Strategy". In: *The Palgrave Encyclopedia of Strategic Management*. Ed. by M. Augier and D. J. Teece. London: Palgrave Macmillan UK, pp. 1–9.

Parker, G., M. Van Alstyne, and S. Choundary (2016). *Platform revolution: how networked markets are transforming the economy and how to make them work for you*. New York: W. W. Norton & Company.

Parker, G., M. Van Alstyne, and X. Jiang (2017). "Platform Ecosystems: How Developers Invert the Firm". In: *MIS Quarterly* 41.1, pp. 255–266.

Parker, G. and M. Van Alstyne (2018). "Innovation, openness, and platform control". In: *Management Science* 64.7, pp. 3015–3032.

Patel, S. K., V. Rathod, and J. B. Prajapati (2011). "Performance Analysis of Content Management Systems Joomla, Drupal and WordPress". In: *International Journal of Computer Applications* 21.4, pp. 39–43.

Patni, S. (2017). "API Management and API Client". In: *Pro RESTful APIs*. Ed. by S. Patni. Berkeley, CA: Apress, pp. 97–106.

Patton, M. Q. (1990). *Qualitative Evaluation and Research Methods*. 2nd ed. Thousand Oaks, CA: Sage Publications.

Paul, J. and A. R. Criado (2020). "The art of writing literature review: What do we know and what do we need to know?" In: *International Business Review* 29.4, pp. 1–17.

Pauli, T., E. Marx, S. Dunzer, and M. Matzner (2021). "Modeling platform ecosystems". In: *Proceedings of the 16th Internationale Tagung Wirtschaftsinformatik (WI2021)*. Essen, Gremany, pp. 1–15.

Pavlou, P. A., H. Liang, and Y. Xue (2007). "Understanding and Mitigating Uncertainty in Online Exchange Relationships: A Principal-Agent Perspective". In: *MIS Quarterly* 31.1, pp. 105–136.

Peffers, K., T. Tuunanen, M. A. Rothenberger, and S. Chatterjee (2007). "A design science research methodology for information systems research". In: *Journal of Management Information Systems* 24.3, pp. 45–77.

Pelizza, A. (2021). "Towards a sociomaterial approach to inter-organizational boundaries: How information systems elicit relevant knowledge in government outsourcing". In: *Journal of Information Technology* 36.2, pp. 94–108.

Peng, L., W. Zhang, X. Wang, and S. Liang (2019). "Moderating effects of time pressure on the relationship between perceived value and purchase intention in social E-commerce sales promotion: Considering the impact of product involvement". In: *Information and Management* 56.2, pp. 317–328.

Penrose, E. (1989). “History, the social sciences and economic “theory”, with special reference to multi-national enterprise”. In: *The Economic Theory of the Multinational Enterprise*. Ed. by A. Teichova, M. Levy-Leboyer, and H. Nussbaum. Cambridge, UK: Cambridge University Press, pp. 7–14.

Peppard, J. and A. Rylander (2006). “From Value Chain to Value Network”. In: *European Management Journal* 24.2-3, pp. 128–141.

Pereira, C. M. and P. Sousa (2004). “A method to define an enterprise architecture using the zachman framework”. In: *Proceedings of the ACM Symposium on Applied Computing* 2, pp. 1366–1371.

Perera, C., A. Zaslavsky, P. Christen, and D. Georgakopoulos (2014). “Sensing as a service model for smart cities supported by Internet of Things”. In: *Transactions on Emerging Telecommunications Technologies* 25.1, pp. 81–93.

Perks, H., C. Kowalkowski, L. Witell, and A. Gustafsson (2017). “Network orchestration for value platform development”. In: *Industrial Marketing Management* 67.7, pp. 106–121.

Peteraf, M., G. Di Stefano, and G. Verona (2013). “The elephant in the room of dynamic capabilities: Bringing two diverging conversations together”. In: *Strategic Management Journal* 34.12, pp. 1389–1410.

Petersen, K. J., J. A. Ogden, and P. L. Carter (2007). “B2B e-marketplaces: A typology by functionality”. In: *International Journal of Physical Distribution & Logistics Management* 37.1, pp. 4–18.

Petrik, D. (2022). *Management der Zufriedenheit der Wertschöpfungspartner auf Basis der Boundary Resources im IIoT*. Wiesbaden, Germany: Springer Fachmedien Wiesbaden, Germany.

Petrik, D. and G. Herzwurm (2019). “iIoT ecosystem development through boundary resources: a Siemens MindSphere case study”. In: *Proceedings of the 2nd ACM SIGSOFT International Workshop on Software-Intensive Business: Start-ups, Platforms, and Ecosystems (IWSiB 2019)*. New York, New York, USA: ACM Press, pp. 1–6.

Petrik, D. and G. Herzwurm (2020). “Boundary resources for IIoT platforms—A complementor satisfaction study”. In: *Proceedings of the 41st International Conference on Information Systems (ICIS 2020)*. 9. Hyderabad, India, pp. 1–17.

Petrik, D., K. Model, L. A. Drebinger, and G. Herzwurm (2021). “Exploring the Satisfaction Potential of Technical Boundary Resources in IoT Platforms—The Microsoft Azure Case”. In: *Proceedings of the 42nd International Conference on Information Systems (ICIS2021)*. Austin, Texas, pp. 1–17.

Petrik, D. and F. Schüler (2021). “Einfluss plattformbasierter Ökosysteme auf unternehmensübergreifende integrierte Wertschöpfungsnetzwerke”. In: *Wertsch ö pfungsnetzwerke mit digitalisierten Dienstleistungen etablieren*. Ed. by T. Schulz. Berlin, Germany: Beuth Verlag. Chap. 7, pp. 85–103.

Petter, S., W. DeLone, and E. McLean (2012). “The Past, Present, and Future of “IS Success” ”. In: *Journal of the Association for Information Systems* 13.5, pp. 341–362.

Pfeffer, J. and G. Salancik (1978). *The External Control of Organizations: A Resource Dependence Perspective*. New York City, USA: Harper & Row Publishers.

Pfeifer, W. (1993). *Plattform*. URL: https://www.dwds.de/wb/wb-etymwb (visited on 11/22/2019).

Phang, C. W., A. Kankanhalli, K. Ramakrishnan, and K. S. Raman (2010). "Customers preference of online store visit strategies: An investigation of demographic variables". In: *European Journal of Information Systems* 19.3, pp. 344–358.

Phillips, N. and T. B. Lawrence (2012). "The turn to work in organization and management theory: Some implications for strategic organization". In: *Strategic Organization* 10.3, pp. 223–230.

Pickering, A. (1993). "The Mangle of Practice : Agency and Emergence in the Sociology of Science". In: *American Journal of Sociology* 99.3, pp. 559–589.

Pickering, A. (1995). *The Mangle of Practice: Time, Agency and Science*. Chicago, USA: The Univeristy of Chicago Press.

Picot, A. (1982). "Transaktionskostenansatz in der Organisationstheorie. Stand der Diskussion und Aussagewert". In: *Die Betriebswirtschaft* 42.2, pp. 267–284.

Picot, A. (1986). "Transaktionskosten im Handel. Zur Notwendigkeit einer flexiblen Strukturentwicklung in der Distribution." In: *Betriebsberater. Zeitrschrift für Recht und Wirtschaft* 13.2, pp. 1–16.

Picot, A., R. Reichwald, and R. T. Wigand (2001). *Die grenzenlose Unternehmung*. 4th ed. Wiesbaden, Germany: Gabler Verlag.

Pierce, L. (2009). "Big losses in ecosystem niches: How core firm decisions drive complementary product shakeouts". In: *Strategic Management Journal* 30.3, pp. 323–347.

Pirklbauer, G., R. Ramler, and R. Zeilinger (2009). "An integration-oriented model for application lifecycle management". In: *Proceedings of the 11th International Conference on Enterprise Information Systems (ICEIS2009)*. Milan, Italy, pp. 399–402.

Plateaux, A. et al. (2018). "A comparative study of card-not-present e-commerce architectures with card schemes: What about privacy?" In: *Journal of Information Security and Applications* 40, pp. 103–110.

Polkinghorne, D. E. (2005). "Language and meaning: Data collection in qualitative research." In: *Journal of Counseling Psychology* 52.2, pp. 137–145.

Pon, B., T. Seppälä, and M. Kenney (2014). "Android and the demise of operating system-based power: Firm strategy and platform control in the post-PC world". In: *Telecommunications Policy* 38.11, pp. 979–991.

Pop, E. and D. Gîfu (2018). "A Service Oriented Architecture Approach for an E-Business Platform". In: *Proceedings of the 12th International Conference on Communications (COMM2018)*. Bucharest, Romania, pp. 407–412.

Popp, K. and R. Meyer (2010). *Profit from Software Ecosystems: Business Models, Ecosystems and Partnerships in the Software Industry*. Norderstedt, Germany: Books on Demand, p. 242.

Popper, K. R. (1978). *Three Worlds*. Ann Arbor, Michigan, USA: The Tanner Lecture on Human Values Delivered at the University of Michigan.

Porch, C., G. T. Timbrell, and M. Rosemann (2015). "Platforms: A systematic review of the literature using algorithmic historiography". In: *Proceedings of the 22rd European Conference on Information Systems (ECIS2015)*. Münster, Germany, pp. 1–17.

Porter, M. E. and J. E. Heppelmann (2014). "How Smart, Connected Products Are Transforming Competition". In: *Harvard Business Review* 92.11, pp. 1–23.

Porter, M. E. (1980). *Competitive Strategy: Techniques for Analyzing Industries and Competitors*. New York: The Free Press.

Porter, M. E. (2001). "Strategy and the Internet". In: *Harvard Business Review* 79.3, pp. 62–78.

Pousttchi, K. (2008). "A modeling approach and reference models for the analysis of mobile payment use cases". In: *Electronic Commerce Research and Applications* 7.2, pp. 182–201.
Pousttchi, K. and A. Gleiss (2019). "Surrounded by middlemen—how multi-sided platforms change the insurance industry". In: *Electronic Markets* 29.4, pp. 609–629.
Pouti, N., M. T. Taghavifard, M. R. Taghva, and M. Fathian (2020). "A comprehensive literature review of acceptance and usage studies in the social commerce field". In: *International Journal of Electronic Commerce Studies* 11.2, pp. 119–166.
Power, T. and G. Jerjian (2001). *Ecosystem: Living the 12 principles of networked business*. London: Pearson Education, p. 392.
Prat, N., I. Comyn-Wattiau, and J. Akoka (2015). "A Taxonomy of Evaluation Methods for Information Systems Artifacts". In: *Journal of Management Information Systems* 32.3, pp. 229–267.
Preston-Werner, T. (2022). *Semantic Versioning 2.0.0*. URL: https://semver.org/#semantic-versioning-200
Priem, R. L., J. E. Butler, and S. Li (2013). "Toward reimagining strategy research: Retrospective and prospection on the 2011 AMR decard award article". In: *Academy of Management Review* 38.4, pp. 471–489.
Pritchard, A. (1969). "Statistical bibliography or bibliometrics". In: *Journal of Documentation* 25.4, pp. 348–349.
Priyadarsini, K., E. F. I. Raj, A. Y. Begum, and V. Shanmugasundaram (2020). "Comparing DevOps procedures from the context of a systems engineer". In: *Proceedings of the International Con- ference on Emerging Trends in Materials Science, Technology and Engineering (ICMSTE2K21)*. Ramanathapuram, India, pp. 1–6.
Priyanka, S. (2016). "Internet Marketing: The New Era of novelty in E-commerce". In: *Internet Marketing: The New Era of novelty in E-commerce* 6.2, pp. 1–22.
Prothmann, O. (2017). *Marketplaces across the World: Analyses and Rankings of Online Marketplaces Worldwide*. Berlin, Germany: p.digital.
Purao, S. (2002). "Design Science Research in the Technology of Information Systems: Truth or Dare". State College, USA.
Puschmann, T. (2017). "CATCHWORD Fintech". In: *Business & Information Systems Engineering* 59.1, pp. 69–76.
Puschmann, T. and S. Shiba (2021). "Developing a Taxonomy for Sustainable Digital Finance". In: *Proceedings of the 42nd International Conference on Information Systems (ICIS2021)*. Austin, Texas, USA, pp. 1–17.
Qin, X., Z. Liu, and L. Tian (2020). "The strategic analysis of logistics service sharing in an e-commerce platform". In: *Omega* 92, pp. 181–192.
Qu, Y. et al. (2015). "A vehicle routing problem with time windows for attended home distribution". In: *Proceedings of the International Conference on Data Science (ICDS2015)*. Ningbo, China, pp. 127–134.
Rach, S. and M. Tschöpel (2011). *Handelsplattformen im Internet: Eine Literaturstudie zur empirischen Evidenz*. URL: https://www.econstor.eu/bitstream/10419/55724/1/687809894.pdf%7D (visited on 07/28/2023).
Rädiker, S. and U. Kuckartz (2019). *Analyse qualitativer Daten mit MAXQDA. Text, Audio und Video*. Wiesbaden, Germany: Springer.
Ragin, C. C. (2008). *Redesigning Social Inquiry*. Chicago, USA: University of Chicago Press.

Rahrovani, Y. (2020). "Platform drifting: When work digitalization hijacks its spirit". In: *The Journal of Strategic Information Systems* 29.2, pp. 1–26.

Rai, A., R. Patnayakuni, and N. Seth (2006). "Firm performance impacts of digitally enabled supply chain integration capabilities". In: *MIS Quarterly* 30.2, pp. 225–246.

Rakuten (2021). *Rakuten Group to Acquire Mobile Industry Innovator Altiostar—Rakuten Group, Inc.* URL: https://global.rakuten.com/corp/news/press/2021/0804_02.html (visited on 06/27/2023).

Rakuten (2022). *Rakuten to Acquire SKY ESTATE—Rakuten Group, Inc.* URL: https://global.rakuten.com/corp/news/press/2022/0420_03.html (visited on 06/27/2023).

Ramachandran, J., A. Pant, and S. K. Pani (2012). "Building the BoP Producer Ecosystem: The Evolving Engagement of Fabindia with Indian Handloom Artisans". In: *Journal of Product Innovation Management* 29.1, pp. 33–51.

Ramakrishnan, G. et al. (2020). "Collaborative Filtering for Book Recommendation System". In: *Advances in Intelligent Systems and Computing* 1057, pp. 325–338.

Ramos-Rodríguez, A.-R. and J. Ruíz-Navarro (2004). "Changes in the intellectual structure of strategic management research: A bibliometric study of theStrategic Management Journal, 1980–2000". In: *Strategic Management Journal* 25.10, pp. 981–1004.

Randolph, J. J. (2009). "A guide to writing the dissertation literature review". In: *Practical Assessment, Research and Evaluation* 14.13, pp. 1–13.

Ranganthan, V. P. et al. (2018). "A decentralized marketplace application on the ethereum blockchain". In: *Proceedings of the 4th IEEE International Conference on Collaboration and Internet Computing (CIC2018)*. Philadelphia, Pennsylvania, USA, pp. 90–97.

Rani, M. S. and S. Sumathy (2018). "Online social networking services and spam detection approaches in opinion mining—A review". In: *International Journal of Web Based Communities* 14.4, pp. 353–378.

Razavi, A. R., P. J. Krause, and A. Strommen-Bakhtiar (2010). "From business ecosystems towards digital business ecosystems". In: *Proceedings of the 4th IEEE International Conference on Digital Ecosystems and Technologies*. Dubai, UAE, pp. 290–295.

Recker, J. (2013). *Scientific Research in Information Systems: A Beginner's Guide*. Berlin, Heidelberg: Springer.

Recker, J., R. Lukyanenko, B. M. Samuel, and A. Castellanos (2021). "From Representation to Mediation: A New Agenda for Conceptual Modeling Research in a Digital World". In: *MIS Quarterly* 45.1, pp. 269–300.

Reillier, L. C. and B. Reillier (2017). *Platform Strategy. How to Unlock the Power of Communities and Networks to Grow Your Business*. London, UK: Routledge.

Reinartz, W., N. Wiegand, and M. Imschloss (2019). "The impact of digital transformation on the retailing value chain". In: *International Journal of Research in Marketing* 36.3, pp. 350–366.

Reisberger, P., T. Reisberger, and S. Strahringer (2008). "Measuring success of comerce". In: *Proceeding of the 6th International Workshop for Technical, Economic and Legal Aspects of Business Models for Virtual Goods*. Warsaw, Pland, pp. 117–138.

Reiss, M. and A. Günther (2010). "Mehrseitige Märkte: Paradigmenwechsel vom Markt- zum Netzwerk-Ansatz". In: *WiSt—Wirtschaftswissenschaftliches Studium* 39.4, pp. 176–181.

Rescher, N. (2000). *Realistic pragmatism. An introduction to pragmatic philosophy*. Albany: SUNY Press.

Resnick, M. L. and J. Sanchez (2004). "Effects of organizational scheme and labeling on task performance in product-centered and user-centered retail Web sites". In: *Human factors* 46.1, pp. 104–117.

Reussner, R. H., H. W. Schmidt, and I. H. Poernomo (2003). "Reliability prediction for component-based software architectures". In: *Journal of Systems and Software* 66.3, pp. 241–252.

Reverb (2023a). *Integrations*. URL: https://reverb.com/page/integrations (visited on 06/27/2023).

Reverb (2023b). *Reverb Market Insights*. URL: https://reverb.com/page/reverb-market-insights (visited on 06/27/2023).

Riasanow, T., F. Burckhardt, et al. (2018). "The Generic Blockchain Ecosystem and its Strategic Implications". In: *Proceedings of the 22nd Americas Conference on Information Systems (AMCIS2018)*. San Diego, USA, pp. 1–10.

Riasanow, T., R. J. Flötgen, et al. (2018). "The Generic Ecosystem and Innovation Patterns of the Digital Transformation in the Financial Industry". In: *Proceedings of the 22nd Pacific Asia Conference on Information Systems (PACIS2018)*. Yokohama, Japan, pp. 1–14.

Riasanow, T., L. Jäntgen, et al. (2021). "Core, intertwined, and ecosystem-specific clusters in platform ecosystems: analyzing similarities in the digital transformation of the automotive, blockchain, financial, insurance and IIoT industry". In: *Electronic Markets* 31.1, pp. 89–104.

Rickmann, T., S. Wenzel, and K. Fischbach (2014). "Software Ecosystem Orchestration: The Perspective of Complementors". In: *Proceedings of the 20th Americas Conference on Information Systems (AMCIS2014)*. Savannah, USA, pp. 1–14.

Riemensperger, F. and S. Falk (2020). "How to capture the B2B platform opportunity". In: *Electronic Markets* 30.1, pp. 61–63.

Rietveld, J. and M. A. Schilling (2021). "Platform Competition: A Systematic and Interdisciplinary Review of the Literature". In: *Journal of Management* 47.6, pp. 1528–1563.

Riggins, F. J. (1999). "A Framework for Identifying Web-Based Electronic Commerce Opportunities". In: *Journal of Organizational Computing and Electronic Commerce* 9.4, pp. 297–310.

Riggins, F. J. and H. S. Rhee (1998). "Toward a Unified View of Electronic Commerce". In: *Communications of the ACM* 41.10, pp. 88–95.

Risch, D. (2008). *Nutzung von Kundenprofilen im E-Commerce*. Cologne, Germany: Josef Eul Verlag.

Ritala, P., V. Agouridas, D. Assimakopoulos, and O. Gies (2013). "Value creation and capture mechanisms in innovation ecosystems: A comparative case study". In: *International Journal of Technology Management* 63.3-4, pp. 244–267.

Ritala, P. and A. Almpanopoulou (2017). "In defense of 'eco' in innovation ecosystem". In: *Technovation* 60-61, pp. 39–42.

Rivard, S. and L. Lapointe (2012). "Information Technology Implementers' Responses to User Resistance: Nature and Effects". In: *MIS Quarterly* 36.3, pp. 897–920.

Robertson, D. and K. Ulrich (1998). "Platform Product Development". In: *MIT Sloan Management Review* 39.4, pp. 1–19.

Rochet, J. and J. Tirole (2003). "Platform competition in two-sided markets". In: *Journal of the European Economic Association* 1.4, pp. 990–1029.

Rochet, J.-C. and J. Tirole (2006). "Two-sided markets: a progress report". In: *The RAND journal of economics* 37.3, pp. 645–667.

Rohrbeck, R., K. Hölzle, and H. G. Gemünden (2009). "Opening up for competitive advantage—How Deutsche Telekom creates an open innovation ecosystem". In: *R&D Management* 39.4, pp. 420–430.

Rosario, A. B., F. Sotgiu, K. de Valck, and T. Bijmolt (2016). "The effect of electronic word of mouth on sales: A meta-analytic review of platform, product, and metric factors". In: *Journal of Marketing Research* 53.3, pp. 297–318.

Rosemann, M. and I. Vessey (2008). "Toward improving the relevance of information systems research to practice: The role of applicability checks". In: *MIS Quarterly* 32.1, pp. 7–22.

Rossberg, J. (2014). *Beginning Application Lifecycle Management*. Berkeley, CA: Apress.

Rotar, A. (2020). *eCommerce Report 2020*. URL: https://www.statista.com/study/42335/ecommerce-report/ (visited on 08/21/2023).

Rowe, F. (2014). "What literature review is not: diversity, boundaries and recommendations". In: *European Journal of Information Systems* 23, pp. 241–255.

Rowley, T. J. (1997). "Moving beyond Dyadic Ties: A Network Theory of Stakeholder Influences". In: *The Academy of Management Review* 22.4, pp. 887–910.

Rückeshäuser, N. (2017). "Typology of distributed ledger-based business models". In: *Proceedings of the 25th European Conference on Information Systems (ECIS2017)*. Vol. 2017. Guimaraes in Portugal, pp. 2202–2217.

Ruddenklau, A. (2022). *Pulse of Fintech H2'21*. URL: https://assets.kpmg/content/dam/kpmg/xx/pdf/2022/02/pulse-of-fintech-h2-21.pdf (visited on 05/02/2022).

Rudolph, T., L. Nagengast, B. Melanie, and D. Bouteiller (2015). "Die Nutzung mobiler Shopping Apps im Kaufprozess". In: *Marketing Review St. Gallen* 3.32, pp. 42–49.

Ruutu, S., T. Casey, and V. Kotovirta (2017). "Development and competition of digital service platforms: A system dynamics approach". In: *Technological Forecasting and Social Change* 117.November 2016, pp. 119–130.

Rysman, M. (2009). "The economics of two-sided markets". In: *Journal of economic perspectives* 23.3, pp. 125–143.

Ryu, S. and J. Park (2020). "The effects of benefit-driven commitment on usage of social media for shopping and positive word-of-mouth". In: *Journal of Retailing and Consumer Services* 55, pp. 1–25.

Saarikko, T., K. Jonsson, and T. Burström (2014). "Towards an understanding of entrepreneurial alertness in the formation of platform ecosystems". In: *Proceedings of the European Conference on Information Systems (ECIS2014)*. Tel Aviv, Israel, pp. 1–15.

Sachse, S., T. Puschmann, and R. Alt (2012). "Towards customer-oriented electronic markets—A survey among digital natives in the financial industry". In: *Proceedings of the BLED Conference (BLED2012)*. Bled, Slovenia, pp. 1–22.

Saghafi, A. and Y. Wand (2014). "Do ontological guidelines improve understandability of conceptual models? A meta-analysis of empirical work". In: *Proceedings of the 47th Annual Hawaii International Conference on System Sciences (HICSS2014)*. Waikoloa, Hawaii: IEEE, pp. 4609–4618.

Saghiri, S., R. Wilding, C. Mena, and M. Bourlakis (2017). "Toward a three-dimensional framework for omni-channel". In: *Journal of Business Research* 77.4, pp. 53–67.

Sahai, A., V. Machiraju, and K. Wurster (2001). “Monitoring and controlling Internet based e-services”. In: *Proceedings of the 2nd IEEE Workshop on Internet Applications (WIAPP2001)*. San Jose, California, USA, pp. 41–48.

Salesforce (2021a). *API End-of-Life*. URL: https://developer.salesforce.com/docs/atlas.en-us.api_rest.meta/api_rest/api_rest_eol.htm (visited on 07/28/2021).

Salesforce (2021b). *Dreamforce 2021*. URL: https://www.salesforce.com/dreamforce/ (visited on 07/27/2021).

Salesforce (2021c). *Salesforce AppExchange—Leading Enterprise Cloud Marketplace*. URL: https://appexchange.salesforce.com/ (visited on 07/26/2021).

Salesforce (2021d). *Salesforce Extensions for Visual Studio Code*. URL: https://developer.salesforce.com/tools/vscode/ (visited on 07/28/2021).

Salesforce (2021e). *Technology Partners*. URL: https://www.salesforce.com/products/commerce-cloud/partner-marketplace/ (visited on 07/26/2021).

Salesforce (2021f). *Trailhead*. URL: https://trailhead.salesforce.com/en/ (visited on 07/27/2021).

Salesforce (2021g). *Trailhead Playground Management*. URL: https://trailhead.salesforce.com/en/content/learn/modules/trailhead_playground_management/create-a-trailhead-playground (visited on 07/27/2021).

Salesforce (2022a). *Build apps and automate with low-code magic*. URL: https://www.salesforce.com/products/platform/overview/?bc=OTH (visited on 10/21/2022).

Salesforce (2022b). *Introduction to Marketing Cloud Platform SDKs*. URL: https://developer.salesforce.com/docs/marketing/marketing-cloud/guide/sdks-overview.html (visited on 10/21/2022).

Salesforce (2022c). *Metadata Coverage*. URL: https://developer.salesforce.com/docs/metadata-coverage/56 (visited on 10/21/2022).

Salesforce (2022d). *Salesforce Roadmap Explorer*. URL: https://architect.salesforce.com/roadmaps/roadmap-explorer (visited on 10/21/2022).

Salesforce (2022e). *Salesforce Winter ’23 Release Notes*. URL: https://help.salesforce.com/s/articleView?id=release-notes.salesforce_release_notes.htm&type=5&release=240 (visited on 10/21/2022).

Salminen, J. and J. Teixeira (2013). “Fool’s gold? developer dilemmas in a closed mobile application market platform”. In: *Lecture Notes in Business Information Processing* 155 LNBIP, pp. 121–132.

Saloner, G. (1990). “Economic issues in computer interface standardization”. In: *Economics of Innovation and New Technology* 1.1-2, pp. 135–156.

Sambarmurthy, V., A. Bhradawaj, and V. Grover (2003). “Shaping Agility through Digital Options: Reconceptualizing the Role of Information Technology in Contemporary Firms”. In: *MIS Quarterly* 27.2, pp. 237–263.

Samila, S. and O. Sorenson (2010). “Venture capital as a catalyst to commercialization”. In: *Research Policy* 39.10, pp. 1348–1360.

Sanchez, L., L. Muñoz, et al. (2014). “SmartSantander: IoT experimentation over a smart city testbed”. In: *Computer Networks* 61, pp. 217–238.

Sanchez, R. and J. Mahoney (1996). “Modularity, flexibility, and knowledge management in product and organization design”. In: *Strategic Management Journal* 17, pp. 63–76.

Sánchez, V., I. Angulo, and S. Bilbao (2009). “E-Business in the construction sector: A service oriented approach”. In: *Business Information Systems Workshops*. Poznan, Poland, pp. 171–182.

Sangiovanni-Vincentelli, A. and G. Martin (2001). “Platform-Based Design and Software Design Methodology for Embedded Systems”. In: *IEEE Design & Test of Computers* 18.6, pp. 23–33.

Santos, F. M. and K. M. Eisenhardt (2005). “Organizational boundaries and theories of organization”. In: *Organization Science* 16.5, pp. 491–508.

SAP (2021a). *All Questions in SAP Commerce Cloud—SAP Community*. URL: https://answers.sap.com/tags/73555000100800001224 (visited on 07/27/2021).

SAP (2021b). *Builds and Cloud Environment Setups—SAP Commerce Cloud*. URL: https://microlearning.opensap.com/playlist/dedicated/178317971/1_akh12tp0/1_0lrtis02 (visited on 07/27/2021).

SAP (2021c). *Prototypes Archive—SAP Customer Experience Labs*. URL: https://cxlabs.sap.com/prototype/ (visited on 07/29/2021).

SAP (2021d). *SAP Road Map Explorer*. URL: https://roadmaps.sap.com/board?PRODUCT=089E017A62AB1EDA94C15F5EDB33E0E1%5C&range=CURRENT-LAST#Q3%202021 (visited on 07/27/2021).

SAP (2021e). *SAP Store*. URL: https://store.sap.com/dcp/en/ (visited on 07/26/2021).

SAP (2022a). *openSAP Microlearning: SAP Commerce Cloud*. URL: https://microlearning.opensap.com/category/SAP+Commerce+Cloud/178315851 (visited on 10/21/2022).

SAP (2022b). *Platform, Services, and Utilities: Logging*. URL: https://help.sap.com/docs/SAP_COMMERCE/d0224eca81e249cb821f2cdf45a82ace/8c07853c866910148a00baf81ea1669e.html?locale=en-US (visited on 10/21/2022).

SAP (2022c). *Platform, Services, and Utilities: Testing*. URL: https://help.sap.com/docs/SAP_COMMERCE/d0224eca81e249cb821f2cdf45a82ace/8c6f059d866910149b6aede7ab7a0f36.html?locale=en-US (visited on 10/21/2022).

SAP (2022d). *SAP API Business Hub*. URL: https://api.sap.com/search?searchterm=commerce%20cloud&tab=APIs&top=12 (visited on 10/21/2022).

SAP (2022e). *SAP Roadmap Explorer*. URL: https://www.sap.com/germany/products/roadmaps.html?gl=https%5C%3A%5C%2F%5C%2Froadmaps.sap.com%5C%2Findex.html%5C%23%5C%2F (visited on 10/21/2022).

Sarkar, S., G. Rama, and A. Kak (2007). “API-Based and Information-Theoretic Metrics for Measuring the Quality of Software Modularization”. In: *IEEE Transactions on Software Engineering* 33.1, pp. 14–32.

Satariano, A. (2020). *Big Fines and Strict Rules Unveiled Against ‘Big Tech’ in Europe*. URL: https://www.nytimes.com/2020/12/15/technology/big-tech-regulation-europe.html (visited on 02/02/2022).

Schaefer, M. (2002). *Ö kologie*. 3rd ed. Jena, Germany: Gustav Fischer.

Schalowski, J. and C. Barrot (2019). “The Long-term Diffusion of Digital Platforms—An Agent-based Model”. In: *Proceedings of the 40th International Conference on Information Systems (ICIS2019)*. Munich, Germany, pp. 1–9.

Schekkerman, J. (2004). *How to Survive in the Jungle of Enterprise Architecture Frameworks: Creating or Choosing an Enterprise Architecture Framework*. 2nd ed. Victoria, British Columbia: Trafford Publishing.

Schenkel, P., P. Osl, and H. Oesterle (2013). "Towards an electronic marketplace for bricks-and-mortar services". In: *Proceedings of the 24th Australasian Conference on Information Systems*. Melbourne, Australia, pp. 1–10.

Schilling, M. A. (2000). "Toward a general modular systems theory and its application to interfirm product modularity". In: *Academy of Management Review* 25.2, pp. 312–334.

Schirrmacher, N. B., J. Ondrus, and T. Kude (2017). "Launch strategies of digital platforms: Platforms with switching and non-switching users". In: *Proceedings of the 25th European Conference on Information Systems (ECIS 2017)*. Guimaraes, Portugal, pp. 658–673.

Schmeken, G. M. (2007). *Erfolgreiche Strategien für E-Commerce. Integrierte Kosten- und Leistungsführerschaft als Orientierungsmuster*. Wiesbaden, Germany: Deutscher Universitäts-Verlag.

Schmid, B. (1993). "Elektronische Märkte". In: *Wirtschaftsinformatik* 35, pp. 465–480.

Schmid, B. (1997). "Requirements For Electronic Markets Architecture". In: *Electronic Markets* 7.1, pp. 3–6.

Schmid, B. F. (1999). "Elektronische Märkte—Merkmale, Organisation und Potentiale". In: *Electronic Commerce*. Ed. by A. Hermanns and M. Sauter. Munich, Germany, Germany: Vahlen, pp. 31–48.

Schmid, M. et al. (2021). "Platform over Market—when is joining a Platform Beneficial?" In: *Proceedings of the 28th European Conference on Information Systems (ECIS2020)*. Marrakech, Morocco, pp. 1–16.

Schmidt, J., P. Drews, and I. Schirmer (2018a). "Charting the emerging business ecosystem of fintechs and banks: Seven types of collaborative business models". In: *Proceedings of the European Conference on Information Systems (ECIS2018)*. Portsmouth, UK, pp. 1–15.

Schmidt, J., P. Drews, and I. Schirmer (2018b). "Charting the Emerging Financial Services Ecosystem of Fintechs and Banks: Six Types of Data-Driven Business Models in the Fintech Sector". In: *Proceedings of the 51st Hawaii International Conference on System Sciences (HICSS2018)*. Maui, Hawaii, USA, pp. 5004–5013.

Schöbel, S. M. and F. Tingelhoff (2023). "Overcoming Challenges to Enable the Potential of Metaverse Platforms: A Qualitative Approach to Understand Value Creation". In: *AIS Transactions on Human-Computer Interaction* 15.1, pp. 1–21.

Schoormann, T., F. Möller, M. Di Maria, and N. Große (2023). "Guiding Design Principle Projects: A Canvas for Young Design Science Researchers". In: *Journal of Information Systems Education* 34.3, pp. 307–325.

Schreieck, M., M. Wiesche, and H. Krcmar (2016). "Design and Governance of Platform Ecosystems-Key Concepts and Issues for Future Research". In: *Proceedings of the 24th European Conference on Information Systems (ECIS2016)*. Istanbul, Turkey, pp. 1–20.

Schreier, M. (2014). "Qualitative Content Analysis". In: *The Sage handbook of qualitative data analysis*. Ed. by U. Flick. Los Angeles, USA: Sage Publications, pp. 170–183.

Schröder, H. (2005). *Multichannel-Retailing. Marketing in Mehrkanalsysemen des Einzelhandels*. Berlin/Heidelberg: Springer-Verlag.

Schroeder, P., K. Anggraeni, and U. Weber (2019). "The Relevance of Circular Economy Practices to the Sustainable Development Goals". In: *Journal of Industrial Ecology* 23.1, pp. 77–95.

Schroth, C., T. Janner, P. Mayer, and A. Stage (2007). "A holistic architecture for collaborative and highly automatized e-Business platforms". In: *Proceedings of the International Conference on Data Engineering*. Istanbul, Turkey, pp. 355–362.

Schryen, G., G. Wagner, and A. Benlian (2015). "Theory of Knowledge for Literature Reviews: An Epistemological Model, Taxonomy and Empirical Analysis of IS Literature". In: *Proceedings of the 36th International Conference on Information Systems (ICIS2015)*. Fort Worth, Texas, USA, pp. 1–22.

Schubert, P. (1999). *Virtuelle Transaktionsgemeinscahften im Electronic Commerce: Management, Marketing und Soziale Umwelt*. Köln: Josef Eul Verlag.

Schuh, G. and S. Rudolf (2015). *Marktspiegel Business Software PLM/PDM 2015/2016*. 3rd ed. Aachen: Trovarit AG.

Schüler, F. and D. Petrik (2021). "Objectives of Platform Research: A Co-citation and Systematic Literature Review Analysis". In: *Management Digitaler Plattformen*. Ed. by M. Seiter, L. Grünert, and A. J. Steur. Wiesbaden, Germany: Springer Fachmedien Wiesbaden, Germany. Chap. 1, pp. 1–33.

Schultz, C. D. (2004). "Consumer Trust in E-Commerce. An Analysis of Means Communicating Trustworthiness From a Buying Transaction Life Cycle Perspective". PhD thesis. Univeristy of Skövde.

Schultze, U. and M. Avital (2011). "Designing interviews to generate rich data for information systems research". In: *Information and Organization* 21.1, pp. 1–16.

Schüritz, R., S. Seebacher, and R. Dorner (2017). "Capturing Value from Data: Revenue Models for Data-Driven Services". In: *Proceedings of the 50th Hawaii International Conference on System Sciences 2017 (HICSS2017)*. Maui, Hawaii, USA, pp. 5348–5357.

Schüßler, E., W. Attwood-Charles, S. Kirchner, and J. B. Schor (2021). "Between mutuality, autonomy and domination: rethinking digital platforms as contested relational structures". In: *Socio-Economic Review* 19.4, pp. 1217–1243.

Schütte, R. (2011). *Modellierung von Handelsinformationssystemen. Kumulative Habilitationsschrift*. Münster: Westfälische Wilhelms-Universität Münster.

Schütte, R. (2018). "Retailing in Zeiten der Digitalisierung. Die Plattformen dominieren den Wettbewerb". In: *IM+io Best & Next Practices aus Digitalisierung—Management—Wissenschaft* Heft 4, pp. 46–51.

Schütte, R. and T. Vetter (2016). "Die Analyse des Digitalisierungspotenzials von Handelsunternehmen". In: *Handel 4.0*. Ed. by R. Gläß and B. Leukert. Berlin, Germany: Springer-Verlag, pp. 75–116.

Schütte, R. (1998). *Grundsätze ordnungsm ä ßiger Referenzmodellierung. Konstruktion konfigurations- und anpassungsorientierter Modelle. Neue betriebswirtschaftliche Forschung (Bd. 233)*. Wiesbaden, Germany: Gabler.

Schütte, R. (2017). "Information systems for retail companies". In: *International Conference on Advanced Information Systems Engineering*. Ed. by E. Dubois and K. Pohl. Essen, Deutschland: Springer, pp. 13–25.

Schütte, R., F. Ahlemann, et al. (2022). "Quo Vadis Information Systems Research in Times of Digitalization?" In: *Business & Information Systems Engineering* 4, pp. 529–540.

Schütte, R., M. Muschkiet, and T. Wulfert (2022). "Smart City Experience: von isolierten Services zu Smart Service-Ökosystemen". In: *Forum Dienstleistungsmanagement*. Ed. by M. Bruhn and K. Hadwich. Heidelberg: Springer, pp. 1–25.

Schütte, R., S. Seufert, and T. Wulfert (2022). *IT-Systeme wirtschaftlich verstehen und gestalten. Methoden—Paradoxien—Grundsätze*. Wiesbaden, Germany: Springer Gabler.

Schütte, R. and T. Wulfert (2022a). "Digital platforms and trading companies: the evolution of traditional business ecosystems to integrated digital business ecosystems". In: *Handbook on Digital Business Ecosystems: Technologies, Markets, Business Models, Management, and Societal Challenges*. Ed. by S. Baumann. Cheltenham, UK: Edward Elgar. Chap. 15, pp. 212–231.

Schütte, R. and T. Wulfert (2022b). “Referenzmodelle und Referenzarchitekturen—Wiederverwendbare Modelle zur Unterstützung der Digitalen Transformation”. In: *Handbuch Digitalisierung*. Ed. by S. Roth and H. Corsten. Munich, Germany: Vahlen, pp. 479–504.

Schwalbe, M. et al. (2000). “Generic Processes in the Reproduction of Inequality: An Interactionist Analysis”. In: *Social Forces* 79.2, pp. 419–452.

Schwarz, J. S. and C. Legner (2020). “Business model tools at the boundary: Exploring communities of practice and knowledge boundaries in business model innovation”. In: *Electronic Markets* 30.3, pp. 421–445.

Scott, W. R. (2014). *Institutions and organizations: ideas, interests, and identities*. 4th ed. Los Angeles [u. a.]: SAGE, p. 345.

Secinaro, S., V. Brescia, D. Calandra, and P. Biancone (2021). “Data quality for health sector innovation and accounting man-agement: a twenty-year bibliometric analysis”. In: *Economia Aziendale Online* 12.4, pp. 407–431.

Secinaro, S., F. Dal Mas, V. Brescia, and D. Calandra (2021). “Blockchain in the accounting, auditing and accountability fields: a bibliometric and coding analysis”. In: *Accounting, Auditing & Accountability Journal*.

Sedlmeir, J., H. Buhl, G. Fridgen, and R. Keller (2020). “The Energy Consumption of Blockchain Technology: Beyond Myth”. In: *Business & Information Systems Engineering* 62.6, pp. 599–608.

Seidman, I. (2019). *Interviewing as qualitative research: A guide for researchers in education and the social sciences*. 5th ed. New York, USA: Teachers college press.

Sein, M. et al. (2011). “Action design research”. In: *MIS Quarterly* 35.1, pp. 37–56.

Selander, L., O. Henfridsson, and F. Svahn (2013). “Capability Search and Redeem across Digital Ecosystems”. In: *Journal of Information Technology* 28.3, pp. 183–197.

Semeijn, J., A. C. van Riel, M. J. van Birgelen, and S. Streukens (2005). “E-services and physical fulfillment: How e-loyalty is created”. In: *Managing Service Quality: An International Journal* 15.2, pp. 182–194.

Sensuse, D. I. et al. (2020). “Challenges and Recommended Solutions for Change Management in Indonesian E-Commerce”. In: *Proceedings of the International Conference on Information Technology Systems and Innovation (ICITSI2020)*. Padang, Indonesia, pp. 250–255.

Senyo, P. K., K. Liu, and J. Effah (2019). “Digital business ecosystem: Literature review and a framework for future research”. In: *International Journal of Information Management* 47, pp. 52–64.

Seppänen, M., S. Hyrynsalmi, K. Manikas, and A. Suominen (2017). “Yet another ecosystem literature review: 10+1 Research communities”. In: *Proceedings of the 2017 IEEE European Technology and Engineering Management Summit (E-TEMS2017)*. Munich, Germany, pp. 1–8.

Seÿffert, R. (1972). *Wirtschaftslehre des Handels*. Ed. by E. Sundhoff. Wiesbaden, Germany: VS Verlag für Sozialwissenschaften.

Shaikh, M. and E. Vaast (2016). "Folding and Unfolding: Balancing Openness and Transparency in Open Source Communities". In: *Information Systems Research* 27.4, pp. 813–833.

Shanks, G. (2002). "Guidelines for Conducting Positivist Case Study Research in Information Systems". In: *Australasian Journal of Information Systems* 10.1, pp. 76–85.

Shannon-Baker, P. (2015). " "But I Wanted to Appear Happy": How using Arts-Informed and Mixed Methods Approaches Complicate Qualitatively Driven Research on Culture Shock". In: *International Journal of Qualitative Methods* 14.2, pp. 34–52.

Shapiro, C. and H. Varian (1998). *Information rules: a strategic guide to the network economy.* Boston, USA: Harvard Business School Press.

Shapiro, C. and H. R. Varian (1999). "The Art of Standards Wars". In: *California Management Review* 41.2, pp. 8–32.

Shaw, M. and D. Garlan (1996). *Software Architecture. Perspectives on an Emerging Discipline.* Upper Saddle River, NJ: Prentice-Hall.

Shen, C.-W., Min Chen, and C.-C. Wang (2019). "Analyzing the trend of O2O commerce by bilingual text mining on social media". In: *Computers in Human Behavior* 101, pp. 474–483.

Shenton, A. K. (2004). "Strategies for ensuring trustworthiness in qualitative research projects". In: *Education for Information* 22.2, pp. 63–75.

Sherif, N. and G. Zhang (2017). "Collaborative filtering using probabilistic matrix factorization and a Bayesian nonparametric model". In: *Proceedings of the 2nd IEEE International Conference on Big Data Analysis (ICBDA2017).* Beijing, China, pp. 391–396.

Shiau, W.-L., Y. K. Dwivedi, and H. S. Yang (2017). "Co-citation and cluster analyses of extant literature on social networks". In: *International Journal of Information Management* 37.5, pp. 390–399.

Shin, D.-H. (2013). "User experience in social commerce: In friends we trust". In: *Behaviour and Information Technology* 32.1, pp. 52–67.

Shopee (2023). *Documentation—Shopee Open Platform.* URL: https://open.shopee.com/documents/v2/v2.shop.get_shop_info?module=92&type=1 (visited on 06/28/2023).

Shopify (2022a). *API version release notes.* URL: https://shopify.dev/api/release-notes (visited on 10/21/2022).

Shopify (2022b). *App Design Guidelines.* URL: https://shopify.dev/apps/design-guidelines (visited on 10/21/2022).

Shopify (2022c). *Fulfillment API Deprecation.* URL: https://community.shopify.com/c/fulfillment-api-deprecation/bd-p/fulfillment-api (visited on 10/21/2022).

Shopify (2022d). *Partners and Developers.* URL: https://community.shopify.com/c/partners-and-developers/ct-p/appdev?shpxid=c60642d1-83D0-4E6D-FA2C-72EDF37CA21A (visited on 10/21/2022).

Shopify (2022e). *Shopify API versioning.* URL: https://shopify.dev/api/usage/versioning (visited on 10/21/2022).

Shopify (2022f). *Shopify Status.* URL: https://www.shopifystatus.com/ (visited on 10/21/2022).

Shopware (2021a). *API Versioning.* URL: https://developer.shopware.com/docs/guides/integrations-api/general-concepts/api-versioning (visited on 07/28/2021).

Shopware (2021b). *Erweiterungen.* URL: https://store.shopware.com/erweiterungen/ (visited on 07/29/2021).

Shopware (2021c). *Setting up PhpStorm.* URL: https://developers.shopware.com/developers-guide/phpstorm/ (visited on 07/28/2021).

Shopware (2022a). *Debugging Shopware.* URL: https://developers.shopware.com/developers-guide/debugging/ (visited on 10/21/2022).

Shopware (2022b). *Shopware 5 upgrade guide.* URL: https://developers.shopware.com/developers-guide/shopware-5-upgrade-guide-for-developers/ (visited on 10/21/2022).

Shopware (2022c). *Shopware Roadmap.* URL: https://www.shopware.com/en/roadmap/ (visited on 10/21/2022).

Shopware (2022d). *Versioning & Dependencies.* URL: https://developer.shopware.com/docs/guides/hosting/installation-updates/composer (visited on 10/21/2022).

Shore, J., D. Larsen, G. Klitgaard, and S. Warden (2022). *The Art of Agile Development.* 2nd ed. Sebastopol, USA: O'Reilly.

Shrivastava, M. and S. Kumar (2021). "A pragmatic and intelligent model for sarcasm detection in social media text". In: *Technology in Society* 64.4, pp. 1–17.

Siering, M., J. Muntermann, and M. Grčar (2021). "Design principles for robust fraud detection: The case of stock market manipulations". In: *Journal of the Association for Information Systems* 22.1, pp. 156–178.

Simon, H. (1996). *The Sciences of the Artificial.* 3rd ed. Cambridge, Massachusetts: MIT Press.

Simon, H. A. (1962). "The architecture of complexity". In: *Proceedings of the American Philosophical Society* 106.6, pp. 467–482.

Simon, P. (2013). *The Age of the Plattform. How Amazon, Facebook and google Have Redefined Business.* Revised Ed. Henderson, Nevada: Motion Publishing.

Sirmon, D. G., M. A. Hitt, R. D. Ireland, and B. A. Gilbert (2011). "Resource Orchestration to Create Competitive Advantage". In: *Journal of Management* 37.5, pp. 1390–1412.

Smolander, K. and M. Rossi (2008). "Conflicts, compromises, and political decisions: Methodological challenges of enterprise-wide e-business architecture creation". In: *Journal of Database Management* 19.1, pp. 19–40.

Soh, C. and M. L. Markus (2002). "Business-to-Business Electronic Marketplaces : A Strategic Archetypes Approach". In: *Proceedings of the International Conference on Information Systems (ICIS2002).* Barcelona, Spain, pp. 835–845.

Sokiyna, M. and M. Aqel (2020). "The role of e-business applications software in driving operational excellence: Impact of departments collaboration using sustainable software". In: *Sustainable Computing: Informatics and Systems* 28, pp. 1–35.

Solaimani, S. and H. Bouwman (2012). "A framework for the alignment of business model and business processes". In: *Business Process Management Journal* 18.4, pp. 655–679.

Song, P., D. Huang, Q. Yang, and Y. Zhang (2013). "Research on financial coordinated supervision platform and supervision strategy for online payment under paperless trade". In: *International Journal of Services, Technology and Management* 19.4-6, pp. 219–239.

Song, T., Q. Tang, and J. Huang (2019). "Triadic Closure, Homophily, and Reciprocation: An Empirical Investigation of Social Ties Between Content Providers". In: *Information Systems Research* 30.3, pp. 912–926.

Soto Setzke, D., M. C. Kavili, and M. Böhm (2020). "On the Use of Qualitative Comparative Analysis in Information Systems Research—A Critical Rerview". In: *Proceedings of the 28th European Conference on Information Systems (ECIS2020).* Marrakesh, Morocco, pp. 1–20.

Spaeth, S. and S. Niederhöfer (2022). "Compatibility promotion between platforms: The role of open technology standards and giant platforms". In: *Electronic Markets* 32, pp. 1891–1915.

Spagnoletti, P., A. Resca, and G. Lee (2015). "A Design Theory for Digital Platforms Supporting Online Communities: A Multiple Case Study". In: *Journal of Information Technology* 30.4, pp. 364–380.

Specht, G. (1998). *Distributionsmanagement*. 3rd ed. Stuttgart, Germany [u.a.]: Kohlhammer.

Srnicek, N. (2017). *Platform Capitalism*. Cambridge, UK: Polity.

Stahl, B. C. (2009). "The Ideology of Design: A Critical Appreciation of the Design Science Discourse in Information Systems and Wirtschaftsinformatik". In: *Wissenschaftstheorie und gestaltungsorientierte Wirtschaftsinformatik*. Heidelberg, Germany: Physica-Verlag, pp. 111–132.

Stähler, P. (2002). *Geschäftsmodelle in der digitalen Ö konomie: Merkmale, Strategien und Auswirkungen*. 2. Aufl. Reihe: Electronic Commerce 7. Lohmar: Eul Verlag.

Stallkamp, M. and A. P. Schotter (2021). "Platforms without borders? The international strategies of digital platform firms". In: *Global Strategy Journal* 11.1, pp. 58–80.

Stanley, J. and G. Briscoe (2010). "The ABC of Digital Business Ecosystems". In: *Communications Law—Journal of Computer, Media and Telecommunications Law* 15.1, pp. 1–24.

Star, S. L. (2010). "This is not a boundary object: Reflections on the origin of a concept". In: *Science Technology and Human Values* 35.5, pp. 601–617.

Star, S. L. and J. R. Griesemer (1989). "Institutional ecology, 'translations' and boundary objects amateurs and professionals in Berkeley's museum of vertebrate zoology". In: *Social Studies of Science* 19, pp. 387–420.

Stark, D. and I. Pais (2020). "Algorithmic management in the platform economy". In: *Sociologica* 14.3, pp. 47–72.

Stark, J. (2011). *Product Lifecycle Management. 21st Century Paradigm for Product Realisation*. Decision Engineering. London, UK: Springer.

Stark, J. (2015). *Product Lifecycle Management: The Devil is in the Details*. 2nd ed. London: Springer, p. 351.

Staub, N., S. Aier, and R. Winter (2021). "Taxonomy of Digital Platforms: A Business Model Perspective". In: *Proceedings of the 54th Hawaii International Conference on System Sciences (HICSS 2021)*. Maui, Hawaii, USA, pp. 6163–6172.

Staudenmayer, N., M. Tripsas, and C. L. Tucci (2005). "Interfirm Modularity and Its Implications for Product Development". In: *Journal of Product Innovation Management* 22.4, pp. 303–321.

Stauss, B. (2006). "Plattformstrategien im Dienstleistungshandel". In: *Service Engineering. Entwicklung und Gestaltung innovativer Dienstleistungen*. Ed. by H.-J. Bullinger and A.-W. Scheer. Vol. 2. Berlin, Germany: Springer.

Staykova, K. S. and J. Damsgaard (2015). "A Typology of Multi-sided Platforms: The Core and the Periphery". In: *Proceedings of the 22rd European Conference on Information Systems (ECIS2015)*. Münster, Germany, pp. 1–16.

Staykova, K. and J. Damsgaard (2021). "How Digital Platforms Compete Against Diverse Rivals". In: *MIS Quarterly Executive* 20.4, pp. 275–295.

Steininger, D. M., P. Mikalef, A. Pateli, and A. Ortiz-de-Guinea (2022). "Dynamic Capabilities in Information Systems Research: A Critical Review, Synthesis of Current Knowledge, and

Recommendations for Future Research". In: *Journal of the Association for Information Systems* 22.2, pp. 447–490.

Strader, T. J. and M. J. Shaw (1997). "Characteristics of electronic markets". In: *Decision Support Systems* 21.3, pp. 185–198.

StrategyR (2022). *Application Lifecycle Management (ALM) World Market Report*. URL: https://www.strategyr.com/market-report-application-lifecycle-management-alm-forecasts-global-industry-analysts-inc.asp (visited on 10/25/2022).

Straub, D. and D. Gefen (2004). "Validation Guidelines for IS Positivist Research". In: *Communications of the Association for Information Systems* 13.1, pp. 380–427.

Street, C. T. and K. W. Ward (2012). "Improving validity and reliability in longitudinal case study timelines". In: *European Journal of Information Systems* 21.2, pp. 160–175.

Strobel, G. (2021). "Information systems in the era of the internet of things: A domain-specific modeling language". In: *Proceedings of the Annual Hawaii International Conference on System Sciences (HICSS2021)*. Maui, Hawaii, pp. 4816–4823.

Stüber, E., C. Leyendecker, and H. Heinick (2019). *Gatekeeper Amazon—Vom Suchen und Finden des eigenen Erfolgswegs*. Tech. rep. Köln: IFH Köln.

Subrahmanian, E. et al. (2003). "Boundary Objects and Prototypes at the Interfaces of Engineering Design". In: *Computer Supported Cooperative Work (CSCW)* 12.2, pp. 185–203.

Subramanian, H. (2017). "Decentralized blockchain-based electronic marketplaces". In: *Communications of the ACM* 61.1, pp. 78–84.

Sun, M. and E. Tse (2009). "The Resource-Based View of Competitive Advantage in Two-Sided Markets". In: *Journal of Management Studies* 46.1, pp. 45–64.

Sun, R., S. Gregor, and B. Keating (2015). "Information technology platforms: conceptualisation and a review of emerging research in the IS discipline". In: *Proceedings of the Australasian Conference on Information Systems (ACIS2015)*. Adelaide, Australia, pp. 1–17.

Sun, W., C. J. Guo, et al. (2010). "Design aspects of Software as a Service to enable e-Business through Cloud platform". In: *Proceedings of the IEEE International Conference on E-Business Engineering (ICEBE2010)*. Shanghai, China, pp. 456–461.

Sunyaev, A. et al. (2021). "Token Economy". In: *Business & Information Systems Engineering* 63.4, pp. 457–478.

Suominen, A., M. Seppänen, and O. Dedehayir (2019). "A bibliometric review on innovation systems and ecosystems: a research agenda". In: *European Journal of Innovation Management* 22.2, pp. 335–360.

Suryono, R. R., B. Purwandari, and I. Budi (2019). "Peer to Peer (P2P) Lending Problems and Potential Solutions: A Systematic Literature Review". In: *Procedia Computer Science* 161, pp. 204–214.

Svátek, V., P. Kompuš, M. Dudáš, and M. Nečaský (2015). "Procurement Notice Enrichment Using Product Ontologies". In: *Proceedings of the 11th International Conference on Semantic Systems (SEMANTICS2015)*. Vienna, Austria, pp. 200–203.

Szanto, K. (2021). *PSR-12: Extended Coding Style—PHP-FIG*. URL: https://www.php-fig.org/psr/psr-12/ (visited on 07/26/2021).

Szyjewski, G. (2019). "Expanding an open source e-commerce with a separate ICT system". In: *Procedia Computer Science* 159, pp. 2091–2101.

Takealot (2023). *Seller Central APIs*. URL: https://seller.takealot.com/api/seller-api (visited on 06/28/2023).

Takeda, H., P. Veerkamp, T. Tomiyama, and H. Yoshikawa (1990). "Modeling design processes". In: *AI Magazine* 11.4, pp. 37–48.

Tan, B., X. Lu, S. L. Pan, and L. Huang (2015). "The Role of IS Capabilities in the Development of Multi-Sided Platforms: The Digital Ecosystem Strategy of Alibaba.com". In: *Journal of the Association for Information Systems* 16.4, pp. 248–280.

Tan, C.-W., I. Benbasat, and R. T. Cenfetelli (2016). "An Exploratory Study of the Formation and Impact of Electronic Service Failures". In: *MIS Quarterly* 40.1, pp. 1–29.

Tan, T. and T. Phan (2016). "Social Media-Driven Credit Scoring: The Predictive Value of Social Structures". In: *Proceedings of the 37th International Conference on Information Systems (ICIS2016)*. Dublin, Ireland, pp. 1–10.

Tandon, A., P. Kaur, M. Mäntymäki, and A. Dhir (2021). "Blockchain applications in management: A bibliometric analysis and literature review". In: *Technological Forecasting and Social Change* 166, pp. 1–19.

Tang, K., J. M. Zhang, and C. H. Feng (2011). "Application Centric Lifecycle Framework in Cloud". In: *Proceedings of the 8th IEEE International Conference on e-Business Engineering (ICEBE2011)*. Beijing, China: IEEE, pp. 329–334.

Tansley, A. G. (1935). "The Use and Abuse of Vegetational Concepts and Terms". In: *Ecology* 16.3, pp. 284–307.

Tapscott, D., D. Ticoll, and A. Lowy (2000). *Digital Capital: Harnessing the Power of Business Webs*. Paperback. Boston, Mass.: Havard Business School Press.

Tatnall, A. (2005). "Actor-Network Theory in Information Systems Research". In: *Encyclopedia of Information Science and Technology, First Edition*. Vol. 19. 2. IGI Global, pp. 42–46.

Taudes, A., M. Feurstein, and A. Mild (2000). "Options analysis of software platform decisions: A case study". In: *MIS Quarterly* 24.2, pp. 227–243.

Täuscher, K. and S. Laudien (2018). "Understanding platform business models: A mixed methods study of marketplaces". In: *European Management Journal* 36.3, pp. 319–329.

Taylor, R. N., N. Medvidovic, and E. M. Dashofy (2010). *Software Architecture. Foundations, Theory and Practice*. Hoboken, NJ, USA: John Wiley & Sons.

Teddie, C. and A. Tashakkori (2009). *Foundations of mixed methods research: integrating quantitative and qualitative approaches in the social and behavioral sciences*. London, UK: Sage Publications.

Teece, D. (1986). "Profiting from technological innovation: Implications for integration, collaboration, licensing and public policy". In: *Research policy* 15.6, pp. 285–305.

Teece, D. J. (2016). "Profiting from Innovation in the Digital Economy: Standards, Complementary Assets, and Business Models In the Wireless World". In: *Working paper Series No. 16* 16, pp. 1–40.

Teece, D. J., G. Pisano, and A. Shuen (1997). "Dynamic capabilities and strategic management". In: *Strategic management journal* 18.7, pp. 509–533.

Teece, D. J. (2007). "Explicating dynamic capabilities: the nature and microfoundations of (sustainable) enterprise performance". In: *Strategic Management Journal* 28.13, pp. 1319–1350.

Teece, D. J. (2014). "A dynamic capabilities-based entrepreneurial theory of the multinational enterprise". In: *Journal of International Business Studies* 45.1, pp. 8–37.

Teece, D. J. (2017). "Dynamic Capabilities and (Digital) Platform Lifecycles". In: *Entrepreneurship, Innovation, and Platforms*. Ed. by J. Furman, A. Gawer, B. Silverman, and S. Stern. Vol. 37. Advances in Strategic Management. Bingley, West Yorkshire, UK: Emerald Publishing Limited, pp. 211–225.

Teece, D. J. (2018). "Profiting from innovation in the digital economy: Enabling technologies, standards, and licensing models in the wireless world". In: *Research Policy* 47.8, pp. 1367–1387.

Teller, C. and J. Elms (2010). "Managing the attractiveness of evolved and created retail agglomerations formats". In: *Marketing Intelligence & Planning* 28.1, pp. 25–45.

Teubner, T. and D. Dann (2018). "How Platforms Build Trust". In: *SSRN Electronic Journal*, pp. 1–5.

Thapa, D. and Haj-Bolouri (2023). "Demystifying Philosophy for Information Systems Researchers". In: *Proceedings of the 29th Americas Conference on Information Systems (AMCIS2023)*. Panama City, Panama, pp. 1–10.

Thelwall, M. (2008). "Social networks, gender, and friending: An analysis of MySpace member profiles". In: *Journal of the American Society for Information Science and Technology* 59.8, pp. 1321–1330.

Thomas, L. D. W., E. Autio, and D. M. Gann (2014). "Architectural Leverage: Putting Platforms in Context". In: *The Academy of Management Perspectives* 28.2, pp. 198–219.

Thompson, J. D. (1967). *Organizations in Action*. New York [u. a.]: McGraw-Hill.

Thoring, K., R. M. Mueller, and P. Badke-Schaub (2020). "Workshops as a research method: Guidelines for designing and evaluating artifacts through workshops". In: *Proceedings of the 53rd Hawaii International Conference on System Sciences (HICSS2020)*. Maui, Hawaii, pp. 5036–5045.

Tian, L., A. J. Vakharia, Y. Tan, and Y. Xu (2018). "Marketplace, Reseller, or Hybrid: Strategic Analysis of an Emerging E-Commerce Model". In: *Production and Operations Management* 27.8, pp. 1595–1610.

Tian, W., L. Wang, and H. Zhang (2012). "Using cloud computing to build E-commerce recommendation platform". In: *International Journal of Digital Content Technology and its Applications* 6.12, pp. 391–398.

Tilson, D., K. Lyytinen, and C. Sørensen (2010). "Digital infrastructures: The missing IS research agenda". In: *Information Systems Research* 21.4, pp. 748–759.

Tilson, D., C. Sorensen, and K. Lyytinen (2012). "Change and Control Paradoxes in Mobile Infrastructure Innovation: The Android and iOS Mobile Operating Systems Cases". In: *Proceedings of the 45th Hawaii International Conference on System Sciences (HICSS2012)*. Maui, Hawaii, pp. 1324–1333.

Timmers, P. (1998). "Business models for electronic markets". In: *Electronic markets* 8.2, pp. 3–8.

Timmers, P. (1999). *Electronic commerce*. Hoboken, New Jersey, United States: John Wiley & Sons, Inc.

Timmers, P. (2000). *Electronic commerce: Strategies and models for business-to-business trading*. Paperback ed. Management/information systems. Chichester and Weinheim: Wiley.

Tiwana, A. (2014). *Platform Ecosystems. Aligning Architecture, Governance, and Strategy*. Waltham: Morgan Kaufman.

Tiwana, A. (2015a). "Evolutionary Competition in Platform Ecosystems". In: *Information Systems Research* 26.2, pp. 266–281.

Tiwana, A., B. Konsynski, and A. A. Bush (2010). "Platform Evolution: Coevolution of Platform Architecture, governance, and Environmental Dynamics". In: *Information Systems Research* 21.4, pp. 675–687.

Tiwana, A. (2015b). "Platform desertion by app developers". In: *Journal of management information systems* 32.4, pp. 40–77.

Tiwana, A. and B. Konsynski (2010). "Complementarities between organizational IT architecture and governance structure". In: *Information Systems Research* 21.2, pp. 288–304.

Tokopedia (2023). *Report Infringement.* URL: https://www.tokopedia.com/intellectual-property-protection/report (visited on 06/28/2023).

Töpfer, A. (2009). *Erfolgreich forschen. Ein Leitfaden für Bachelor-, Master-Studierende und Doktoranden.* Berlin, Germany: Springer.

Topkis, D. M. (1998). *Supermodularity and complementarity.* New York City, USA: Princeton University Press.

Torno, A. et al. (2021). "More than Mobile Banking—A Taxonomy-based Analysis of Mobile Personal Finance Applications". In: *Proceedings of the 25th Pacific Asia Conference on Information Systems (PACIS2021).* Dubai, UAE, pp. 1–14.

Tran, L. (2021). "Managing the effectiveness of e-commerce platforms in a pandemic". In: *Journal of Retailing and Consumer Services* 58, pp. 1–9.

Treutner, M. F. and H. Ostermann (2011). "Evolution of Standard Web Shop Software Systems: A Review and Analysis of Literature and Market Surveys". In: *The Open Information Systems Journal* 5, pp. 8–18.

Tsai, C.-L. and U.-C. Lin (2011). "Information security of cloud computing for enterprises". In: *Advances in Information Sciences and Service Sciences* 3.1, pp. 132–142.

Tsai, C. H., J. Zdravkovic, and J. Stirna (2022). "Modeling Digital Business Ecosystems: A Systematic Literature Review". In: *Complex Systems Informatics and Modeling Quarterly* 30, pp. 1–30.

Tsay, A. A. (2001). "Managing retail channel overstock: markdown money and return policies". In: *Journal of Retailing* 77.4, pp. 457–492.

Tsujimoto, M., Y. Kajikawa, J. Tomita, and Y. Matsumoto (2018). "A review of the ecosystem concept: Towards coherent ecosystem design". In: *Technological Forecasting and Social Change* 136, pp. 49–58.

Tucker, C. and J. Zhang (2010). "Growing two-sided networks by advertising the user base: A field experiment". In: *Marketing Science* 29.5, pp. 805–814.

Turban, E., E. McLean, and J. Wetherbe (2004). *Information Technology for Management: Transforming Organizations in the Digital Economy.* Hoboken, New Jersey, United States: John Wiley & Sons, Inc.

Turban, E., J. Whiteside, D. King, and J. Outland (2017). *Introduction to Electronic Commerce and Social Commerce.* 4th ed. Springer Texts in Business and Economics. Cham: Springer International Publishing.

Tushman, M. L. (1977). "Special Boundary Roles in the Innovation Process". In: *Administrative Science Quarterly* 22.4, pp. 587–605.

Tüzün, E., B. Tekinerdogan, Y. Macit, and K. İnce (2019). "Adopting integrated application lifecycle management within a large-scale software company: An action research approach". In: *Journal of Systems and Software* 149, pp. 63–82.

Tyworth, M. (2014). "Organizational identity and information systems: How organizational ICT reflect who an organization is". In: *European Journal of Information Systems* 23.1, pp. 69–83.

U.S. Department of Commcerce (2020). *Quarterly Retail E-Commerce Sales 2nd Quarter 2020*. URL: https://www.census.gov/retail/mrts/www/data/pdf/ec_current.pdf (visited on 08/21/2023).

Ulrich, K. (1995). "The role of product architecture in the manufacturing firm". In: *Research Policy* 24.3, pp. 419–440.

Uluyol, B., S. Secinaro, D. Calandra, and F. Lanzalonga (2021). "Mapping waqf research: a thirty-year bibliometric analysis". In: *Journal of Islamic Accounting and Business Research* 12.5, pp. 748–767.

Urbaczewski, A., L. M. Jessup, and B. Wheeler (2002). "Electronic Commerce Research: A Taxonomy and Synthesis". In: *Journal of Organizational Computing* 12.4, pp. 263–305.

Vaishnavi, V., B. Kuechler, and S. Petter (2019). *Design science research in information systems*. URL: http://www.desrist.org/design-research-in-information-systems/ (visited on 10/15/2021).

Vaishnavi, V. K., V. K. Vaishnavi, and W. Kuechler (2015). *Design Science Research Methods and Patterns*. Boca Raton, USA: CRC Press.

Van Aken, J. E. (2004). "Management Research Based on the Paradigm of the Design Sciences: The Quest for Field-Tested and Grounded Technological Rules". In: *Journal of Management Studies* 41.2, pp. 219–246.

Van Alstyne, M., G. Parker, and S. P. Choudary (2016). "6 Reasons Platforms Fail". In: *Havard Business Review* 31.6, pp. 1–6.

Van Alstyne, M. W., G. G. Parker, and S. Paul Choudary (2016). "Pipelines, platforms, and the new rules of strategy". In: *Harvard Business Review* 2016.4, pp. 54–62.

Van den Berk, I., S. Jansen, and L. Luinenburg (2010). "Software ecosystems". In: *Proceedings of the Fourth European Conference on Software Architecture: Companion Volume*. New York, NY, USA, pp. 127–134.

Van der Aalst, W., O. Hinz, and C. Weinhardt (2019). "Big Digital Platforms: Growth, Impact, and Challenges". In: *Business and Information Systems Engineering* 61.6, pp. 645–648.

van Eck, N. J. and L. Waltman (2014). "Visualizing Bibliometric Networks". In: *Measuring Scholarly Impact*. Ed. by Y. Ding, R. Rousseau, and D. Wolfram. Cham: Springer International Publishing, pp. 285–320.

Vargo, S. L. and R. F. Lusch (2004). "Evolving to a New Dominant Logic for Marketing". In: *Journal of Marketing* 68.1, pp. 1–17.

Vargo, S. L. and R. F. Lusch (2006). "Service-dominant logic: what it is, what it is not, what it might be". In: *The Service-Dominant Logic of Marketing: Dialog, Debate, and Directions*. Ed. by R. F. Lusch and S. L. Vargo. Armonk: M. E. Sharpe Inc., pp. 43–56.

Vargo, S. L. and R. F. Lusch (2008). "Service-dominant logic: Continuing the evolution". In: *Journal of the Academy of Marketing Science* 36.1, pp. 1–10.

Vargo, S. L. and R. F. Lusch (2014). "Evolving to a new dominant logic for marketing". In: *Journal of Marketing* 68.1, pp. 3–28.

Vargo, S. L. and R. F. Lusch (2016). "Institutions and axioms: an extension and update of service-dominant logic". In: *Journal of the Academy of Marketing Science* 44.1, pp. 5–23.

Vargo, S. L., P. P. Maglio, and M. A. Akaka (2008). "On value and value co-creation: A service systems and service logic perspective". In: *European Management Journal* 26.3, pp. 145–152.

Väyrynen, K. and J. Iivari (2015). "The competitive potential of IT applications-An analytical-Argumentative evaluation". In: *Proceedings of the 36th International Conference on Information Systems (ICIS 2015)*. Fort Worth, Texas, USA, pp. 1–20.

Venable, J., J. Pries-Heje, and R. Baskerville (2016). "FEDS: a Framework for Evaluation in Design Science Research". In: *European Journal of Information Systems* 25.1, pp. 77–89.

Venable, J. R. (2006). "The Role of Theory and Theorising in Design Science Research". In: *Proceedings of the 1st International Conference on Design Science in Information Systems and Technology (DESRIST 2006)*. Claremont, California, USA, pp. 1–18.

Venkatesh, Viswanath, et al. (2003). "User Acceptance of Information Technology: Toward a Unified View". In: *MIS Quarterly* 27.3, p. 425.

Venkatesh, V., J. Y. Thong, and X. Xu (2016). "Unified Theory of Acceptance and Use of Technology: A Synthesis and the Road Ahead". In: *Journal of the Association for Information Systems* 17.5, pp. 328–376.

Verband der Hochschullehrerinnen und Hochschullehrer für Betriebswirtschaft e.V. (VHB) (2022). *Complete List*. URL: https://vhbonline.org/en/vhb4you/vhb-jourqual/vhb-jourqual-3/complete-list (visited on 08/18/2023).

Verborgh, R. and M. Dumontier (2018). "A Web API Ecosystem through Feature-Based Reuse". In: *IEEE Internet Computing* 22.3, pp. 29–37.

Verbovetska, A. (2019). "The Impact of Financial Technology on Customer Intention to Use Financial Services through the Lenses of Process Virtualization Theory". In: *Proceedings of the 40th International Conference on Information Systems (ICIS2019)*. Munich, Germany, pp. 1–17.

Verhagen, T., S. Meents, and Y.-H. Tan (2006). "Perceived risk and trust associated with purchasing at electronic marketplaces". In: *European Journal of Information Systems* 15.6, pp. 542–555.

Verhoef, P. C., P. K. Kannan, and J. J. Inman (2015). "From Multi-Channel Retailing to Omni-Channel Retailing. Introduction to the Special Issue on Multi-Channel Retailing." In: *Journal of Retailing* 91.2, pp. 174–181.

Verhoef, P. C., K. N. Lemon, et al. (2009). "Customer Experience Creation: Determinants, Dynamics and Management Strategies". In: *Journal of Retailing* 85.1, pp. 31–41.

Verschuren, P. and R. Hartog (2005). "Evaluation in design-oriented research". In: *Quality and Quantity* 39.6, pp. 733–762.

Vester, J. (2018). *RESTful API Lifecycle Management*. URL: https://dzone.com/refcardz/restful-api-lifecycle-management (visited on 10/18/2022).

Vetter, T. and R. Morasch (2019). "Integrierte Plattformen im Handel". In: *Handel mit Mehrwert*. Ed. by G. Heinemann, H. M. Gehrckens, M. Täuber, and A. GmbH. Heidelberg: Springer, pp. 321–343.

Vial, G. (2019). "Understanding digital transformation: A review and a research agenda". In: *Journal of Strategic Information Systems* 28.2, pp. 118–144.

Viardot, E., I. Mccarthy, and J. Chen (2021). "Editorial: Standardization in a Digital and Global World: State-of-the-Art and Future Perspectives". In: *IEEE Transactions on Engineering Management* 68.1, pp. 11–17.

Voigt, S. and O. Hinz (2015). "Network effects in two-sided markets: why a 50/50 user split is not necessarily revenue optimal". In: *Business Research* 8.1, pp. 139–170.

Völker, R. and E. Voit (2000). "Planung und Bewertung von Produktplattformen". In: *Controlling & Management* 44.3, pp. 137–143.

Vom Brocke, J. (2015). *Referenzmodellierung: Gestaltung und Verteilung von Konstruktionsprozessen*. Ed. by J. Becker et al. 2. Berlin, Germany: Logos Verlag.

Vom Brocke, J., A. Simons, et al. (2009). "Reconstructing the Giant: On the Importance of Rigour in Documenting the Literature Search Process". In: *Proceedings of the 17th European Conference on Information Systems (ECIS2017)*. Verona, Italy, pp. 2206–2217.

Vom Brocke, J., R. Winter, A. Hevner, and A. Maedche (2020). "Special issue editorial—accumulation and evolution of design knowledge in design science research: A journey through time and space". In: *Journal of the Association for Information Systems* 21.3, pp. 520–544.

Vom Brocke, J. and A. Maedche (2019). "The DSR grid: six core dimensions for effectively planning and communicating design science research projects". In: *Electronic Markets* 29.3, pp. 379–385.

Vom Brocke, J., A. Simons, et al. (2015). "Standing on the Shoulders of Giants: Challenges and Recommendations of Literature Search in Information Systems Research". In: *Communications of the Association for Information Systems* 37.1, pp. 205–224.

Von Krogh, G., S. Spaeth, and K. R. Lakhani (2003). "Community , joining , and specialization in open source software innovation : a case study". In: *Research Policy* 32, pp. 1217–1241.

Von Reden, S. (2015). *Antike Wirtschaft*. Berlin, Germany: Walter de Gruyter GmbH & Co KG.

Von Bertalanffy, L. (1950). "An Outline of General System Theory". In: *The British Journal for the Philosophy of Science* 1.2, pp. 134–165.

Von Engelhardt, S., L. Wangler, and S. Wischmann (2017). *Eigenschaften und Erfolgsfaktoren digitaler Plattformen*. URL: https://www.digitale-technologien.de/DT/Redaktion/DE/Downloads/Publikation/autonomik-studie-digitale-plattformen.pdf?__blob=publicationFile%5C&v=6 (visited on 03/09/2020).

Wallentowitz, H., A. Freilandenhoven, and I. Olschewski (2009). *Strategien in der Automobilindustrie. Technologietrends und Marktentwicklungen*. Wiesbaden, Germany: Vieweg + Teubner.

Walls, J. G., G. R. Widmeyer, and O. A. El Sawy (1992). "Building an Information System Design Theory for Vigilant EIS". In: *Information Systems Research* 3.1, pp. 36–59.

Walmart (2021a). *Sandbox—Walmart Developer Portal*. URL: https://developer.walmart.com/doc/sandbox/ (visited on 07/29/2021).

Walmart (2021b). *Walmart Affiliates Mobile SDK*. URL: https://affiliates.walmart.com/#!/mobilesdk (visited on 07/26/2021).

Walmart (2021c). *What's New—Walmart Developer Portal*. URL: https://developer.walmart.com/category/us/whats-new/ (visited on 07/28/2021).

Walsh, J. P. (1995). "Managerial and Organizational Cognition: Notes from a Trip Down Memory Lane". In: *Organization Science* 6.3, pp. 280–321.

Walsham, G. (1995). "Interpretive case studies in IS research: nature and method". In: *European Journal of Information Systems* 4.2, pp. 74–81.

Walsham, G. (2006). "Doing interpretive research". In: *European Journal of Information Systems* 15.3, pp. 320–330.

Walunj, S. G. and K. Sadafale (2013). "An Online Recommendation System for E-Commerce Based on Apache Mahout Framework". In: *Proceedings of the Annual Conference on Computers and People Research (SIGMIS-CPR2013)*. New York, NY, USA, pp. 153–158.

Wamser, C. (2001). *Strategisches Electronic Commerce. Wettbewerbsvorteile auf elektronischen Märkten*. Munich, Germany: Vahlen.

Wang, C. and P. Zhang (2012). "The evolution of social commerce: The people, management, technology, and information dimensions". In: *Communications of the Association for Information Systems* 31.1, pp. 105–127.

Wang, J., P. Huang, et al. (2018). "Billion-scale commodity embedding for E-commerce recommendation in alibaba". In: *Proceedings of the ACM SIGKDD International Conference on Knowledge Discovery and Data Mining (KDD2018)*. London, UK, pp. 839–848.

Wang, L. (2013). "Post-crisis era of SMEs management innovation in E-commerce". In: *Proceedings of 6th International Conference on Information Management, Innovation Management and Industrial Engineering (ICIII2013)*. Xi'an, China, pp. 471–474.

Wang, M., H. Deng, and K. G. Leong (2022). "Innovation investment and subsidy strategy in two-sided market". In: *Information Technology and Management* X.X, pp. 1–15.

Wang, N., D. Xu, and B. Xu (2010). "Collaborative integration and management of community information in the cloud". In: *Proceedings of the International Conference on E-Business and E-Government (ICEE 2010)*. Guangzhou, China, pp. 1406–1409.

Wang, P., X. Liu, et al. (2018). "QoS-aware service composition using blockchain-based smart contracts". In: *Proceedings of the 40th International Conference on Software Engineering*. Gothenburg, Sweden, pp. 296–297.

Wang, Q. and M. Gao (2012). "The research of E-commerce recommendation system based on cloud computing". In: *International Journal of Advancements in Computing Technology* 4.17, pp. 393–400.

Wang, S., S. Zheng, et al. (2008). "A literature review of electronic marketplace research: Themes, theories and an integrative framework". In: *Information Systems Frontiers* 10.5, pp. 555–571.

Wang, S. and N. P. Archer (2007). "Electronic marketplace definition and classification: literature review and clarifications". In: *Enterprise Information Systems* 1.1, pp. 89–112.

Wang, X., X. Lin, and M. K. Spencer (2019). "Exploring the effects of extrinsic motivation on consumer behaviors in social commerce: Revealing consumers' perceptions of social commerce benefits". In: *International Journal of Information Management* 45, pp. 163–175.

Wang, Z. (2011). "Research on cloud computing-based online payment mode". In: *Proceedings of the 3rd International Conference on Multimedia Information Networking and Security (MINES2011)*. Shanghai, China, pp. 559–563.

Wanner, J., C. Bauer, and C. Janiesch (2019). "Two-sided digital markets: Disruptive chance meets chicken or egg causality dilemma". In: *Proceedings of the 21st IEEE Conference on Business Informatics (CBI2019)*. Moscow, Russia, pp. 335–344.

Wareham, J., P. B. Fox, and J. L. C. Giner (2014). "Technology ecosystem governance". In: *Organization Science* 25.4, pp. 1195–1215.

Wassermann, S. and J. Galaskiewicz (1994). *Advances in social network analysis: Research in the social and behavioral sciences*. Thousand Oaks, CA: Sage.

Weber, C. E. (2005). "Edgeworth on Complementarity, or Edgeworth, Auspitz-Lieben, and Pareto De-Homogenized". In: *History of Political Economy* 37.2, pp. 293–307.

Weber, F. and R. Schütte (2019a). "Markdown Optimization: A Taxonomy and Realization of the Economic Decision Problem". In: *Intelligent Systems Applications in Software Engineering*. Ed. by R. Silhavy, P. Silhavy, and Z. Prokopova. Cham: Springer International Publishing, pp. 449–460.

Weber, F. D. and R. Schütte (2019b). "State-of-the-art and adoption of artificial intelligence in retailing". In: *Digital Policy, Regulation and Governance* 21.3, pp. 264–279.

Weber, M. (1947). *The Theory of Social and Economic Organization, Translated and edited by A. M. Henderson and Talcott Parsons*. New York City, USA: Oxford University Press.

Webster, J. and R. T. Watson (2002). "Analyzing the Past to Prepare for the Future: Writing a Literature Review." In: *MIS Quarterly* 26.2, pp. xiii–xxiii.

Wei, B., Y. Li, L. Deng, and N. Visalli (2020). "Teaching Distributed Software Architecture by Building an Industrial Level E-Commerce Application". In: *Studies in Computational Intelligence*. Ed. by R. Lee. Vol. 845. Cham, Switzerland: Springer Nature, pp. 43–54.

Weick, K. (1995). *Sensemaking in Organizations*. London, UK: Sage Publications.

Weir, L. (2019). *Enterprise API Management: Design and deliver valuable business APIs*. Birmingham, UK: Packt Publishing.

Wells, J. (2021). *Kroger maps its new digital 'ecosystem'*. URL: https://www.grocerydive.com/news/kroger-maps-its-new-digital-ecosystem/597693/ (visited on 06/28/2023).

Wenke, C., F. Lang, and M. Buschle (2022). "Distributed Ledger Technology in the Financial Industry: Managerial, Organizational, and Technological Challenges". In: *Proceedings of the 30th European Conference on Information Systems (ECIS2022)*. Timisoara, Romania, pp. 1–15.

Wernerfelt, B. (1984). "A resource-based view of the firm". In: *Strategic Management Journal* 5.2, pp. 171–180.

West, J. (2003). "How open is open enough? Melding proprietary and open source platform strategies". In: *Research policy* 32, pp. 1259–1285.

West, J. and S. O'Mahony (2008). "The role of participation architecture in growing sponsored open source communities". In: *Industry and Innovation* 15.2, pp. 145–168.

Wheelwright, S. and K. Clark (1992). *Revolutionizing Product Development: Quantum Leap in Speed, Efficiency and Quality*. New York City, USA: Free Press.

Whetten, D. A. (1989). "What Constitutes a Theoretical Contribution? The Academy of Management Review What Constitutes a Theoretical Contribution?" In: *The Academy of Management ReviewAcaderyof Management Review* 14.4, pp. 490–495.

White, A., E. Daniel, J. Ward, and H. Wilson (2007). "The adoption of consortium B2B e-marketplaces: An exploratory study". In: *The Journal of Strategic Information Systems* 16.1, pp. 71–103.

Wiesche, M., M. Schermann, and H. Krcmar (2011). "Understanding the Role of Information Technology for Organizational Control Design: Risk Control as New Control Mechanism". In: *Proceedings of the International Working Conference on Transfer and Diffusion of IT (TDIT2011)*. Hamburg, Germany, pp. 135–152.

Wigand, R. T. (1997). "Electronic commerce: Definition, theory, and context". In: *Information Society* 13.1, pp. 1–16.

Willcocks, L. P., D. Feeny, and G. Islei (1997). *Managing information technology as a strategic resource*. New York City: McGraw-Hill.

Williamson, O. E. (1985). *The Economic Institutions of Capitalism. Firms, Markets, Relational Contracting*. New York [u.a.]: The Free Press/MacMillan.

Wilson, N. (2017). *Modernizing Application Development Primer*. URL: https://www.gartner.com/en/documents/3571618 (visited on 10/20/2022).

Wind, Y., V. Fung, and W. Fung (2009). "Network Orchestration: Creating and Managing Global Supply Chains Without Owning Them". In: *The Network Challenge: Strategy, Profit, and Risk in an Interlinked World*. Ed. by P. R. Kleindorfer and Y. Wind. Upper Saddle River, NJ: FT Press, pp. 299–315.

Winter, R. (2009). "Was ist eigentlich Grundlagenforschung in der Wirtschaftsinformatik?" In: *Wirtschaftsinformatik* 51.2, pp. 223–231.

Winter, R. and R. Fischer (2006). "Essential layers, artifacts, and dependencies of enterprise architecture". In: *Proceedings of the 10th IEEE International Enterprise Distributed Object Computing Conference Workshops (EDOCW2006)*. Hong Kong, China, pp. 30–38.

Winter, R. and A. Gericke (2006). "Teradata University Network—Ein Portal zur Unterstützung der Lehre in den Bereichen Business Intelligence, Data Warehousing und Datenbanken". In: *Wirtschaftsinformatik* 48.4, pp. 276–281.

Wirtz, B. W. (2020). *Electronic Business*. 7th ed. Wiesbaden, Germany: Springer Gabler.

Witte, A. and R. Zarnekow (2018). "Is Open Always Better?—A Taxonomy-based Analysis of Platform Ecosystems for Fitness Trackers". In: *Proceedings of the Multikonferenz Wirtschaftsinformatik*. Lüneburg, Germany, pp. 732–742.

Wittig, M., A. Wittig, and B. Whaley (2016). *Amazon web services in action*. Shelter Island, NY: Manning.

WooCommerce (2022a). *Become a certified WooExpert*. URL: https://woocommerce.com/partner-program/ (visited on 10/21/2022).

WooCommerce (2022b). *Store Editing Roadmap Update—Q3 2022*. URL: https://developer.woocommerce.com/category/roadmap-insights/ (visited on 10/21/2022).

WooCommerce (2022c). *Woo Rest API*. URL: https://woocommerce.github.io/woocommerce-rest-api-docs/#introduction (visited on 10/21/2022).

WooCommerce (2022d). *WooCommerce Beta Tester*. URL: https://wordpress.org/plugins/woocommerce-beta-tester/ (visited on 10/21/2022).

WooCommerce (2022e). *WooCommerce Developer Blog*. URL: https://developer.woocommerce.com/blog/ (visited on 10/21/2022).

WooCommerce (2022f). *WooCommerce Documentation*. URL: https://woocommerce.com/documentation/ (visited on 10/21/2022).

Woods, E. and N. Rozanski (2013). "Relating Enterprise, Application, and Infrastructure Architects". In: *Aliging Enterprise, System, and Software Architectures*. Ed. by I. Mistrik, A. Tang, R. Bahsoon, and J. A. Stafford. Heershey: Business Science Reference, pp. 1–21.

WordPress (2022). *Support WooCommerce Plugin*. URL: https://wordpress.org/support/plugin/woocommerce/ (visited on 10/21/2022).

Woroch, R., G. Strobel, and T. Wulfert (2022). "Four Shades of Customer: How Value Flows in Fintech Ecosystems". In: *Proceedings of the 43rd International Conference on Information Systems (ICIS2022)*. Copenhagen, Denmark, pp. 1–17.

Wright, J. (2003). *One-Sided Logic in Two-Sided Markets*. Singapore: National University of Singapore (NUS)—Department of Economics.

Wu, J., S. Fan, and J. L. Zhao (2018). "Community engagement and online word of mouth: An empirical investigation". In: *Information and Management* 55.2, pp. 258–270.

Wu, J. H. and T. L. Hisa (2008). "Developing E-business dynamic capabilities: An analysis of E-commerce innovation from I-, M-, to U-commerce". In: *Journal of Organizational Computing and Electronic Commerce* 18.2, pp. 95–111.

Wu, M., X. Ding, and R. Hou (2019). "Design and implementation of B2B E-commerce platform based on microservices architecture". In: *Proceedings of the 2nd International Conference on Computer Science and Software Engineering (CSSE2019)*. Xi'an China, pp. 30–34.

Wu, S.-y. and K.-C. Lin (2011). "Structured design, consistency analysis and failure reasoning of business workflows with activity-control templates and causal ordering". In: *Expert Systems with Applications* 38.7, pp. 8000–8013.

Wu, Y.-L. and X.-K. Cheng (2010). "The design and implementation of special market-orientated integrated E-business portal system". In: *Proceedings of the International Conference on E-Business and E-Government (ICEE2010)*. Guangzhou, China, pp. 148–151.

Wulf, J. and I. Blohm (2020). "Fostering Value Creation with Digital Platforms: A Unified Theory of the Application Programming Interface Design". In: *Journal of Management Information Systems* 37.1, pp. 251–281.

Wulfert, T., J. H. Betzing, and J. Becker (2019). "Eliciting Customer Preferences for Shopping Companion Apps: A Service Quality Approach". In: *Proceedings of the 14th Internationale Tagung Wirtschaftsinformatik (WI2019)*. Siegen, Deutschland, pp. 1220–1234.

Wulfert, T. (2023). "Boundary Resource Management in Innovation Ecosystems: The Case of E-Commerce". In: *Electronic Markets* 33.27, pp. 1–27.

Wulfert, T. and J. Busch (2022). "Reference Architectures Facilitating a Retailer's Dual Role on Digital Marketplaces: A Literature Review". In: *Proceedings of the 24th International Conference on Enterprise Information Systems (ICEIS2022) Volume 2*. Online Event, pp. 494–505.

Wulfert, T. and B. Dennhardt (2023). "Establishing Successful E-Commerce Ecosystems: Evluating Strategies for Reaching a Critical Mass of Participants". In: *Corporate Ownership & Control* 20.3, pp. 130–145.

Wulfert, T. and E. Karger (2022a). "A Bibliometric Analysis of Platform Research in E-Commerce: Past, Present, and Future Research Agenda". In: *Corporate Ownership & Control* 20.1, pp. 185–200.

Wulfert, T. and E. Karger (2022b). "Shaping Digital Platforms in E-Commerce: Developing an Architecture Framework". In: *Proceedings of the 26th Pacific Asia Conference on Information Systems (PACIS2022)*. Virtual Conference, pp. 1–17.

Wulfert, T. and R. Schütte (2021). "Retailer's Dual Role in Digital Marketplaces: Towards Architectural Patterns for Retail Information Systems". In: *Proceedings of the 23rd International Conference on Enterprise Information Systems (ICEIS 2021)*, pp. 601–612.

Wulfert, T. and R. Schütte (2022). "Retailer's Dual Role in Digital Marketplaces—Reference Architecture for Retail Information Systems". In: *SN Computer Science* 3.206, pp. 1–17.

Wulfert, T., S. Seufert, and C. Leyens (2021). "Developing Multi-Sided Markets in Dynamic Electronic Commerce Ecosystems—Towards a Taxonomy of Digital Marketplaces". In: *Proceedings of the 54th Hawaii International Conference on System Sciences (HICSS 2021)*. Maui, Hawaii, pp. 6133–6142.

Wulfert, T., G. Strobel, and H. Hoang (2024). "Let's Join Forces: Boundary Resource as Enablers of Value Cocreation in E-Commerce Ecosystems". In: *Electronic Commerce Research* Forthcoming.

Wulfert, T., R. Woroch, and G. Strobel (2024). “How Giants Rule the World: A Qualitative Comparative Analysis of Dynamic Capabilities for E-Commerce Ecosystem Orchestration”. In: *Journal of Strategic Information Systems* Forthcoming.

Wulfert, T., R. Woroch, G. Strobel, T. Schoormann, et al. (2023). “Unboxing the Role of E-Commerce Ecosystems to Address Grand Challenges”. In: *Proceedings of the 31st European Conference on Information Systems (ECIS2023).* Kristiansand, Norway, pp. 1–21.

Wulfert, T., R. Woroch, G. Strobel, S. Seufert, et al. (2022). “Developing Design Principles to Standardize E-Commerce Ecosystems: A Systematic Literature Review and Multi-Case Study of Boundary Resources”. In: *Electronic Markets* 32, pp. 1813–1842.

Wunck, C. and S. Baumann (2017). “Towards a process reference model for the information value chain in IoT applications”. In: *Proceedings of the IEEE European Technology and Engineering Management Summit (E-TEMS2017).* Munich, Germany, pp. 1–6.

Xebia (2023). *Supporting bol.com as a Frontrunner in Data & AI.* URL: https://customers.xebia.com/data-ai-supporting-bol.com-as-a-frontrunner-in-data-ai (visited on 06/28/2023).

Xia, Y., Q. Li, and L. Wang (2010). “Research on decentralized E-commerce architecture in P2P environment”. In: Wuhan, China, pp. 2929–2932.

Xiang, Z., Q. Du, Y. Ma, and W. Fan (2017). “A comparative analysis of major online review platforms: Implications for social media analytics in hospitality and tourism”. In: *Tourism Management* 58, pp. 51–65.

Xie, L., W. Zhou, and Y. Li (2016). “Application of improved recommendation system based on spark platform in big data analysis”. In: *Cybernetics and Information Technologies* 16.6, pp. 245–255.

Xie, J., L. Ye, W. Huang, and M. Ye (2021) “Understanding FinTech Platform Adoption: Impacts of Perceived Value and Perceived Risk”. In: *Journal of Theoretical and Applied Electronic Commerce Research* 16.5, pp. 1893–1911.

Xu, X., V. Venkatesh, K. Y. Tam, and S.-J. Hong (2010). “Model of migration and use of platforms: role of hierarchy, current generation, and complementarities in consumer settings”. In: *Management Science* 56.8, pp. 1304–1323.

Xu, X., X. Chen, et al. (2018). “Supply chain finance: A systematic literature review and bibliometric analysis”. In: *International Journal of Production Economics* 204, pp. 160–173.

Yadav, R. and T. Mahara (2019). “An Empirical Study of Consumers Intention to Purchase Wooden Handicraft Items Online: Using Extended Technology Acceptance Model”. In: *Global Business Review* 20.2, pp. 479–497.

Yaddarabullah, M. F. Muttaqin, and M. Rafiansyah (2019). “Service-Oriented Architecture for E-Marketplace Model Based on Multi-Platform Distributed System”. In: *IOP Conference Series: Materials Science and Engineering* 662.4, pp. 1–8.

Yan, N., Y. Liu, X. Xu, and X. He (2020). “Strategic dual-channel pricing games with e-retailer finance”. In: *European Journal of Operational Research* 283.1, pp. 138–151.

Yang, F. (2012). “The research of E-commerce recommendation system based on Cloud Computing”. In: *International Journal of Advancements in Computing Technology* 4.16, pp. 256–263.

Yang, L., Y. Li, J. Wang, and R. S. Sherratt (2020). “Sentiment Analysis for E-Commerce Product Reviews in Chinese Based on Sentiment Lexicon and Deep Learning”. In: *IEEE Access* 8, pp. 23522–23530

Yang, L. and X. Tang (2008). "Research of C2C e-business trust evaluation model based on entropy method". In: *Proceedings of the International Symposium on Electronic Commerce and Security (ISECS2008)*. Guangzhou, China, pp. 599–602.

Yang, X., T. Pan, and J. Shen (2010). "On 3G mobile E-commerce platform Based on Cloud Computing". In: *Proceedings of the 3rd IEEE International Conference on Ubi-Media Computing (U-Media2010)*. Zhejiang, China, pp. 198–201.

Yang, Z., C. Ou, and Z. Zhou (2017). "Investigating the Impact of Recommendation Agents on E-commerce Ecosystem". In: *23rd Americas Conference on Information Systems (AMCIS 2017)*. Boston, USA, pp. 1–5.

Ye, F., Z.-J. Wang, et al. (2013). "Cloud-based big data mining & analyzing services platform integrating R". In: *Proceedings of the International Conference on Advanced Cloud and Big Data (CBD2013)*. Nanjing, China, pp. 147–151.

Ye, G., R. L. Priem, and A. A. Alshwer (2012). "Achieving demand-side synergy from strategic diversification: How combining mundane assets can leverage consumer utilities". In: *Organization Science* 23.1, pp. 207–224.

Yi, X. and X. Ming (2011). "Study on Competitive Advantage and Construction Strategy of E-Commerce Ecosystem". In: *Proceedings of the 4th International Conference on Business Intelligence and Financial Engineering*. Wuhan, China: IEEE, pp. 207–210.

Yin, R. K. (1981). "The Case Study as a Serious Research Strategy". In: *Science Communication* 3.1, pp. 97–114.

Yin, R. K. (2018). *Case study research: design and methods*. 6th ed. Los Angeles [u. a.]: Sage Publications.

Ying, W., S. Jia, and W. Du (2018). "Digital enablement of blockchain: Evidence from HNA group". In: *International Journal of Information Management* 39, pp. 1–4.

Yoo, B. and M. Jang (2019). "A bibliographic survey of business models, service relationships, and technology in electronic commerce". In: *Electronic Commerce Research and Applications* 33, pp. 1–21.

Yoo, Y. (2013). "The tables have turned: How can the information systems field contribute to technology and innovation management research?" In: *Journal of the Association for Information Systems* 14.5, pp. 227–236.

Yoo, Y., R. J. Boland, K. Lyytinen, and A. Majchrzak (2012). "Organizing for innovation in the digitized world". In: *Organization Science* 23.5, pp. 1398–1408.

Yoo, Y., O. Henfridsson, and K. Lyytinen (2010). "The new organizing logic of digital innovation: An agenda for information systems research". In: *Information Systems Research* 21.4, pp. 724–735.

Young, J. (2022). *The poi-katsu life: Earn Rakuten Points in ways you never imagined.* URL: https://rakuten.today/blog/poi-katsu-earning-rakuten-points-in-surprising-ways.html (visited on 06/27/2023).

Yu, D.-F., C.-Y. Chang, H. C. Jiau, and K. F. Ssu (2017). "Which API Lifecycle Model is the Best for API Removal Management?" In: *Proceedings of the 12th International Conference on Software Engineering Advances (ICSEA2017)*. Athens, Greece, pp. 219–224.

Yu, H., Y. Zhao, et al. (2021). "Research on the financing income of supply chains based on an E-commerce platform". In: *Technological Forecasting and Social Change* 169.

Yu, J. and J. Ni (2013). "Development strategies for SME E-commerce based on cloud computing". In: *Proceedings of the 7th International Conference on Internet Computing for Engineering and Science (ICICSE2013)*. Shanghai, China, pp. 1–8.

Zalando (2021). *Download Center: ZALANDO SE—Annual Report 2017*. URL: https://geschaeftsbericht.zalando.de/2017/service/download-center/ (visited on 06/27/2023).
Zalando (2023). *Zalando Tech Hubs*. URL: https://jobs.zalando.com/en/tech/locations/?gh_src=22377bdd1us (visited on 06/27/2023).
Zapadka, P., A. Hanelt, and S. Firk (2022). "Digital at the edge—antecedents and performance effects of boundary resource deployment". In: *Journal of Strategic Information Systems* 31.1, pp. 1–22.
Zelewski, S. (2007). "Kann Wissenschaftstheorie behilflich für die Publikationspraxis sein?" In: *Wissenschaftstheoretische Fundierung und wissenschaftliche Orientierung der Wirtschaftsinformatik*. Ed. by F. Lehner and S. Zelewski. Berlin, Germany: GITO. Chap. 5, pp. 71–120.
Zelewski, S. (2008). "Grundlagen". In: *Betriebswirtschaftslehre, Band 1. Grundlagen, Internes Rechnungswesen, Externes Rechnungswesen, Beschaffung*. Ed. by H. Corsten and M. K. Reiß. Munich, Germany: Vahlen, pp. 1–97.
Zeljko, D., B. Jakovic, I. Strugar, and Katalinic B. (2018). "New methods of online advertising: Social media influencers". In: *Annals of DAAAM* 29.1, pp. 41–50.
Zeng, L., B. Benatallah, et al. (2004). "QoS-aware middleware for Web services composition". In: *IEEE Transactions on Software Engineering* 30.5, pp. 311–327.
Zeng, Y., Y. Cheng, G. Xie, and R. Wang (2021). "Design of Mobile Application Lifecycle Security Management Platform". In: *Proceedings of the International Conference on Computer Network, Electronic and Automation (ICCNEA2021)*. Xi'an, China, pp. 26–30.
Zentes, J. (2006). "Dynamik des Handels: Perspektiven und Zukunftsstrategien". In: *Handbuch Handel*. Ed. by J. Zentes. Wiesbaden, Germany: Betriebswirtschaftlicher Verlag Dr. Th. Gabler, pp. 3–22.
Zhang, C., S. Viswanathan, and J. W. Henke (2011). "The boundary spanning capabilities of purchasing agents in buyer-supplier trust development". In: *Journal of Operations Management* 29.4, pp. 318–328.
Zhang, L., Y. Yuan, Z. Wu, and J. Cao (2017). "Semi-SGD: Semi-Supervised Learning Based Spammer Group Detection in Product Reviews". In: *Proceedings of the 5th International Conference on Advanced Cloud and Big Data (CBD2017)*. Shanghai, China, pp. 368–373.
Zhang, X., Y. Yu, H. Li, and Z. Lin (2016). "Sentimental interplay between structured and unstructured user-generated contents". In: *Online Information Review* 40.1, pp. 119–145.
Zhang, Y. and J. Wen (2017). "The IoT electric business model: Using blockchain technology for the internet of things". In: *Peer-to-Peer Networking and Applications* 10.4, pp. 983–994.
Zhang, Z. (2016). "Who boost your sales, somebody or nobody?" In: *Proceedings of the 37th International Conference on Information Systems (ICIS2016)*. Dublin, Ireland, pp. 1–10.
Zhao, J. L. and H. K. Cheng (2005). "Web services and process management: A union of convenience or a new area of research?" In: *Decision Support Systems* 40.1, pp. 1–8.
Zhao, W. X., S. Li, et al. (2016). "Exploring demographic information in social media for product recommendation". In: *Knowledge and Information Systems* 49.1, pp. 61–89.
Zhao, X. and P. Luo (2022). "Hybrid Platform Operation Decision of Retail Enterprises". In: *Journal of Theoretical and Applied Electronic Commerce Research* 17.2, pp. 809–835.
Zhao, Y., L. Wen, et al. (2020). "How managerial responses to online reviews affect customer satisfaction: An empirical study based on additional reviews". In: *Journal of Retailing and Consumer Services* 57, pp. 1–11.

Zhao, Y., S. von Delft, A. Morgan-Thomas, and T. Buck (2020). "The evolution of platform business models: Exploring competitive battles in the world of platforms". In: *Long Range Planning* 53.4, pp. 1–24.

Zheng, Q. et al. (2009). "Fundamentals of E-commerce". In: *Introduction to E-commerce*. Ed. by Q. Zheng. Berlin, Heidelberg: Springer Berlin Heidelberg. Chap. 1, pp. 3–76.

Zhong, J. (2009). "A Comparison of Mobile Payment Procedures in Finnish and Chinese Markets". In: *Proceedings of the BLED Conference (BLED2009)*. Bled, Slovenia, pp. 79–96.

Zhu, F. and Q. Liu (2018). "Competing with complementors: An empirical look at Amazon.com". In: *Strategic Management Journal* 39.10, pp. 2618–2642.

Zhu, K. (2004). "Information transparency of business-to-business electronic markets: A game-theoretic analysis". In: *Management Science* 50.5, pp. 670–685.

Zittrain, J. (2006). "The generative internet". In: *Havard Law Review* 119.7, pp. 1974–2040.

Zittrain, J. (2008). *The Future of the Internet: And How to Stop It*. New Haven, USA: Yale University Press.

Zittrain, J. (2009). "Law and technology—The end of the generative internet". In: *Communications of the ACM* 52.1, pp. 18–20.

Zollet, R. (2014). "Interactivity of Corporate Websites: An Integrative Review of the Literature". In: *IEEE Transactions on Professional Communication* 57.1, pp. 2–16.

Zott, C., R. Amit, and L. Massa (2011). "The Business Model: Recent Developments and Future Research". In: *Journal of Management* 37.4, pp. 1019–1042.

Zulian, A. and A. Bouza (2021). *API Product Management: Product Strategy and Execution for the Digital Economy*. Victoria, Canada: leanpub.

Zutshi, A. and A. Grilo (2019). "The Emergence of Digital Platforms: A Conceptual Platform Architecture and impact on Industrial Engineering". In: *Computers and Industrial Engineering* 136.7, pp. 546–555.

Zwass, V. (1996). "Electronic commerce: Structures and issues". In: *International Journal of Electronic Commerce* 1.1, pp. 3–23.

Zwass, V. (2010). "Co-creation: Toward a taxonomy and an integrated research perspective". In: *International Journal of Electronic Commerce* 15.1, pp. 11–48.